Advances of Remote Sensing in Land Cover and Land Use Mapping

Advances of Remote Sensing in Land Cover and Land Use Mapping

Guest Editors

Sébastien Gadal
Gintautas Mozgeris

MDPI

Basel • Beijing • Wuhan • Barcelona • Belgrade • Novi Sad • Cluj • Manchester

Guest Editors

Sébastien Gadal
CNRS ESPACE UMR 7300
Aix-Marseille University
Aix-en-Provence
France

Gintautas Mozgeris
Department of Forest Sciences
Vytautas Magnus University
Akademija
Lithuania

Editorial Office
MDPI AG
Grosspeteranlage 5
4052 Basel, Switzerland

This is a reprint of the Special Issue, published open access by the journal *Remote Sensing* (ISSN 2072-4292), freely accessible at: https://www.mdpi.com/journal/remotesensing/special_issues/00249M7PBL.

For citation purposes, cite each article independently as indicated on the article page online and as indicated below:

Lastname, A.A.; Lastname, B.B. Article Title. *Journal Name* **Year**, *Volume Number*, Page Range.

ISBN 978-3-7258-5919-1 (Hbk)
ISBN 978-3-7258-5920-7 (PDF)
https://doi.org/10.3390/books978-3-7258-5920-7

Contents

About the Editors

Sébastien Gadal

Sébastien Gadal is a Full Professor of Geography with a focus on geomatic and spatial analysis. His area of research concerns the analysis of the processes of urbanization and territorial development in the context of globalization, climate changes, and adaptation, especially in the Nordic, The Arctic, post-Soviet territories and Africa. Geographic analysis is based on the development and implementation of earth observation remote sensing monitoring systems and geo-simulations, including the following: multispectral, hyperspectral imagery, artificial intelligence, massive data image processing, multi-temporal geographic knowledge systems, and GIS.

Gintautas Mozgeris

Gintautas Mozgeris is a Professor in the Department of Forest Resource Sciences and is also Head of this Department. His research interests focus on GIS, spatial modelling, mapping, photogrammetry and remote sensing, with a special focus on forestry, agricultural, urban and hydrological applications.

Preface

This Special Issue of *Remote Sensing* is focused on the latest advances in the mapping of land use and land cover (LULC). It provides an overview of the progress made in the methodologies of mapping LULC applications, approaches, and methods. Due to climate variability, territorial dynamics, and societal changes, efficient planning strategies cannot be developed without considering processes related to land cover and land use changes. This requires solutions that support the observation, mapping, monitoring, analysis, and modelling of land-related activities. Remote sensing applications for exploring land covers and land uses have demonstrated notable progress with the development of current approaches, especially advances in artificial intelligence, the proliferation of remote sensing sensors, and the availability of space and airborne data with ancillary databases and massive geodata processing.

The aim of this Special Issue, related to the mapping of land covers and land uses via remote sensing, is to review the latest methods and to increase methodological soundness: that is, the consistency, comparability, accuracy, and transparency of assessing, monitoring, and predicting land uses, land covers and their changes using spatial analysis, artificial intelligence, and remote sensing techniques. New advances in the spatial modelling of territorial patterns, their dynamics and their evolution, achieved by remote sensing, introduce new conceptual questions related to the actual environmental and territorial processes: the mapping of multi-scalar dynamics as well as different land use and land cover temporalities and changes. The included articles focus on new methodologies for mapping land use patterns, land covers, their evolution, and their dynamics based on successful operational examples from geography, environmental sciences, climate change mitigation, spatial planning and other fields, with land use and land cover information being an important factor.

Sébastien Gadal and Gintautas Mozgeris
Guest Editors

remote sensing

MDPI

Editorial

Advances of Remote Sensing in Land Cover and Land Use Mapping

Sébastien Gadal [1,*] and Gintautas Mozgeris [2]

[1] Aix Marseille Univ, Université Côte d'Azur, Avignon Université, CNRS, ESPACE, UMR 7300, Avignon, Technopôle de l'Environnement Arbois Méditerranée, Avenue Louis Philibert, Bâtiment Laennec Hall C, BP 80, CEDEX 04, 13545 Aix-en-Provence, France

[2] Vytautas Magnus University, Agriculture Academy, Faculty of Forest Sciences and Ecology, Department of Forest Sciences, Studentų Str. 11, LT-53361 Akademija, Kaunas Region, Lithuania; gintautas.mozgeris@vdu.lt

* Correspondence: sebastien.gadal@univ-amu.fr

1. Introduction

Understanding and monitoring land use and land cover (LULC) dynamics have become increasingly critical in the face of mounting socio-economic, political, and environmental challenges, including climate change, urbanization, and resource degradation [1,2]. Effective spatial planning and sustainable development strategies necessitate the integration of up-to-date information on land use and land cover changes, which are driven by complex interactions among climatic variability, demographic shifts, and economic transformations [3,4]. Addressing these challenges requires robust methods for observing, mapping, monitoring, analyzing, and modeling land surface processes. Remote sensing has emerged as a fundamental tool in this domain, benefiting from recent advances in artificial intelligence and machine learning, the expansion of Earth observation platforms, and the increasing availability of high-resolution satellite and airborne data, often supplemented with ancillary geospatial datasets [5,6]. These developments have significantly enhanced the capacity to map, characterize, and interpret land cover and land use changes at various spatial and temporal scales for a variety of applications.

The aim of this Special Issue is to present recent advances in the methodologies for mapping land cover and land use using remote sensing technologies, with a particular focus on enhancing methodological consistency, comparability, accuracy, and transparency in the assessment, monitoring, and prediction of land use and land cover changes. The contributions emphasize the integration of spatial analysis, machine learning, and remote sensing techniques to improve the detection and understanding of spatial and temporal land dynamics. Recent progress in the spatial modeling of territorial patterns and their evolution through remote sensing has also raised important conceptual questions concerning the interpretation of environmental and territorial processes, including the representation of multi-scalar dynamics and the diverse temporalities associated with land cover and land use changes.

Altogether, this Special Issue features 13 papers, including 12 research articles and 1 literature review. The esteemed first authors contributing to this Special Issue represent seven countries, namely China (5 publications), the United States (2), Ethiopia (2), Brazil, Egypt, Germany, and South Africa. In addition, their co-authors are affiliated with institutions in seven other countries beyond those of the first authors, reflecting the international and collaborative nature of the research. More detailed information on the individual articles published in this Special Issue is provided below, listed alphabetically by the first author's name.

2. An Overview of Published Articles

We begin with a comprehensive review paper by Ajibola and Cabral analyzing semantic segmentation models in land cover mapping (Contribution 1). They conducted a systematic literature review and bibliometric analysis of semantic segmentation models used in land cover mapping from 2017 to 2023. By analyzing 106 articles, they identified dominant research themes (e.g., urban areas, agriculture, forests), key geographic focuses (notably China and the USA), and widely used data sources like Sentinel-2 and Landsat. Most models rely on encoder–decoder and CNN-based architectures due to their strong performance, while transformer-based models remain underutilized because of their high computational demands. The study also highlights major research gaps, providing a solid foundation for future advancements in deep learning-based land cover mapping.

Admas et al. (Contribution 2) evaluated the combined and individual impacts of LULC and climate change on runoff and sediment flows in Ethiopia's Megech watershed using the GeoWEPP model and CA-ANN-based land use/land cover projections. Their results indicate that cropland and settlements are expanding while forests and rangelands decline, and their climate projections (RCP4.5 and RCP8.5) suggest increased temperatures and shifts in rainfall patterns. These changes are projected to significantly elevate soil loss and sediment yield—up to 41.01%, especially under RCP8.5—posing a threat to reservoir capacity. The study emphasizes the urgent need for climate adaptation and land management strategies to mitigate these impacts.

Bhungeni et al. (Contribution 3) evaluated the effectiveness of four machine learning algorithms—Random Forest (RF), Support Vector Machine (SVM), Artificial Neural Network (ANN), and Naive Bayes (NB)—for classifying LULC in the uMngeni Catchment, South Africa, using Landsat 8 imagery. Among the tested models, RF demonstrated the highest accuracy (OA = 97.02%, Kappa = 0.96), outperforming SVM and ANN, while NB showed the lowest performance, likely due to its sensitivity to limited training data. The findings underscore RF's suitability for high-dimensional, remote sensing-based LULC mapping in complex watershed environments.

De Barros de Sousa et al. (Contribution 4) demonstrated that rainfall in a semi-arid Brazilian basin exhibited strong spatial dependence and increased with elevation, as revealed through geostatistical analysis of 55 years of CHIRPS and station data. The Gaussian semi variogram model effectively captured this spatial structure, and the remote sensing indices indicated that rainfall variability significantly influenced land use and vegetation dynamics. The findings highlight geostatistics as a robust tool for understanding precipitation patterns and guiding land and water resource management in semi-arid environments.

Khan and Chen (Contribution 5) examined the impact of LULC change on environmental conditions in Khyber Pakhtunkhwa, Pakistan, using Landsat imagery from 2000 to 2020, along with future projections up to 2100. Urban areas increased by 138%, significantly contributing to rising land surface temperatures, which in turn negatively affected key environmental indices such as NDVI, MNDWI, and NDMI, indicating declines in vegetation, water resources, and soil moisture. Future scenarios project continued urban expansion, vegetation loss, and intensifying environmental stress, underscoring the urgent need for sustainable land use planning, green infrastructure, and climate adaptation measures to mitigate ecological degradation and heat-related risks.

The Region-Specific Model Adaptation (RSMA) method was introduced by Li et al. (Contribution 6) to improve training data extraction for long-term, large-scale land cover mapping in Alaska—a region with limited historical datasets and complex land cover variability. RSMA combines data from NLCD, LANDFIRE EVT, and NWI with machine learning to generate high-quality training samples, accounting for regional spectral variations while preserving generalizable land cover traits. Validation using vegetation plot data showed that RSMA improved classification accuracy by approximately 30% over models using only NLCD-derived training data, demonstrating its effectiveness for large-scale, data-sparse regions.

Liu et al. (Contribution 7) introduced an Explainability Feature Bands Adaptive Selection Network (EFBASN) approach for hyperspectral image classification, addressing challenges such as poor interpretability, high dimensionality, and classification accuracy. The method identified and visualized salient spectral bands that contributed most to classification, enhancing both performance and interpretability of deep learning models. On benchmark datasets like Pavia University, EFBASN achieved state-of-the-art accuracy (97.68%), outperforming existing methods. Importantly, it assigned significantly higher attention weights to physically meaningful bands (e.g., 672 nm for iron oxide), offering scientifically grounded insights into spectral feature relevance.

The study by Lucas et al. (Contribution 8) evaluated past and predicted future LULC changes in Ethiopia's Upper Omo–Gibe River basin from 1991 to 2067 using satellite imagery and a hybrid Cellular Automata–Artificial Neural Network (CA-ANN) model implemented through QGIS's MOLUSCE plugin. Classification was performed using the random forest algorithm, achieving high accuracy (kappa = 0.82). The results showed that agricultural land and shrubland are projected to expand significantly between 2022 and 2037, while forest and grassland decline. From 2037 to 2067, urban expansion and continued agricultural growth are expected to further reduce forest cover. These trends highlight the need for targeted land management strategies to address environmental concerns stemming from ongoing LULC changes.

Selmy et al. (Contribution 9) applied a CA–Markov hybrid model, along with Landsat imagery and GIS techniques, to detect, analyze, and predict LULC changes in the arid Sohag Governorate of Egypt from 1984 to 2050. The analysis revealed significant urban expansion and growth in cultivated lands—primarily at the expense of desert lands—driven by urbanization and land reclamation. The model demonstrated high predictive accuracy (Kappa: 0.84–0.93), indicating its strong utility for forecasting LULC trends and informing sustainable land management and policy decisions in arid regions.

Shastry et al. (Contribution 10) demonstrated that machine learning (ML) could significantly improve the accuracy of bedrock outcrop mapping in the Sierra Nevada Mountains compared to the existing 30 m resolution USGS NLCD, which underrepresents barren land. By training a model on high-resolution 0.6 m NAIP imagery and labeled data from 83 km^2 of the region, the ML approach achieved 90% overall accuracy, 83% precision, and 78% recall in detecting bedrock. The results showed that NLCD captured only about 40% of bedrock mapped by the ML model, suggesting that ML-based methods provide enhanced detail and reliability for environmental and geological applications like erosion modeling and land–atmosphere interaction studies.

Sun et al. (Contribution 11) introduced a novel, high-quality 1 km monthly NDVI time series for China, reconstructed using a Random Forest (RF) model to overcome contamination issues (e.g., clouds, snow) in MODIS NDVI data. By integrating climate, geographic, and auxiliary datasets, the RF model accurately predicted NDVI values, achieving superior performance (MAE = 0.024, RMSE = 0.034, R^2 = 0.974) compared to existing methods like STSG and LSTM. The resulting product showed improved spatial–temporal continuity and robustness in complex environmental conditions, offering a valuable tool for monitoring vegetation dynamics in China.

Sun et al. (Contribution 12) presented a hybrid method combining Otsu thresholding and Random Forest classification to produce high-resolution (10 m) LULC maps using Sentinel-2 imagery for Wensu County, focusing on seasonal changes, particularly in water and snow/ice dynamics. The approach achieved strong classification accuracy—93.77%

for vegetation and 85.50% for water/snow/ice—highlighting the significance of terrain features in improving model performance. Seasonal analysis revealed notable expansion of snow/ice in winter, covering large areas of bare land and grassland, with most changes occurring in areas with moderate winter temperatures (>−8 °C). The method offers a robust framework for monitoring environmental changes and supporting resource and agricultural management.

Zhang et al. (Contribution 13) introduced a High-Performance Automated Large-Area Land-Cover Mapping Framework (HALF), designed to address the inefficiencies and computational demands of large-scale, multi-temporal land cover mapping. By integrating Docker, workflow technologies, and high-performance computing, HALF streamlined model deployment, automated sample generation, and significantly accelerated image processing and classification—achieving a tenfold increase in speed during sample–image matching and the rapid mosaicking of classification outputs. The framework's modular, scalable design demonstrates strong potential for producing high-resolution land cover maps efficiently across extensive regions, advancing environmental monitoring and sustainable land management.

3. Conclusions

This Special Issue presents a broad array of methodological approaches and applications in land use and land cover mapping, monitoring, and analysis using remote sensing. A central theme across the contributions is the integration of machine learning and deep learning techniques—particularly Random Forest, Convolutional Neural Networks, and novel explainable AI architectures—which have notably enhanced classification accuracy, scalability, and interpretability. These methods are especially effective in handling high-dimensional and heterogeneous remote sensing datasets, highlighting their increasing significance in operational LULC monitoring. Notably, traditional supervised classification methods appeared in only one of the studies.

Several contributions extend beyond mapping current and historical LULC patterns, focusing instead on long-term monitoring and scenario-based projections of future changes. These studies often employ hybrid models such as Cellular Automata–Artificial Neural Networks and CA-Markov, typically using remote sensing products as primary inputs. Such approaches enable detailed assessments of the environmental impacts of land transformations, including vegetation change, soil erosion, and urban expansion, frequently in the context of climate change scenarios (e.g., RCP pathways).

A distinguishing feature of the contributions is their geographical breadth, with case studies spanning diverse ecological and socio-economic contexts—from the semi-arid regions of Brazil and Egypt to the mountainous landscapes of Alaska and the Sierra Nevada. These region-specific analyses highlight the importance of context-sensitive model adaptation, and the development of methodologies suited to data-scarce and environmentally complex settings.

Most studies utilized imagery from Landsat sensors (TM, ETM+, and OLI), with additional cases drawing on Sentinel-2 and MODIS-derived products. Thematic databases derived from remote sensing also played a key role in supporting these analyses. Advances in data fusion—particularly the integration of satellite imagery with climatic and ancillary datasets—were shown to contribute to the production of more detailed and accurate LULC classifications.

Finally, many of the studies exhibited a clear alignment with policy-relevant issues, including urbanization, climate adaptation, and resource management. This highlights the practical utility of remote sensing-based LULC monitoring not only for advancing scientific understanding but also for informing evidence-based decision-making in environmental and land governance.

In summary, the contributions to this Special Issue reflect a rapidly evolving field characterized by increasing technical sophistication and growing applied relevance. Nevertheless, the contributions also highlight key directions for future research in the application of remote sensing to LULC mapping. Future progress will depend on expanding access to high-quality training data, enhancing model generalizability and interpretability, and fostering deeper integration of LULC science with broader climate, ecological, and socio-economic systems. The increasing availability and diversity of Earth observation data present new opportunities for near-real-time LULC monitoring, emphasizing the need for robust pipelines capable of automated change detection and alert generation. While there remains significant scope for advancing explainable AI techniques that provide transparent insights into classification and prediction processes, future work should also focus on adapting interpretable AI frameworks to address the spatiotemporal complexities of LULC dynamics. In parallel, research should prioritize the development of scalable approaches for multimodal data assimilation, integrating optical, radar, thermal, and ancillary socio-environmental datasets to improve predictive performance. The creation, thorough documentation, and open dissemination of large, labeled datasets would further support the development and validation of generalizable AI models. Additionally, extending research on transfer learning and domain adaptation to LULC applications—particularly in data-scarce regions or environmentally heterogeneous isolated ecosystems—remains a critical priority. Finally, a closer coupling of LULC modeling efforts with socio-economic forecasting and Earth system models could enable more comprehensive scenario analyses under varying climate trajectories, thereby enhancing both scientific insight and the utility of remote sensing for evidence-based environmental planning and policy-making.

Author Contributions: S.G. and G.M. contributed equally to this work. All authors have read and agreed to the published version of the manuscript.

Funding: This research received no external funding.

Acknowledgments: As the Guest Editors, we would like to thank all the authors who accepted the challenge of sharing their research results and ideas in this Special Issue. Special thanks to the reviewers and the editorial staff of *Remote Sensing*.

Conflicts of Interest: The authors declare no conflicts of interest.

Abbreviations

The following abbreviations are used in this manuscript:

AI	Artificial Intelligence
CA-ANN	Cellular Automata–Artificial Neural Network
CHIRPS	Climate Hazards Group InfraRed Precipitation with Stations
CNN	Convolutional Neural Networks
EFBASN	Explainability Feature Bands Adaptive Selection Network
EVT	LANDFIRE's Existing Vegetation Type
HALF	High-Performance Automated Large-Area Land Cover Mapping Framework
LSTM	Long Short-Term Memory
LULC	Land Use/Land Cover
MAE	Mean Absolute Error
ML	Machine Learning
NB	Naive Bayes
NDVI	Normalized Difference Vegetation Index

NLCD	National Land Cover Database
NWI	National Wetlands Inventory
OA	Overall Accuracy
R^2	Coefficient of Determination
RCP	Representative Concentration Pathway
RF	Random Forest
RMSE	Root Mean Square Error
RSMA	Region-Specific Model Adaptation
STSG	Spatial–Temporal Savitzky–Golay
SVM	Support Vector Machine
USGS	United States Geological Survey

List of Contributions:

1. Ajibola, S.; Cabral, P. A Systematic Literature Review and Bibliometric Analysis of Semantic Segmentation Models in Land Cover Mapping. *Remote Sens.* **2024**, *16*, 2222. https://doi.org/10.3390/rs16122222.
2. Admas, M.; Melesse, A.M.; Tegegne, G. Predicting the Impacts of Land Use/Cover and Climate Changes on Water and Sediment Flows in the Megech Watershed, Upper Blue Nile Basin. *Remote Sens.* **2024**, *16*, 2385. https://doi.org/10.3390/rs16132385.
3. Bhungeni, O.; Ramjatan, A.; Gebreslasie, M. Evaluating Machine-Learning Algorithms for Mapping LULC of the uMngeni Catchment Area, KwaZulu-Natal. *Remote Sens.* **2024**, *16*, 2219. https://doi.org/10.3390/rs16122219.
4. de Barros de Sousa, L.; de Assunção Montenegro, A.A.; da Silva, M.V.; Almeida, T.A.B.; de Carvalho, A.A.; da Silva, T.G.F.; de Lima, J.L.M.P. Spatiotemporal Analysis of Rainfall and Droughts in a Semiarid Basin of Brazil: Land Use and Land Cover Dynamics. *Remote Sens.* **2023**, *15*, 2550. https://doi.org/10.3390/rs15102550.
5. Khan, M.; Chen, R. Assessing the Impact of Land Use and Land Cover Change on Environmental Parameters in Khyber Pakhtunkhwa, Pakistan: A Comprehensive Study and Future Projections. *Remote Sens.* **2025**, *17*, 170. https://doi.org/10.3390/rs17010170.
6. Li, C.; Xian, G.; Jin, S. A "Region-Specific Model Adaptation (RSMA)"-Based Training Data Method for Large-Scale Land Cover Mapping. *Remote Sens.* **2024**, *16*, 3717. https://doi.org/10.3390/rs16193717.
7. Liu, J.; Lan, J.; Zeng, Y.; Luo, W.; Zhuang, Z.; Zou, J. Explainability Feature Bands Adaptive Selection for Hyperspectral Image Classification. *Remote Sens.* **2025**, *17*, 1620. https://doi.org/10.3390/rs17091620.
8. Lukas, P.; Melesse, A.M.; Kenea, T.T. Prediction of Future Land Use/Land Cover Changes Using a Coupled CA-ANN Model in the Upper Omo–Gibe River Basin, Ethiopia. *Remote Sens.* **2023**, *15*, 1148. https://doi.org/10.3390/rs15041148.
9. Selmy, S.A.H.; Kucher, D.E.; Mozgeris, G.; Moursy, A.R.A.; Jimenez-Ballesta, R.; Kucher, O.D.; Fadl, M.E.; Mustafa, A.-R.A. Detecting, Analyzing, and Predicting Land Use/Land Cover (LULC) Changes in Arid Regions Using Landsat Images, CA-Markov Hybrid Model, and GIS Techniques. *Remote Sens.* **2023**, *15*, 5522. https://doi.org/10.3390/rs15235522.
10. Shastry, A.; Cerovski-Darriau, C.; Coltin, B.; Stock, J.D. Mapping Bedrock Outcrops in the Sierra Nevada Mountains (California, USA) Using Machine Learning. *Remote Sens.* **2025**, *17*, 457. https://doi.org/10.3390/rs17030457.
11. Sun, M.; Gong, A.; Zhao, X.; Liu, N.; Si, L.; Zhao, S. Reconstruction of a Monthly 1 km NDVI Time Series Product in China Using Random Forest Methodology. *Remote Sens.* **2023**, *15*, 3353. https://doi.org/10.3390/rs15133353.
12. Sun, X.; Li, X.; Tan, B.; Gao, J.; Wang, L.; Xiong, S. Integrating Otsu Thresholding and Random Forest for Land Use/Land Cover (LULC) Classification and Seasonal Analysis of Water and Snow/Ice. *Remote Sens.* **2025**, *17*, 797. https://doi.org/10.3390/rs17050797.
13. Zhang, J.; Fu, Z.; Zhu, Y.; Wang, B.; Sun, K.; Zhang, F. A High-Performance Automated Large-Area Land Cover Mapping Framework. *Remote Sens.* **2023**, *15*, 3143. https://doi.org/10.3390/rs15123143.

References

1. Lambin, E.F.; Turner, B.L.; Geist, H.J.; Agbola, S.B.; Angelsen, A.; Bruce, J.W.; Coomes, O.T.; Dirzo, R.; Fischer, G.; Folke, C.; et al. The causes of land-use and land-cover change: Moving beyond the myths. *Glob. Environ. Change* **2001**, *11*, 261–269. [CrossRef]
2. Foley, J.A.; DeFries, R.; Asner, G.P.; Barford, C.; Bonan, G.; Carpenter, S.R.; Chapin, F.S.; Coe, M.T.; Daily, G.C.; Gibbs, H.K.; et al. Global consequences of land use. *Science* **2005**, *309*, 570–574. [CrossRef] [PubMed]
3. Seto, K.C.; Güneralp, B.; Hutyra, L.R. Global forecasts of urban expansion to 2030 and direct impacts on biodiversity and carbon pools. *Proc. Natl. Acad. Sci. USA* **2012**, *109*, 16083–16088. [CrossRef] [PubMed]
4. d'Amour, C.B.; Reitsma, F.; Baiocchi, G.; Barthel, S.; Güneralp, B.; Erb, K.; Haberl, H.; Creutzig, F.; Seto, K.C. Future urban land expansion and implications for global croplands. *Proc. Natl. Acad. Sci. USA* **2016**, *114*, 8939–8944. [CrossRef] [PubMed]
5. Ma, L.; Liu, Y.; Zhang, X.; Ye, Y.; Yin, G.; Johnson, B.A. Deep learning in remote sensing applications: A meta-analysis and review. *ISPRS J. Photogramm. Remote Sens.* **2019**, *152*, 166–177. [CrossRef]
6. Wu, S.; Cao, J.; Zhao, X. Land cover classification of high-resolution remote sensing images based on improved spectral clustering. *PLoS ONE* **2025**, *20*, e0316830. [CrossRef] [PubMed]

remote sensing MDPI

Review

A Systematic Literature Review and Bibliometric Analysis of Semantic Segmentation Models in Land Cover Mapping

Segun Ajibola [1,2,*] and Pedro Cabral [2,3]

[1] Afridat UG (haftungsbeschränkt), Sebastianstrasse 38, 53115 Bonn, Germany
[2] NOVA Information Management School (NOVA IMS), Universidade Nova de Lisboa, Campus de Campolide, 1070-312 Lisbon, Portugal; cabral@nuist.edu.cn
[3] School of Remote Sensing and Geomatics Engineering, Nanjing University of Information Science & Technology, Nanjing 210044, China
* Correspondence: d20210969@novaims.unl.pt

Abstract: Recent advancements in deep learning have spurred the development of numerous novel semantic segmentation models for land cover mapping, showcasing exceptional performance in delineating precise boundaries and producing highly accurate land cover maps. However, to date, no systematic literature review has comprehensively examined semantic segmentation models in the context of land cover mapping. This paper addresses this gap by synthesizing recent advancements in semantic segmentation models for land cover mapping from 2017 to 2023, drawing insights on trends, data sources, model structures, and performance metrics based on a review of 106 articles. Our analysis identifies top journals in the field, including MDPI *Remote Sensing*, IEEE *Journal of Selected Topics in Earth Science*, and IEEE *Transactions on Geoscience and Remote Sensing*, IEEE *Geoscience and Remote Sensing Letters*, and ISPRS *Journal Of Photogrammetry And Remote Sensing*. We find that research predominantly focuses on land cover, urban areas, precision agriculture, environment, coastal areas, and forests. Geographically, 35.29% of the study areas are located in China, followed by the USA (11.76%), France (5.88%), Spain (4%), and others. Sentinel-2, Sentinel-1, and Landsat satellites emerge as the most used data sources. Benchmark datasets such as ISPRS Vaihingen and Potsdam, LandCover.ai, DeepGlobe, and GID datasets are frequently employed. Model architectures predominantly utilize encoder–decoder and hybrid convolutional neural network-based structures because of their impressive performances, with limited adoption of transformer-based architectures due to its computational complexity issue and slow convergence speed. Lastly, this paper highlights existing key research gaps in the field to guide future research directions.

Keywords: remote sensing; semantic segmentation; land cover mapping; deep learning; land cover classification

1. Introduction

Semantic segmentation models are crucial in land cover mapping especially in generating precise Land Cover (LC) maps [1]. LC maps show various types of land cover, such as forests, grasslands, wetlands, urban areas, and bodies of water. These maps are typically created using Remote Sensing (RS) data like satellite imagery or aerial photography [2]. Land cover maps serve for different purposes, including land use management [3], disaster management, urban planning [4], precision agriculture [5], forestry [6], building infrastructure development [7], climate changes problems [8], and others.

Due to advancements in Deep Convolutional Neural Network (DCNN) models, the domain of land cover mapping has progressively evolved [9]. DCNN are potentially successful for extracting information from high-resolution RS data [10]. They possess deep layers and hierarchical architectures, aiming to automatically identify high-level patterns in data [11]. Although DCNN have shown impressive performance in the image classification task, the conventional models still struggle to capture comprehensive global information as well as long-range dependencies inherent to RS data [12]. Consequently, they may not achieve precise image segmentation because they could potentially overlook certain edge details of objects [13]. However, they have successfully contributed valuably to semantic segmentation methodologies.

The semantic segmentation model allocates every pixel in an image to a predefined class [14]. Its effectiveness in landcover segmentation has established it as a mainstream method [15]. It has demonstrated an enhanced performance leading to more accurate segmentation outcomes. A prominent example of a state-of-the-art (SOTA) semantic segmentation model is UNet [16]. Recently, many novel semantic segmentation models tailored for land cover mapping have been proposed, including DFFAN [12], MFANet [17], Sgformer [18], UNetFormer [19], and CSSwin-unet [20]. Accordingly, these models have demonstrated exceptional segmentation accuracy in this domain.

In recent years, literature reviews have predominantly explored Deep Learning (DL) semantic segmentation models. The review by [21] presented significant methods, their origins, and contributions, including insights into datasets, performance metrics, execution time, and memory consumption relevant to DL-based segmentation projects. Similar reviews by [14,22] categorized existing semantic segmentation with DL methods based on criteria like supervision degree during training and architectural design. In addition, ref. [23] summarized various semantic segmentation models for RS Imagery. These reviews offer comprehensive overviews of DL-based semantic segmentation models but have not specifically examined their application to land cover mapping. To address this gap, this literature review focuses on emerging semantic segmentation models in land cover mapping, aiming to answer predefined research questions quantitatively and qualitatively. Our objective is to identify knowledge gaps in semantic segmentation models applied to land cover mapping and understand the evolution of these models in relation to domain-specific studies, data sources, model structures, and performance metrics. Furthermore, this review offers insights for future research directions in land cover mapping.

The next sections of this review are structured as follows: Section 2 provides the research questions and method used in conducting the review. Section 3 delves into the results obtained from the performed systematic literature review, discusses the evolution and trends, the domain study, the data, and semantic segmentation methodologies. Section 4 challenges and provides future insights into land cover mapping. Finally, Section 5 summarizes the highlights of the review.

2. Materials and Methods

The methodology used in conducting this comprehensive literature review follows the PRISMA framework as outlined by [24] of identification, eligibility, screening, and data extraction. A search strategy was developed to identify the literature for this review (Figure 1). Peer-reviewed papers published in relevant journals between 2017 and 2023 are reviewed. In this section, the research questions are formulated, thereafter, the search strategy was defined using the Scopus database (https://scopus.com accessed on 23 February 2024). The selection of the inclusion and exclusion criteria was explained as well as the eligibility criteria defined to assess articles utilized as the final record for the bibliometric analysis.

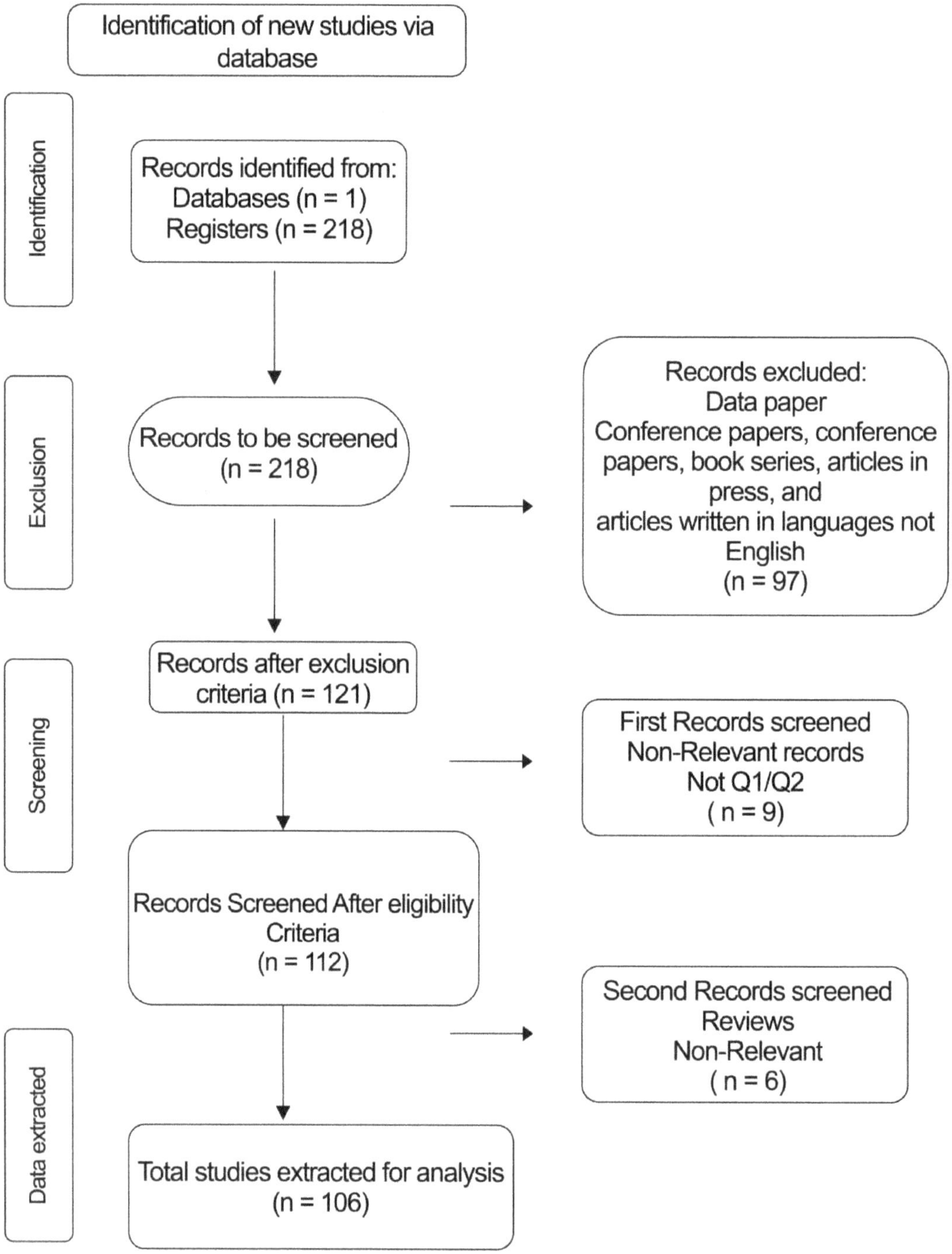

Figure 1. Flow chart of peer-review procedure.

2.1. Research Questions (RQs)

The objectives of the study are addressed based on 4 main research questions. These RQs are specifically selected to elicit trends, benchmark datasets, the state-of-the-art architecture, and performances of semantic segmentation in land cover mapping. This review is built around the following RQs:

RQ1. What are the emerging patterns in land cover mapping?
RQ2. What are the domain studies of semantic segmentation models in land cover mapping?
RQ3. What are the data used in semantic segmentation models for land cover mapping?
RQ4. What are the architecture and performances of semantic segmentation methodologies used in land cover mapping?

2.2. Search Strategy

The search strategy was carried out using Scopus (Elsevier) database. This database is renowned in the scientific community for its high-quality journals, extensive, multidisciplinary abstract collection, and it is an excellent fit for the purposes of review articles [25]. The search keywords used as search criteria to identify articles from the Scopus database are 'semantic AND segmentation', AND 'land AND use AND land AND cover', AND 'deep AND learning', AND 'land AND cover AND classification'. The search string is used to find relevant papers that link deep learning semantic segmentation to land use and land cover.

2.3. Study Selection Criteria

After the defined search strings were entered on Scopus, a total of 218 articles were initially retrieved from the scientific database. To further process the data retrieved, the papers are filtered between 2017 and 2023 excluding conference papers, conference reviews, data paper, books, and book chapters, articles in the press, conference proceedings, book series and articles that are not in English language. The stated period 2017–2023 is selected to provide us with the recent developments in the field and during search, there were almost no notable articles published before 2017 on the subject matter. Consequently, the retrieved record was reduced to 121 articles.

2.4. Eligibility and Data Analysis

To determine eligibility and quality assessment of the extracted papers, we remove articles published in journals that are not Q1 or Q2. This is to ensure that the articles with a rigorous peer-review process and quality research output in this field are selected and synthesized for a top-quality review. Following this criteria, 9 articles are excluded, and 112 records are retrieved. For further assessment, titles and abstracts are assessed regarding their relevance to the study. The relevant study focused on articles that implemented various deep learning semantic segmentation models focused on land use/land cover classification, and/or various satellite datasets extracted through semantic segmentation in different applications. At the end of this step, 6 articles are further excluded bringing down the record to a total number of 106 articles between the period of 2017 to 2023, which are eligible for bibliometric analysis.

2.5. Data Synthesis

Data synthesis is an important way to the answer the RQs. The data are visualized in such a way that it presents findings and synthesizes through quantitative and qualitative analysis. Categorization and visualization were carried out to draw important trends and findings of deep learning semantic segmentation models in land cover mapping in relation to datasets, applications, architecture, and performance. The study's synthesizing themes were developed by full-text content analysis of the 106 articles. Vosviewer software version 1.6.18 (https://vosviewer.com accessed on 24 February 2024) was used to provide graphical visualizations of occurrence of key terms, taking title, abstract, and keywords as input.

3. Results and Discussion

This section investigates the results derived from the performed systematic literature review. Using 106 extracted articles, the results are presented and discussed based on the RQs.

3.1. RQ1. What Are the Emerging Patterns in Land Cover Mapping?

- Annual distribution of research studies

Figure 2 shows the number of research articles published annually from 2017 to 2023. The year 2017 saw a modest output of merely two articles. There is a surge in the number of articles to 12 in 2020, 17 articles by the year 2021, 30 articles by 2022 and 40 articles by 2023. This observation aligns with the understanding that the adoption of deep learning semantic segmentation models on satellite imagery gained significant momentum in 2020 and subsequent years.

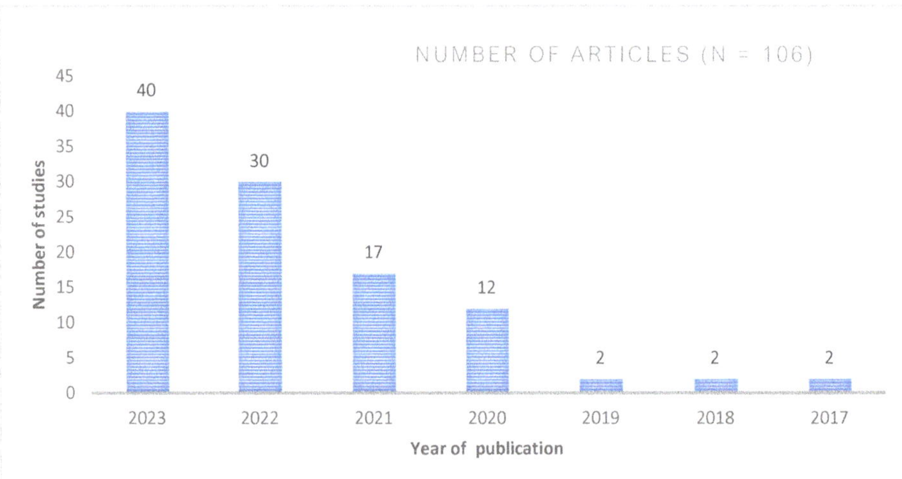

Figure 2. Annual distribution of research studies.

- Leading Journals

Figure 3 depicts the number of articles published in academic journals of this domain. The top 13 journals produced over 81% of the number of research studies of semantic segmentation in land cover mapping. MDPI Remote Sensing (30) has the highest number of published articles in this domain, followed by IEEE Journal of selected topics in Earth Science (15), IEEE Transactions on Geoscience And Remote Sensing (12), IEEE Geoscience and Remote Sensing Letters (5), ISPRS Journal Of Photogrammetry And Remote Sensing (4), and so on, while 20 other journals have one article each published and are grouped as the "other" category.

- Geographic distribution of studies

In terms of geographical distribution of studies extracted from the Scopus database, 35 countries contributed to the study domain. Almost all continents have contributions, except for the African continent. Figure 4 shows that China published 63 articles of the total 106 articles, the second is the United States with 10 articles, followed by India (6), Italy (5), South Korea (4), the United Kingdom(4), Canada, Finland, and Germany (3), then Austria, Australia, Brazil, France, Greece, Turkey, the Netherlands, and Japan (2) while the rest of the countries have one each.

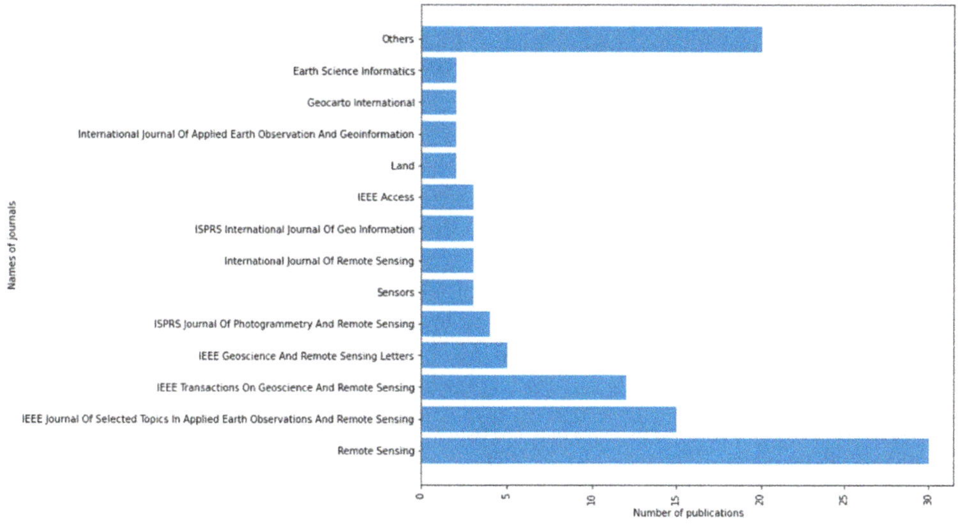

Figure 3. Number of relevant publications in this field distributed per relevant journals.

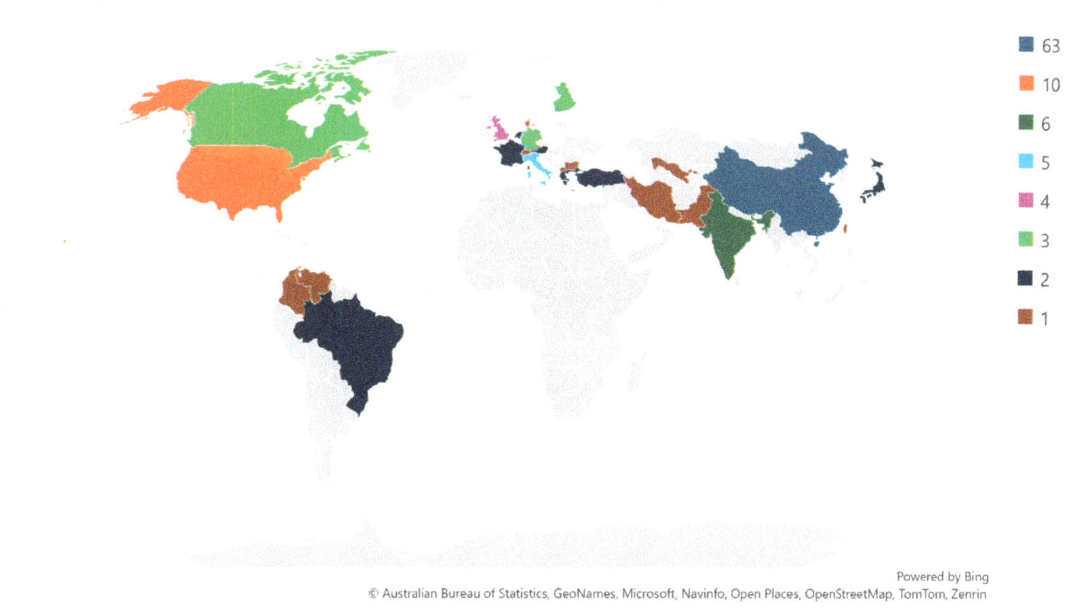

Figure 4. Geographic distribution of studies per country.

- Leading Themes and Timelines

The significant keyword occurrences are obtained from the titles and abstracts of the extracted articles. Figure 5 shows relevant and leading keywords. A threshold of five was set, which means the minimum number of occurrences of a keyword. Only 68 out of 897 keywords met the threshold. Bibliometric analysis reveals that keywords such as "high-resolution RS images", "Remote Sensing", "satellite imagery" and "very high resolution" exhibit prominence, showing strong associations with neural network-related terms including "Semantic Segmentation", "Deep Learning,", "Machine Learning", and "Neural Network". The Semantic Segmentation has “attention mechanisms” and “transformer” as different model’s architectural component. These learning models are further linked to various application domains, evident in their connections to terms like "Land Cover Classification", "Image Classification", "Image Segmentation", "Land Cover", "Land Use “, “Change Detection”, and "Object Detection." In 2020, the research revolved around network architectures, object detection, and image processing. In the later part of 2021, there was a notable shift in research domains, predominantly towards image segmentation, image classification, and land cover segmentation. In 2022 and 2023, there was a pronounced shift in research focusing more on semantic segmentation employing satellite high-resolution images for change detection, land use, and land cover classification and segmentation.

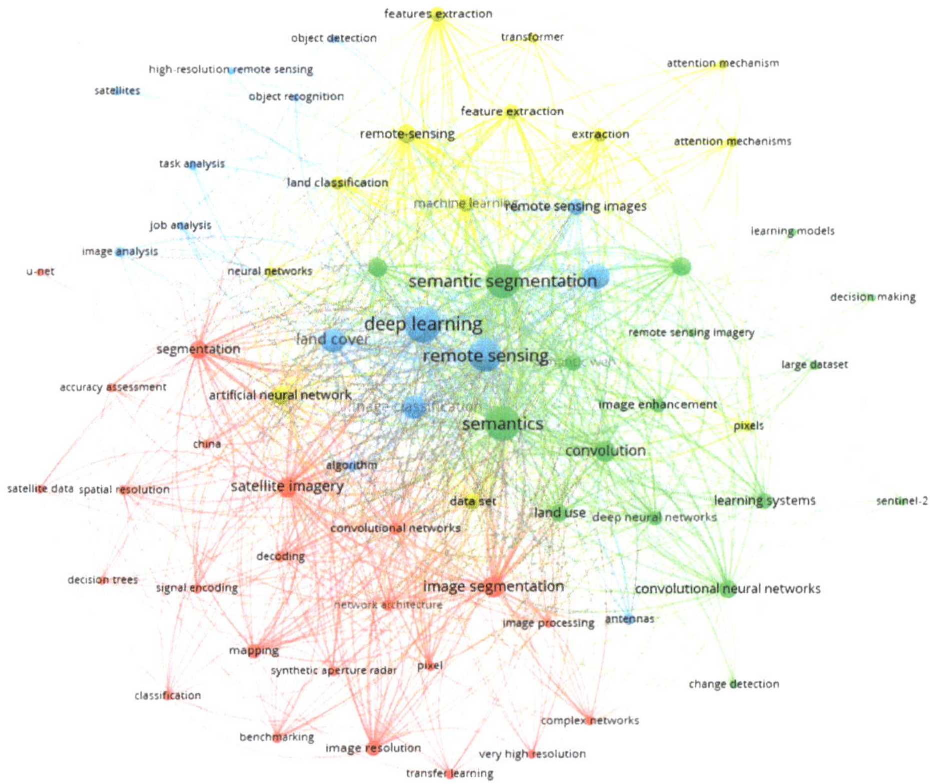

Figure 5. Significant keywords occurrence driving domain theme.

3.2. *RQ2. What Are Domain Studies of Semantic Segmentation Models in Land Cover Mapping?*

In this section, each extracted paper is clustered based on similar study domain areas. Figure 6 shows the overall mind map of the domain studies. Land cover, urban, precision agriculture, environment, coastal areas, and forest are the mostly studied domain areas.

Figure 7 shows that among the 106 articles, 36 studies cover both land cover classification and segmentation, 24 specifically focus on land cover classification, 15 concentrate on urban applications, 9 address land cover segmentation alone, 8 addresses environment issues, 5 center around precision agriculture, 5 are oriented toward coastal applications, 3 articles addressing land cover change detection and 2 focus on forestry.

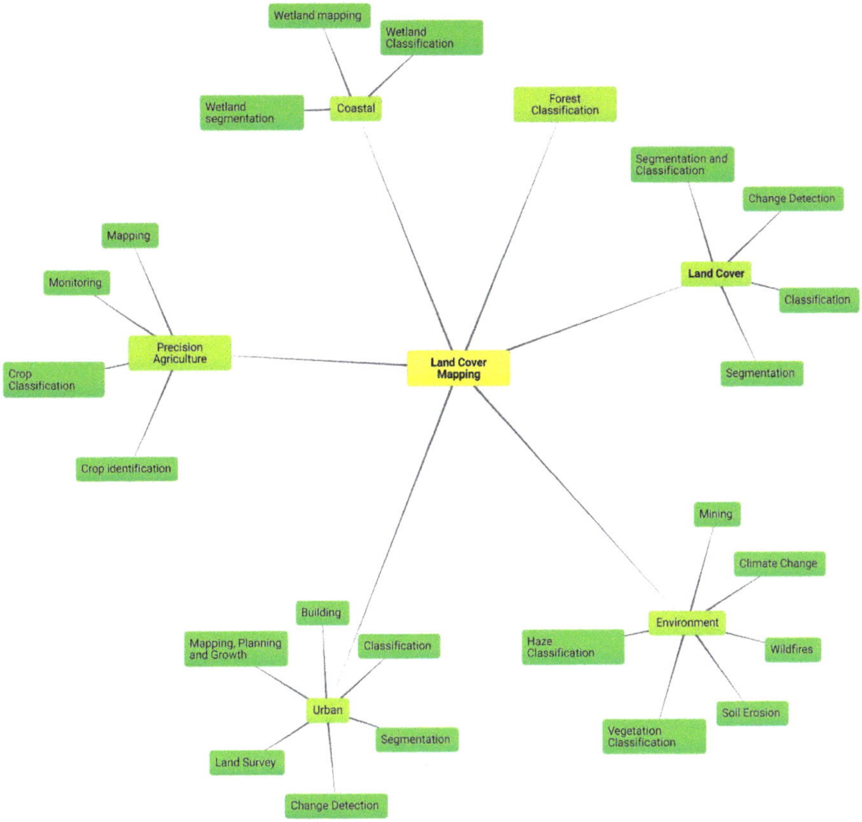

Figure 6. Land cover mapping domain studies.

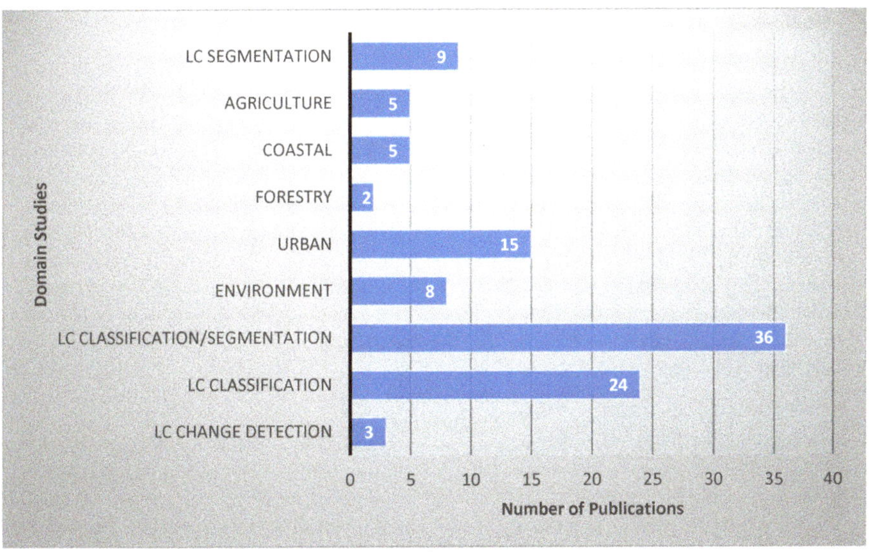

Figure 7. Number of publications per domain studies.

- Land Cover Studies

In the 72 out of 106 studies related to land cover, research activities encompass land cover classification (33.3%), land cover segmentation (12.5%), change detection (4.2%), and the combined application of land cover classification and segmentation (50%). These areas of study are extensively documented and represent the most widely researched applications within land cover studies. Land cover classification involves assigning an image to one of several classes of land use and land cover (LULC), while land cover segmentation entails assigning a semantic label to each pixel within an image [26]. Land cover refers to various classes of biophysical Earth cover, while land use describes how human activities modify land cover. On the other hand, change detection plays a crucial role in monitoring LULC changes by identifying changes over time periods, which can help predict future events or environmental impacts. Change detection methods employing DL have attained remarkable achievements [27] across various domains, including urban change detection [4], agriculture, forestry, wildfire management, and vegetation monitoring [28].

- Urban

Among the 15 publications related to urban studies, 38% focus on segmentation applications, including urban scene segmentation [19,29,30], while 31% address urban change detection for mapping, planning, and growth [31,32]. For instance, change detection techniques provide insights into urban dynamics by identifying changes from remote sensing imagery [7], including changes in settlement areas. Additionally, these studies involve predicting urban trends and growth over time, managing land use [3], and monitoring urban densification [33] as well as mapping built-up areas to assess human activities across large regions [34].

Publications in urban studies also cover 6% in land survey management [35] and 25% in urban classification and detection [36–40], particularly in building applications and for urban land-use Classification [40].

- Precision Agriculture

Out of the five publications concerning precision agriculture studies, the research focuses on various aspects such as crop mapping [41], identification [42,43], classification [44], and monitoring [45]. For example, mapping large-scale rice farms [5], monitoring crops to analyze different growth stages [42], and classifying Sentinel data for creating an oil palm land cover map [46].

- Environment

Among the eight environmental studies analyzed, the research spans various applications. These include 25% focusing on soil erosion applications [47,48], which involves rapid monitoring of ground covers to mitigate soil erosion risks. Additionally, 12.5% of the studies center around wildfire applications [49], encompassing burned area mapping, wildfire detection [50], and smoke monitoring [51], along with initiatives for preventing wildfires through sustainable land planning [8]. Another 12.5% of the studies involve haze classification [52], specifically cloud classification using Sentinel-2 imagery. Climate change research [53] accounts for another 12.5% of the studies, focusing on aspects such as the urban thermal environment. Furthermore, 12.5% of the studies are dedicated to vegetation classification [54] and an additional 25% address mining applications [55] including the detection of changes in mining areas [56].

- Forest

Research in this domain encompasses forest classification, including the classification of landscapes affected by deforestation [57]. More so, change detection in vegetation and forest areas enables decision-makers, conservationists, and policymakers to make informed decisions through forest monitoring initiatives [6] and mapping strategies tailored to tropical forests [58].

- Coastal Areas

Within this field, 5 studies out of the 106 articles focus on wetland mapping, classification, and segmentation [10,59–62]. However, the exploration and study of coastal area remote sensing image segmentation remains a relatively underexplored research area, as noted by [60]. This challenge is primarily attributed to the significant complexities associated with coastal land categories, including issues such as homogeneity, multiscale features, and class imbalance, as highlighted [63].

3.3. RQ3. What Are the Data Used in Semantic Segmentation Models for Land Cover Mapping?

In this section, this paper synthesizes extensively employed, particularly for the study location, data source and benchmark datasets used for land cover mapping.

- Study Locations

Figure 8 illustrates the countries where the study areas were located and where in-depth research was conducted among the extracted articles. Among the 22 countries represented in 51 studies, 35.29% of the study areas are located in China, with 11.76% in the USA, 5.88% in France, and 3.92% in Spain, Italy, Brazil, South Korea, and Finland each. Other countries in the chart each account for 1.96% of the study areas.

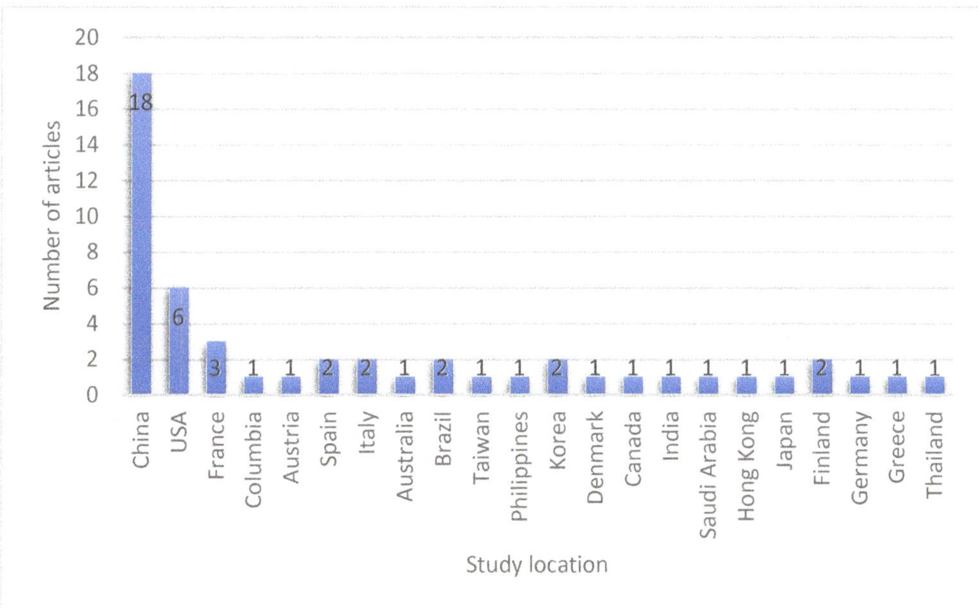

Figure 8. Number of publications per study location.

- Data Sources

Table 1 presents the identified data sources along with the number of articles and their corresponding references. The data sources identified in the literature include RS satellites, RS Unmanned Aerial Vehicles (UAVs) and Unmanned Aircraft Systems (UAS), mobile phones, Google Earth, Synthetic Aperture Radar (SAR), and Light Detection and Ranging (LiDAR) sources. Among these, Sentinel-2, Sentinel-1, and Landsat satellites are the most frequently utilized data sources. It is important to recognize that the primary RS technologies include RS satellite imagery, SAR, and LiDAR.

Table 1. Data sources identified in review.

Data Sources	Number of Articles	References
RS Satellites		
Sentinel-2	7	[45,64–67]
Landsat	5	[33,68–70]
Worldview-03	2	[71,72]
Rapid eye	1	[73]
Worldview-02	1	[74]
Quickbird	1	[74]
ZY-3	1	[48]
PlanetScope	1	[49]
GF-2	2	[48,75]
Aerial images		
Phantom m multi-rotor AUS	1	[59]
Quadcopter drone	1	[61]
Vexcel Ultracam Eagle Camera	1	[76]
DJI-Phantom 4 pro UAV	1	[47]
SAR SAT		
RADARSAT-2	1	[77]
Sentinel-1	6	[10,41,43,64,65,78]
GF-3	1	[79]
ALOS-2	1	[80]
Others		
Earth digitalglobe	2	[44,60]
Mobile phone	1	[35]
Lidar Sources	1	[37]

RS Imagery data are the most extensively utilized in land cover mapping. In RS, data are collected from satellite sources such as Sentinel-1, Landsat, Sentinel-2, WorldView-2, and QuickBird at certain time step intervals for a period. The data capture products of these satellite include Panchromatic (1 Channel – 2D), Multispectral, or Hyperspectral images [1]. RSI can be represented as aerial images [2]; these are taken using Drone or UAVs. These data usually possess spatial resolution and spectral resolution of certain image sizes.

SAR is among the radar systems used in Land Cover Mapping; SAR stands out as a notable data source [81]. SAR utilizes radio detection technology and constitutes an essential tool in this field. SAR data carry distinct advantages, especially in scenarios where optical imagery faces limitations such as cloud cover or limited visibility. SAR can penetrate through cloud cover and offer Earth surface imaging even in the presence of clouds or unfavorable weather conditions. This is one of the key advantages of SAR technology, unlike electromagnetic spectrum which is obstructed by clouds [82].

There are different types of SAR identified in the literature based on different parameters harnessed for the purpose of land cover mapping, such as polarimetric synthetic aperture radar (PolSAR) [80], E-SAR, AIRSAR, and Interferometric Synthetic Aperture Radar [6]. Notable benchmark SAR datasets include Gaofen-3 and RADARSAT-2 datasets [83]. At present, semantic segmentation of PolSAR images holds significant utility in the interpretation of SAR imagery, particularly within agricultural contexts [84]. Similarly, the High-Resolution GaoFen-3 SAR Dataset is useful for the Semantic Segmentation of Building [34,79,85]. The benchmark dataset Gaofen-3 (GF-3), comprised of single-polarization SAR images, holds significant importance [86]. This dataset is derived from China's pioneering civilian C-band polarimetric SAR satellite, designed for high-resolution RS. Notably,

FUSAR-Maps are generated from extensive semantic segmentation efforts utilizing high-resolution GF-3 single-polarization SAR images [87], while the GID dataset is collected from the Gaofen-2 satellite.

LiDAR holds a significant role within the sphere of land cover mapping and climate change [88]. LiDAR involves the emission of laser pulses and the measurement of their return times to precisely gauge distances, creating highly accurate and detailed elevation models of the Earth's surface. It provides detailed information, including topographic features, terrain variations, and the vertical structure of vegetation. It stands as an indispensable data source for land cover mapping endeavors. Notable examples include the utilization of multispectral LiDAR [54], an advanced RS technology merging conventional LiDAR principles with the capacity to concurrently capture multiple spectral bands. There's the Follo 2014 LiDAR data, a dataset that specifically captures Light Detection and Ranging (LiDAR) data in the Follo region during 2014. Additionally, the NIBIO AR5 (Norwegian Institute of Bioeconomy Research—Assessment Report 5) Land Resources dataset, developed by the Norwegian Institute of Bioeconomy Research, represents a comprehensive evaluation of land resources. This dataset encompasses a range of attributes including land cover, land use, and pertinent environmental factors [89].

- Benchmark datasets

The benchmark datasets used for evaluation in this domain as identified in the review are shown in Figure 9. ISPRS Vaihingen and Potsdam are the widely used benchmark datasets, followed by GID, Landcover.ai, DeepGlobe, and WHDLD.

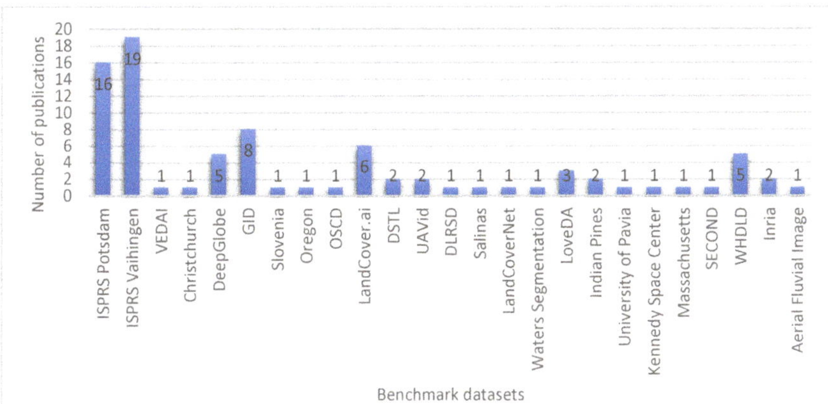

Figure 9. Benchmark datasets identified in the literature.

The ISPRS Vaihingen comprises 33 aerial image patches in IRRGB format along with their associated Digital Surface Model (DSM) data, each with a size of around 2500×2500 pixels at 9 cm spatial resolution [90,91]. Similarly, the publicly accessible ISPRS Potsdam dataset encompasses Potsdam city, Germany. It is composed of 38 aerial IR-RGB images measuring 6000×6000 pixels each at a spatial resolution of 5 centimeters [92].

The Gaofen Image Dataset (GID) dataset [93] contains 150 images of GaoFen-2 data (GF-2 images) [94,95] collected from different cities in China. The GaoFen-2 (GF-2) is high-resolution images acquired from the GaoFen-2 satellite, which captures Panchromatic (PAN) and Multispectral (MS) Spectral Bands. It offers valuable insights into land cover characteristics and patterns. Every image is composed of pixels measuring 6908×7300 and R, G, B, and NIR bands, each with a spatial resolution of 4 meters. The GID dataset consists of five land-use categories: farmland, meadow, forest, water, and built-up [96].

The LandCover.ai dataset [97] comprises images chosen from aerial photographs encompassing 216.27 square kilometers of Poland. The dataset includes 41 RS images, with 33 images having an approximate resolution of 25 cm, measuring around 9000 × 9000 pixels, and 8 images with a resolution of approximately 50 cm, spanning about 4200 × 4700 pixels [12,17,98]. The dataset was manually categorized into four types of objects such as buildings, woodland, and water, as well as the background.

DeepGlobe Data [99] is another important dataset in land cover mapping. The dataset stands out as the inaugural publicly available collection of high-resolution satellite imagery primarily emphasizing urban and rural regions. This dataset comprises a total of 1146 satellite images, each with dimensions of 20448 × 20448 [100]. It is of great importance to the Land Cover Classification Challenge. Likewise, the Inria dataset [101] consists of aerial visual images encompassing 10 regions in the United States and Austria, collected at a 30 cm resolution, with RGB bands [102]. It is organized by five cities in both in training and test data. Every city includes 36 image tiles, each sized at 5000 × 5000 pixels, and these tiles are divided into two semantic categories: buildings and non-building classes.

In addition, the Disaster Reduction and Emergency Management Building dataset exhibits a notable similarity to the Inria dataset. It has image tiles size of 5000 × 5000 with a spatial resolution of 30 cm, all the tiles contain R, G, and B bands [96]. The building dataset from Wuhan University comprises an aerial dataset encompassing 8189 image patches captured at a 30 cm resolution. These images are in RGB format, with each patch measuring 512 × 512 pixels [103]. The Aerial Imagery for Roof Segmentation dataset [104] is composed of aerial images that encompass the Christchurch city area in New Zealand. These images are captured at a resolution of 7.5 cm and include RGB bands [102]. It was captured following the seismic event that impacted the town of Christchurch in New Zealand. Four images, each with dimensions of 5000 × 4000 pixels, were labeled to include the following categories: buildings, cars, and vegetation [105]. Other benchmark datasets include the Massachusetts building and road datasets [106], Dense labeling RS dataset [107], VEhicle Detection in Aerial Imagery (VEDAI) dataset, and LoveDA dataset [108].

3.4. RQ4. What Are the Architecture and Performances of Semantic Segmentation Methodologies Used in Land Cover Mapping?

This section investigates the design and effectiveness of recent advancements in novel semantic segmentation methodologies applied to land cover mapping. In this paper, the methodologies employed in land cover mapping are classified based on similarities in their structural components. We have identified three primary architectural structures: (i) encoder-decoder structures; (ii) transformer structures; and (iii) hybrid structures. Figure 10 shows that among 80 articles employing different model structures, 59% utilized hybrid structures, 36% utilized encoder-decoder structures, and 5% utilized transformer-based structures.

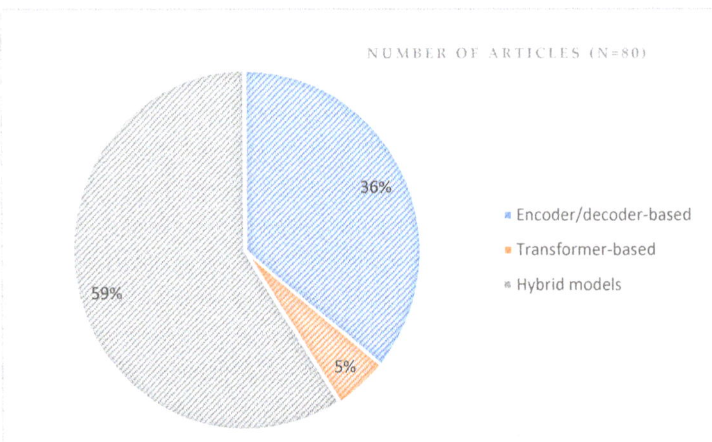

Figure 10. Percentage number of articles employing different model structures in data retrieved.

- Encoder-Decoder based structure

Encoder-decoder structures consist of two main parts: an encoder that processes the input data and extracts high-level features, and a decoder that generates the output (e.g., segmentation map) based on the encoder's representations [16]. The authors [109] suggested an innovative encoding-to-decoding technique known as the Full Receptive Field network, which utilizes two varieties of attention mechanisms, with ResNet-101 serving as the fundamental backbone. Similarly, a different DL segmentation framework known as the DGFNET Dual-gate fusion network. Paper [110] adopts an encoder-decoder architecture design. Typically, encoder-decoder architectures encounter difficulties with the semantic gap. To address this, the DGFNET framework comprises two modules: the Feature Enhancement Module as well as the Dual Gate Fusion Module to mitigate the impact of semantic gaps in deep convolutional neural networks, leading to improved performance in land cover classification. The model underwent evaluation using both the landcover dataset and the Potsdam dataset, achieving MIoU scores of 88.87% and 72.25%, respectively.

The article [111] proposed U-Net incorporating asymmetry and fusion coordination. It is an encoder and decoder architecture with an integrated coordinated attention mechanism, a non-symmetric convolution block refinement fusion block that gets long term dependencies and intricate information from RS data. It was reported that the method was evaluated on DeepGlobe datasets and performed best with an MIoU of 85.54% as reported compared to other models like UNet, MAResU-Net, PSPNet, DeepLab v3+, etc. However, the model has low network efficiency and is not recommended for mobile applications. Also, ref. [112] suggested the Attention dilation-LinkNet neural network, which contains an encoder-decoder structure. It takes advantage of serial-parallel combination dilated convolution and two channel-wise attention mechanisms, as well as a pretrained encoder to be useful for satellite image segmentation particularly road extraction. The best performance of an ensemble of the model achieved an IoU of 64.49% on DeepGlobe road extraction dataset. Table 2 tabulates some semantic segmentation models using the encoder-decoder structure. It shows that these models have relatively impressive generalization performances on different data; however, accuracy can be further enhanced through parameter optimization.

Table 2. Encoder-decoder-based semantic segmentation models for land cover segmentation.

Models	Datasets	Performance Metrics	Limitation/Future Work
RAANet [108]	LoveDA, ISPRS Vaihingen	MIoU = 77.28, MIoU = 73.47	Accuracy can be improved with optimization.
PSE-UNet Model [113]	Salinas Dataset	MIoU = 88.50	Inaccurate segmentation of land cover features with low frequencies, superfluous parameter redundancy, and unvalidated generalization capabilities.
SEG-ESRGAN [114]	Sentinel-2 and WorldView-2 image pairs.	MIoU = 62.78	The assessment of utilizing medium-resolution images has not been tested
Class-wise FCN [26]	Vaihingen, Potsdam	MIoU = 72.35, MIoU = 76.88	Enhancements in performance can be achieved through class-wise considerations for multiple classes, along with improved and more efficient implementations.
MARE [115]	Vaihingen	MIoU = 81.76	Improve performance through parameter optimization and extend approach incorporating other self-supervised algorithms.
Feature fusion with dual attention and flexible contextual adaptation [94]	Vaihingen, GaoFen-2	MIoU = 70.51, MIoU = 56.98	Computational complexity issue.

Table 2. *Cont.*

Models	Datasets	Performance Metrics	Limitation/Future Work
Deanet [100]	LandCover.ai, DSTL dataset, DeepGlobe	MIoU = 90.28, MIoU = 52.70, MIoU = 71.80	Suboptimal performance. Future efforts involve incorporating the spatial attention module into a single unified backbone network.
An encoder-decoder framework featuring attention-guided multi-scale context integration [116]	GF-2 images	MIoU = 62.3%	Reduced accuracy on imbalance data.

- Transformer-based structure

Transformer-based architectures are neural network structures originally designed for natural language processing (NLP), utilizing transformer modules as their fundamental building components. In the context of land cover mapping tasks, transformer-based architectures such as the Swin-S-GF [117], BANet [30], DWin-HRFormer [29], spectral spatial transformer [118], Sgformer [18], and Parallel Swin Transformer have been developed. Table 3 presents various transformer-based structures alongside their performance metrics and limitations. Researchers have noted that while these architectures achieve effective segmentation accuracy with an average OA of approximately 89%, transformers can exhibit slow convergence and be computationally expensive, particularly in land cover mapping tasks. This limitation contributes to their relatively low adoption in land cover segmentation applications.

Table 3. Transformer-based semantic segmentation models for land cover segmentation.

Models	Data	Performance	Limitation
Swin-S-GF [117],	GID	OA = 89.15 MIoU = 80.14	Computational complexity issue and slow convergence speed.
CG-Swin [119]	Vaihingen, Potsdam	OA = 91.68 MIoU = 83.39, OA = 91.93 MIoU = 87.61	Extending CG-Swin to accommodate multi-modal data sources for more comprehensive and robust classification.
BANet [30]	Vaihingen, Potsdam, UAVid dataset	MIoU = 81.35, MIoU = 86.25, MIoU = 64.6	Combine convolution and Transformer as a hybrid structure to improve performance.
Spectral spatial transformer [118]	Indian dataset	OA = 0.94	Computational complexity issue
Sgformer [18]	Landcover dataset	MIOU = 0.85	Computational complexity issue and slow convergence speed.
Parallel Swin Transformer [120]	Postdam, GID WHDLD	OA = 89.44, OA = 84.67, OA = 84.86	Performance can be improved.

- Hybrid-based structure

A hybrid-based structure combines elements from different neural network architectures or techniques to create a unified model for semantic segmentation. Traditional convolutional neural network methods face limitations in accurately capturing boundary details and small ground objects, potentially leading to the loss of crucial information. While deep convolutional neural networks are applied for classifying land use covers, results often show suboptimal performance in the land cover segmentation task [73]. However, this result can be tackled by a hybrid through introduction of encoder-decoder style semantic segmentation models, leverage existing deep learning backbone [68], and explore diverse data settings and parameters in their experimentation [121]. Other methods of the structure's enhancement include architectural modifications through the integration of attention mechanisms, transformer architecture, module fusion, and multi-scale feature fusion [122,123]. An example is the SOCNN framework [124], which addresses the limitation faced by CNN through module integration: (i) a module for semantic segmentation; (ii) a module for superpixel optimization; and (iii) a module for fusion. While the evaluated performance of the framework demonstrated improvement, further enhancement in boundary retrieval can be achieved by incorporating superior boundary adhesion and integrating it into the boundary optimization module.

Moreso, paper [125] proposed multi-level context-guided classification method object-based CNN. It involves high-level feature-fusing and employed a Conditional Random Field for better classification performance. The model attained a comparable overall accuracy with DeepLabV3+ at various segmentation scale parameters on the Vaihingen dataset and suboptimal overall accuracy compared to DeepLabV3+ on the Potsdam dataset. Another approach identified is utilizing a Generative Adversarial Network-based approach for domain adaptation, such as the Full Space Domain Adaptation Network [103], as well as leveraging domain adaptation and transfer learning [126]. It has been proven to enhance accuracy in scenarios where source and target images originate from distinct domains. However, the domain adaption segmentation using RS images remains largely underexplored [127]. The authors [92] presented a CNN-based SegNet model that classifies terrain features using 3D geospatial data; the model did better on building classification than other natural objects. The model was validated on the Vaihingen dataset and tested on the Potsdam dataset, achieving an IoU of 84.90%.

In addition, ref. [128] proposed SBANet, which stands for Semantic Boundary Awareness Network used to extract sharp boundaries, and ResNet was employed as the backbone. Subsequently, it was enhanced by introducing a boundary attention module and applying adaptive weights in multitask learning to incorporate both low- and high-level features, with the goals of improving land-cover boundary precision and expediting network convergence. The method was evaluated on the Potsdam and Vaihingen semantic labelling datasets; they reported that SBANet performed best compared to models like UNet, FCN, SegNet, PSPNet, Deeplab3+, and others. The DenseNet-based model [129], a proposed method, modified one of the DL backbones DenseNet by adding two novel fusions that is the unit fusion and cross-level fusion. The unit fusion is detailed-oriented fusion, and the other integrates different information levels. This model with both fusions performed best on the DeepGlobe dataset.

Furthermore, paper [93] suggested a bidirectional grid fusion network, a two-way fusion architecture for classifying land in very high-resolution RS data. It encourages bidirectional information flow with mutual benefits of feature propagation, and a grid fusion architecture is attached for further improvement. The best refined model was tested on ISPRS and GID datasets, achieving MIoU performances of 68.88% and 64.01%, respectively. Table 4 shows some identified hybrid semantic segmentation models and performance metric in land cover mapping. These models have demonstrated effective performances, with an average overall accuracy of 91.3% across presented datasets.

Table 4. Hybrid-based semantic segmentation models for land cover segmentation.

Models	Datasets	Performance Metrics	Limitation
RSI-Net [95]	Vaihingen, Potsdam, GID	OA = 91.83, OA = 93.31, OA = 93.67	Limitation in segmentation of pixel-wise semantics. Enhanced feature map fusion decoders can lead to performance improvements.
HMRT [32]	Potsdam	OA = 85.99 MIoU = 74.14	Parameter complexity issue, decrease in segmentation accuracy due to a lot of noise. Optimization is required.

Table 4. *Cont.*

Models	Datasets	Performance Metrics	Limitation
UNetFormer [19]	UAVid, Vaihingen, Potsdam, LoveDA	MIoU = 67.8, OA = 91.0 MIoU = 82.7, OA = 91.3 MIoU = 86.8, MIoU = 52.4	Investigate the Transformer's potential and practicality in addressing geospatial vision tasks is open for research.
(TL-ResUNet) model [130]	DeepGlobe	IoU = 0.81	Improve classification performance is open for research, and developing real time and automated solution for land use land cover.
CNN-enhanced heterogeneous GCN [131]	Beijing dataset, Shenzhen dataset.	MIoU = 70.48, MIoU = 62.45	Future endeavor is to optimize the utilization of pretrained deep CNN features and GCN features across various segmentation scales.
HFENet [132]	MZData, LandCover Dataset, WHU Building Dataset	MIoU = 87.19, MIoU = 89.69, MIoU = 92.12	Time and space complexity issues. Future work can be to automatically fine-tune the parameters to attain the optimal performance of the model.

- Performance analysis of semantic segmentation model structures on ISPRS 2-D labelling Potsdam and Vaihingen datasets

Our study examines the performance of various semantic segmentation model structures using the most widely utilized datasets in land cover mapping. The models are categorized based on the three primary architectural structures, and their performances are averaged across ISPRS benchmark datasets.

Figures 11 and 12 show that semantic segmentation models demonstrate impressive generalization capabilities across all ISPRS land cover categories. Among the different structures, the encoder-decoder-based models performed the least well across ISPRS land cover categories. On the Potsdam dataset, hybrid models outperformed others across all landcover categories. For the Vaihingen data, Transformer-based models slightly outperformed other structures in classifying buildings, low vegetation, and trees. Notably, all three model structures showed the highest effectiveness in the building category, with the least generalization performance observed in either the cars or low vegetation categories. Semantic segmentation models perform better in categorizing buildings compared to natural objects such as low vegetation and forests in remote sensing (RS) applications. This supports the authors' assertion [92] that their model is more effective in man-made features due to its geometric properties than in natural features.

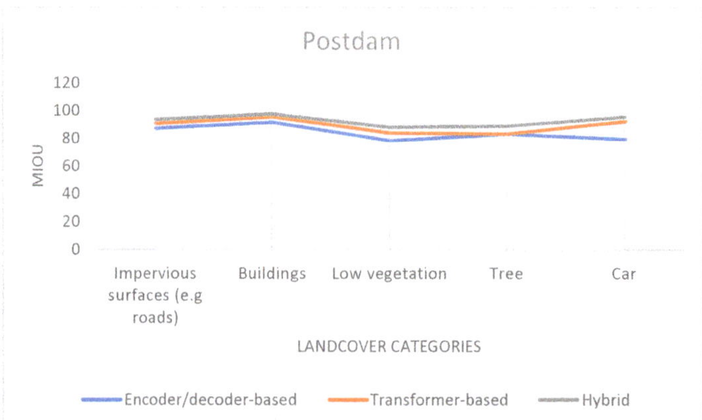

Figure 11. Semantic segmentation model structures analysis on ISPRS 2-D labelling Potsdam.

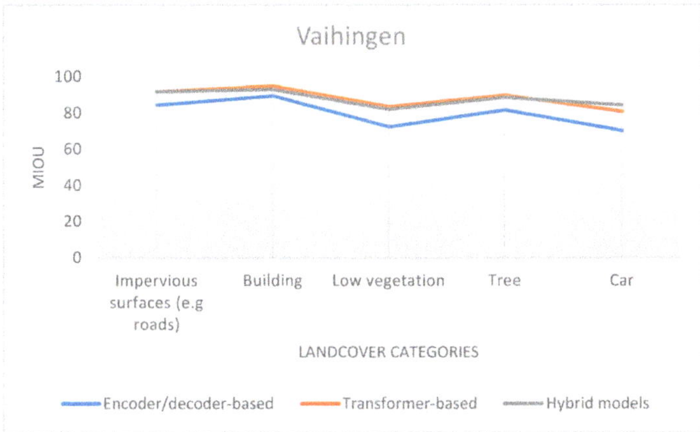

Figure 12. Semantic segmentation model structures analysis on ISPRS 2-D labelling Vaihingen.

- Common experimental training settings

The common experimental training settings found in studies are detailed in Table 5.

Table 5. Frequently used and common hyper-parameter values for segmentation models structures.

Model's Structures	Batch Size	Epochs	Learning Rate	Data Augmentation	Backbone	Popular Optimizer	Parameters	Evaluation Metrics
Encoder/ decoder-based	4, 8, 16, 64	100–500	0.01	Yes	ResNet	SGD	Low–High	MIoU, OA, F1
Transformer-based	6, 8	100–200	0.0006	Yes	ResNet/ Swintiny	Adam	High	MIoU, OA, F1
Hybrid models	8, 16	40–100	0.0006	Yes	ResNet	Adam	Low–High	MIoU, OA, F1

Table 5 shows that all three model structures typically utilize data augmentation during pre-processing. Both the encoder-decoder and hybrid structures use the ResNet model as their backbone, while the Transformer-based structure employs either the ResNet or Swintiny model. For evaluation, commonly used metrics include Mean Intersection over Union (MIoU), Overall Accuracy (OA), and F1 score, among others. It is notable that the Adam optimizer is frequently used for both Transformer-based and hybrid structures, as it helps to improve convergence speed. This is particularly important for Transformer-based structures, which have a high number of parameters that tend to slow down convergence.

4. Challenges, Future Insights and Directions

In this section, we highlight the challenges in land cover mapping and semantic segmentation methodologies in the context of land cover mapping, as revealed through an extensive review of the existing literature. We found persistent gaps requiring targeted attention and innovative solutions in forthcoming research efforts. This section examines these challenges and outlines potential avenues for future investigation.

4.1. Land Cover Mapping

In land cover mapping, the key challenges include extracting land cover boundary information and generating precise land cover maps.

- Extracting boundary information

The precise delineation of sharp and well-defined boundaries [133], the refinement of object edges [134], the extraction of boundary details [128], and the acquisition of content details [135] from RS Imagery, all aimed at achieving accurate land cover segmentation. This research gap remains an area that needs further exploration.

Defining clear land cover boundaries is a crucial aspect of RS and geospatial analysis. It involves precisely delineating the borders that divide distinct land cover categories on the Earth's surface. Semantic segmentation encounters performance degradation due to the loss of crucial boundary information [124]. The challenge of delineating precise boundaries in land cover maps is exacerbated by the heightened probability of prediction errors occurring at borders and within smaller segments [19,136]. This holds particularly true in scenarios like the segmentation and classification of vegetation land covers [54]. An illustrative example of this complexity is encountered when attempting to segment the boundaries among ecosystems characterized by a combination of both forest and grassland in regions with semi-arid to semi-humid climates [137].

Unclear boundaries, and loss of essential and detailed information at boundaries decreases the chance of producing fine segmentation results [128]. The task of effectively capturing entire and well-defined boundaries in intricate RS images of very high resolution remains open for further research and improvement.

- Generating Precise Land Cover Maps

The demand for accurate and timely high-resolution land cover maps is high and are of immediate importance to various sectors and communities [125]. The creation of precise land cover maps holds significant value for subsequent applications. These applications encompass a diverse range of tasks, including vehicle detection [105], the extraction of building footprints [138], building segmentation [32], road extraction, surface classification [112], determining optimal seamlines for orthoimage mosaicking within settlements [139], and land consumption [140]. This offers valuable assistance in the monitoring and reporting of data within rapidly changing urban regions. However, it has been reported that generating and automating accurate land cover maps still present a formidable challenge [141,142].

4.2. Semantic Segmentation Methodologies

The key challenges of semantic segmentation methodologies in land cover mapping, as identified in reviewed articles, includes enhancing semantic segmentation models performances, improving RS Images analysis using semantic segmentation models, addressing imbalance and unlabeled RS data problem.

- Enhancing deep learning model performance

Improving deep learning semantic segmentation architecture is a notable research gap in land cover mapping. This gap is characterized by the ongoing need to advance the capabilities of these models to address complex challenges emanating in land cover mapping such as improved performance on natural images and convergence speed. Various studies have critically compared the accuracy assessment and effectiveness of models, both on natural images and RS imagery, aiming to enhance their generalization performance [10] and convergence speed. Moreover, multiple studies such as [143] have modified the semantic segmentation model to improve generalization performance for land cover mapping.

- Analysis of RS images

In terms of RS data, extracting information from RS imagery data remains a challenge. The following factors contribute to inaccurate RS classification, namely the complexities inherent in deciphering intricate spatial and spectral patterns within RS Imagery [13,125], the challenge of handling diverse distributions of ground objects with variations within the same class [26] and significant intra-class and limited inter-class pixel differences [144–146]. Other contributing factors include data complexity, geographical time difference [147], foggy conditions [148], and data acquisition errors [141]. More so, several studies have implemented works for very high RS resolution images. However, only few studies have focused on low and medium-resolution images [68,75,114,149]. As a future insight, it is recommended to conduct more research using fast and efficient DL methods for low and medium resolution RS.

Another area is the analysis of SAR Images. SAR images is extremely important for many applications especially in Agriculture. Researchers find it very challenging classifying SAR data and the segmentation is poorly understood [84]. Some studies have undertaken land cover classification and segmentation tasks across diverse categories of SAR data [86], including polarimetric SAR imagery [77,84,85], single-polarization SAR images [87], and multi-temporal SAR data [81]. As a consideration, we recommend that a roadmap for simplified and automated semantic segmentation of SAR images should be investigated.

In LiDAR data analysis, ref. [89] pioneered novel deep learning architectures designed specifically for land cover classification and segmentation, which were extensively validated using airborne LiDAR data. Additionally, ref. [150] developed DL models that synergistically leverage airborne LiDAR data and high-resolution RS Imagery to achieve enhanced generalization performance. Beyond RS Imagery and LiDAR, semantic segmentation is applied to high-resolution SAR images [79,83] and aerial images of high resolution [2,151].

- Unlabeled and Imbalance RS data

A large majority of RS images lack high-quality or are largely unlabeled [152], with weak and missing annotations that reduce the model's performances. The efforts to obtain well-annotated RS data are expensive, laborious, and highly time-consuming. This affects generating accurate land use maps, thereby impacting the generation of precise land cover maps negatively. To mitigate this challenge, exploration into study areas like domain adaptation techniques [153] and the application of DL models can be considered to facilitate the creation of accurate land cover maps [154]. Another suggested approach for addressing the shortage of well-annotated data is to utilize networks capable of utilizing training labels derived from lower resolution land cover data [155]. Furthermore, it is advisable to harness the benefits of multi-modal RS data, as its potential to enhance model performance particularly in situations with limited training samples has not been fully realized [156]. To overcome the laborious and time-consuming process of manually labeling data, certain studies such as [148,157] introduced DL models that addressed this issue. The authors [135] proposed the effectiveness of semi-supervised adversarial learning methods for handling limited and unannotated high-resolution satellite images. Furthermore, the investigations by [39,149,158,159] sought remedies for challenges stemming from the scarcity of well-annotated pixel-level data and as well as other studies proposed steps on how to tackle instances of class imbalance [64,154,160].

Class and sample imbalance of RS images, and data with limited annotation cause performance degradation, which can lead to poor performance especially in minority classes [161,162].

5. Conclusions

This study conducted an analysis of emerging patterns, performance, applications, and data sources related to semantic segmentation in land cover mapping. The objective is to identify knowledge gaps within this domain and offer readers a roadmap and detailed insights into semantic segmentation for land use/land cover mapping. Employing the PRISMA methodology, a comprehensive review is undertaken to address predefined research questions.

The results reveal a substantial increase in publications between 2020 and 2023, with 81% appearing in the top 13 journals. These studies originate from diverse global institutions, with over 59% attributed to Chinese institutions, followed by the USA and India. Research focuses primarily on land cover, urban areas, precision agriculture, environment, coastal areas, and forests, particularly in tasks such as land use change detection, land cover classification, and segmentation of forests, buildings, roads, agriculture, and urban areas. RS satellites, UAV and UAS, mobile phones, Google Earth, SAR, and LiDAR sources are the major data sources, with Sentinel-2, Sentinel-1, and Landsat satellites being the most utilized. Many studies use publicly available benchmark datasets for semantic segmentation model evaluation. ISPRS Vaihingen and Potsdam being widely employed, followed by GID, Landcover.ai, DeepGlobe, and WHDLD datasets.

In terms of semantic segmentation models, three primary architectural structures are identified and grouped as encoder-decoder structures, transformer structures, and hybrid structures. Based on our research, all semantic segmentation model's structures demonstrate effective performance in land cover mapping. Hybrid and encoder-decoder structures are most popular due to their impressive generalization capabilities and speed. Transformer-based structures show good generalization but slower convergence. Current research directions and expanding frontiers in land cover mapping emphasize the introduction and implementation of innovative semantic segmentation techniques for satellite imagery in RS. Furthermore, key research gaps identified include the need to enhance model accuracy on RS data, improve existing model architectures, extract precise boundaries in land cover maps, address scarcity of well-labeled datasets, and tackle challenges associated with low and medium-resolution RS data. This study provides useful domain specific information. However, there are some threats to review validity, which include database choice, searched keywords, and classification bias.

Author Contributions: Conceptualization, S.A.; methodology, S.A.; software, S.A.; validation, S.A., and P.C.; investigation, S.A.; data curation, S.A.; writing—original draft preparation, S.A.; writing—review and editing, S.A. and P.C.; visualization, S.A.; supervision, P.C. All authors have read and agreed to the published version of the manuscript.

Funding: This research was funded by the European Union's Horizon 2020 research and innovation program under the Marie Sklodowska-Curie-RISE, Project SUSTAINABLE (https://www.projectsustainable.eu) with grant number 101007702". This work was partially supported by national funds through FCT (Fundação para a Ciência e a Tecnologia), under the project—UIDB/04152/2020 (DOI:10.54499/UIDB/04152/2020)—Centro de Investigação em Gestão de Informação (MagIC)/NOVA IMS.

Data Availability Statement: Data sharing is not applicable to this article.

Acknowledgments: The authors are grateful to the editors and anonymous reviewers for their informative suggestions.

Conflicts of Interest: The author Segun Ajibola was employed by Afridat UG. The authors declare no conflicts of interest.

Abbreviations

The following abbreviations are used in this manuscript:

BANet	Bilateral Awareness Network
CNN	Convolutional Neural Networks
DCNN	Deep Convolutional Neural Network
DEANET	Dual Encoder with Attention Network
DGFNET	Dual-Gate Fusion Network
DL	Deep Learning
DSM	Digital Surface Model
FCN	Fully Convolutional Networks
GF-2	GaoFen-2
GF-3	GaoFen-3
GID	GaoFen Image Data
HFENet	Hierarchical Feature Extraction Network
HMRT	Hybrid Multi-resolution and Transformer semantic extraction Network
IEEE	Institute of Electrical and Electronics Engineers
IoU	Mean Intersection over Union
ISPRS	International Society for Photogrammetry and Remote Sensing
LC	Land Cover
LiDAR	Light Detection and Ranging data

LoveDA	Land-cOVEr Domain Adaptive
LULC	Land Use and Land Cover
MARE	Multi-Attention REsu-Net
MDPI	Multidisciplinary Digital Publishing Institute
MIoU	Mean Intersection over Union
NLP	Natural Language Processing
OA	Overall Accuracy
PolSAR	Polarimetric Synthetic Aperture Radar
RAANET	Residual ASPP with Attention Net
RQ	Research Question
RS	Remote Sensing
RSI	Remote Sensing Imaginary
SAR	Synthetic Aperture Radar
SBANet	Semantic Boundary Awareness Network
SEG-ESRGAN	Segmentation Enhanced Super-Resolution Generative Adversarial Network
SOCNN	Superpixel-Optimized convolutional neural network
SOTA	State-Of-The-Art
UAS	Unmanned Aircraft System
UAV	Unmanned Aerial Vehicle
VEDAI	VEhicle Detection in Aerial Imagery
WHDLD	Wuhan Dense Labeling Dataset

References

1. Vali, A.; Comai, S.; Matteucci, M. Deep Learning for Land Use and Land Cover Classification Based on Hyperspectral and Multispectral Earth Observation Data: A Review. *Remote Sens.* **2020**, *12*, 2495. [CrossRef]
2. Ma, J.; Wu, L.; Tang, X.; Liu, F.; Zhang, X.; Jiao, L. Building Extraction of Aerial Images by a Global and Multi-Scale Encoder-Decoder Network. *Remote Sens.* **2020**, *12*, 2350. [CrossRef]
3. Pourmohammadi, P.; Adjeroh, D.A.; Strager, M.P.; Farid, Y.Z. Predicting Developed Land Expansion Using Deep Convolutional Neural Networks. *Environ. Model. Softw.* **2020**, *134*, 104751. [CrossRef]
4. Di Pilato, A.; Taggio, N.; Pompili, A.; Iacobellis, M.; Di Florio, A.; Passarelli, D.; Samarelli, S. Deep Learning Approaches to Earth Observation Change Detection. *Remote Sens.* **2021**, *13*, 4083. [CrossRef]
5. Wei, P.; Chai, D.; Lin, T.; Tang, C.; Du, M.; Huang, J. Large-Scale Rice Mapping under Different Years Based on Time-Series Sentinel-1 Images Using Deep Semantic Segmentation Model. *ISPRS J. Photogramm. Remote Sens.* **2021**, *174*, 198–214. [CrossRef]
6. Dal Molin Jr., R.; Rizzoli, P. Potential of Convolutional Neural Networks for Forest Mapping Using Sentinel-1 Interferometric Short Time Series. *Remote Sens.* **2022**, *14*, 1381. [CrossRef]
7. Sun, Y.; Zhang, X.; Huang, J.; Wang, H.; Xin, Q. Fine-Grained Building Change Detection from Very High-Spatial-Resolution Remote Sensing Images Based on Deep Multitask Learning. *IEEE Geosci. Remote Sens. Lett.* **2022**, *19*, 8000605. [CrossRef]
8. Trenčanová, B.; Proença, V.; Bernardino, A. Development of Semantic Maps of Vegetation Cover from UAV Images to Support Planning and Management in Fine-Grained Fire-Prone Landscapes. *Remote Sens.* **2022**, *14*, 1262. [CrossRef]
9. Zhang, X.; Wang, Z.; Zhang, J.; Wei, A. MSANet: An Improved Semantic Segmentation Method Using Multi-Scale Attention for Remote Sensing Images. *Remote Sens. Lett.* **2022**, *13*, 1249–1259. [CrossRef]
10. Scepanovic, S.; Antropov, O.; Laurila, P.; Rauste, Y.; Ignatenko, V.; Praks, J. Wide-Area Land Cover Mapping with Sentinel-1 Imagery Using Deep Learning Semantic Segmentation Models. *IEEE J. Sel. Top. Appl. Earth Obs. Remote Sens.* **2021**, *14*, 10357–10374. [CrossRef]
11. Guo, Y.; Liu, Y.; Oerlemans, A.; Lao, S.; Wu, S.; Lew, M.S. Deep Learning for Visual Understanding: A Review. *Neurocomputing* **2016**, *187*, 27–48. [CrossRef]
12. Huang, J.; Weng, L.; Chen, B.; Xia, M. DFFAN: Dual Function Feature Aggregation Network for Semantic Segmentation of Land Cover. *ISPRS Int. J. Geoinf.* **2021**, *10*, 125. [CrossRef]
13. Chen, S.; Wu, C.; Mukherjee, M.; Zheng, Y. Ha-Mppnet: Height Aware-Multi Path Parallel Network for High Spatial Resolution Remote Sensing Image Semantic Seg-Mentation. *ISPRS Int. J. Geoinf.* **2021**, *10*, 672. [CrossRef]
14. Hao, S.; Zhou, Y.; Guo, Y. A Brief Survey on Semantic Segmentation with Deep Learning. *Neurocomputing* **2020**, *406*, 302–321. [CrossRef]
15. Chen, G.; Zhang, X.; Wang, Q.; Dai, F.; Gong, Y.; Zhu, K. Symmetrical Dense-Shortcut Deep Fully Convolutional Networks for Semantic Segmentation of Very-High-Resolution Remote Sensing Images. *IEEE J. Sel. Top. Appl. Earth Obs. Remote Sens.* **2018**, *11*, 1633–1644. [CrossRef]
16. Ronneberger, O.; Fischer, P.; Brox, T. U-Net: Convolutional Networks for Biomedical Image Segmentation. In Proceedings of the Lecture Notes in Computer Science (including subseries Lecture Notes in Artificial Intelligence and Lecture Notes in Bioinformatics), Munich, Germany, 5–9 October 2015; Volume 9351.

17. Chen, B.; Xia, M.; Huang, J. Mfanet: A Multi-Level Feature Aggregation Network for Semantic Segmentation of Land Cover. *Remote Sens.* **2021**, *13*, 731. [CrossRef]
18. Weng, L.; Pang, K.; Xia, M.; Lin, H.; Qian, M.; Zhu, C. Sgformer: A Local and Global Features Coupling Network for Semantic Segmentation of Land Cover. *IEEE J. Sel. Top. Appl. Earth Obs. Remote Sens.* **2023**, *16*, 6812–6824. [CrossRef]
19. Wang, L.; Li, R.; Zhang, C.; Fang, S.; Duan, C.; Meng, X.; Atkinson, P.M. UNetFormer: A UNet-like Transformer for Efficient Semantic Segmentation of Remote Sensing Urban Scene Imagery. *ISPRS J. Photogramm. Remote Sens.* **2022**, *190*, 196–214. [CrossRef]
20. Xiao, D.; Kang, Z.; Fu, Y.; Li, Z.; Ran, M. Csswin-Unet: A Swin-Unet Network for Semantic Segmentation of Remote Sensing Images by Aggregating Contextual Information and Extracting Spatial Information. *Int. J. Remote Sens.* **2023**, *44*, 7598–7625. [CrossRef]
21. Garcia-Garcia, A.; Orts-Escolano, S.; Oprea, S.; Villena-Martinez, V.; Martinez-Gonzalez, P.; Garcia-Rodriguez, J. A Survey on Deep Learning Techniques for Image and Video Semantic Segmentation. *Appl. Soft Comput. J.* **2018**, *70*, 41–65. [CrossRef]
22. Lateef, F.; Ruichek, Y. Survey on Semantic Segmentation Using Deep Learning Techniques. *Neurocomputing* **2019**, *338*, 321–348. [CrossRef]
23. Yuan, X.; Shi, J.; Gu, L. A Review of Deep Learning Methods for Semantic Segmentation of Remote Sensing Imagery. *Expert. Syst. Appl.* **2021**, *169*, 114417. [CrossRef]
24. Page, M.J.; McKenzie, J.E.; Bossuyt, P.M.; Boutron, I.; Hoffmann, T.C.; Mulrow, C.D.; Shamseer, L.; Tetzlaff, J.M.; Akl, E.A.; Brennan, S.E.; et al. The PRISMA 2020 Statement: An Updated Guideline for Reporting Systematic Reviews. *BMJ* **2021**, *372*, n71. [CrossRef] [PubMed]
25. Manley, K.; Nyelele, C.; Egoh, B.N. A Review of Machine Learning and Big Data Applications in Addressing Ecosystem Service Research Gaps. *Ecosyst. Serv.* **2022**, *57*, 101478. [CrossRef]
26. Tian, T.; Chu, Z.; Hu, Q.; Ma, L. Class-Wise Fully Convolutional Network for Semantic Segmentation of Remote Sensing Images. *Remote Sens.* **2021**, *13*, 3211. [CrossRef]
27. Wan, L.; Tian, Y.; Kang, W.; Ma, L. D-TNet: Category-Awareness Based Difference-Threshold Alternative Learning Network for Remote Sensing Image Change Detection. *IEEE Trans. Geosci. Remote Sens.* **2022**, *60*, 5633316. [CrossRef]
28. Picon, A.; Bereciartua-Perez, A.; Eguskiza, I.; Romero-Rodriguez, J.; Jimenez-Ruiz, C.J.; Eggers, T.; Klukas, C.; Navarra-Mestre, R. Deep Convolutional Neural Network for Damaged Vegetation Segmentation from RGB Images Based on Virtual NIR-Channel Estimation. *Artif. Intell. Agric.* **2022**, *6*, 199–210. [CrossRef]
29. Zhang, Z.; Huang, X.; Li, J. DWin-HRFormer: A High-Resolution Transformer Model With Directional Windows for Semantic Segmentation of Urban Construction Land. *IEEE Trans. Geosci. Remote Sens.* **2023**, *61*, 5400714. [CrossRef]
30. Wang, L.; Li, R.; Wang, D.; Duan, C.; Wang, T.; Meng, X. Transformer Meets Convolution: A Bilateral Awareness Network for Semantic Segmentation of Very Fine Resolution Urban Scene Images. *Remote Sens.* **2021**, *13*, 3065. [CrossRef]
31. Akcay, O.; Kinaci, A.C.; Avsar, E.O.; Aydar, U. Semantic Segmentation of High-Resolution Airborne Images with Dual-Stream DeepLabV3+. *ISPRS Int. J. Geoinf.* **2022**, *11*, 23. [CrossRef]
32. Sun, Z.; Zhou, W.; Ding, C.; Xia, M. Multi-Resolution Transformer Network for Building and Road Segmentation of Remote Sensing Image. *ISPRS Int. J. Geoinf.* **2022**, *11*, 165. [CrossRef]
33. Chen, T.-H.K.; Qiu, C.; Schmitt, M.; Zhu, X.X.; Sabel, C.E.; Prishchepov, A.V. Mapping Horizontal and Vertical Urban Densification in Denmark with Landsat Time-Series from 1985 to 2018: A Semantic Segmentation Solution. *Remote Sens. Environ.* **2020**, *251*, 112096. [CrossRef]
34. Wu, F.; Wang, C.; Zhang, H.; Li, J.; Li, L.; Chen, W.; Zhang, B. Built-up Area Mapping in China from GF-3 SAR Imagery Based on the Framework of Deep Learning. *Remote Sens. Environ.* **2021**, *262*, 112515. [CrossRef]
35. Xu, S.; Zhang, S.; Zeng, J.; Li, T.; Guo, Q.; Jin, S. A Framework for Land Use Scenes Classification Based on Landscape Photos. *IEEE J. Sel. Top. Appl. Earth Obs. Remote Sens.* **2020**, *13*, 6124–6141. [CrossRef]
36. Xu, L.; Shi, S.; Liu, Y.; Zhang, H.; Wang, D.; Zhang, L.; Liang, W.; Chen, H. A Large-Scale Remote Sensing Scene Dataset Construction for Semantic Segmentation. *Int. J. Image Data Fusion* **2023**, *14*, 299–323. [CrossRef]
37. Sirous, A.; Satari, M.; Shahraki, M.M.; Pashayi, M. A Conditional Generative Adversarial Network for Urban Area Classification Using Multi-Source Data. *Earth Sci. Inf.* **2023**, *16*, 2529–2543. [CrossRef]
38. Vasavi, S.; Sri Somagani, H.; Sai, Y. Classification of Buildings from VHR Satellite Images Using Ensemble of U-Net and ResNet. *Egypt. J. Remote Sens. Space Sci.* **2023**, *26*, 937–953. [CrossRef]
39. Kang, J.; Fernandez-Beltran, R.; Sun, X.; Ni, J.; Plaza, A. Deep Learning-Based Building Footprint Extraction with Missing Annotations. *IEEE Geosci. Remote Sens. Lett.* **2022**, *19*, 3002805. [CrossRef]
40. Yu, J.; Zeng, P.; Yu, Y.; Yu, H.; Huang, L.; Zhou, D. A Combined Convolutional Neural Network for Urban Land-Use Classification with GIS Data. *Remote Sens.* **2022**, *14*, 1128. [CrossRef]
41. Wei, P.; Chai, D.; Huang, R.; Peng, D.; Lin, T.; Sha, J.; Sun, W.; Huang, J. Rice Mapping Based on Sentinel-1 Images Using the Coupling of Prior Knowledge and Deep Semantic Segmentation Network: A Case Study in Northeast China from 2019 to 2021. *Int. J. Appl. Earth Obs. Geoinf.* **2022**, *112*, 102948. [CrossRef]
42. Liu, S.; Peng, D.; Zhang, B.; Chen, Z.; Yu, L.; Chen, J.; Pan, Y.; Zheng, S.; Hu, J.; Lou, Z.; et al. The Accuracy of Winter Wheat Identification at Different Growth Stages Using Remote Sensing. *Remote Sens.* **2022**, *14*, 893. [CrossRef]

43. Bem, P.P.D.; de Carvalho Júnior, O.A.; Carvalho, O.L.F.D.; Gomes, R.A.T.; Guimarães, R.F.; Pimentel, C.M.M. Irrigated Rice Crop Identification in Southern Brazil Using Convolutional Neural Networks and Sentinel-1 Time Series. *Remote Sens. Appl.* **2021**, *24*, 100627. [CrossRef]
44. Niu, B.; Feng, Q.; Su, S.; Yang, Z.; Zhang, S.; Liu, S.; Wang, J.; Yang, J.; Gong, J. Semantic Segmentation for Plastic-Covered Greenhouses and Plastic-Mulched Farmlands from VHR Imagery. *Int. J. Digit. Earth* **2023**, *16*, 4553–4572. [CrossRef]
45. Sykas, D.; Sdraka, M.; Zografakis, D.; Papoutsis, I. A Sentinel-2 Multiyear, Multicountry Benchmark Dataset for Crop Classification and Segmentation With Deep Learning. *IEEE J. Sel. Top. Appl. Earth Obs. Remote Sens.* **2022**, *15*, 3323–3339. [CrossRef]
46. Descals, A.; Wich, S.; Meijaard, E.; Gaveau, D.L.A.; Peedell, S.; Szantoi, Z. High-Resolution Global Map of Smallholder and Industrial Closed-Canopy Oil Palm Plantations. *Earth Syst. Sci. Data* **2021**, *13*, 1211–1231. [CrossRef]
47. He, J.; Lyu, D.; He, L.; Zhang, Y.; Xu, X.; Yi, H.; Tian, Q.; Liu, B.; Zhang, X. Combining Object-Oriented and Deep Learning Methods to Estimate Photosynthetic and Non-Photosynthetic Vegetation Cover in the Desert from Unmanned Aerial Vehicle Images with Consideration of Shadows. *Remote Sens.* **2023**, *15*, 105. [CrossRef]
48. Wan, L.; Li, S.; Chen, Y.; He, Z.; Shi, Y. Application of Deep Learning in Land Use Classification for Soil Erosion Using Remote Sensing. *Front. Earth Sci.* **2022**, *10*, 849531. [CrossRef]
49. Cho, A.Y.; Park, S.-E.; Kim, D.-J.; Kim, J.; Li, C.; Song, J. Burned Area Mapping Using Unitemporal PlanetScope Imagery With a Deep Learning Based Approach. *IEEE J. Sel. Top. Appl. Earth Obs. Remote Sens.* **2023**, *16*, 242–253. [CrossRef]
50. Bergado, J.R.; Persello, C.; Reinke, K.; Stein, A. Predicting Wildfire Burns from Big Geodata Using Deep Learning. *Saf. Sci.* **2021**, *140*, 105276. [CrossRef]
51. Wang, Z.; Yang, P.; Liang, H.; Zheng, C.; Yin, J.; Tian, Y.; Cui, W. Semantic Segmentation and Analysis on Sensitive Parameters of Forest Fire Smoke Using Smoke-Unet and Landsat-8 Imagery. *Remote Sens.* **2022**, *14*, 45. [CrossRef]
52. Liu, C.-C.; Zhang, Y.-C.; Chen, P.-Y.; Lai, C.-C.; Chen, Y.-H.; Cheng, J.-H.; Ko, M.-H. Clouds Classification from Sentinel-2 Imagery with Deep Residual Learning and Semantic Image Segmentation. *Remote Sens.* **2019**, *11*, 119. [CrossRef]
53. Ji, W.; Chen, Y.; Li, K.; Dai, X. Multicascaded Feature Fusion-Based Deep Learning Network for Local Climate Zone Classification Based on the So2Sat LCZ42 Benchmark Dataset. *IEEE J. Sel. Top. Appl. Earth Obs. Remote Sens.* **2023**, *16*, 449–467. [CrossRef]
54. Ayhan, B.; Kwan, C. Tree, Shrub, and Grass Classification Using Only RGB Images. *Remote Sens.* **2020**, *12*, 1333. [CrossRef]
55. Maxwell, A.E.; Bester, M.S.; Guillen, L.A.; Ramezan, C.A.; Carpinello, D.J.; Fan, Y.; Hartley, F.M.; Maynard, S.M.; Pyron, J.L. Semantic Segmentation Deep Learning for Extracting Surface Mine Extents from Historic Topographic Maps. *Remote Sens.* **2020**, *12*, 4145. [CrossRef]
56. Zhou, G.; Xu, J.; Chen, W.; Li, X.; Li, J.; Wang, L. Deep Feature Enhancement Method for Land Cover With Irregular and Sparse Spatial Distribution Features: A Case Study on Open-Pit Mining. *IEEE Trans. Geosci. Remote Sens.* **2023**, *61*, 4401220. [CrossRef]
57. Lee, S.-H.; Han, K.-J.; Lee, K.; Lee, K.-J.; Oh, K.-Y.; Lee, M.-J. Classification of Landscape Affected by Deforestation Using High-resolution Remote Sensing Data and Deep-learning Techniques. *Remote Sens.* **2020**, *12*, 3372. [CrossRef]
58. Yu, T.; Wu, W.; Gong, C.; Li, X. Residual Multi-Attention Classification Network for a Forest Dominated Tropical Landscape Using High-Resolution Remote Sensing Imagery. *ISPRS Int. J. Geoinf.* **2021**, *10*, 22. [CrossRef]
59. Pashaei, M.; Kamangir, H.; Starek, M.J.; Tissot, P. Review and Evaluation of Deep Learning Architectures for Efficient Land Cover Mapping with UAS Hyper-Spatial Imagery: A Case Study over a Wetland. *Remote Sens.* **2020**, *12*, 959. [CrossRef]
60. Fang, B.; Chen, G.; Chen, J.; Ouyang, G.; Kou, R.; Wang, L. Cct: Conditional Co-Training for Truly Unsupervised Remote Sensing Image Segmentation in Coastal Areas. *Remote Sens.* **2021**, *13*, 3521. [CrossRef]
61. Buchsteiner, C.; Baur, P.A.; Glatzel, S. Spatial Analysis of Intra-Annual Reed Ecosystem Dynamics at Lake Neusiedl Using RGB Drone Imagery and Deep Learning. *Remote Sens.* **2023**, *15*, 3961. [CrossRef]
62. Wang, Z.; Mahmoudian, N. Aerial Fluvial Image Dataset for Deep Semantic Segmentation Neural Networks and Its Benchmarks. *IEEE J. Sel. Top. Appl. Earth Obs. Remote Sens.* **2023**, *16*, 4755–4766. [CrossRef]
63. Chen, J.; Chen, G.; Wang, L.; Fang, B.; Zhou, P.; Zhu, M. Coastal Land Cover Classification of High-Resolution Remote Sensing Images Using Attention-Driven Context Encoding Network. *Sensors* **2020**, *20*, 7032. [CrossRef] [PubMed]
64. Li, Y.; Zhou, Y.; Zhang, Y.; Zhong, L.; Wang, J.; Chen, J. DKDFN: Domain Knowledge-Guided Deep Collaborative Fusion Network for Multimodal Unitemporal Remote Sensing Land Cover Classification. *ISPRS J. Photogramm. Remote Sens.* **2022**, *186*, 170–189. [CrossRef]
65. Tzepkenlis, A.; Marthoglou, K.; Grammalidis, N. Efficient Deep Semantic Segmentation for Land Cover Classification Using Sentinel Imagery. *Remote Sens.* **2023**, *15*, 2027. [CrossRef]
66. Billson, J.; Islam, M.D.S.; Sun, X.; Cheng, I. Water Body Extraction from Sentinel-2 Imagery with Deep Convolutional Networks and Pixelwise Category Transplantation. *Remote Sens.* **2023**, *15*, 1253. [CrossRef]
67. Bergamasco, L.; Bovolo, F.; Bruzzone, L. A Dual-Branch Deep Learning Architecture for Multisensor and Multitemporal Remote Sensing Semantic Segmentation. *IEEE J. Sel. Top. Appl. Earth Obs. Remote Sens.* **2023**, *16*, 2147–2162. [CrossRef]
68. Yang, X.; Zhang, B.; Chen, Z.; Bai, Y.; Chen, P. A Multi-Temporal Network for Improving Semantic Segmentation of Large-Scale Landsat Imagery. *Remote Sens.* **2022**, *14*, 5062. [CrossRef]
69. Yang, X.; Chen, Z.; Zhang, B.; Li, B.; Bai, Y.; Chen, P. A Block Shuffle Network with Superpixel Optimization for Landsat Image Semantic Segmentation. *Remote Sens.* **2022**, *14*, 1432. [CrossRef]
70. Boonpook, W.; Tan, Y.; Nardkulpat, A.; Torsri, K.; Torteeka, P.; Kamsing, P.; Sawangwit, U.; Pena, J.; Jainaen, M. Deep Learning Semantic Segmentation for Land Use and Land Cover Types Using Landsat 8 Imagery. *ISPRS Int. J. Geoinf.* **2023**, *12*, 14. [CrossRef]

71. Bergado, J.R.; Persello, C.; Stein, A. Recurrent Multiresolution Convolutional Networks for VHR Image Classification. *IEEE Trans. Geosci. Remote Sens.* **2018**, *56*, 6361–6374. [CrossRef]
72. Karila, K.; Matikainen, L.; Karjalainen, M.; Puttonen, E.; Chen, Y.; Hyyppä, J. Automatic Labelling for Semantic Segmentation of VHR Satellite Images: Application of Airborne Laser Scanner Data and Object-Based Image Analysis. *ISPRS Open J. Photogramm. Remote Sens.* **2023**, *9*, 100046. [CrossRef]
73. Zhang, X.; Du, L.; Tan, S.; Wu, F.; Zhu, L.; Zeng, Y.; Wu, B. Land Use and Land Cover Mapping Using Rapideye Imagery Based on a Novel Band Attention Deep Learning Method in the Three Gorges Reservoir Area. *Remote Sens.* **2021**, *13*, 1225. [CrossRef]
74. Zhu, Y.; Geis, C.; So, E.; Jin, Y. Multitemporal Relearning with Convolutional LSTM Models for Land Use Classification. *IEEE J. Sel. Top. Appl. Earth Obs. Remote Sens.* **2021**, *14*, 3251–3265. [CrossRef]
75. Fan, Z.; Zhan, T.; Gao, Z.; Li, R.; Liu, Y.; Zhang, L.; Jin, Z.; Xu, S. Land Cover Classification of Resources Survey Remote Sensing Images Based on Segmentation Model. *IEEE Access* **2022**, *10*, 56267–56281. [CrossRef]
76. Clark, A.; Phinn, S.; Scarth, P. Pre-Processing Training Data Improves Accuracy and Generalisability of Convolutional Neural Network Based Landscape Semantic Segmentation. *Land* **2023**, *12*, 1268. [CrossRef]
77. Mohammadimanesh, F.; Salehi, B.; Mahdianpari, M.; Gill, E.; Molinier, M. A New Fully Convolutional Neural Network for Semantic Segmentation of Polarimetric SAR Imagery in Complex Land Cover Ecosystem. *ISPRS J. Photogramm. Remote Sens.* **2019**, *151*, 223–236. [CrossRef]
78. Wenger, R.; Puissant, A.; Weber, J.; Idoumghar, L.; Forestier, G. Multimodal and Multitemporal Land Use/Land Cover Semantic Segmentation on Sentinel-1 and Sentinel-2 Imagery: An Application on a MultiSenGE Dataset. *Remote Sens.* **2023**, *15*, 151. [CrossRef]
79. Xia, J.; Yokoya, N.; Adriano, B.; Zhang, L.; Li, G.; Wang, Z. A Benchmark High-Resolution GaoFen-3 SAR Dataset for Building Semantic Segmentation. *IEEE J. Sel. Top. Appl. Earth Obs. Remote Sens.* **2021**, *14*, 5950–5963. [CrossRef]
80. Kotru, R.; Turkar, V.; Simu, S.; De, S.; Shaikh, M.; Banerjee, S.; Singh, G.; Das, A. Development of a Generalized Model to Classify Various Land Covers for ALOS-2 L-Band Images Using Semantic Segmentation. *Adv. Space Res.* **2022**, *70*, 3811–3821. [CrossRef]
81. Mehra, A.; Jain, N.; Srivastava, H.S. A Novel Approach to Use Semantic Segmentation Based Deep Learning Networks to Classify Multi-Temporal SAR Data. *Geocarto Int.* **2022**, *37*, 163–178. [CrossRef]
82. Pešek, O.; Segal-Rozenhaimer, M.; Karnieli, A. Using Convolutional Neural Networks for Cloud Detection on VENμS Images over Multiple Land-Cover Types. *Remote Sens.* **2022**, *14*, 5210. [CrossRef]
83. Jing, H.; Wang, Z.; Sun, X.; Xiao, D.; Fu, K. PSRN: Polarimetric Space Reconstruction Network for PolSAR Image Semantic Segmentation. *IEEE J. Sel. Top. Appl. Earth Obs. Remote Sens.* **2021**, *14*, 10716–10732. [CrossRef]
84. Zhang, R.; Chen, J.; Feng, L.; Li, S.; Yang, W.; Guo, D. A Refined Pyramid Scene Parsing Network for Polarimetric SAR Image Semantic Segmentation in Agricultural Areas. *IEEE Geosci. Remote Sens. Lett.* **2022**, *19*, 4014805. [CrossRef]
85. Garg, R.; Kumar, A.; Bansal, N.; Prateek, M.; Kumar, S. Semantic Segmentation of PolSAR Image Data Using Advanced Deep Learning Model. *Sci. Rep.* **2021**, *11*, 15365. [CrossRef]
86. Zheng, N.-R.; Yang, Z.-A.; Shi, X.-Z.; Zhou, R.-Y.; Wang, F. Land Cover Classification of Synthetic Aperture Radar Images Based on Encoder—Decoder Network with an Attention Mechanism. *J. Appl. Remote Sens.* **2022**, *16*, 014520. [CrossRef]
87. Shi, X.; Fu, S.; Chen, J.; Wang, F.; Xu, F. Object-Level Semantic Segmentation on the High-Resolution Gaofen-3 FUSAR-Map Dataset. *IEEE J. Sel. Top. Appl. Earth Obs. Remote Sens.* **2021**, *14*, 3107–3119. [CrossRef]
88. Yoshida, K.; Pan, S.; Taniguchi, J.; Nishiyama, S.; Kojima, T.; Islam, M.T. Airborne LiDAR-Assisted Deep Learning Methodology for Riparian Land Cover Classification Using Aerial Photographs and Its Application for Flood Modelling. *J. Hydroinformatics* **2022**, *24*, 179–201. [CrossRef]
89. Arief, H.A.; Strand, G.-H.; Tveite, H.; Indahl, U.G. Land Cover Segmentation of Airborne LiDAR Data Using Stochastic Atrous Network. *Remote Sens.* **2018**, *10*, 973. [CrossRef]
90. Xu, Z.; Su, C.; Zhang, X. A Semantic Segmentation Method with Category Boundary for Land Use and Land Cover (LULC) Mapping of Very-High Resolution (VHR) Remote Sensing Image. *Int. J. Remote Sens.* **2021**, *42*, 3146–3165. [CrossRef]
91. Liu, M.; Zhang, P.; Shi, Q.; Liu, M. An Adversarial Domain Adaptation Framework with KL-Constraint for Remote Sensing Land Cover Classification. *IEEE Geosci. Remote Sens. Lett.* **2022**, *19*, 3002305. [CrossRef]
92. Lee, D.G.; Shin, Y.H.; Lee, D.C. Land Cover Classification Using SegNet with Slope, Aspect, and Multidirectional Shaded Relief Images Derived from Digital Surface Model. *J. Sens.* **2020**, *2020*, 8825509. [CrossRef]
93. Wang, Y.; Shi, H.; Zhuang, Y.; Sang, Q.; Chen, L. Bidirectional Grid Fusion Network for Accurate Land Cover Classification of High-Resolution Remote Sensing Images. *IEEE J. Sel. Top. Appl. Earth Obs. Remote Sens.* **2020**, *13*, 5508–5517. [CrossRef]
94. Shi, H.; Fan, J.; Wang, Y.; Chen, L. Dual Attention Feature Fusion and Adaptive Context for Accurate Segmentation of Very High-Resolution Remote Sensing Images. *Remote Sens.* **2021**, *13*, 3715. [CrossRef]
95. He, S.; Lu, X.; Gu, J.; Tang, H.; Yu, Q.; Liu, K.; Ding, H.; Chang, C.; Wang, N. RSI-Net: Two-Stream Deep Neural Network for Remote Sensing Images-Based Semantic Segmentation. *IEEE Access* **2022**, *10*, 34858–34871. [CrossRef]
96. Yang, N.; Tang, H. Semantic Segmentation of Satellite Images: A Deep Learning Approach Integrated with Geospatial Hash Codes. *Remote Sens.* **2021**, *13*, 2723. [CrossRef]
97. Boguszewski, A.; Batorski, D.; Ziemba-Jankowska, N.; Dziedzic, T.; Zambrzycka, A. LandCover.Ai: Dataset for Automatic Mapping of Buildings, Woodlands, Water and Roads from Aerial Imagery. In Proceedings of the IEEE Computer Society Conference on Computer Vision and Pattern Recognition Workshops, Nashville, TN, USA, 20–25 June 2021.

98. Gao, J.; Weng, L.; Xia, M.; Lin, H. MLNet: Multichannel Feature Fusion Lozenge Network for Land Segmentation. *J. Appl. Remote Sens.* **2022**, *16*, 016513. [CrossRef]
99. Demir, I.; Koperski, K.; Lindenbaum, D.; Pang, G.; Huang, J.; Basu, S.; Hughes, F.; Tuia, D.; Raska, R. DeepGlobe 2018: A Challenge to Parse the Earth through Satellite Images. In Proceedings of the IEEE Computer Society Conference on Computer Vision and Pattern Recognition Workshops, Salt Lake City, UT, USA, 18–23 June 2018; Volume 2018.
100. Wei, H.; Xu, X.; Ou, N.; Zhang, X.; Dai, Y. Deanet: Dual Encoder with Attention Network for Semantic Segmentation of Remote Sensing Imagery. *Remote Sens.* **2021**, *13*, 3900. [CrossRef]
101. Maggiori, E.; Tarabalka, Y.; Charpiat, G.; Alliez, P. Can Semantic Labeling Methods Generalize to Any City? The Inria Aerial Image Labeling Benchmark. In Proceedings of the International Geoscience and Remote Sensing Symposium (IGARSS), Fort Worth, TX, USA, 23–28 July 2017; Volume 2017.
102. Li, W.; He, C.; Fang, J.; Zheng, J.; Fu, H.; Yu, L. Semantic Segmentation-Based Building Footprint Extraction Using Very High-Resolution Satellite Images and Multi-Source GIS Data. *Remote Sens.* **2019**, *11*, 403. [CrossRef]
103. Ji, S.; Wang, D.; Luo, M. Generative Adversarial Network-Based Full-Space Domain Adaptation for Land Cover Classification from Multiple-Source Remote Sensing Images. *IEEE Trans. Geosci. Remote Sens.* **2021**, *59*, 3816–3828. [CrossRef]
104. Chen, Q.; Wang, L.; Wu, Y.; Wu, G.; Guo, Z.; Waslander, S.L. Aerial Imagery for Roof Segmentation: A Large-Scale Dataset towards Automatic Mapping of Buildings. *ISPRS J. Photogramm. Remote Sens.* **2019**, *147*, 42–55. [CrossRef]
105. Audebert, N.; Le Saux, B.; Lefèvre, S. Segment-before-Detect: Vehicle Detection and Classification through Semantic Segmentation of Aerial Images. *Remote Sens.* **2017**, *9*, 368. [CrossRef]
106. Abdollahi, A.; Pradhan, B.; Shukla, N.; Chakraborty, S.; Alamri, A. Multi-Object Segmentation in Complex Urban Scenes from High-Resolution Remote Sensing Data. *Remote Sens.* **2021**, *13*, 3710. [CrossRef]
107. Khan, S.D.; Alarabi, L.; Basalamah, S. Deep Hybrid Network for Land Cover Semantic Segmentation in High-Spatial Resolution Satellite Images. *Information* **2021**, *12*, 230. [CrossRef]
108. Liu, R.; Tao, F.; Liu, X.; Na, J.; Leng, H.; Wu, J.; Zhou, T. RAANet: A Residual ASPP with Attention Framework for Semantic Segmentation of High-Resolution Remote Sensing Images. *Remote Sens.* **2022**, *14*, 3109. [CrossRef]
109. Sang, Q.; Zhuang, Y.; Dong, S.; Wang, G.; Chen, H. FRF-Net: Land Cover Classification from Large-Scale VHR Optical Remote Sensing Images. *IEEE Geosci. Remote Sens. Lett.* **2020**, *17*, 1057–1061. [CrossRef]
110. Guo, Y.; Wang, F.; Xiang, Y.; You, H. Article Dgfnet: Dual Gate Fusion Network for Land Cover Classification in Very High-Resolution Images. *Remote Sens.* **2021**, *13*, 3755. [CrossRef]
111. Niu, X.; Zeng, Q.; Luo, X.; Chen, L. FCAU-Net for the Semantic Segmentation of Fine-Resolution Remotely Sensed Images. *Remote Sens.* **2022**, *14*, 215. [CrossRef]
112. Wu, M.; Zhang, C.; Liu, J.; Zhou, L.; Li, X. Towards Accurate High Resolution Satellite Image Semantic Segmentation. *IEEE Access* **2019**, *7*, 55609–55619. [CrossRef]
113. Li, J.; Wang, H.; Zhang, A.; Liu, Y. Semantic Segmentation of Hyperspectral Remote Sensing Images Based on PSE-UNet Model. *Sensors* **2022**, *22*, 9678. [CrossRef]
114. Salgueiro, L.; Marcello, J.; Vilaplana, V. SEG-ESRGAN: A Multi-Task Network for Super-Resolution and Semantic Segmentation of Remote Sensing Images. *Remote Sens.* **2022**, *14*, 5862. [CrossRef]
115. Marsocci, V.; Scardapane, S.; Komodakis, N. MARE: Self-Supervised Multi-Attention REsu-Net for Semantic Segmentation in Remote Sensing. *Remote Sens.* **2021**, *13*, 3275. [CrossRef]
116. Wang, S.; Mu, X.; Yang, D.; He, H.; Zhao, P. Attention Guided Encoder-Decoder Network with Multi-Scale Context Aggregation for Land Cover Segmentation. *IEEE Access* **2020**, *8*, 215299–215309. [CrossRef]
117. Feng, D.; Zhang, Z.; Yan, K. A Semantic Segmentation Method for Remote Sensing Images Based on the Swin Transformer Fusion Gabor Filter. *IEEE Access* **2022**, *10*, 77432–77451. [CrossRef]
118. Bai, J.; Wen, Z.; Xiao, Z.; Ye, F.; Zhu, Y.; Alazab, M.; Jiao, L. Hyperspectral Image Classification Based on Multibranch Attention Transformer Networks. *IEEE Trans. Geosci. Remote Sens.* **2022**, *60*, 3196661. [CrossRef]
119. Meng, X.; Yang, Y.; Wang, L.; Wang, T.; Li, R.; Zhang, C. Class-Guided Swin Transformer for Semantic Segmentation of Remote Sensing Imagery. *IEEE Geosci. Remote Sens. Lett.* **2022**, *19*, 6517505. [CrossRef]
120. Wang, D.; Yang, R.; Zhang, Z.; Liu, H.; Tan, J.; Li, S.; Yang, X.; Wang, X.; Tang, K.; Qiao, Y.; et al. P-Swin: Parallel Swin Transformer Multi-Scale Semantic Segmentation Network for Land Cover Classification. *Comput. Geosci.* **2023**, *175*, 105340. [CrossRef]
121. Dong, R.; Fang, W.; Fu, H.; Gan, L.; Wang, J.; Gong, P. High-Resolution Land Cover Mapping through Learning with Noise Correction. *IEEE Trans. Geosci. Remote Sens.* **2022**, *60*, 4402013. [CrossRef]
122. Shen, X.; Weng, L.; Xia, M.; Lin, H. Multi-Scale Feature Aggregation Network for Semantic Segmentation of Land Cover. *Remote Sens.* **2022**, *14*, 6156. [CrossRef]
123. Luo, Y.; Wang, J.; Yang, X.; Yu, Z.; Tan, Z. Pixel Representation Augmented through Cross-Attention for High-Resolution Remote Sensing Imagery Segmentation. *Remote Sens.* **2022**, *14*, 5415. [CrossRef]
124. Yuan, X.; Chen, Z.; Chen, N.; Gong, J. Land Cover Classification Based on the PSPNet and Superpixel Segmentation Methods with High Spatial Resolution Multispectral Remote Sensing Imagery. *J. Appl. Remote Sens.* **2021**, *15*, 034511. [CrossRef]
125. Zhang, C.; Yue, P.; Tapete, D.; Shangguan, B.; Wang, M.; Wu, Z. A Multi-Level Context-Guided Classification Method with Object-Based Convolutional Neural Network for Land Cover Classification Using Very High Resolution Remote Sensing Images. *Int. J. Appl. Earth Obs. Geoinf.* **2020**, *88*, 102086. [CrossRef]

126. Van den Broeck, W.A.J.; Goedemé, T.; Loopmans, M. Multiclass Land Cover Mapping from Historical Orthophotos Using Domain Adaptation and Spatio-Temporal Transfer Learning. *Remote Sens.* **2022**, *14*, 5911. [CrossRef]
127. Zhang, B.; Chen, T.; Wang, B. Curriculum-Style Local-to-Global Adaptation for Cross-Domain Remote Sensing Image Segmentation. *IEEE Trans. Geosci. Remote Sens.* **2022**, *60*, 5611412. [CrossRef]
128. Li, A.; Jiao, L.; Zhu, H.; Li, L.; Liu, F. Multitask Semantic Boundary Awareness Network for Remote Sensing Image Segmentation. *IEEE Trans. Geosci. Remote Sens.* **2022**, *60*, 5400314. [CrossRef]
129. Shan, L.; Wang, W. DenseNet-Based Land Cover Classification Network with Deep Fusion. *IEEE Geosci. Remote Sens. Lett.* **2022**, *19*. [CrossRef]
130. Safarov, F.; Temurbek, K.; Jamoljon, D.; Temur, O.; Chedjou, J.C.; Abdusalomov, A.B.; Cho, Y.-I. Improved Agricultural Field Segmentation in Satellite Imagery Using TL-ResUNet Architecture. *Sensors* **2022**, *22*, 9784. [CrossRef] [PubMed]
131. Liu, Z.-Q.; Tang, P.; Zhang, W.; Zhang, Z. CNN-Enhanced Heterogeneous Graph Convolutional Network: Inferring Land Use from Land Cover with a Case Study of Park Segmentation. *Remote Sens.* **2022**, *14*, 5027. [CrossRef]
132. Wang, D.; Yang, R.; Liu, H.; He, H.; Tan, J.; Li, S.; Qiao, Y.; Tang, K.; Wang, X. HFENet: Hierarchical Feature Extraction Network for Accurate Landcover Classification. *Remote Sens.* **2022**, *14*, 4244. [CrossRef]
133. Zhang, H.; Liu, M.; Wang, Y.; Shang, J.; Liu, X.; Li, B.; Song, A.; Li, Q. Automated Delineation of Agricultural Field Boundaries from Sentinel-2 Images Using Recurrent Residual U-Net. *Int. J. Appl. Earth Obs. Geoinf.* **2021**, *105*, 102557. [CrossRef]
134. Maggiolo, L.; Marcos, D.; Moser, G.; Serpico, S.B.; Tuia, D. A Semisupervised CRF Model for CNN-Based Semantic Segmentation with Sparse Ground Truth. *IEEE Trans. Geosci. Remote Sens.* **2022**, *60*, 5606315. [CrossRef]
135. Barthakur, M.; Sarma, K.K.; Mastorakis, N. Modified Semi-Supervised Adversarial Deep Network and Classifier Combination for Segmentation of Satellite Images. *IEEE Access* **2020**, *8*, 117972–117985. [CrossRef]
136. Wang, Q.; Luo, X.; Feng, J.; Li, S.; Yin, J. CCENet: Cascade Class-Aware Enhanced Network for High-Resolution Aerial Imagery Semantic Segmentation. *IEEE J. Sel. Top. Appl. Earth Obs. Remote Sens.* **2022**, *15*, 6943–6956. [CrossRef]
137. Zhang, C.; Zhang, L.; Zhang, B.Y.J.; Sun, J.; Dong, S.; Wang, X.; Li, Y.; Xu, J.; Chu, W.; Dong, Y.; et al. Land Cover Classification in a Mixed Forest-Grassland Ecosystem Using LResU-Net and UAV Imagery. *J. Res.* **2022**, *33*, 923–936. [CrossRef]
138. Xu, Y.; Wu, L.; Xie, Z.; Chen, Z. Building Extraction in Very High Resolution Remote Sensing Imagery Using Deep Learning and Guided Filters. *Remote Sens.* **2018**, *10*, 144. [CrossRef]
139. Li, L.; Yao, J.; Liu, Y.; Yuan, W.; Shi, S.; Yuan, S. Optimal Seamline Detection for Orthoimage Mosaicking by Combining Deep Convolutional Neural Network and Graph Cuts. *Remote Sens.* **2017**, *9*, 701. [CrossRef]
140. Cecili, G.; De Fioravante, P.; Congedo, L.; Marchetti, M.; Munafò, M. Land Consumption Mapping with Convolutional Neural Network: Case Study in Italy. *Land* **2022**, *11*, 1919. [CrossRef]
141. Abadal, S.; Salgueiro, L.; Marcello, J.; Vilaplana, V. A Dual Network for Super-Resolution and Semantic Segmentation of Sentinel-2 Imagery. *Remote Sens.* **2021**, *13*, 4547. [CrossRef]
142. Henry, C.J.; Storie, C.D.; Palaniappan, M.; Alhassan, V.; Swamy, M.; Aleshinloye, D.; Curtis, A.; Kim, D. Automated LULC Map Production Using Deep Neural Networks. *Int. J. Remote Sens.* **2019**, *40*, 4416–4440. [CrossRef]
143. Shojaei, H.; Nadi, S.; Shafizadeh-Moghadam, H.; Tayyebi, A.; Van Genderen, J. An Efficient Built-up Land Expansion Model Using a Modified U-Net. *Int. J. Digit. Earth* **2022**, *15*, 148–163. [CrossRef]
144. Dong, X.; Zhang, C.; Fang, L.; Yan, Y. A Deep Learning Based Framework for Remote Sensing Image Ground Object Segmentation. *Appl. Soft Comput.* **2022**, *130*, 109695. [CrossRef]
145. Guo, X.; Chen, Z.; Wang, C. Fully Convolutional Densenet with Adversarial Training for Semantic Segmentation of High-Resolution Remote Sensing Images. *J. Appl. Remote Sens.* **2021**, *15*, 016520. [CrossRef]
146. Zhang, B.; Wan, Y.; Zhang, Y.; Li, Y. JSH-Net: Joint Semantic Segmentation and Height Estimation Using Deep Convolutional Networks from Single High-Resolution Remote Sensing Imagery. *Int. J. Remote Sens.* **2022**, *43*, 6307–6332. [CrossRef]
147. Li, W.; Chen, K.; Shi, Z. Geographical Supervision Correction for Remote Sensing Representation Learning. *IEEE Trans. Geosci. Remote Sens.* **2022**, *60*, 5411520. [CrossRef]
148. Shi, W.; Qin, W.; Chen, A. Towards Robust Semantic Segmentation of Land Covers in Foggy Conditions. *Remote Sens.* **2022**, *14*, 4551. [CrossRef]
149. Zhang, W.; Tang, P.; Zhao, L. Fast and Accurate Land Cover Classification on Medium Resolution Remote Sensing Images Using Segmentation Models. *Int. J. Remote Sens.* **2021**, *42*, 3277–3301. [CrossRef]
150. Dechesne, C.; Mallet, C.; Le Bris, A.; Gouet-Brunet, V. Semantic Segmentation of Forest Stands of Pure Species Combining Airborne Lidar Data and Very High Resolution Multispectral Imagery. *ISPRS J. Photogramm. Remote Sens.* **2017**, *126*, 129–145. [CrossRef]
151. Zhang, Z.; Lu, W.; Cao, J.; Xie, G. MKANet: An Efficient Network with Sobel Boundary Loss for Land-Cover Classification of Satellite Remote Sensing Imagery. *Remote Sens.* **2022**, *14*, 4514. [CrossRef]
152. Li, W.; Chen, K.; Chen, H.; Shi, Z. Geographical Knowledge-Driven Representation Learning for Remote Sensing Images. *IEEE Trans. Geosci. Remote Sens.* **2022**, *60*, 5405516. [CrossRef]
153. Liu, W.; Liu, J.; Luo, Z.; Zhang, H.; Gao, K.; Li, J. Weakly Supervised High Spatial Resolution Land Cover Mapping Based on Self-Training with Weighted Pseudo-Labels. *Int. J. Appl. Earth Obs. Geoinf.* **2022**, *112*, 102931. [CrossRef]
154. Liu, Q.; Kampffmeyer, M.; Jenssen, R.; Salberg, A.-B. Dense Dilated Convolutions Merging Network for Land Cover Classification. *IEEE Trans. Geosci. Remote Sens.* **2020**, *58*, 6309–6320. [CrossRef]

155. Li, Z.; Zhang, H.; Lu, F.; Xue, R.; Yang, G.; Zhang, L. Breaking the Resolution Barrier: A Low-to-High Network for Large-Scale High-Resolution Land-Cover Mapping Using Low-Resolution Labels. *ISPRS J. Photogramm. Remote Sens.* **2022**, *192*, 244–267. [CrossRef]
156. Yuan, Q.; Mohd Shafri, H.Z. Multi-Modal Feature Fusion Network with Adaptive Center Point Detector for Building Instance Extraction. *Remote Sens.* **2022**, *14*, 4920. [CrossRef]
157. Mboga, N.; D'aronco, S.; Grippa, T.; Pelletier, C.; Georganos, S.; Vanhuysse, S.; Wolff, E.; Smets, B.; Dewitte, O.; Lennert, M.; et al. Domain Adaptation for Semantic Segmentation of Historical Panchromatic Orthomosaics in Central Africa. *ISPRS Int. J. Geoinf.* **2021**, *10*, 523. [CrossRef]
158. Zhang, Z.; Doi, K.; Iwasaki, A.; Xu, G. Unsupervised Domain Adaptation of High-Resolution Aerial Images via Correlation Alignment and Self Training. *IEEE Geosci. Remote Sens. Lett.* **2021**, *18*, 746–750. [CrossRef]
159. Simms, D.M. Fully Convolutional Neural Nets In-the-Wild. *Remote Sens. Lett.* **2020**, *11*, 1080–1089. [CrossRef]
160. Liu, C.; Du, S.; Lu, H.; Li, D.; Cao, Z. Multispectral Semantic Land Cover Segmentation from Aerial Imagery with Deep Encoder-Decoder Network. *IEEE Geosci. Remote Sens. Lett.* **2022**, *19*, 5000105. [CrossRef]
161. Sun, P.; Lu, Y.; Zhai, J. Mapping Land Cover Using a Developed U-Net Model with Weighted Cross Entropy. *Geocarto Int.* **2022**, *37*, 9355–9368. [CrossRef]
162. Chen, J.; Sun, B.; Wang, L.; Fang, B.; Chang, Y.; Li, Y.; Zhang, J.; Lyu, X.; Chen, G. Semi-Supervised Semantic Segmentation Framework with Pseudo Supervisions for Land-Use/Land-Cover Mapping in Coastal Areas. *Int. J. Appl. Earth Obs. Geoinf.* **2022**, *112*, 102881. [CrossRef]

remote sensing

MDPI

Article

Predicting the Impacts of Land Use/Cover and Climate Changes on Water and Sediment Flows in the Megech Watershed, Upper Blue Nile Basin

Mulugeta Admas [1], Assefa M. Melesse [2,*] and Getachew Tegegne [1]

[1] Department of Civil Engineering, Addis Ababa Science and Technology University, Addis Ababa 1000, Ethiopia; mulugeta.admas@aastu.edu.et (M.A.); getachew.tegegne@aastu.edu.et (G.T.)
[2] Institute of Environment, Department of Earth and Environment, Florida International University, Miami, FL 33199, USA
* Correspondence: melessea@fiu.edu

Abstract: This study assessed the impacts of the land use/cover (LULC) and climate changes on the runoff and sediment flows in the Megech watershed. The Geospatial Water Erosion Prediction Project (GeoWEPP) was used to assess LULC and climate changes' impact on runoff, soil loss, and sediment yield. The QGIS 2.16.3 plugin module for land use change evaluation (MOLUSCE) tool with the cellular automata artificial neural network (CA-ANN) was used for LULC prediction based on historical data and exploratory maps. Two commonly used representative concentration pathways (RCPs)—4.5 and 8.5—were used for climate projection in the 2030s, 2050s, and 2070s. The LULC prediction analysis showed an expansion of cropland and settlement areas, with the reduction in the forest and rangelands. The climate projections indicated an increase in maximum temperatures and altered precipitation patterns, particularly with increased wet months and reduced dry periods. The average annual soil loss and sediment yield rates were estimated to increase under both the RCP4.5 and RCP8.5 climate scenarios, with a more noticeable increase under RCP8.5. By integrating DEM, soil, land use, and climate data, we evaluated runoff, soil loss, and sediment yield changes on only land use/cover, only climate, and the combined impacts in the watershed. The results revealed that, under all combined scenarios, the sediment yield in the Megech Reservoir was projected to substantially increase by 23.28–41.01%, showing a potential loss of reservoir capacity. This study recommends strong climate adaptation and mitigation measures to alleviate the impact on land and water resources. It is possible to lessen the combined impacts of climate and LULC change through implementing best-management practices and adaptation strategies for the identified scenarios.

Keywords: climate change; land use/cover change; GeoWEPP; sediment yield; flow

1. Introduction

Soil erosion is a natural process that has many impacts throughout the world, and its rate is influenced by many factors. Climate change and land use/cover are considered the two most influential factors that accelerate soil erosion and increase the rates of sediment flow to reservoirs [1–3] and affect the overall hydrology of a watershed [4]. Climate change and land use/cover interact with each other, and their combined impacts disturb biodiversity [5,6] and the hydrology of the watershed [6].

Globally, land use/cover (LULC) has altered dramatically because of anthropogenic activities and population growth [7,8], which have increased watershed soil loss, sediment yield, and runoff flow [9]. Soil erosion rates are frequently impacted by land use change, and it has been found that the least developed nations, primarily those found in sub-Saharan Africa, are experiencing a faster pace of increase [10]. In Ethiopia, more than 80% of the population depends on agriculture and lives in the highlands of the country [11,12], which contributes to land dynamism. Studies in Ethiopia have also explored whether the severity of soil erosion has increased due to the expansion of cultivated lands [13,14]. This expansion has occurred at the expense of forests, shrubs, and grazing lands [15,16], which are facilitated by deforestation. For example, Tewabe and Fentahun [17] reported that agricultural and residential lands increased by 13% and 9.1%, respectively, in the last 30 years, with diminishing forest, bush land, and grassland in the Lake Tana Basin. Although the Ethiopian government has been implementing an afforestation initiative for the past ten years, it has not been very successful in bringing about positive changes [18].

As various studies indicate, landscape changes, including land use/cover alterations, are one of the major factors for variations in water and sediment fluxes in watersheds. Understanding the land use/cover changes and predictions can benefit the sediment and water transport studies. Results of such studies can be useful in watershed management planning and intervention measures. Similarly, climate change is responsible for changes in precipitation regime changes as well as for increased air temperatures, which can directly change the hydrologic cycle negatively by increasing the rainfall volume, variability, and intensity. These changes can increase sediment production and transport to receiving water bodies. Berihun et al. [19] reported that population growth and related changes in farming demand are considered major drivers of LULC changes in the Upper Blue Nile Basin. Another major driver of LULC change is the increase in built-up areas due to the expansion of urbanization [20], which impacts the hydrology of the watershed. The LULC changes, for any reason, increased the sediment yields of the watershed through onsite and offsite effluence [21–23]. Negese [9] reported that LULC changes and impact studies provide little information on Ethiopian watersheds and in their focus on past land use changes, little attention has been paid to future prediction and planning.

Climate change has had an impact on natural and human systems on all continents in recent years. It is triggered by human activities and is one of the main driving forces for increases in flooding and drought, loss of productive soils and sediment yield, and related pollution. Globally, future climate projections are increasing soil erosion rates [24], sediment yield, deposition, and distribution in watersheds [25]. However, the degree to which climate change impacts the hydrology and soil erosion of a watershed depends on the scenario selected [26]. An increase in precipitation amounts and intensity is expected to impact soil erosion and sediment production in watersheds [25,27,28]. Previous studies on climate change impact assessments used different downscaling methods and models available in the Coupled Model of Inter-comparison Projects (CMIP5) World Climate database. Climate change studies in the Upper Blue Nile River Basin (UBNRB) showed a projected increase in rainfall intensity and maximum and minimum temperatures, while they reported a decrease in annual rainfall depth and river flow [29,30] and an increase in flood and erosivity factors that led to accelerated soil loss and sediment yield [27,31,32]. Moreover, some climate change impact studies have indicated that soil erosion will increase under all RCPs in croplands, with more changes under RCP8.5 [32,33]. According to [24,33], soil erosion under climate change is increasing in Europe and Asia, but in Africa, only a few studies are available.

Climate change studies in the Upper Blue Nile Basin of Ethiopia revealed increases in temperature and precipitation in the rainy season and in runoff in the future due to climate change [34,35]. However, according to Mengistu et al. [36], an increase in surface runoff does not mean an increase in total water yield; rather, the water yield in the basin declines due to climate change under different climate scenarios.

Climate change studies have also shown temperature increases from 0.3 to 0.7 °C. Precipitation and temperature will increase in the future in East Africa under the RCP4.5 and RCP8.5 climate scenarios of the CMIP5 model domain, with a greater relative change under RCP8.5 [37]. Several studies in the Blue Nile Basin (BNB), based on the review by Gelete et al. [38], showed that hydrology will be highly affected in the future by climate change, with a more adverse impact on soil erosion. The authors of [39] concluded that the annual average rainfall will increase in the 2020s, 2050s, and 2080s using RCPs in East Africa. Conversely, Getachew and Manjunatha [40] reported that precipitation did not show a clear trend in the Lake Tana Basin and could be described as either increasing at some stations or decreasing at others.

The high rate of soil erosion in the Lake Tana Basin poses a major problem for reservoir sedimentation and land use/cover, and climate change exacerbates this problem [41]. A modeling approach is common and important for forecasting soil erosion and runoff under future LULC and climate change conditions. Belay and Mengistu [3] reported that climate change has greater impacts on soil erosion than LULC. The combined impacts of LULC and climate change have been studied in many parts of the world, but less attention has been paid to the Megech watershed in the Upper Blue Nile, Ethiopia.

It is necessary to know how LULC and climate changes hasten soil loss and sediment yield rates for proper planning of erosion reduction and offsite sediment impact alleviation measures. The combined land use/cover and climate change studies suggested that, under various scenarios, the watershed hydrology and soil erosion would be significantly changed in the planned future time periods. The combined effects of those changes were highlighted by the studies conducted around different parts of the globe [42–44]. To examine the dynamics of LULC in the future scenarios, the QGIS plugin MOLUSCE for LULC change using cellular automata artificial neural network (CA-ANN) algorithms can be applied [45–47].

For the robustness of climate change impacts on soil erosion analysis, ensembles of five climate models are recommended [24]. Different numbers of GCM/RCM combinations have been used in climate change studies by different researchers [3,30,34,48–51]. For this specific study, combinations of GCM/RCM that were previously used in the study region (CanESM2/RCA4, EC-EARTH-RACMO22T, GFDL-ESM2M/RCA4 MPI-ESM-PI/REMO 2009, and NorESM1-M/RCA4) were used to download 0.440 X 0.440-resolution data from Coordinated Experimental (CORDEX) Africa CMIP5 data sources, downscaled and bias-corrected for impact assessment. RCP4.5 and RCP8.5 are commonly used climate emission scenarios for climate change impacts on soil erosion assessments in the Upper Blue Nile River Basin [3,27].

Therefore, the purpose of this study was to understand the impacts of land use/cover, climate change, and combined land use/cover and climate change on soil loss, sediment yield, and river flow using the process and physically based GeoWEPP model. The land use/cover and climate change under the RCP4.5 and 8.5 scenarios was conducted for the 2030s, 2050s, and 2070s timeframes. These timeframes were set by considering the shelf-life of the Megech Dam, which is 50 years. We aimed to (1) predict and detect the land use/cover changes in the study, (2) project climate change under the RCP4.5 and 8.5 climate scenarios, with the ensemble of five climate models for CORDEX Africa data, and (3) identify the impacts of land use and climate changes on soil loss, sediment yield, and runoff using the GeoWEPP soil erosion model.

2. Materials and Methods

2.1. Study Area

The research area watershed, named the Megech watershed, is located at 11°45′40″ and 12°44′57″N latitude and 37°24′2″ and 37°37′21″E longitude in the Upper Blue Nile Basin of the tributary of Lake Tana, the largest lake in Ethiopia (Figure 1). The Megech watershed covers 424 km^2 at the dam site, with a mean annual flow of 5.6 m^3/s. The watershed is characterized as a mountainous, wedge-shaped, and steeply sloping watershed. The study site has two distinctive seasons: a rainy season from May to October and a dry season that extends from November to April. The monthly rainfall varies from 67 mm in October to 306 mm in July, and the mean annual precipitation is approximately 1100 mm. The mean monthly minimum and maximum temperatures are approximately 14.16 °C and 27.63 °C, respectively. The Angereb watershed at the gauge site is located at 37°28′5″ to 37°32′56″E longitude and 12°38′26″ to 12°43′31″N latitude (WGS_1984 Lat/Long). The watershed area at the gauge site covers 49.74 km^2.

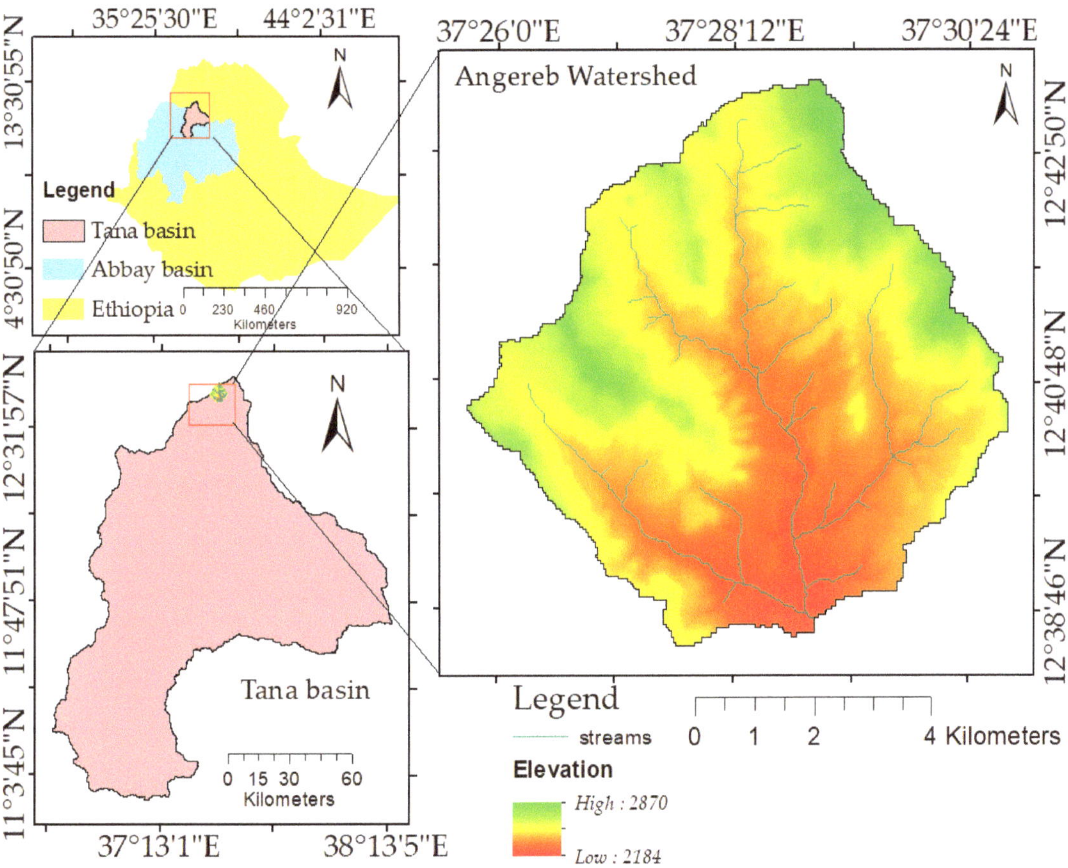

Figure 1. Study area map.

2.2. Datasets

2.2.1. Land Use/Land Cover Data

The data used in this research were the digital elevation model (DEM), soil maps, land use/cover maps, and climate data for the baseline period. The 12.5 m high-resolution, terrain-adjusted ALOS PALSAR geotiff DEM data were downloaded from https://search.asf.alaska.edu/ (accessed on 5 January 2023) and resampled to 30 m resolution for the modeling. This high-resolution DEM is radiometric-terrain-corrected, which removes geometric and radiometric distortion and has very good vertical and horizontal accuracy. The FAO-based soil raster file was collected from the Ministry of Water and Energy (MoWE). Land use/land cover (LULC) data for the years 2012, 2017, 2019, and 2022 were collected from the Ethiopian Mapping Agency, the Water and Land Resources Centre (WLRC), and the Ministry of Water and Energy (MoWE), and more than 50% of the area was covered with cropland for all four time periods of land use data (Table 1). The land use/cover data of 2017, 2019, and 2022 gathered from those organizations were acquired from the Sentinel-2, 10 m resolution map, and ground truthing was performed to adjust the cloud cover errors, while the 2012 land use map was of 30 m resolution, sourced from the FAO catalog resourced from the Ethiopian Mapping Agency (EMA). Phiri et al. [52] stated that because Sentinel-2 data employs machine-learning classifiers, such as random forest (RF) and support vector machine (SVM), the classification accuracy is high (>80%).

Table 1. Percent of area coverage by different land use/cover types for 2012, 2017, 2019, and 2022.

Land Use/Land Cover Class	Area (Hectare)				Percentage Distribution			
	2012	2017	2019	2022	2012 (%)	2017 (%)	2019 (%)	2022 (%)
Water bodies	38.97	42.57	41.4	73.98	0.092	0.100	0.098	0.174
Forest	1743.3	1042.65	1132.02	1230.12	4.106	2.456	2.666	2.898
Wetland/Vegetative	5.85	0.72	0.27	0.72	0.014	0.002	0.001	0.002
Cropland	20,880.54	29,998.08	30,804.57	31,252.41	49.185	70.662	72.562	73.617
Settlement	1806.66	3347.28	4112.46	4677.75	4.256	7.885	9.687	11.019
Bare land	291.06	0.18	1.35	0.18	0.686	0.000	0.003	0.000
Rangeland	17,686.44	8021.34	6360.75	5217.66	41.661	18.895	14.983	12.290

2.2.2. Climate Data

Climate data, such as precipitation and maximum and minimum temperatures for the baseline period, were collected from the Ethiopian National Meteorological Agency (NMA), and the number of wet days was calculated from daily precipitation, with a 1 mm depth as a threshold value.

The observed precipitation and maximum and minimum temperature data were collected from the National Meteorological Agency (NMA) for 28 years (1986–2014) of records for the stations in the study areas. The precipitation and temperature (maximum and minimum) climate data under the RCP4.5 and RCP8.5 climate scenarios and historical data for 1986 to 2005, with an overlap of more than 70%, were downloaded from: https://esgfdata.dkrz.de/search/esgf-dkrz/ (accessed on 20 January 2023). The model domain was AFR-44, the daily time step data resembled those of r1i1p1, and the selected experiments included historical, RCP4.5, and RCP8.5 data.

2.3. Methods

The methods employed in this study included predicting land use/cover using the QGIS2.16.3 software plugin, named the MOLUSCE, based on the cellular automata artificial neural network (CA-ANN) algorithm. Land use/cover prediction was performed for the years of 2035, 2055, and 2075 when the near, middle, and far future impact assessment of flow and sediment was performed. Then, using GCM/RCM climate models, future climatic projections were performed for the periods that were necessary to assess soil erosion and sediment output of the impact analysis. Finally, soil loss, sediment yield, and flow estimation were conducted using the GeoWEPP model. The impacts were identified by comparing the near, middle, and far future results with the baseline results for all metrics.

2.3.1. Modeling Approach

This study utilized modeling of land use/cover prediction, climate projection, soil loss, sediment yield, and runoff simulation. Changes in soil loss, sediment yield, and flow were evaluated by comparing the baseline time results with projected simulations, and impacts were identified. To do this, the data, such as climate, flow, sediment, land use/cover, soil, and road and stream networks, were collected, checked for errors, and corrected before application. The overall conceptual framework for the methods employed for this study is presented below (Figure 2).

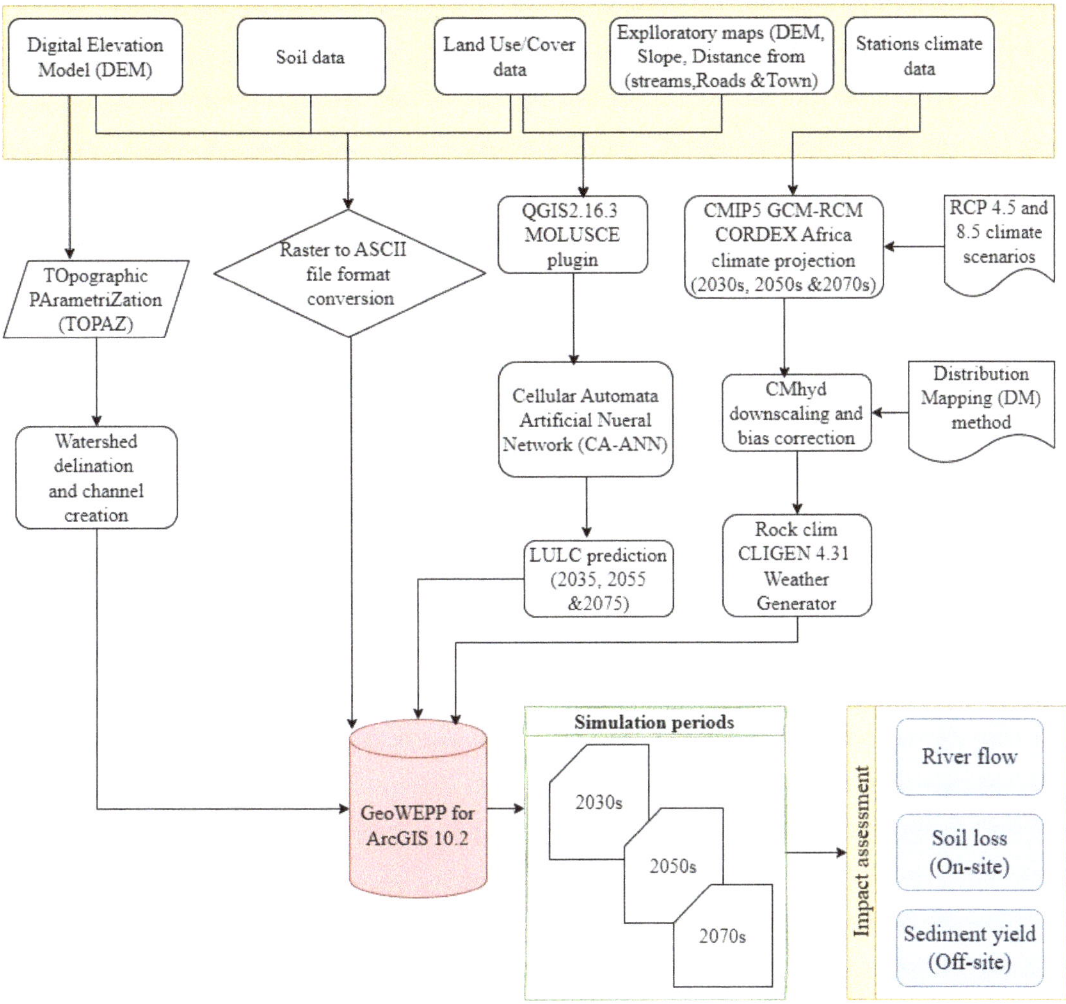

Figure 2. Land use/cover and climate change impact framework diagram.

2.3.2. Land Use/Cover Forecasting

We employed the Quantum Geographic Information System (QGIS 2.16.3) spatial analysis software with a wide range of applications, including the plugin Modules for Land Use Change Evaluation (MOLUSCE), which provides a set of algorithms for land use change simulations, such as ANN (artificial neural network), LR (logistic regression), WoE (weight of evaluation), and multi-criteria evaluation for modeling and simulation of land use change. The MOLUSCE plugin uses either Pearson's correlation or Cramer's coefficient to evaluate the correlation between land use/cover and exploratory variables. After creating the correlation land use data with spatial variables, a transition matrix table was generated, land use change maps were created, and finally, the future land use map was predicted.

The future LULC spatial data for the 2035, 2055, and 2075 datasets were predicted using the cellular automata artificial neural network (CA-ANN) algorithm. Setting the frames of 2030s, 2050s, and 2070s was selected by considering the useful life of the Megech Dam in the research watershed. The CA-ANN algorithm was chosen because of its ability to simulate complicated land use change patterns and a range of land covers, as well as its high efficiency with minimal data and ease of calibration [7,45]. For this study, generation of the transition matrix and model validation was performed for the years 2017–2022. Then, the processes were repeated for the three timeframes to generate the future land use/cover changes. In the MLP-CA-ANN algorithm, neurons or nodes are interconnected to create a complex neural network that is composed of layers, including an input layer, hidden layers, and an output layer (Figure 3).

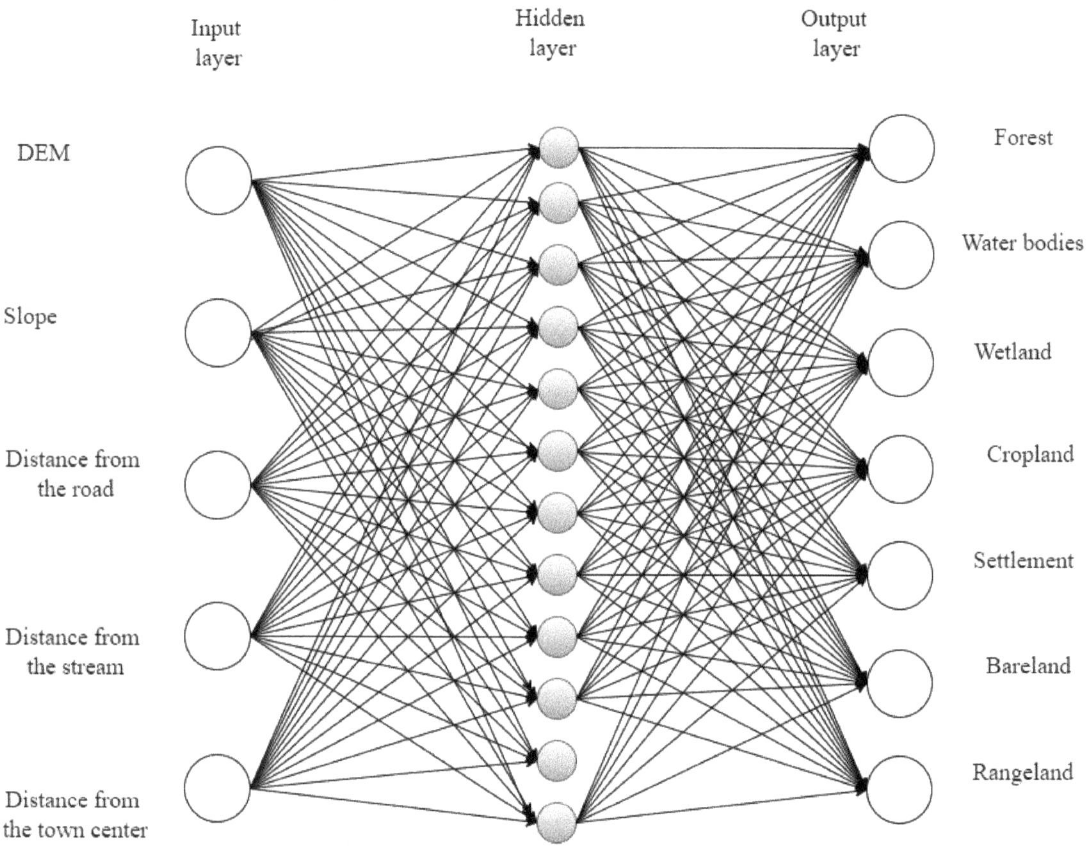

Figure 3. Structural map of the multilayer neural network.

2.3.3. Climate Modeling

Five climate models were selected based on their appropriateness in the Upper Blue Nile Basin. The selected GCMs included CanESM2, EC-EARTH, GFDL-ESM2M, MPI-ESM-PI, and NorESM1-M. Three regional climate models (RCMs), named RCA4, REMO2009, and RACMO22T, were used from the Coordinated Regional Climate Downscaling Experiment (CORDEX), Africa. The combinations of GCM/RCM (CanESM2/RCA4, EC-EARTH/RACMO22T, GFDL-ESM2M/RCA4, MPI-ESM-PI/REMO2009, and NorESM1-M/RCA4) were considered to download daily precipitation, and maximum and minimum temperature climate variables, with a resolution of 0.440 × 440, or an approximately 50 × 50 km grid size. Note that the middle-level scenario (RCP4.5) and the worst-case scenario in terms of greenhouse gas emissions (RCP8.5), with no clear climate policy that shows a radiative forcing pathway leading to 8.5 w/m^2 (greater than 1370 ppm CO_2-

equivalent) in 2100 [53], were used to examine the impacts of climate change on the soil erosion and sediment yield in the study area.

2.3.4. Bias Correction Method

Using the climate model for the hydrological modeling tool (CMhyd) at the station level, bias correction of the RCM simulations was carried out in this study. The distribution mapping [53] bias correction method, which can better transfer the observed precipitation and temperature statistics to the raw ensemble GCM/RCM CORDEX-Africa RCP4.5 and RCP8.5 climate scenarios, was employed. The climate model data for the hydrologic modeling (CMhyd) tool were used for future data extraction and bias correction.

2.3.5. Soil Erosion Model Description

The geospatial water erosion prediction project (GeoWEPP) model is an erosion model from the continuous simulation and process-based water erosion prediction project (WEPP) [54,55]. This model was employed as an impact model to identify the changes in flow and soil erosion due to LULC and climate change individually, as well as in combination. The GeoWEPP model was developed to integrate the advanced features of geographical information systems (GISs) within the WEPP model, such as processing digital data sources and generating digital outputs [56]. The movement of the sediment along the hill slope is described based on the steady-state sediment continuity equation, which applies flow conditions [57]:

$$\frac{dG}{dX} = D_i + D_r \quad (1)$$

where G is the sediment load (kg s^{-1} m^{-1}), X is the distance downslope, D_i is the interrill erosion rate (kg s^{-1} m^{-1}), and D_r is the rill erosion rate (kg s^{-1} m^{-1}):

$$D_i = K_i I_e \, \sigma_{ir} SDR_{RR} F_{nozele} \left(\frac{R_s}{W} \right) \quad (2)$$

where K_i is the adjusted interrill erodibility (kg s/m^4), I_e is the effective rainfall intensity (mm/h), σ_{ir} is the interrill runoff rate (mm/h), SDR_{RR} is the interrill sediment delivery ratio, Fnozzle is the adjustment factor for the sprinkler irrigation nozzle impact energy variation, R_s is the rill spacing, and W is the width of the rill. Since there is no sprinkler irrigation in the study areas, the factor was taken as a unit:

$$D_r = K_r \left(\tau_f - \tau_c \right) \left(1 - \frac{G}{T_c} \right) \quad (3)$$

where K_r is the adjusted soil erodibility parameter (s/m), τ_f is the flow shear stress (kg/m/s^2), τ_c is the adjusted critical shear stress of the rill surface (kg/m/s^2), and T_c is the sediment transport capacity of the rill flow (kg/m/s). T_c was calculated as follows:

$$T_c = K_{tr} q_w s \quad (4)$$

where K_{tr} is a constant parameter, q_w is the flow discharge per unit width (m^2/s), and S is the slope (%). The GeoWEPP model can be used to capture changes in soil loss, runoff, and sediment delivery rates in watersheds caused by changes in land use/cover [58]. According to Luetzenburg et al. [59], the model is a useful tool for depicting the impacts on soil erosion, flow, and sediment delivery under climate and land use/cover change scenarios to design possible mitigation measures at the watershed level.

2.4. Scenario Development

Three future time window scenarios for the 2030s, 2050s, and 2070s were selected to identify the impacts of LULC changes and climate change on flow, soil loss, and sediment yield. These years were selected based on the life of the Megech Dam and considering time intervals for significant changes in land use and climate. Input data for the GeoWEPP model were collected from the Ministry of Water and Energy [60], Abay Basin Development [61], National Meteorological Agency (NMA), and the Water and Land Resources Center (WLRC). Based on the collected data, future LULC forecasting was performed using the CA-ANN algorithm.

Scenario 1: time window of the 2030s—only land use/cover change (climate is constant), only climate change (land use/cover is constant), and both climate and LULC changes (synchronizing both).

Scenario 2: time window of the 2050s—only land use/land cover change (climate is constant), only climate change (land use/cover is constant), and both climate and LULC changes (synchronizing both).

Scenario 3: time window of the 2070s—only land use/cover change (climate is constant), only climate change (land use/land cover is constant), and both climate and LULC changes (synchronizing both).

3. Results

3.1. Precipitation and Temperature

According to [62], the Intergovernmental Panel on Climate Change (IPCC) recently revealed that the global surface air temperature has risen considerably, by about 1.53 °C on average, for the time period of 1850–1900. In the 2050s, the warming may even increase by 2 °C (medium confidence) and 3 °C (low confidence) [62]. The projected climate data were based on five global circulation models (GCMs) (CanESM2/RCA4, EC-EARTH/RACMO22T, GFDL-ESM2M/RCA4, MPI-ESM-PI/REMO 2009, and NorESM1-M/RCA4), and downscaling and bias correction were performed using the distribution mapping [53] algorithm method for the RCP4.5 and RCP8.5 scenarios at representative stations. In Megech watershed, the average increases in maximum temperature in the 2050s were 1.94 °C and 2.63 °C for the RCP4.5 and RCP8.5, respectively. In this study, a maximum of 5.80 °C temperature change was identified for the far future time period for the worst-case scenario. The multi-model ensemble average of the downscaled and bias-corrected future climate data from five climate models was used to feed the RockClime (CLIGEN4.31) stochastic weather generator model to produce a climate PAR file, which was used as climate input for the GeoWEPP impact assessment erosion model.

3.2. Climate Change

The projected ensemble maximum temperatures under the RCP4.5 climate scenario for the 2030s, 2050s, and 2070s timeframes increased from 1.860 °C to 3.720 °C, 0.80 °C to 3.08 °C, and 0.90 °C to 3.89 °C, respectively. For the RCP8.5 scenario, the maximum temperatures increased for all months, with ranges of 0.45 °C to 2.15 °C, 1.78 °C to 3.48 °C, and 1.71 °C to 5.80 °C for the 2030s, 2050s, and 2070s, respectively. The relative percentage change in maximum temperature for RCP4.5 compared with the baseline climate varied from 6.79 to 14.28%, 3 to 11.01%, and 3.85 to 13.52% for the 2030s, 2050s, and 2070s, respectively. For the RCP8.5 climate scenario, relative percentage increases of 3.80–7.70%, 6.48–12.42%, and 6.98–24.57% were observed for the 2030s, 2050s, and 2070s, respectively (Figure 4A,B).

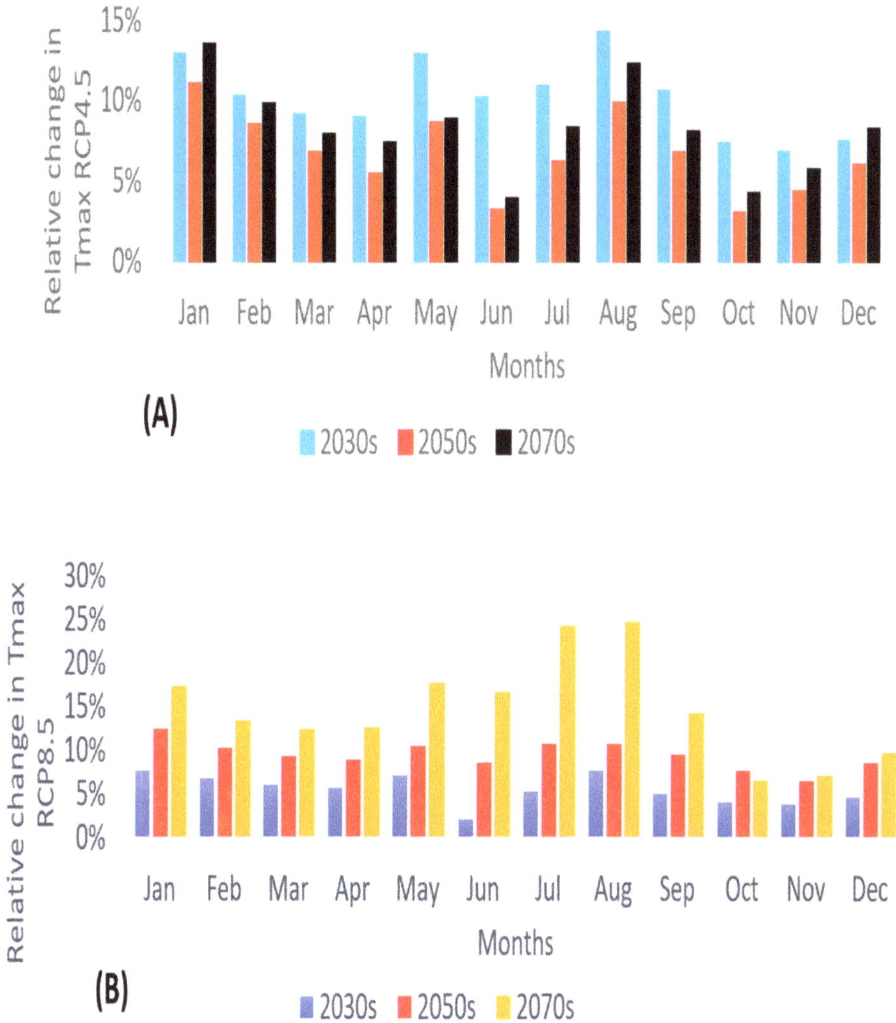

Figure 4. Future seasonal maximum temperature changes under the (**A**) RCP4.5 and (**B**) RCP8.5 scenarios.

The relative changes in the minimum temperature showed that forecasted monthly values decreased for the 2030s, 2050s, and 2070s compared to the base period for all RCP4.5 and RCP8.5 climate scenarios (Figure 5).

The projected maximum temperatures in the 2030s, 2050s, and 2070s at the Gondar Station increased compared with those in the baseline period (1985–2014), while the minimum temperature decreased under both RCP4.5 and RCP8.5 under all the time window scenarios (Figure 6).

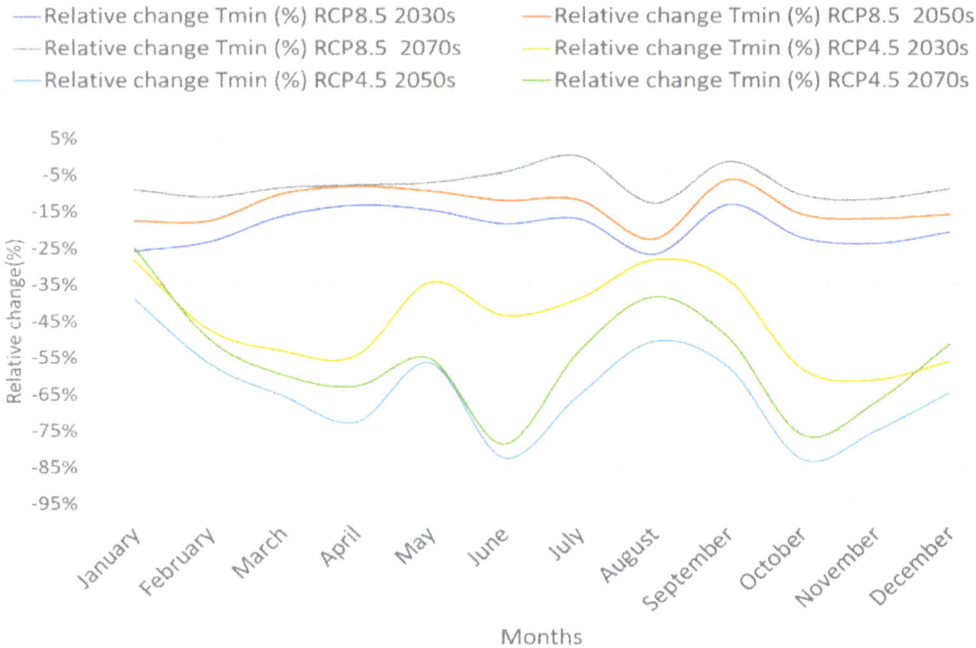

Figure 5. Relative changes in minimum temperature for the 2030s, 2050s, and 2070s for both the RCP4.5 and RCP8.5 scenarios.

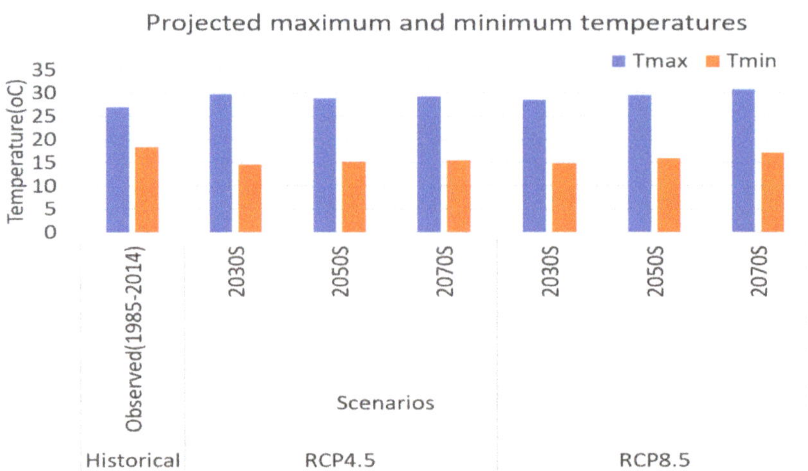

Figure 6. Projected mean annual maximum and minimum temperatures compared with historical records.

The projection of the ensemble precipitation results from the five GCM/RCM model combinations did not show a consistent pattern for all three timeframes of the 2030s, 2050s, and 2070s. This irregular pattern was observed for both the RCP4.5 and RCP8.5 climate scenarios. Figure 7 demonstrates that precipitation was as high as two-fold for some months, or 1.5 times the observed baseline value for other months. In contrast, for other months, precipitation was lower compared with the baseline, as verified by the negative relative change. This characteristic was observed for all time domains and RCP scenarios, which indicated that precipitation did not have consistent characteristics due to climate change in the watershed.

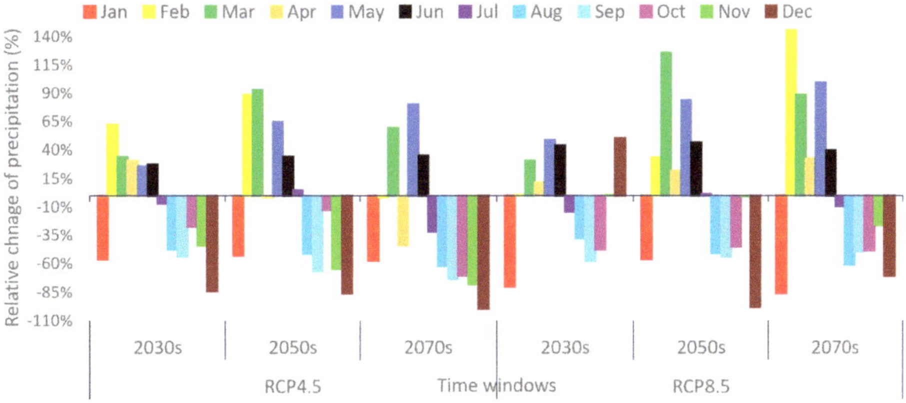

Figure 7. Precipitation relative changes for the RCP4.5 and RCP8.5 climate scenarios.

The monthly number of wet days, calculated based on the criterion of a depth greater than 1 mm [63] in the study region from the projected ensemble precipitation, indicated fewer wet days than in the base period, especially for the rainy months (Figure 8). The mean monthly depth of precipitation and the number of wet days were used to generate weather/climate inputs to the GeoWEPP model.

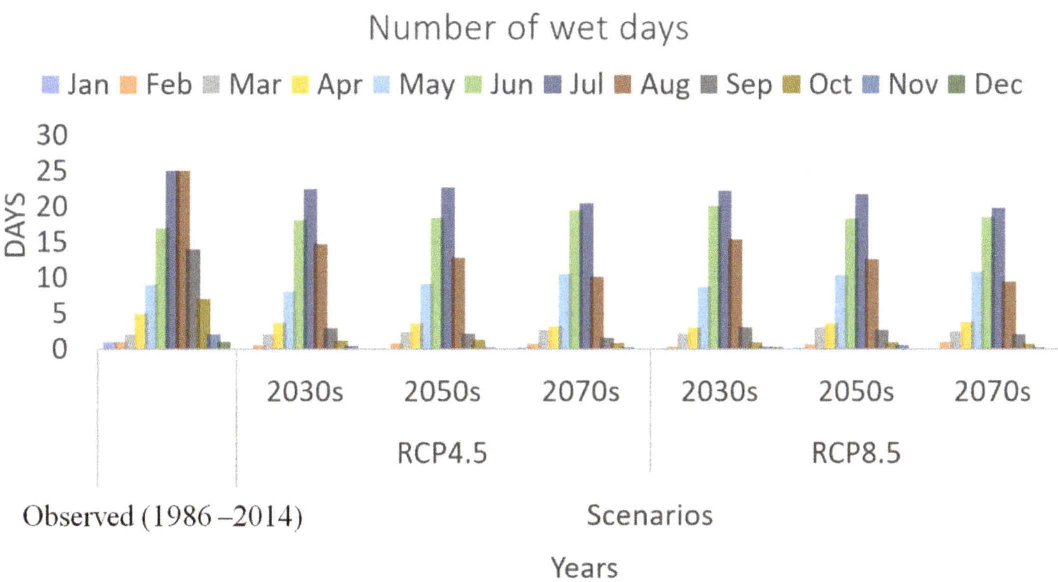

Figure 8. Projected wet days for the 2030s, 2050s, and 2070s for RCP4.5 and RCP8.5.

In the RCP4.5 and RCP8.5 climatic scenarios, the ensemble annual precipitation forecasts for the 2030s, 2050s, and 2070s decreased relative to those in the base period, with ranges from 6 to 28%, even though the monthly values occasionally showed increasing and decreasing trends (Figure 9).

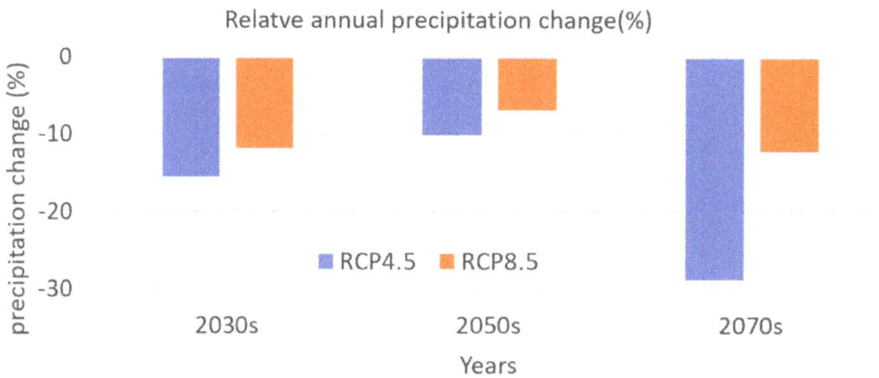

Figure 9. Relative change percentage of precipitation.

3.3. *Land Use/Cover Prediction*

The land use/cover change prediction was performed using QGIS 2.16.3 software with the modules for land use change evaluation (MOLUSCE-version 3.0.13) plugin software. The plugin provides a set of algorithms for land use/cover change simulations, such as the cellular automata artificial neural network (CA-ANN). Because of its dynamic simulation capacity, high efficiency with limited data, simple calibration, and ability to reproduce different land covers and complicated patterns, CA-ANN was chosen to simulate land use in the future [45]. The learning algorithm analyzes the reached accuracy on training and validation sets of samples and stores the best neural network in memory. The training process finishes when the best accuracy is reached (Figure 10).

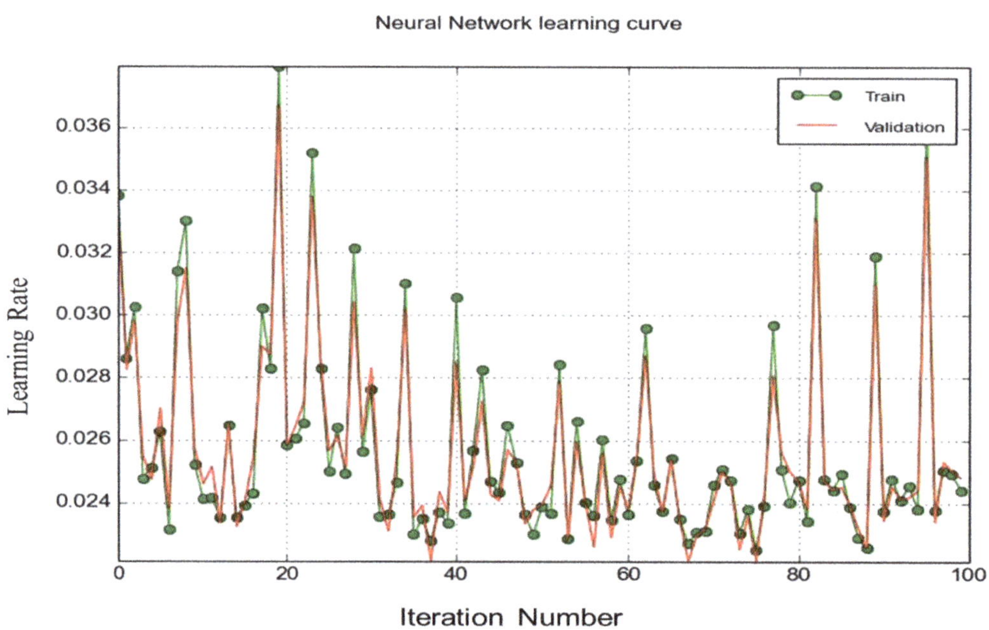

Figure 10. Cellular automata artificial neural network learning curve for 2012 and 2017 to predict the land use/cover map of 2022.

During the validation process, a multiple-resolution budget method was used in the MOLUSCE plugin. By selecting one hundred iterations, the reference and simulated maps of the same period were validated (Figure 10). In the model, performance measures, such as kappa, the overall accuracy (% of correctness), kappa (overall), kappa (historical), and kappa (location) can be calculated.

Cellular Automata Simulation

Future land use/cover was forecasted using the cellular automata simulation (CA-ANN) method based on historical land use data for 2012, 2017, 2019, and 2022. Spatial exploratory maps of elevation, aspect, slope, distance from the city center, distance from streams, and distance from the road were generated, and then a transition matrix was generated. Model validation for 2022 between the observed and simulated land use/cover maps of the Megech watershed resulted in a kappa value of 0.76 and an overall correctness of 82.64%.

The predicted land use/cover maps for 2035 were produced after generating the 2026 maps from the transitions of 2018 and 2022 and the explanatory variables, with a kappa of 0.82 rom the transition matrix of the 2017 and 2026 maps. The 2055 map was predicted with the same procedure using the simulated map of 2037 and the observed map of 2019, with a kappa of 0.78, and the spatial variables and the transition matrix were used to simulate this map. To forecast the land use/cover map of 2075, the predicted maps of 2035 and 2055 were used, and spatial variables and a transition matrix with a kappa value of 0.68 were used. Cropland and settlement/built-up areas were shown to be increasing for all time projections, with diminishing forest and rangelands. Compared with the baseline data, other land use/cover classes, such as wetlands, bare lands, and waterbodies, exhibited non-significant changes under all the prediction scenarios (Table 2 and Figure 11).

Table 2. Class statistics of land use/cover for 2017, 2035, 2055, and 2075.

Year Land Use/Cover	Area (ha)				% of Coverage				Change Percentages		
	2017	**2035**	**2055**	**2075**	**2017%**	**2035%**	**2055%**	**2075%**	**Δ %**	**Δ %**	**Δ %**
Water bodies	42.57	70.92	38.7	51.12	0.10	0.17	0.09	0.12	0.07	−0.01	0.02
Forest	1042.65	349.47	317.52	381.87	2.46	0.82	0.75	0.90	−1.63	−1.71	−1.56
Wetland/ Vegetative	0.72	1.35	0.63	0.63	0.001	0.00	0.00	0.00	0.00	0.00	0.00
Cropland	29,998.08	33,428.88	33,480.36	36,397.26	70.68	78.74	78.86	85.7	8.06	8.19	15.06
Settlement	3347.28	4702.14	5115.6	4104.9	7.87	11.08	12.05	9.67	3.21	4.18	1.80
Bare land	0.18	0.27	0.18	1.62	0.001	0.00	0.00	0.00	0.00	0.00	0.00
Rangeland	8021.34	3899.79	3499.83	1515.42	18.90	9.19	8.24	3.57	−9.71	−10.65	−15.33

From the land use/cover prediction to future time scenarios, cropland and settlement areas increased, while range lands and forests decreased in the Megech watershed. For 2035, 2055, and 2075, the cropland and settlement areas increased by 8.07%, 8.18%, and 15.06% and 3.21%, 4.18%, and 1.80%, respectively. For 2035, 2055, and 2075, the percentage of rangeland decreased by 9.71%, 10.65%, and 15.33%, respectively, whereas that of forests decreased by 1.63%, 1.71%, and 1.56%, respectively (Figure 12). The relative changes in other land use/cover classes, such as water bodies, wetlands, and bare lands, were insignificant. To ascertain the relative changes, data from the chosen baseline period of 2022 land use/cover data were compared with predicted land use/cover data.

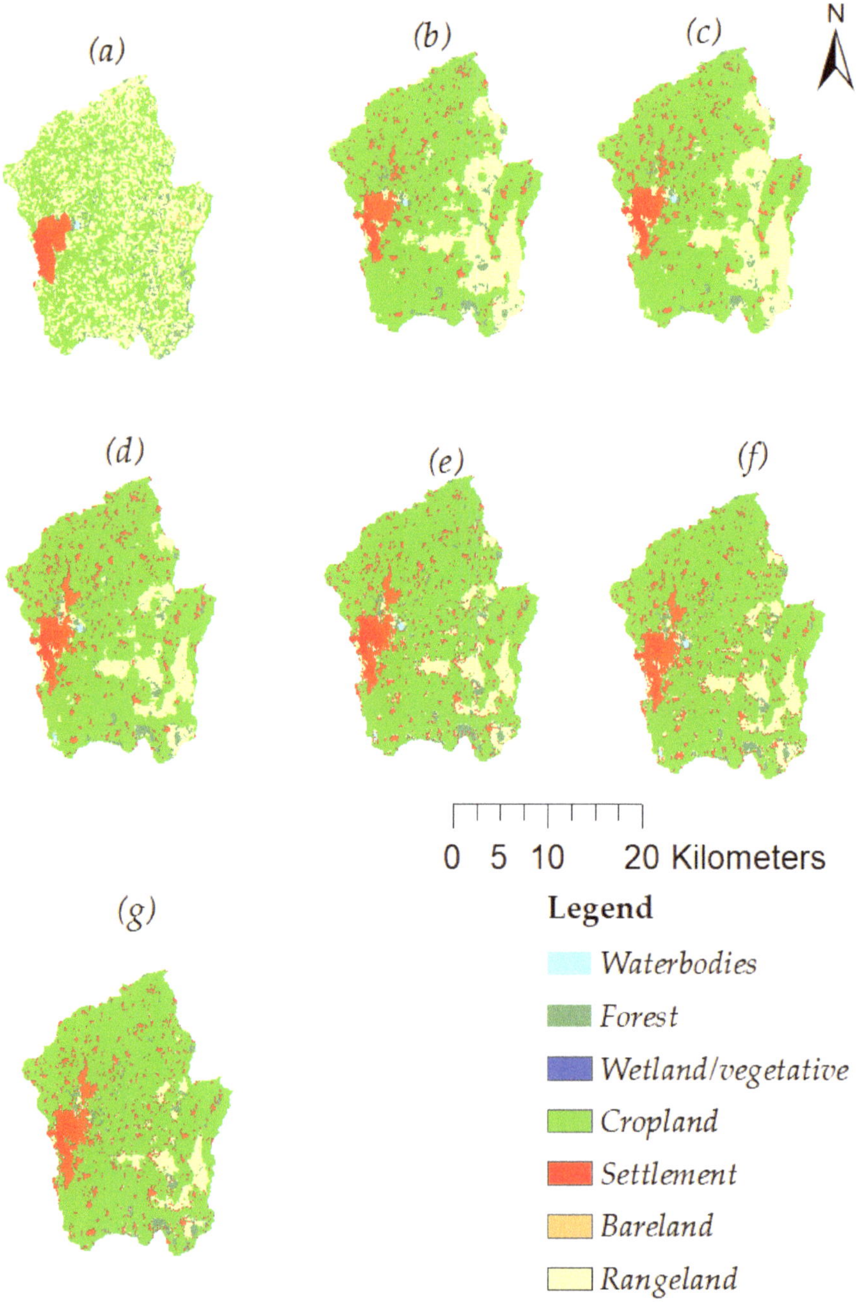

Figure 11. Land use/land cover maps observed in the Megech Dam watershed in (**a**) 2012, (**b**) 2017, (**c**) 2019, and (**d**) 2022 and in (**e**) 2035, (**f**) 2055, and (**g**) 2075.

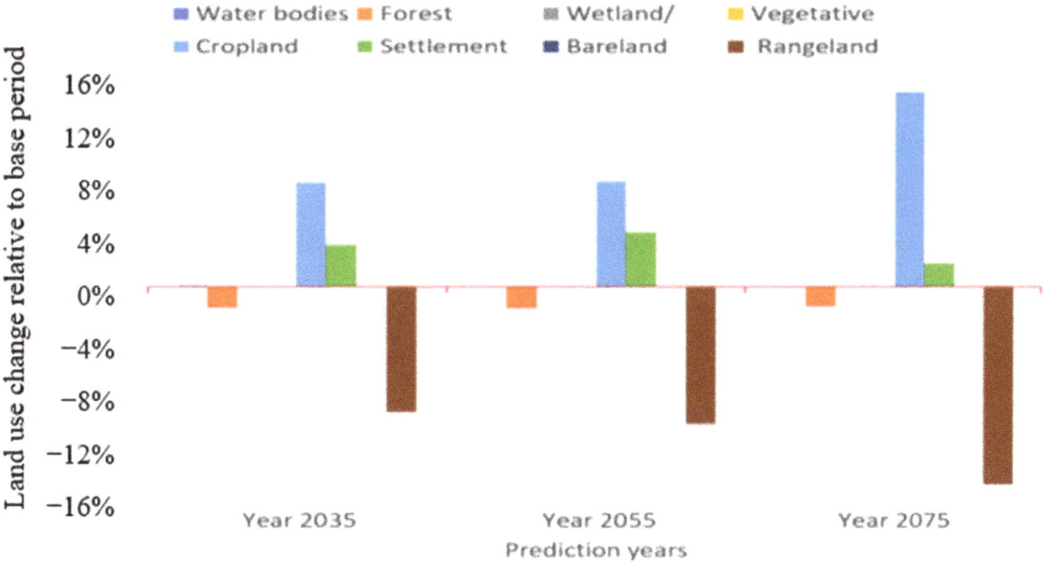

Figure 12. Relative change in predicted land use/cover.

However, after the dam commenced, the land use/cover in the reservoir area changed to water bodies, and the relative percentage shift changed with the change in value throughout the season and throughout the year due to the variation in the reservoir level. In this study, such variation is not included since the land use/land cover impact study focused on the upstream sub-watersheds. Another factor contributing to the study's limitations is the dynamic nature of the reservoir size, which is roughly 8.9 km2 when the water reservoir is at the NWL and lowers with the water level decline during the dry and high-water-intake seasons.

3.4. GeoWEPP Model Simulations

Through sensitivity analysis of parameters controlling surface runoff, soil loss, and sediment yield rates from the hill slope, such as hydraulic conductivity (K), interrill erodibility (Ki), rill erodibility, and critical shear stress (τc), the performance of the GeoWEPP model simulation in the studied watershed was verified. Following the selection of the sensitive factors, the statistical metrics of coefficient of determination, Nash–Sutcliffe, and percent of bias were calibrated and validated for runoff and sediment yield [64,65]. According to Moriasi et al. [66], the aforementioned statistical indicators' results demonstrated how effectively the simulations reproduced the reported values (Table 3). Model calibration and validation were performed at three and two years, respectively, for both water and sediment flows.

Table 3. Statistical metric values for calibration and validation periods.

	Runoff			Sediment Yield		
	R2	NSE	PBIAS	R2	NSE	PBIAS
Calibration	0.942	0.941	−2.41%	0.69	0.70	+3.77%
Validation	0.75	0.65	−38.77%	0.75	0.54	+53.01%

The GeoWEPP model simulation was based on climate data generated using Rock-Clime (CLIGEN 4.31) stochastic weather generator software for the data collected from stations for the baseline period and projected climate data for the future scenario assessment. Weather generated by long-duration simulation data, which is the input format for the GeoWEPP model in the daytime step and includes precipitation depth, intensity, time to peak, peak intensity, dew point, maximum and minimum temperature, radiation, wind velocity, and direction, can provide us with a long-term assessment of runoff, soil loss, and sediment yield in the study watershed.

3.4.1. Impacts of Land Use/Cover Change on Flow, Soil Loss, and Sediment Yield

Land use/cover change may increase or decrease runoff, soil loss, and sediment delivery rates in watersheds. The value of the effective hydraulic conductivity (K_e) changed, while other soil parameters, such as the critical shear stress (τ_c), rill erodibility [53], and interrill erodibility (K_i), and climate factors, such as the effective intensity (I_e) and interrill flow rate (σ_r), remained unchanged during the evaluation of land use/cover change effects on soil erosion and flow using the GeoWEPP model. Compared with those in the baseline period (2017), cultivated land and built-up areas increased by 8.1–15.1% and 1.8 to 4.18%, respectively, for 2035, 2055, and 2075 due to the reduction in forest and rangelands. The land use/cover change sediment alteration model simulated 4.06–7.67%, 8.26–18.86%, and 5.23–15.94% of the runoff, hill slope soil loss, and yield, respectively (Table 4). The changes in land use/cover varied the distribution of effective hydraulic conductivity and impacted the model results. Figure 13 illustrates that the Angereb sub-watershed sediment yield distribution was greater than the tolerable soil loss (T = 10 t/ha/year) for the baseline year of 2017 and for the future (2035, 2055, and 2075) predicted land use/cover changes. According to the distribution of soil loss (flow path technique) and sediment yield (watershed method) model simulation results, the severity level and its spatial extent increased from the near to far future land use/cover change analysis, in comparison with the acceptable loss (T = 10 t/ha/yr.) as a reference for comparison (Figure 13).

Table 4. Watershed method simulation results under land use/cover change.

Simulated Parameters	Simulated Model Results				% Relative Change by LULC Compared with the Baseline		
	Base	2030s	2050s	2070s	2035	2055	2075
Total contributing area to an outlet (ha)	4974.02	4974.02	4974.02	4974.02	-	-	-
Avg. Ann. Precipitation volume in contributing area (m^3/yr)	52,330,679	52,330,679	52,330,679	52,330,679	-	-	-
Avg. Ann. water discharge from an outlet (m^3/yr)	14,001,599	15,075,338	14,570,401.0	14,572,299.00	7.67	4.06	4.08
Avg. Ann. total hillslope soil loss (tons/yr)	265,085.50	295,608.1	315,079.7	320,035.7	8.26	18.86	20.73
Avg. Ann. Sed. delivery per unit area of the watershed (T/ha/year)	32.69	34.4	36.60	37.90	5.23	11.96	15.94

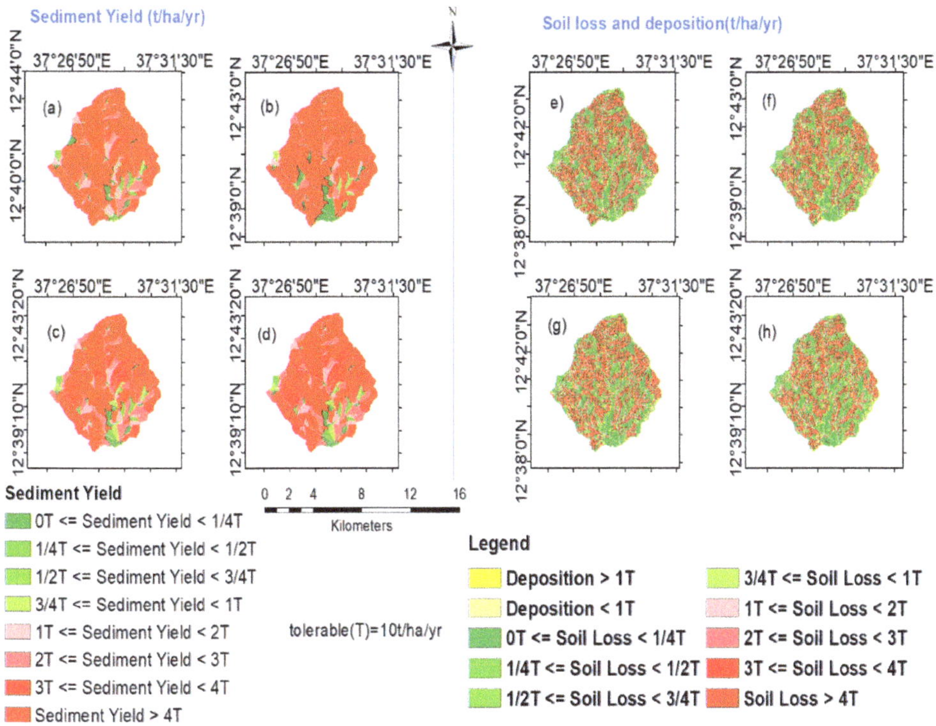

Figure 13. Under land use/cover change, watershed sediment yield for (**a**) 2017, (**b**) 2035, (**c**) 2055, and (**d**) 2075 and soil loss for (**e**) 2017, (**f**) 2035, (**g**) 2055, and (**h**) 2075.

3.4.2. Impacts of Climate Change on Flow, Soil Loss, and Sediment Yield

RockClime or CLIGEN version 4.31 was used to create climate PAR files to analyze the effects of climate change on runoff, soil loss, and sediment yield. The monthly maximum and minimum temperatures, monthly precipitation, and wet days projected for the 2030s, 2050s, and 2070s under the RCP4.5 and RCP8.5 climate scenarios were used as inputs to create par files that served as inputs to the GeoWEPP model. The computed water flow for the projected climate for the 2030s, 2050s, and 2070s was compared with the baseline and indicated an increase in RCP4.5 for the 2030s and 2050s and a decrease for the 2070s. For RCP8.5, the decreasing stream flow pattern ranged from 4.01 to 20.65%. The sediment yield rates exhibited an increasing trend for all time windows, with values ranging from 4.1 to 24.95% for RCP4.5 and from 16.51 to 63.55% for RCP8.5, which may be due to an increase in rainfall intensity during the wet season (Figure 14). Climate-change-related increases in sediment yield put the reservoir storage capacity of the Megech Dam under greater stress unless appropriate measures are taken to address them. Onsite soil loss is predicted to increase under the RCP4.5 and RCP8.5 climate scenarios, with ranges of 1.26–5.90% and 11.44–16.86%, respectively.

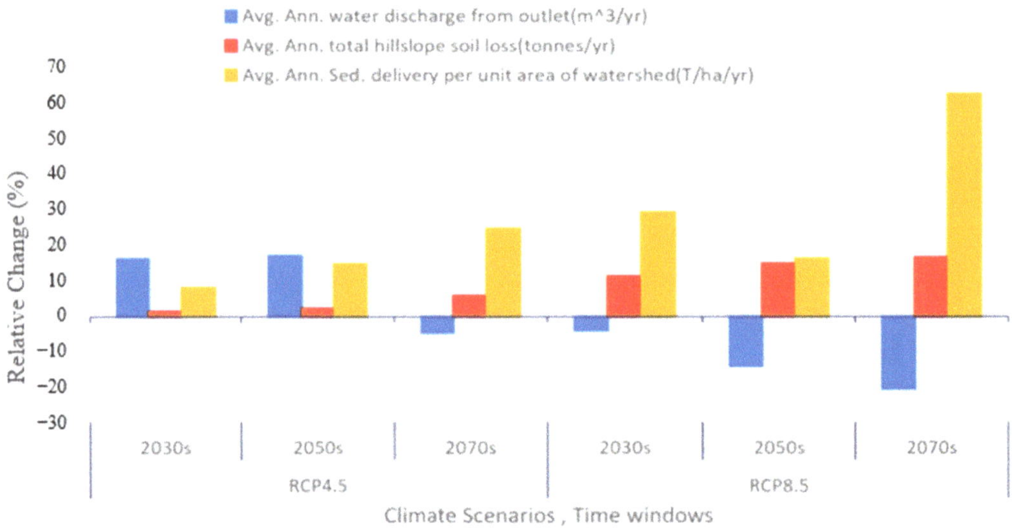

Figure 14. Relative percent change in discharge, hill slope soil loss, and sediment yield rates for the Angereb watershed.

4. Discussion

Megech River is an important source of water supply for Gondar Town and irrigation use in the Megech watershed. Therefore, evaluating soil loss, sediment yield, and flow as well as assessing climate and land use/cover changes' impacts are very important. In mountainous areas, such as the Upper Blue Nile Basin, onsite soil loss and offsite sedimentation are the most complex problems threatening agricultural productivity and dam productivity [67,68]. This study examined isolated land use, climate change, and the combined impacts of changing land use/cover and climate on runoff, soil loss, and sediment yield for the Megech watershed. Study results indicated increasing cropland and settlement areas with reductions in forest and rangelands. Studies by Gashaw et al. [69] and Tadesse et al. [70] showed similarity on land use changes in the region, indicating the intensification of soil erosion and sediment yield rates over time in the Upper Blue Nile, Ethiopia. In addition, this study identified climate change and combined scenarios on the impact assessment of watershed soil loss, sediment yield, and flow. The study results of climate change and combined land use/cover and climate changes on soil loss and sediment yield indicated that soil loss and sediment yield rates had higher impacts than land use/cover change for all 2030, 2050, and 2070 timeframes.

4.1. Impact of Land Use Change on Soil Loss, Sediment Yield, and Flow

The findings of this study showed that land use/cover change highly altered soil loss, sediment yield, and river flow of the Megech watershed. To understand land use change patterns and their impacts on the watershed's hydrology in the Upper Blue Nile Basin, several studies have been performed and reported [13,69,71]. Our study focused on land use/cover change based on the CA-ANN model, and the predictions were for the years 2035, 2055, and 2075, which indicated an increase in cropland and settlement areas with the loss of forest and rangeland in the Megech watershed. Such an increase in cropland and built-up areas and loss of rangeland and forest areas is consistent with other findings reported in the study region [69,72,73] that impacted the soil erosion and flow of the watershed. However, this study integrated land use/cover changes with climate change (RCP4.5 and 8.5) scenarios and employed the process-based GoeWEPP model for the impact analysis. By employing the CA-ANN algorithm for the land use/cover change prediction, this study contributed the technique of applying land use maps and spatial variables to capture the watershed land use dynamics. Additionally, the results of land use change and impacts on soil loss, sediment yield, and flow are useful for sustainable land use planning and implementation of improved soil and water conservation practices to prevent the adverse impacts. Nevertheless, population density data are not directly incorporated into the artificial neural network's learning process for land use/cover prediction. However, the influence of demography on land use/cover change is indirectly accounted for through spatial variables, such as distance from the city and distance from roads. It is important to note that the model's prediction results may undergo significant variation due to changing governmental plans, such as the construction of dams and the expansion of agricultural areas, which should be the focus of future studies.

4.2. Impact of Climate Change on Soil Loss, Sediment Yield, and Runoff

In the Megech Dam watershed, we found that climate change had an overall negative impact on soil loss and sediment yield. The maximum temperature showed an increasing trend, while the minimum temperature decreased compared with that in the base period. The projected precipitation showed an increase for some months (wet months) and a decrease for others (dry months), with time windows of the 2030s, 2050s, and 2070s under the RCP4.5 and 8.5 climate scenarios. The studies by Tariku et al. [74] and Getachew and Manjunatha [40] showed similar patterns of temperature, precipitation, and runoff in the study region. These changes result in amplified intensity, consequently leading to an increase in rill erodibility (Kr) and interrill erodibility (Ki), thus increasing soil loss and sediment yield rates over the predicted timeframe. The annual sediment yield rates of the Angereb watershed showed an increasing trend for the RCP4.5 and RCP8.5 climate scenarios for all time windows of the 2030s, 2050s, and 2070s, which implied that climate change will decrease the live storage of dams. Studies in the region also indicated that the RCP8.5 climate scenario produced higher sediment yield and soil loss rates as compared to RCP4.5 [3]. As we projected further into the future, specifically examining scenarios for the near future (2030s), middle future (2050s), and far future (2070s), the intensification of climate change impacts became increasingly evident [40]. This situation was observed across various simulated metrics of runoff, soil loss, and sediment yield. Therefore, in the far future, particularly by the 2070s, climate change impacts are projected to escalate significantly, underscoring the urgent need for proactive measures to mitigate its consequences. This study provided important insights, though it is limited in that only a single method (distribution mapping) was employed in the climate data downscaling and bias corrections, with the ensemble of the five GCM/RCM CORDEX-Africa data, in which a similar weightage was employed, which should be the main focus area of future studies to enhance the accuracy of climate projections.

4.3. Combined LULC and Climate Change Impacts

Figure 15 shows the soil loss, sediment yield, and river flow changes, in percent, under the impacts of combined climate and land use/cover impacts. The change in flow at Angerb River, the main tributary of the Megech River, showed an increase under the combined land use/cover and climate changes impacts for the case of the RCP4.5 scenario. However, the combined effects of land use and the RCP8.5 climatic scenario showed an increase in the 2030s and a decline for the 2050s and 2070s (Figure 15). The results of this study showed that flow declined in the far future predictions, consistent with those of previous studies for the Upper Blue Nile Basin [75,76]. Unlike river flow, the effects of combined land use and climate changes on sediment yield showed a significant increase in the far future. The rate of increase of sediment yield ranged from 23.28% to 41.01% under the combined effects of land use changes and climate scenarios (RCP4.5 and RCP8.5) across all timeframes. Similarly, the relative changes in percentage of soil loss under the synergetic land use change and climatic scenarios (RCP4.5 and 8.5) depicted an increase for all time periods. The results indicated a higher relative change under the RCP8.5 scenario and for the farthest time window. Previous studies on soil loss and sediment yield have shown some degree of consistency in the basin [41]. Therefore, the study findings showed that the combined land use/cover and climate changes substantially changed the soil loss, sediment yield, and flow metrics of the watershed. However, it should be noticed that research findings under the combined impacts of climate changes and land use/cover changes are subject to various uncertainties. Among the various sources of uncertainty, climate models, downscaling, bias correction techniques, land use/cover classification errors, the LULC prediction algorithm, and the GeoWEPP model scaling can be taken as the study's limitations.

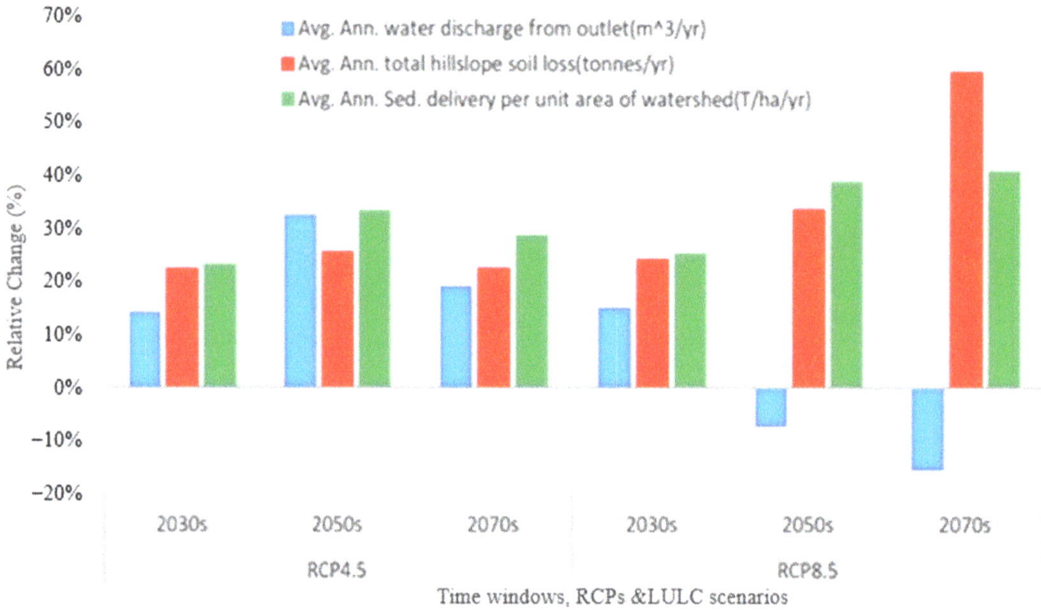

Figure 15. Combined climate and land use impacts on discharge, hill slope soil loss, and sediment yield rates.

The results clearly indicate that the combined impacts of land use/cover and climate (RCP4.5 and 8.5) changes place significant pressure on the sustainable use of water and land resources in the Megech watershed. The focus of future studies will be on reducing the uncertainties and enhancing the accuracy of the results. Future studies should also include climate mitigation and adaptation and soil and water conservation techniques as strategies to address these significant challenges.

5. Conclusions

This study underlined the potential impacts of land use change and climate change on precipitation and temperature, with a focus on runoff, soil loss, and sediment yield in the Megech Dam watershed. Changes in climate and land use/cover have great potential to impose pressure on water availability, sustainable use of dams, and agricultural productivity in the Upper Blue Nile region.

Three temporal window scenarios of the 2030s, 2050s, and 2070s were selected for the impact analysis, which lie within the useful life of the Megech Dam. The impacts of projected land use/cover and climate on sediment yield, soil loss, and runoff were quantified using the physically based GeoWEPP soil erosion model. The generated land use/cover maps for 2035, 2055, and 2075 indicated an increase in cropland and settlement areas and a decrease in forest and rangelands, while other land uses/cover types, such as water, wetland, and bare land, did not show significant changes. As a result, the soil loss and sediment yield of the watershed increased, while flow depicted inconsistent changes across the selected time periods.

Climate downscaling and bias correction were performed using the distribution mapping technique. The analysis showed that the maximum temperature will increase, while the minimum temperature will decrease. Precipitation was projected to increase for wet months and decrease for dry months under both the RCP4.5 and RCP8.5 scenarios, without a clear pattern. As a result, for the three time windows, the river discharge due to climate change alone and the combination of land use/cover and climate change scenarios decreased/increased by −4.67%/17.32% and 14.18–32.63% for the RCP4.5 scenario and decreased by −4.01% to –20.65% and increased/decreased by 15.07%/−15.53% for the RCP8.5 scenario, respectively.

The sediment yield results showed an increase under the combined land/cover and climate change impacts in all scenarios, with ranges of 23.28–33.37% for RCP4.5 and 25.42–41.02% for RCP8.5, while for soil loss, the ranges were 1.26–5.90% and 11.44–16.86% for RCP4.5 and RCP8.5, respectively. Generally, addressing these challenges requires the use of watershed best-management practices (BMP) and proactive soil and water conservation measures. Overall, policymakers and land managers can mitigate and protect the recently constructed dams' sedimentation and land productivity losses by adopting climate adaptations strategies and sustainable management practices in the Upper Blue Nile Basin.

Author Contributions: Conceptualization, M.A. and A.M.M.; methodology, M.A.; software, M.A.; validation, M.A.; formal analysis, M.A.; investigation, M.A., A.M.M. and G.T.; resources, M.A.; data curation, M.A.; writing—original draft preparation, M.A.; writing—review and editing, M.A., A.M.M. and G.T.; visualization, M.A.; supervision, A.M.M.; project administration, M.A., A.M.M. and G.T. All authors have read and agreed to the published version of the manuscript.

Funding: This research has not received any external fund.

Data Availability Statement: Data can be obtained upon request from the corresponding author.

Acknowledgments: The authors would like to thank the data providers, the Ministry of Water and Energy (MoWE), the Abbay Basin Authority, and the National Meteorological Agency. The authors express their sincere thanks to the reviewers and the editor for their comments to improve publication quality.

Conflicts of Interest: The authors declare that there are no known competing financial interests or personal relationships that could have appeared to influence the work reported in this paper.

References

1. Azimi Sardari, M.R.; Bazrafshan, O.; Panagopoulos, T.; Sardooi, E.R. Modeling the impact of climate change and land use change scenarios on soil erosion at the Minab Dam Watershed. *Sustainability* **2019**, *11*, 3353. [CrossRef]
2. Chuenchum, P.; Xu, M.; Tang, W. Predicted trends of soil erosion and sediment yield from future land use and climate change scenarios in the Lancang–Mekong River by using the modified RUSLE model. *Int. Soil Water Conserv. Res.* **2020**, *8*, 213–227. [CrossRef]
3. Belay, T.; Mengistu, D.A. Impacts of land use/land cover and climate changes on soil erosion in Muga watershed, Upper Blue Nile basin (Abay), Ethiopia. *Ecol. Process.* **2021**, *10*, 68. [CrossRef]
4. Tsegaye, L.; Bharti, R. The impacts of LULC and climate change scenarios on the hydrology and sediment yield of Rib watershed, Ethiopia. *Environ. Monit. Assess.* **2022**, *194*, 717. [CrossRef] [PubMed]
5. Dale, V.H. The relationship between land-use change and climate change. *Ecol. Appl.* **1997**, *7*, 753–769. [CrossRef]
6. Mango, L.M.; Melesse, A.M.; McClain, M.E.; Gann, D.; Setegn, S. Land use and climate change impacts on the hydrology of the upper Mara River Basin, Kenya: Results of a modeling study to support better resource management. *Hydrol. Earth Syst. Sci.* **2011**, *15*, 2245–2258. [CrossRef]
7. Baig, M.F.; Mustafa, M.R.U.; Baig, I.; Takaijudin, H.B.; Zeshan, M.T. Assessment of land use land cover changes and future predictions using CA-ANN simulation for selangor, Malaysia. *Water* **2022**, *14*, 402. [CrossRef]
8. Genet, A. Population growth and land use land cover change scenario in Ethiopia. *Int. J. Environ. Prot. Policy* **2020**, *8*, 77–85. [CrossRef]

9. Negese, A. Impacts of land use and land cover change on soil erosion and hydrological responses in Ethiopia. *Appl. Environ. Soil Sci.* **2021**, *2021*, 6669438. [CrossRef]
10. Borrelli, P.; Robinson, D.A.; Fleischer, L.R.; Lugato, E.; Ballabio, C.; Alewell, C.; Meusburger, K.; Modugno, S.; Schütt, B.; Ferro, V. An assessment of the global impact of 21st century land use change on soil erosion. *Nat. Commun.* **2017**, *8*, 2013. [CrossRef]
11. Kebede, B.; Bekele, A.; Kedir, E. Can the urban poor afford modern energy? The case of Ethiopia. *Energy Policy* **2002**, *30*, 1029–1045. [CrossRef]
12. Awulachew, S.B.; Yilma, A.D.; Loulseged, M.; Loiskandl, W.; Ayana, M.; Alamirew, T. *Water Resources and Irrigation Development in Ethiopia*; IWMI: Colombo, Sri Lanka, 2007; Volume 123.
13. Melese, T.; Senamaw, A.; Belay, T.; Bayable, G. The spatiotemporal dynamics of land use land cover change, and its impact on soil erosion in Tagaw Watershed, Blue Nile Basin, Ethiopia. *Glob. Chall.* **2021**, *5*, 2000109. [CrossRef] [PubMed]
14. Moisa, M.B.; Negash, D.A.; Merga, B.B.; Gemeda, D.O. Impact of land-use and land-cover change on soil erosion using the RUSLE model and the geographic information system: A case of Temeji watershed, Western Ethiopia. *J. Water Clim. Change* **2021**, *12*, 3404–3420. [CrossRef]
15. Chimdessa, K.; Quraishi, S.; Kebede, A.; Alamirew, T. Effect of land use land cover and climate change on river flow and soil loss in Didessa River Basin, South West Blue Nile, Ethiopia. *Hydrology* **2018**, *6*, 2. [CrossRef]
16. Tsegaye, B. Effect of land use and land cover changes on soil erosion in Ethiopia. *Int. J. Agric. Sci. Food Technol.* **2019**, *5*, 026–034. [CrossRef]
17. Tewabe, D.; Fentahun, T. Assessing land use and land cover change detection using remote sensing in the Lake Tana Basin, Northwest Ethiopia. *Cogent Environ. Sci.* **2020**, *6*, 1778998. [CrossRef]
18. Takele, A.; Lakew, H.B.; Kabite, G. Does the recent afforestation program in Ethiopia influenced vegetation cover and hydrology? A case study in the upper awash basin, Ethiopia. *Heliyon* **2022**, *8*, e09589. [CrossRef]
19. Berihun, M.L.; Tsunekawa, A.; Haregeweyn, N.; Meshesha, D.T.; Adgo, E.; Tsubo, M.; Masunaga, T.; Fenta, A.A.; Sultan, D.; Yibeltal, M. Exploring land use/land cover changes, drivers and their implications in contrasting agro-ecological environments of Ethiopia. *Land Use Policy* **2019**, *87*, 104052. [CrossRef]
20. Zhai, H.; Lv, C.; Liu, W.; Yang, C.; Fan, D.; Wang, Z.; Guan, Q. Understanding spatio-temporal patterns of land use/land cover change under urbanization in Wuhan, China, 2000–2019. *Remote Sens.* **2021**, *13*, 3331. [CrossRef]
21. Alatorre, L.; Beguería, S.; Lana-Renault, N.; Navas, A.; García-Ruiz, J.M. Soil erosion and sediment delivery in a mountain catchment under scenarios of land use change using a spatially distributed numerical model. *Hydrol. Earth Syst. Sci.* **2012**, *16*, 1321–1334. [CrossRef]
22. Nut, N.; Mihara, M.; Jeong, J.; Ngo, B.; Sigua, G.; Prasad, P.V.; Reyes, M.R. Land use and land cover changes and its impact on soil erosion in Stung Sangkae catchment of Cambodia. *Sustainability* **2021**, *13*, 9276. [CrossRef]
23. Kidane, M.; Bezie, A.; Kesete, N.; Tolessa, T. The impact of land use and land cover (LULC) dynamics on soil erosion and sediment yield in Ethiopia. *Heliyon* **2019**, *5*, e02981. [CrossRef]
24. Eekhout, J.P.; de Vente, J. Global impact of climate change on soil erosion and potential for adaptation through soil conservation. *Earth-Sci. Rev.* **2022**, *226*, 103921. [CrossRef]
25. Chen, C.-N.; Tfwala, S.S.; Tsai, C.-H. Climate change impacts on soil erosion and sediment yield in a watershed. *Water* **2020**, *12*, 2247. [CrossRef]
26. Borrelli, P.; Robinson, D.A.; Panagos, P.; Lugato, E.; Yang, J.E.; Alewell, C.; Wuepper, D.; Montanarella, L.; Ballabio, C. Land use and climate change impacts on global soil erosion by water (2015–2070). *Proc. Natl. Acad. Sci. USA* **2020**, *117*, 21994–22001. [CrossRef] [PubMed]
27. Girmay, G.; Moges, A.; Muluneh, A. Assessment of current and future climate change impact on soil loss rate of Agewmariam Watershed, Northern Ethiopia. *Air Soil Water Res.* **2021**, *14*, 1178622121995847. [CrossRef]
28. Jilo, N.B.; Gebremariam, B.; Harka, A.E.; Woldemariam, G.W.; Behulu, F. Evaluation of the impacts of climate change on sediment yield from the Logiya Watershed, Lower Awash Basin, Ethiopia. *Hydrology* **2019**, *6*, 81. [CrossRef]
29. Alaminie, A.A.; Tilahun, S.A.; Legesse, S.A.; Zimale, F.A.; Tarkegn, G.B.; Jury, M.R. Evaluation of past and future climate trends under CMIP6 scenarios for the UBNB (Abay), Ethiopia. *Water* **2021**, *13*, 2110. [CrossRef]
30. Tegegne, G.; Melesse, A.M. Comparison of trend preserving statistical downscaling algorithms toward an improved precipitation extremes projection in the headwaters of blue nile river in Ethiopia. *Environ. Process.* **2021**, *8*, 59–75. [CrossRef]
31. Chapman, S.; Birch, C.E.; Galdos, M.V.; Pope, E.; Davie, J.; Bradshaw, C.; Eze, S.; Marsham, J.H. Assessing the impact of climate change on soil erosion in East Africa using a convection-permitting climate model. *Environ. Res. Lett.* **2021**, *16*, 084006. [CrossRef]
32. Panagos, P.; Ballabio, C.; Himics, M.; Scarpa, S.; Matthews, F.; Bogonos, M.; Poesen, J.; Borrelli, P. Projections of soil loss by water erosion in Europe by 2050. *Environ. Sci. Policy* **2021**, *124*, 380–392. [CrossRef]
33. Choudhury, B.U.; Nengzouzam, G.; Islam, A. Runoff and soil erosion in the integrated farming systems based on micro-watersheds under projected climate change scenarios and adaptation strategies in the eastern Himalayan mountain ecosystem (India). *J. Environ. Manag.* **2022**, *309*, 114667. [CrossRef]
34. Chakilu, G.G.; Sándor, S.; Zoltán, T.; Phinzi, K. Climate change and the response of streamflow of watersheds under the high emission scenario in Lake Tana sub-basin, upper Blue Nile basin, Ethiopia. *J. Hydrol. Reg. Stud.* **2022**, *42*, 101175. [CrossRef]
35. Roth, V.; Lemann, T.; Zeleke, G.; Subhatu, A.T.; Nigussie, T.K.; Hurni, H. Effects of climate change on water resources in the upper Blue Nile Basin of Ethiopia. *Heliyon* **2018**, *4*, e00771. [CrossRef] [PubMed]

36. Mengistu, D.; Bewket, W.; Dosio, A.; Panitz, H.-J. Climate change impacts on water resources in the upper blue nile (Abay) river basin, ethiopia. *J. Hydrol.* **2021**, *592*, 125614. [CrossRef]
37. Ongoma, V.; Chen, H.; Gao, C. Projected changes in mean rainfall and temperature over East Africa based on CMIP5 models. *Int. J. Climatol.* **2018**, *38*, 1375–1392. [CrossRef]
38. Gelete, G.; Gokcekus, H.; Gichamo, T. Impact of climate change on the hydrology of Blue Nile basin, Ethiopia: A review. *J. Water Clim. Change* **2020**, *11*, 1539–1550. [CrossRef]
39. Gebrechorkos, S.H.; Hülsmann, S.; Bernhofer, C. Long-term trends in rainfall and temperature using high-resolution climate datasets in East Africa. *Sci. Rep.* **2019**, *9*, 11376. [CrossRef]
40. Getachew, B.; Manjunatha, B. Climate change projections and trends simulated from the CMIP5 models for the Lake Tana sub-basin, the Upper Blue Nile (Abay) River Basin, Ethiopia. *Environ. Chall.* **2021**, *5*, 100385. [CrossRef]
41. Getachew, B.; Manjunatha, B.R.; Bhat, G.H. Assessing current and projected soil loss under changing land use and climate using RUSLE with Remote sensing and GIS in the Lake Tana Basin, Upper Blue Nile River Basin, Ethiopia. *Egypt. J. Remote Sens. Space Sci.* **2021**, *24*, 907–918. [CrossRef]
42. Kumar, A.; Singh, A.; Gaurav, K. Assessing the synergic effect of land use and climate change on the upper Betwa River catchment in Central India under present, past, and future climate scenarios. *Environ. Dev. Sustain.* **2023**, *25*, 5163–5184. [CrossRef]
43. Berihun, M.L.; Tsunekawa, A.; Haregeweyn, N.; Tsubo, M.; Fenta, A.A.; Ebabu, K.; Bayabil, H.K.; Dile, Y.T. Predicting runoff and sediment responses to climate-resilient land use and management scenarios. *Environ. Sci. Pollut. Res.* **2023**, *30*, 72262–72283. [CrossRef] [PubMed]
44. Choukri, F.; Raclot, D.; Naimi, M.; Chikhaoui, M.; Nunes, J.P.; Huard, F.; Hérivaux, C.; Sabir, M.; Pépin, Y. Distinct and combined impacts of climate and land use scenarios on water availability and sediment loads for a water supply reservoir in northern Morocco. *Int. Soil Water Conserv. Res.* **2020**, *8*, 141–153. [CrossRef]
45. Ahmad, M.N.; Shao, Z.; Javed, A. Modelling land use/land cover (LULC) change dynamics, future prospects, and its environmental impacts based on geospatial data models and remote sensing data. *Environ. Sci. Pollut. Res.* **2023**, *30*, 32985–33001. [CrossRef] [PubMed]
46. Kamaraj, M.; Rangarajan, S. Predicting the future land use and land cover changes for Bhavani basin, Tamil Nadu, India, using QGIS MOLUSCE plugin. *Environ. Sci. Pollut. Res.* **2022**, *29*, 86337–86348. [CrossRef] [PubMed]
47. Muhammad, R.; Zhang, W.; Abbas, Z.; Guo, F.; Gwiazdzinski, L. Spatiotemporal change analysis and prediction of future land use and land cover changes using QGIS MOLUSCE plugin and remote sensing big data: A case study of Linyi, China. *Land* **2022**, *11*, 419. [CrossRef]
48. Eshete, D.G.; Rigler, G.; Shinshaw, B.G.; Belete, A.M.; Bayeh, B.A. Evaluation of streamflow response to climate change in the data-scarce region, Ethiopia. *Sustain. Water Resour. Manag.* **2022**, *8*, 187. [CrossRef]
49. Tenagashaw, D.Y.; Andualem, T.G.; Ayele, W.T.; Belew, A.Z. Climate change impact on hydro-climatic variables of Ribb watershed, Tana sub-basin, Ethiopia. *Appl. Water Sci.* **2023**, *13*, 35. [CrossRef]
50. Worku, G.; Teferi, E.; Bantider, A.; Dile, Y.T. Statistical bias correction of regional climate model simulations for climate change projection in the Jemma sub-basin, upper Blue Nile Basin of Ethiopia. *Theor. Appl. Climatol.* **2020**, *139*, 1569–1588. [CrossRef]
51. Tigabu, T.B.; Wagner, P.D.; Hörmann, G.; Kiesel, J.; Fohrer, N. Climate change impacts on the water and groundwater resources of the Lake Tana Basin, Ethiopia. *J. Water Clim. Change* **2021**, *12*, 1544–1563. [CrossRef]
52. Phiri, D.; Simwanda, M.; Salekin, S.; Nyirenda, V.R.; Murayama, Y.; Ranagalage, M. Sentinel-2 data for land cover/use mapping: A review. *Remote Sens.* **2020**, *12*, 2291. [CrossRef]
53. Van Vuuren, D.P.; Edmonds, J.; Kainuma, M.; Riahi, K.; Thomson, A.; Hibbard, K.; Hurtt, G.C.; Kram, T.; Krey, V.; Lamarque, J.-F. The representative concentration pathways: An overview. *Clim. Change* **2011**, *109*, 5–31. [CrossRef]
54. Flanagan, D.; Nearing, M. USDA-Water Erosion Prediction Project: Hillslope profile and watershed model documentation. *NSERL Rep.* **1995**, *10*, 1196–47097.
55. Laflen, J.M.; Lane, L.J.; Foster, G.R. WEPP: A new generation of erosion prediction technology. *J. Soil Water Conserv.* **1991**, *46*, 34–38.
56. Renschler, C.S. Designing geo-spatial interfaces to scale process models: The GeoWEPP approach. *Hydrol. Process.* **2003**, *17*, 1005–1017. [CrossRef]
57. Flanagan, D.; Livingston, S.J. *WEPP User Summary*; NSERL Report: West Lafayette, IN, USA, 1995.
58. Maalim, F.K.; Melesse, A.M.; Belmont, P.; Gran, K.B. Modeling the impact of land use changes on runoff and sediment yield in the Le Sueur watershed, Minnesota using GeoWEPP. *Catena* **2013**, *107*, 35–45. [CrossRef]
59. Luetzenburg, G.; Bittner, M.J.; Calsamiglia, A.; Renschler, C.S.; Estrany, J.; Poeppl, R. Climate and land use change effects on soil erosion in two small agricultural catchment systems Fugnitz–Austria, Can Revull–Spain. *Sci. Total Environ.* **2020**, *704*, 135389. [CrossRef]
60. Mwendera, E.; Saleem, M.M.; Dibabe, A. The effect of livestock grazing on surface runoff and soil erosion from sloping pasture lands in the Ethiopian highlands. *Aust. J. Exp. Agric.* **1997**, *37*, 421–430. [CrossRef]
61. Yasir, S.; Crosato, A.; Mohamed, Y.A.; Abdalla, S.H.; Wright, N.G. Sediment balances in the Blue Nile River basin. *Int. J. Sediment Res.* **2014**, *29*, 316–328. [CrossRef]

62. IPCC. Summary for Policymakers. In *Climate Change and Land: An IPCC Special Report on Climate Change, Desertification, Land Degradation, Sustainable Land Management, Food Security, and Greenhouse Gas Fluxes in Terrestrial Ecosystems*; Shukla, P.R., Skea, J., Buendia, E.C., Masson-Delmotte, V., Pörtner, H.-O., Roberts, D.C., Zhai, P., Slade, R., Connors, S., van Diemen, R., et al., Eds.; IPCC: Geneva, Switzerland, 2019. [CrossRef]
63. Tekleab, S.; Mohamed, Y.; Uhlenbrook, S. Hydro-climatic trends in the Abay/upper Blue Nile basin, Ethiopia. *Phys. Chem. Earth Parts A/B/C* **2013**, *61*, 32–42. [CrossRef]
64. Admas, M.; Melesse, A.M.; Abate, B.; Tegegne, G. Soil erosion, sediment yield, and runoff modeling of the megech watershed using the GeoWEPP model. *Hydrology* **2022**, *9*, 208. [CrossRef]
65. Admas, M.; Melesse, A.M.; Abate, B.; Tegegne, G. Impacts of best management practices on runoff, soil loss, and sediment yield in the megech watershed, Ethiopia. *Water* **2023**, *15*, 788. [CrossRef]
66. Moriasi, D.N.; Arnold, J.G.; Van Liew, M.W.; Bingner, R.L.; Harmel, R.D.; Veith, T.L. Model evaluation guidelines for systematic quantification of accuracy in watershed simulations. *Trans. ASABE* **2007**, *50*, 885–900. [CrossRef]
67. Sinshaw, B.G.; Belete, A.M.; Mekonen, B.M.; Wubetu, T.G.; Anley, T.L.; Alamneh, W.D.; Atinkut, H.B.; Gelaye, A.A.; Bilkew, T.; Tefera, A.K. Watershed-based soil erosion and sediment yield modeling in the Rib watershed of the Upper Blue Nile Basin, Ethiopia. *Energy Nexus* **2021**, *3*, 100023. [CrossRef]
68. Bewket, W.; Teferi, E. Assessment of soil erosion hazard and prioritization for treatment at the watershed level: Case study in the Chemoga watershed, Blue Nile basin, Ethiopia. *Land Degrad. Dev.* **2009**, *20*, 609–622. [CrossRef]
69. Gashaw, T.; Tulu, T.; Argaw, M.; Worqlul, A.W. Modeling the impacts of land use–land cover changes on soil erosion and sediment yield in the Andassa watershed, upper Blue Nile basin, Ethiopia. *Environ. Earth Sci.* **2019**, *78*, 679. [CrossRef]
70. Tadesse, L.; Suryabhagavan, K.; Sridhar, G.; Legesse, G. Land use and land cover changes and Soil erosion in Yezat Watershed, North Western Ethiopia. *Int. Soil Water Conserv. Res.* **2017**, *5*, 85–94. [CrossRef]
71. Teshome, D.S.; Moisa, M.B.; Gemeda, D.O.; You, S. Effect of Land Use-Land Cover Change on Soil Erosion and Sediment Yield in Muger Sub-Basin, Upper Blue Nile Basin, Ethiopia. *Land* **2022**, *11*, 2173. [CrossRef]
72. Bogale, A. Review, impact of land use/cover change on soil erosion in the Lake Tana Basin, Upper Blue Nile, Ethiopia. *Appl. Water Sci.* **2020**, *10*, 235. [CrossRef]
73. Getachew, H.E.; Melesse, A.M. The impact of land use change on the hydrology of the Angereb Watershed, Ethiopia. *Int. J. Water Sci.* **2012**, *1*, 6.
74. Tariku, T.B.; Gan, T.Y.; Li, J.; Qin, X. Impact of climate change on hydrology and hydrologic extremes of Upper Blue Nile River Basin. *J. Water Resour. Plan. Manag.* **2021**, *147*, 04020104. [CrossRef]
75. Mekonnen, D.F.; Duan, Z.; Rientjes, T.; Disse, M. Analysis of combined and isolated effects of land-use and land-cover changes and climate change on the upper Blue Nile River basin's streamflow. *Hydrol. Earth Syst. Sci.* **2018**, *22*, 6187–6207. [CrossRef]
76. Girma, M.M. Potential Impact of Climate and Land Use Changes on the Water Resources of the Upper Blue Nile Basin. Ph.D. Dissertation, Ferie Universität Berlin, Berlin, Germany, 2013.

remote sensing

MDPI

Article

Evaluating Machine-Learning Algorithms for Mapping LULC of the uMngeni Catchment Area, KwaZulu-Natal

Orlando Bhungeni *, Ashadevi Ramjatan and Michael Gebreslasie

School of Agricultural, Earth and Environmental Sciences, University of KwaZulu-Natal, Westville Campus, Private Bag X54001, Durban 4000, South Africa; 822820676@stu.ukzn.ac.za (A.R.); gebreslasie@ukzn.ac.za (M.G.)

* Correspondence: 223146923@stu.ukzn.ac.za

Abstract: Analysis of land use/land cover (LULC) in catchment areas is the first action toward safeguarding freshwater resources. LULC information in the watershed has gained popularity in the natural science field as it helps water resource managers and environmental health specialists develop natural resource conservation strategies based on available quantitative information. Thus, remote sensing is the cornerstone in addressing environmental-related issues at the catchment level. In this study, the performance of four machine learning algorithms (MLAs), namely Random Forests (RFs), Support Vector Machines (SVMs), Artificial Neural Networks (ANNs), and Naïve Bayes (NB), were investigated to classify the catchment into nine relevant classes of the undulating watershed landscape using Landsat 8 Operational Land Imager (L8-OLI) imagery. The assessment of the MLAs was based on a visual inspection of the analyst and commonly used assessment metrics, such as user's accuracy (UA), producers' accuracy (PA), overall accuracy (OA), and the kappa coefficient. The MLAs produced good results, where RF (OA = 97.02%, Kappa = 0.96), SVM (OA = 89.74%, Kappa = 0.88), ANN (OA = 87%, Kappa = 0.86), and NB (OA = 68.64%, Kappa = 0.58). The results show the outstanding performance of the RF model over SVM and ANN with a significant margin. While NB yielded satisfactory results, its sensitivity to limited training samples could primarily influence these results. In contrast, the robust performance of RF could be due to an ability to classify high-dimensional data with limited training data.

Keywords: uMngeni river catchment; machine learning; LULC; Landsat 8; remote sensing; South Africa

1. Introduction

The catchment areas are important ecological components that provide various ecosystem services that benefit society and biodiversity [1,2]. These include the availability of freshwater for consumption, supporting societal needs such as food production, sport, and recreation, and providing a natural habitat [3,4]. Such essential services can be sustained when the biophysical environment of the catchment areas is sustainably managed. However, natural land cover and freshwater resources in the water catchment areas are threatened by Land use/Land cover (LULC) changes in response to societal demands [5]. It is essential to map and monitor the water catchment landscape dynamics to provide valuable information on the distribution of land use activities in the catchment, which serve as the tool for initiatives whose mandates aim to strike a balance between development and the natural environment in the catchment areas. In addition, this information will help water resource managers and other environmental health practitioners establish environmental management plans and policies based on available information [6,7]. Moreover, the availability of such information is important as it can be used to troubleshoot water quality issues and prompt the identification of land resource degradation [8].

Mapping the land use activities of the water catchment can be achieved by either traditional field observation or remote sensing methods. Remote sensing emerged as a convenient and cost-effective way to provide the spatiotemporal data required to analyse heterogeneous landscapes on a large scale [9]. In addition, the remote sensing data can be systematically stored, maintained, and openly shared to end users. A variety of remote sensing sensors (optical and active sensors) have been utilized to map the LULC in the catchment areas. Akar et al. [10] assessed the influence of the contextual data features and various dimensionality filters, including Discrete Wavelet Transform (DWT), in mapping LULC using Headwall Hyperspec data in Bergama, Turkey. Random forest (RF) and support vector machine (SVM) were employed, and RF and DWT data obtained higher overall accuracy and kappa scores of 88.13% and 0.88, respectively. In comparison, SVM recorded lower accuracy than RF. Lekka et al. [11] have used Environmental Mapping and Analysis Program (EnMAP) imagery data to map the LULC using pixel-based machine learning algorithms (MLAs), such as RF and SVM image classification. Their results showed that the SVM image classifier returned an overall accuracy of 92.6%, while the RF model obtained an overall accuracy of 88.1%. Delogu et al. [12] used PRISMA hyperspectral data with machine learning methods, such as RF, an artificial neural network (ANN), and a convolutional neural network (CNN), to classify land use activities in Naples, Italy. The CNN achieved the best results at 0.973% and 0.968 overall accuracy and Kappa, respectively. As the second-best model, the ANN recorded 0.963% and 0.956, and RF showed the lowest classification accuracy at 0.887% and 0.867, respectively.

Other studies have demonstrated the potential of very high spatial resolution (VHR) multispectral data in providing the same service as hyperspectral data in different terrestrial settings and achieving higher classification accuracies. For instance, James [13] utilized Pleiades-1 imagery to map the LULC in Côte d'Emeraude, France. Maximum likelihood (ML) and SVM algorithms were used to implement the classification. The ML algorithm yielded a higher overall accuracy of 84.64% compared to the SVM algorithm, which only produced a map with an overall accuracy of 76.13%. In Fayoum City, Egypt, Mahmoud et al. [14] evaluated a variety of classifiers in mapping the city land cover classes using PlanetScope imagery. They found the ANN model more accurate with a kappa coefficient of 0.97 compared to Naïve Bayes (NB), SVM, RF, and decision trees (DTs), which recorded 0.93, 0.90, 0.86, and 0.87, respectively.

Some studies explored the capabilities of moderate-to-low-resolution sensors' capabilities in classifying spatial land use classes. For example, in a comparative study, Ref. [6] employed Landsat-8 (L8-OLI) data and SVM and RF algorithms to map Nkandla Forest cover classes in South Africa. L8-OLI was found to be efficient in mapping the forest cover, and SVM was found to be more accurate, with overall accuracy and kappa records of 95.83% and 0.94 compared to 95.24 and 0.93 achieved by RF, respectively. On the west coast of India, a study by Bayas et al. [15] assessed the efficiency of the MLAs in mapping LULC using Sentinel-2 (S2) imagery. NB, Classification And Regression Trees (CART), gradient tree boost (GTB), and RF MLAs were assessed, and the models achieved an overall accuracy of 81.34%, 90.18%, 91.06%, and 92.14%, respectively. A study by Ouchra et al. [9] also used L8-OLI data to evaluate the capabilities of MLAs in land use analysis in Casablanca, Morocco. The RF, SVM, CART, and GTB algorithms were among the tested methods, and RF showed robust performance with an overall accuracy of 95.42% compared to other classifiers with an overall accuracy of 83%, 91.50%, and 93.46%, respectively. A study by Parashar et al. [16] compared SVM and RF algorithms using S2 datasets with and without spectral indices to analyse the LULC of hilly terrain in Gopeshwar, India. The RF models provided more accurate results in both datasets, with overall accuracy and kappa of 86% and 0.83 for a dataset with only spectral bands and 88% and 0.85 for a dataset with incorporated spectral indices, respectively, while SVM reported the lowest accuracy, with overall accuracy and kappa of 82% and 0.79 for spectral bands and 87% and 0.83 for incorporated indices, respectively. In the Munneru River Basin, India, Loukika et al. [17] explored the performance of the MLAs viz. SVM, RF, and CART were used to map the LULC based on L8-OLI and S2 data. They found S2 and RF to dominate their counterparts. The average overall accuracy for RF, SVM, and CART in L8-OLI was 94.85%, 90.88%, and 82.88%. In comparison, S2 average overall accuracy for RF, SVM, and CART was 95.84%, 93.65%, and

86.48%, respectively. Woldemariam et al. [18] evaluated the MLAs in classifying the land use dynamics using L8-OLI data in the Lake Haramaya catchment, Ethiopia. Object-based image analysis (OBIA), SVM, RF, and ANN were tested, and the SVM dominated other classifiers with an overall accuracy of 94%, while RF and ANN achieved 92% and 89%, respectively, and OBIA reported the lowest overall accuracy at 75%. Dash et al. [19] compared the parametric and non-parametric classifiers in mapping LULC based on L8-OLI data in the Big Sunflower River watershed, United States of America (USA). RF, SVM, and the ML were factored in, and their results showed a robust performance with an overall accuracy (kappa) of 93.5% (0.88) and 88.8% (0.82), with RF displaying the worst performance at 84.6%, (0.72).

The success of land cover mapping is dependent not only on the remote sensing data but also on the classification algorithm utilized. As discussed above, various machine-learning algorithms have been used in the remote sensing field to explain the distribution of spatial features in catchment areas, with varying degrees of accuracy reported. Their performance varies with the biophysical features of the landscapes [15]. Hence, there is no generic approach for LULC classification for all terrestrial landscapes [18]. Therefore, it is vital to compare the MLA models to determine their suitability and sensitivity to different geographic settings [18]. Studies that compare the MLAs in mapping LULC in the case of South African landscapes are still limited, if there are any. Thus, in this study, we have identified four prominent MLAs, namely NB, RF, SVM, and ANN, and conducted a comparative study to map the thematic land use classes of the uMngeni river catchment in KwaZulu Natal using L8-OLI data. The current study aims to (i) map the LULC that defines the uMngeni River catchment area and (ii) compare the performance of four non-parametric pixel-based MLAs in a supervised classification based on L8-OLI data. The study is expected to improve the understanding of LULC configuration over the uMngeni River catchment, a key element in building a LULC change and projection module within the catchment to generate information for water resource management strategies.

2. Materials and Methods

The analysis of the LULC was carried out in a series of steps, as illustrated in Figure 1. The acquired data were preprocessed before implementing the machine learning algorithms, and the LULC was computed. Post-classification analysis was carried out, including accuracy assessment and computing variable importance.

2.1. Study Area

The study area is located in the uMngeni River catchment, KwaZulu-Natal, South Africa (29°29′05.0″S 29°46′52.8″E and 29°48′41.8″S 31°02′27.0″E). The area is situated in the central-western part of KwaZulu-Natal province (Figure 2). It covers 4444 km^2, which is subdivided into three main regions (lower, middle, and upper) [20]. Its biophysical environment is characterized by a complex river network with an undulating landscape in the upper reaches and gentle slopes with hills towards the lower parts of the catchment, shown in dark color in Figure 2.

The river network forms the uMngeni River, which stretches from the Drakensberg hills in the inner lands of KZN and flows down through Pietermaritzburg, finally depositing in the Indian Ocean via the Durban north estuary. The uMngeni River has four main dams (Midmar, Albert Falls, Nagle, and Inanda), which serve as a water source for 45% of KZN province's total population [8], including industrial activities, recreation, and agricultural purposes in the cities along the river path, such as Pietermaritzburg and Durban [20]. Land cover change implications not only affect these services but also have an impact on habitat loss, climate change, and biodiversity. To ensure the continuity of such important services, the uMngeni River Catchment was chosen as the study area of interest.

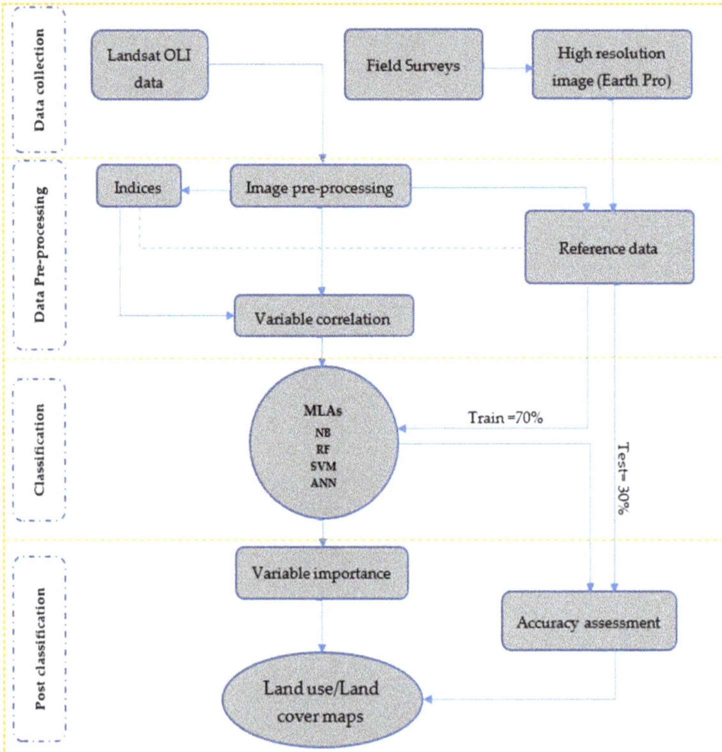

Figure 1. The conceptual framework of the study.

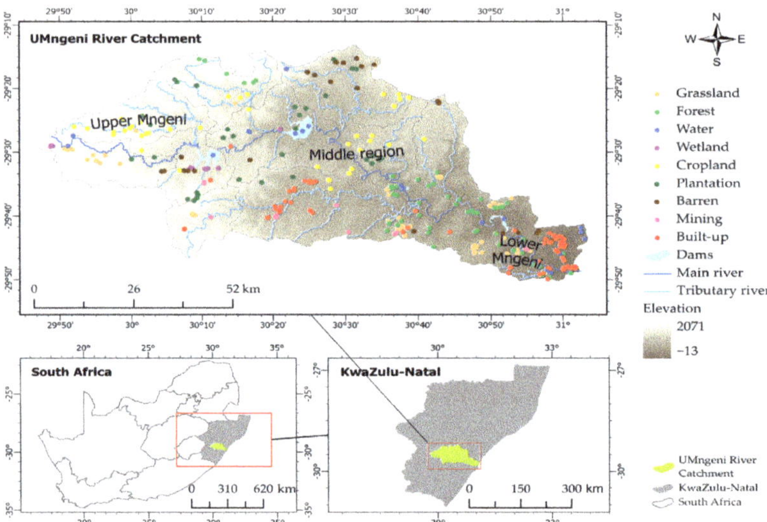

Figure 2. Layout of the uMngeni River catchment with elevation gradients, the river network, and dams.

The catchment has an elevation of 1913 m above sea level [21]. Rainfall occurs during the summer (October to March), ranging from 700 to 550 mL per annum. It has moderate temperatures with an annual average of 12 to 20 °C. Its climate and geographic position play a critical role in the province's economic development. Due to these biophysical characteristics, natural features, and economic activities, the catchment is characterized by a range of land use types.

From our field observation, the land cover types in the catchment include but are not limited to the following: built-up areas, green patches (cropland, plantations, and forests), barren land, and water bodies. Natural vegetation and cropland dominate the uMngeni catchment [8]. Plantations are one of the important land use classes in the catchment and dominate the upper parts of the catchment. The middle part is characterized by heterogeneous patches of different land use types (plantations and built-up areas) [8], while the lower part is mainly dominated by built-up patches with fragments of green vegetated land cover. It is worth unpacking the information in such biodiversity-rich areas to provide updated scientific information.

2.2. Definition of LULC Classes in the Umngeni River Catchment

To assess the LULC in the catchment, a total of nine thematic land use classes were identified, namely built-up areas, cropland, water, plantations, mining, forests, barren, grasslands, and wetlands, as described in Table 1.

Table 1. Shows The Descriptions of Nine Thematic LULC Classes Used in The Current Study.

Class ID	LULC Classes	Description
1	Grassland	Includes grass and shrubs open areas, as well as golf courses, and sport fields grounds
2	Forest	Natural forests or transition foarest areas rangeing from low to highly dense canopy cover, dominated with trees of appraximately 5 m high.
3	Water	This class considers natural or artificial surface fresh waterbodies within the study area, which includes main rivers and tributries, dams, lakes, and ponds
4	Wetland	This class is mainly made up of hydrphytes and herbaceous species which are solely based on the moisture of the soil to survive. The wetlands can be permanent or temporary depending on the local climate or its size and water holding capacity.
5	Cropland	Composed of both subsistance and ecommercial farming, which might be annual crops, or seasonal. During the post harvest, cultivated lands are charecterised as bare land in the case of seasonal crops.
6	Plantation	Range of tree species cultivated for commercial purposes, this class includes greener patches of the mature trees, sub adult, young, jiviniles, and tree stumps.
7	Barren	Partial vegetated areas, bare land due to either erosion, natural degradation or human factors, post harvest crop fields, dry river banks, and stripped rock areas.
8	Mining	Small patches of mining scars, which comprise of open cast pits, sand mining, and the open raw material processing sites.
9	Built-up	Built-up structures, which includes all forms residential areas, economic corridors, industrial, commercial, educational, religious and health infrastructures.

2.3. Remote Sensing Data Acquisition and Preprocessing

In this study, a 30 m resolution image from an L8-OLI image was used for LULC analysis (Table 2). The multispectral image was obtained on 15 May 2023 from the open Google Earth Engine (GEE) data catalog, from the surface reflectance, level 2 category (https://developers.google.com/earth-engine/datasets/catalog/landsat). The surface reflectance category provides images that are ready for analysis with a set of preprocessing requirements that have already been performed to improve the quality of the image [22]. According to [23], surface reflectance images do not require fully blown image preprocessing. Therefore, *ee.filter* functions were used to set the criteria to select data that meet the needs of the study. It must also be noted that scenes from different dates but in the same season have been collected using *ee.Filter.date()*. Images from the same season will not have an impact on the analysis since the atmospheric variation is minimal. A total of four scenes (each with a total area of 185 km by 180 km) from two rows and paths were collected to cover our study area. Then, the *filterMetadata* function was set to return only scenes with a cloud cover of less than 10% [24].

Table 2. Details of L8-OLI Spectral Bands and their Respective Spatial Resolutions.

Bands	Wavelength (λ)	Resolution (m)
Blue	0.45–0.51 µm	30 m
Green	0.53–0.59 µm	30 m
Red	0.64–0.67 µm	30 m
Near Infrared (NIR)	0.85–0.87 µm	30 m
Shortwave Infrared 1 (SWIR 1)	1.56–1.65 µm	30 m
Shortwave Infrared 2 (SWIR 2)	2.10–2.29 µm	30 m

However, the filtering processing alone was infeasible to prepare the image for the analysis. Therefore, image preprocessing was performed as a prerequisite for any satellite image in the GEE interface. The median composite was formed from a series of scenes from the winter season, 01 May to 30 August 2022.

To stretch our specifications, the boundary shape file was used to create a subset of our study area from the composite image using the *Clip()* function. Then, the subset was subject to further processing, as the clouds are unavoidable in the satellite image [25]; the function of the mask from C code, commonly known as CFmasking, was applied to remove cloud-infested pixels in the image, as used by [26]. The approach involves the quality assurance (QA-PIXEL) band, which assigns zero values to the cloud-contaminated pixels and cloud shadows, producing a cloud-free composite [27].

Calculated Spectral Indices

Spectral indices are derived from the pixel-wise ratios of two or more bands of the multispectral image, which serve as both logistics and added explanatory variables for image classification [28]. Many studies have achieved meaningful LULC classification results by supplementing multispectral bands with information from spectral indices [17,28,29]. The indices layers facilitate the discrimination of specific features and assign the pixels to predicted classes. Our study area possesses diverse spatial attributes, making distinguishing the class members from others difficult. Hence, the current study employed various mathematical equations (Table 3) to compute the normalized difference indices (NDI) for vegetation (NDVI) [30] to improve the separation of vegetation classes [31], as well as water indices (NDWI) to distinguish waterbodies from hill shadows [32] and bare soil indices (BSI) [33] to identify bare surfaces [34]. These indices were also used as additional explanatory inputs to L8-OLI visible, NIR, and SWIR bands.

Table 3. Supplementary Indices Used in the Land Use Classification.

Index	Equation	References
NDVI	$\mathrm{NDVI} = \frac{NIR - Red}{Red + NIR}$	[30]
NDWI	$\mathrm{NDWI} = \frac{Green - NIR}{Green + NIR}$	[32]
BSI	$\mathrm{BSI} = \frac{((Red + SWIR) - (NIR + Blue))}{((Red + SWIR) + (NIR + Blue))}$	[33]

2.4. Reference Data Collection

The field survey approach and Google Earth Pro 7.3.6 high-resolution images were used in collecting sampling data for building and measuring the mapping accuracy of the classifiers. The field survey was carried out on 16 August 2023, where coordinates of predefined thematic land use classes (Table 1) were extracted using a portable global positioning system (GPS) through a random sampling method. The same method was also used in Google Earth Pro-based sample collection. This approach widely places the sampling points across the study area (Figure 2) and ensures all the land cover classes are adequately represented in the reference data [35,36]. The GPS coordinates were imported into the Google Earth Pro 7.3.6 platform and overlaid on high-resolution images to locate the land cover classes [37]. Around each coordinate, the irregularly sized polygons were then digitized to represent the corresponding land cover class (Table 4). The digitized reference polygons were imported into GEE and overlayed onto an L8-OLI image. Then, the spectral properties of cover classes were extracted from the L8-OLI image per pixel, as shown in Table 4, and randomly divided into two datasets, a training dataset (70%) and a validation data set (30%) [7]. These two datasets were used to train and assess the accuracy of the MLA models—NB, RF, SVM, and ANN.

Table 4. Training and Test Samples Used in the Study.

LULC Class	No. of Polygons	Training Pixel Count	Test Pixel Count
Grassland	63	6944	1685
Forest	51	5823	1483
Water	38	1865	435
Wetland	11	357	86
Cropland	38	8965	2163
Plantation	32	7820	1957
Barren	30	4492	1075
Built-up	86	2504	598
Mining	13	667	159
Total	362	39437	9641

2.5. Image Classification

Regions with diverse land use classes can make it difficult for machine learning models to discover the relationship between explanatory features [38]. Therefore, before commencing the classification stage, it is important to scrutinize the variables based on their spectral similarities in the given class. To address this, the Pearson correlation coefficient was used to perform variable correlation analysis. It is a commonly used pair-wise technique that helps to identify the collinearity between two variables and eliminate noise [39]. It improves the accuracy of the MLAs by reducing redundancy and improving the computation processing rate. The stepwise correlation analysis was performed on L8-OLI explanatory variables; (1) spectral reflectance information for each variable was extracted from sampling sites, (2) imported into Python to compute variable correlation using the *corr()* algorithm, and (3) Seaborn's package was employed to plot the correlation array as a heat map matrix.

2.5.1. Selected Classifiers and Parameter Tuning

Naïve Bayes

The NB classification approach evolved from the Bayesian and Naive theorem. The Bayesian technique assumes the possibility of a recurrence of the event based on learning from the previous event [40], while Naive is a probabilistic classifier with a strong independent assumption of the feature from the others [29]. In other words, dimensions that define a certain variable are independent of each other. Naïve Bayesian theory can be expressed mathematically as the probability of a pixel being in class k given X as a training sample, which is equivalent to the density of the sample size given the class multiplied by the probability of the pixel being in class *k* and divided by the density of the sample size:

$$P(y = k|X) = \frac{p(X|y = k)(P(y = k))}{p(X)} \quad (1)$$

where y is the unassigned variable, k is the class, and X is the training sample. Therefore, $P(y = k \mid X)$ is the probability of the y and p is the density. NB models can be trained in supervised classification methods either in the form of parametric or non-parametric models by incorporating Gaussian functions.

The NB model is easy to tune in GEE and lambda (λ) is the only parameter that can be adjusted to avoid assigning zero probability to classes that could not be identified in the training stages [38]. The model computes the statistics (the mean and the variance of the variables for each class) required for classification from the provided training data [15]. In LULC analysis, it has been reported as a promising classifier with satisfactory results [29]. Its performance is attributed to the independence of the feature dimensions from the others. For the purpose of our study, the NB model from the "*smileNaiveBayes*" library in GEE was used.

Random Forests

The RF classifier is an ensemble classifier that uses regression decision trees to classify the dataset [41]. It gained popularity in the remote sensing community due to its reduced computational processing time, unique classification approach, and robust performance [28]. The RF proposed by Breiman [41] involves bagging the CART split criterion. Bagging facilitates the building of the bootstrap samples from the original training sample data, a process also known as bootstrapping, forming predictor trees from each bootstrap sample. The class membership of the unknown pixel is determined by the votes from all decision trees [7]. The decision to assign the pixels to the class is based on the aggregate (average of the votes) results from multiple RF decision trees [42,43]. The advantage of the RF model is the ability to handle high-dimensional datasets [15] while being insensitive to outliers and overfitting [44]. The RF model excludes some samples for each newly generated tree, commonly known as out-of-bag samples (OOB), which report on the important features in the classification and misclassification errors of each tree [45]. For the model to meet analysts' needs, requirements must be predefined in the parameter tuning stages.

Two important parameters to be set when building the model are the number of decision trees to be created (*Ntree*) and the number of variables to be selected per split (*Mtry*). The higher value of *Ntree* does not improve the accuracy of the classifier; instead, it brings confusion. A lower *Ntree* increases the confidence in features used to build the decision trees and reduces the correlation between the electorates [28,44]. In contrast, *Ntree* can be as large as possible, while *Mtry* should be equal to the square root of the number of features [46]. There is no standardized way to adjust the parameters for RF models. Therefore, it depends on the discretion of the analyst and the nature of the study. In this study, the RF classifier from the "*smileRandomForest*" library was utilized. To optimize the model, Ntree was adjusted to 25 and *Mtry* was set to three, which is equal to the square root of explanatory features, and each node was adjusted to have 2 leaves, as detailed in Table 5. Each decision tree received a share of 0.5% from the original sample data for OOB, and nodes per tree were set not to exceed 1000. The optimization of the parameters was obtained using 10-fold cross-validation.

Support Vector Machine

The SVM algorithm is a supervised vector-based classifier that has been used for both regression and classification [47]. Different types of SVM models have been widely used for different remote sensing applications which include LULC, water quality, and vegetation mapping. SVM models use discriminative methods by learning the distribution patterns of training data. It uses linear or nonlinear lines (known as hyperplanes) to optimize the boundaries between the vector data. The development of SVM models was designed as a binary classifier that only deals with linear data [48]. The method uses the principle of finding the best marginal line (hyperplane) between the two features in the high-dimensional space. Then, the optimal hyperplane serves as the decision boundary that separates the two features. This principle generally works well when applied to linear data.

Table 5. Hyperparameter Tuning of the MLA Models.

Classifier	Classifiers	Parameters	Parameter Adjustments
NB	*ee.Classifier.smileNaiveBayes()*	λ	0.000001
RF	*ee.Classifier.smileRandomForest()*	Ntree Mtry MinLeafPopulation MaxNodes BagFraction	25.00 null 2.00 1000.00 0.50%
SVM	*ee.Classifier.libsvm()*	kernelType svmType coef0 degree cost	POLY C_SVC 0.3 1.00 10
ANN	*MLPClassifier (ANN)*	activation hidden_layer nuerons	RELU 1.00 10

However, it is challenging to separate nonlinear data using the linear SVM approach because the inherent design of its model tends to incompletely separate the training dataset, leading to classification errors [49]. After numerous attempts to develop SVMs suitable for classifying nonlinear data, combined SVM and kernel functions became a viable way to separate nonlinear data. The primary role of the kernel function in SVM models is to mobilize the training data to a new dimensional space, allowing the marginal lines to find the optimal hyperplane to separate the vectors [50]. The most popular kernel functions for the classification of complex remote-sensed data include the radial basis function (RBF) and the polynomial kernel [51]. These two kernels have been reported in the literature for their ability to produce accurate classification results; for example, [52] used RBF-SVM to monitor dynamic land use changes in Urmia Lake Iran, as did [53] in urban expansion mapping using a polynomial kernel function. In addition, [51] compared the performance of the kernels and found RBF to be slightly more accurate than the polynomial kernel in delineating the sub-catchment.

As there is no rule of thumb in kernel selection, the analyst's discretion is used based on the nature of the study. Therefore, for the current study, the polynomial kernel was selected based on the trial-and-error approach. The applicable SVM type for the polynomial kernel in sensing data analysis is support vector classifier (SVC), which allows for the cost (C) of non-separatable vectors (C-SVC). Thus, it is also known as a soft margin because it requires only the best hyperplane and accepts some separation errors as defined in the C threshold; if value C is high, it can lead to overfitting. In this way, C-SVC guarantees results whose integrity is highly dependent on the training data and parameter tuning of the model (Table 5). The C-SVC model has been used in different studies for LULC classification and achieved meaningful results, such as in [17].

The C-SVC from the "*libsvm*" library in GEE was used. The kernel and C were set to polynomial and 10, respectively, as the main parameters to control marginal lines and error threshold. Additionally, hyperparameters were also tuned, including a degree, gamma, and coef0, to achieve the optimum hyperplane, as shown in Table 5.

Artificial Neural Network

ANN is a non-parametric algorithm with a complex network of interconnected neurons. The network is composed of an input phase, one or more intermediate phases, and an output phase. In remote sensing applications, the input layer has neurons that serve as receptor sites for data from each variable. They pass the data to the intermediate phase via channels. Each channel has a numeric value assigned, known as weight, which is the factor of the inputs to be transmitted to the neurons of the intermediate phase [54]. Neurons in the intermediate phase house the numeric values called bias, which are added to the inputs. Then, the total neuron values are subjected to the activation function, which determines the neurons that are going to contribute to the inputs of the next phase [55]; all the activated neurons transmit the inputs to the next phase in a process called forward propagation. The inputs are propagated through the network to the output phase, where the inputs are assigned to a corresponding class [56]. In addition, the misclassified data are backpropagated through the network model iteratively for revision by adjusting the weights. In this study, the feed-forward propagation ANN model from the *Sci-Kit learn* package applicable in Python (Google Colab cloud computing platform) was used, as defined:

$$\text{Output} = \text{aS}\left(\sum_{i-1}^{n} w_i w z_i + b\right) \tag{2}$$

where w_i and z_i are weights that are carried by the channels, b is the bias as the remainder of the training samples, and i is the number of neurons in the intermediate phase.

According to [57], ANN models, in general, tend to overestimate the classes. The network size and the number of neurons to be used are very important parameters to consider in achieving the best results; a large complex network can lead to overfitting of the input data [58]. The number of intermediates was set to 1 phase with 10 iterations. The nonlinearity was solved by the RELU activation function, as detailed parameter tuning is presented in Table 5.

2.6. Variable Importance

The inherent designs of MLA models are complex in nature, and it is challenging to interpret their interaction with the input data [59]. However, understanding the model's behaviour is important [60] to determine the utility features and gain insight into the mechanisms of the models [61,62]. Thus, the built models (NB, RF, SVM, and ANN) were further investigated to explain the interaction between models and features in classifying given LULC classes (Table 1) using the Shapley Additive exPlanation (SHAP) value. Its approach is based on game theory, which explains the impact of individual features on the prediction of machine learning models [59]. To compute the SHAP value, the *Sci-Kit learn* and *shap* packages in Python were used to build both MLA and SHAP value explainer models, respectively.

2.7. Accuracy Assessment

The most used accuracy assessment standard for land use mapping is the Kappa coefficient [37]. The Kappa coefficient assesses the status of classification results against the randomly assigned values. The nature of the Kappa values ranges from −1 to 1. These kappa values rank the classification accuracy of the classified image; when the value is less than zero, there is no agreement. When the value is ≤0.40, there is a fair amount of accuracy. A value ≤0.75 indicates satisfactory accuracy, and ≤1 indicates good performance [63]. The Kappa coefficient can be defined as follows:

$$K = \frac{P(A) - P(E)}{1 - P(E)} \tag{3}$$

where P(A) is the overall accuracy and P(E) is the chance agreement, which is given by the sum of the predicted values and actual values for each class and agreement by chance [64].

Note: j = land use class of interest (e.g., water bodies) and will be used as an example to determine the accuracies of the class of interest.

The overall accuracy represents the sum of correctly labelled pixels for all the classes in the input image as per the provided test sample sites and is mathematically expressed as a fraction of correctly labelled pixels and the total number of test sample sites [65].

$$OA = \frac{\sum \text{correctly leballed pixels}}{\sum \text{pixels in test sample sites}} \tag{4}$$

PA is the number of pixels that the test sample of the given class (j) agreed to labelled pixels as class j in the input image. It is given by the sum of correctly labelled pixels of a class, j, divided by the sum of correctly labelled pixels and the pixels claimed to not belong to class j, as adapted from [37]:

$$PA_j = \frac{\sum_{\text{correctly leballed pixelsj}}}{\sum_{\text{correct leballed pixelsj + misleballed as classj}}} \quad (5)$$

The user's accuracy is determined by the number of correctly labelled pixels in the class divided by the sum of the pixels in the class claimed to belong to class *j* in the test samples, as proposed by [66].

$$UA_j = \frac{\sum_{\text{correctly leballed pixelsj}}}{\sum_{\text{correct leballed pixelsj + leballed as classj}}} \quad (6)$$

In an effort to assess the performance of the NB, RF, SVM, and ANN models, the aforementioned accuracy metrics will be calculated from the confusion matrix of each model.

3. Results

Figure 3 illustrates the result of the L8-OLI variable collinearity test, as discussed in Section 2.3. One variable of the highly correlated pairs was removed. For example, the ultra-blue and blue bands show a linear spectral relationship of 0.98, as shown in the red and light-red cells in Figure 3. In contrast, variables that showed the least correlation or distinct spectral variation were retained to facilitate the classification process. As a result, from a total of eleven tested variables, six multi-spectral bands, such as visible (blue, green, and red), NIR, and SWIR (1 and 2) bands, were selected as explanatory variables for the LULC classification, along with three spectral indices (NDVI, NWDI, and BSI). To demonstrate the unique spectral information of each variable per class, the spectral variability was computed, as illustrated in Figure 4.

3.1. Mapping and Spatial Extent of LULC Classes

This study successfully mapped nine LULC classes covering about 4444 km^2 of the uMngeni River catchment using four MLA strategies (NB, RF, SVM, and ANN). The classification results are shown in Figure 5, and the graphical representation is in Figure 6. The MLAs show estimates of varying extent but agree to the class geographic position in the study area. The catchment is dominated by grassland vegetation, covering 1819.31 km^2 (RF) and 1206.16 km^2 (ANN). Natural Forest is the second dominant class, with an area cover of 1200.13 km^2 determined by NB and 762.2 km^2 by ANN. This is followed by planted forest, with an estimate of 738.9 km^2 and 296.5 km^2 by NB.

Plantation, as the third dominant class, shows slight variations in area cover in NB, SVM, and RF (Figure 6), while there was a distinct area coverage of 738.88 km^2 recorded by the ANN model. ANN showed a significant area extent of 831.27 km^2 in the cropland class. In contrast, RF and SVM reported statistically insignificant areas for cropland class cover at 372.73 km^2 and 440.29 km^2, respectively. Built-up and bare soil area cover vary across the four MLAs, ranking them in the fifth and sixth dominant classes. The water, wetland, and mining classes showed the least area coverage. In this category, the mining class represents the least coverage of the total study area, with an area cover at a maximum of 45.04 km^2 as determined by ANN (Figure 6).

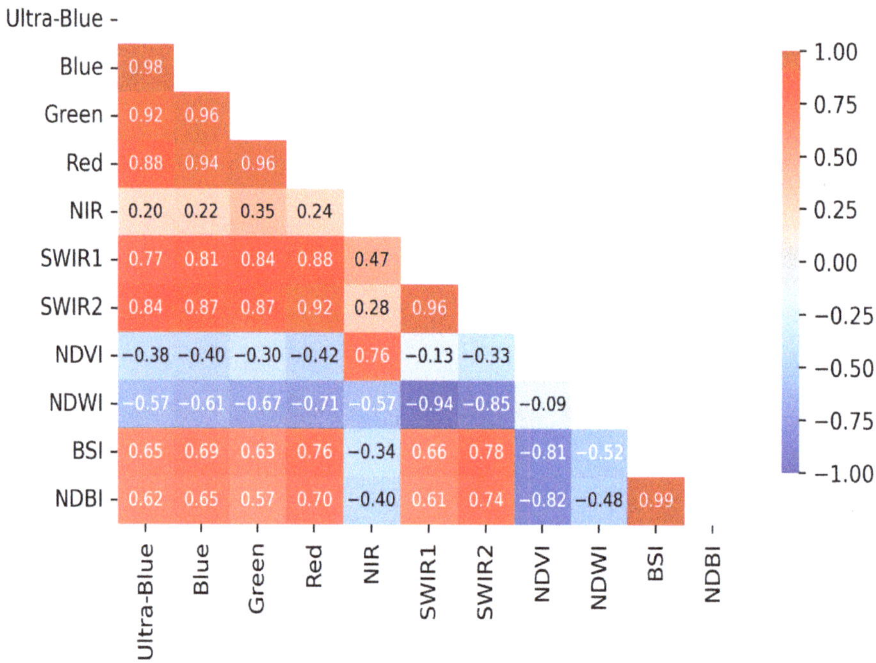

Figure 3. The variable collinearity heatmap. Blue to red shows an increasing correlation relationship and vice versa.

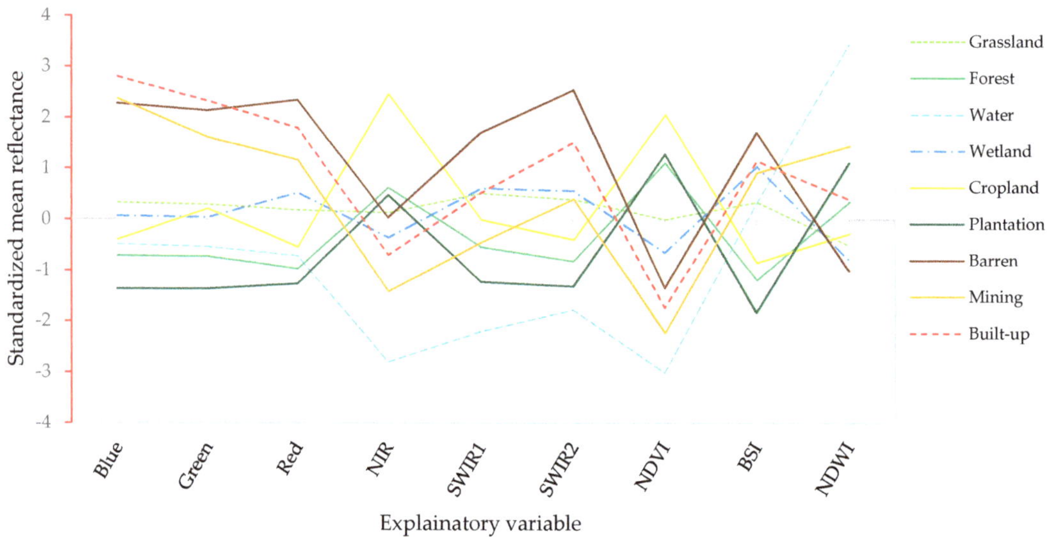

Figure 4. Standardized average reflectance values of each land use class.

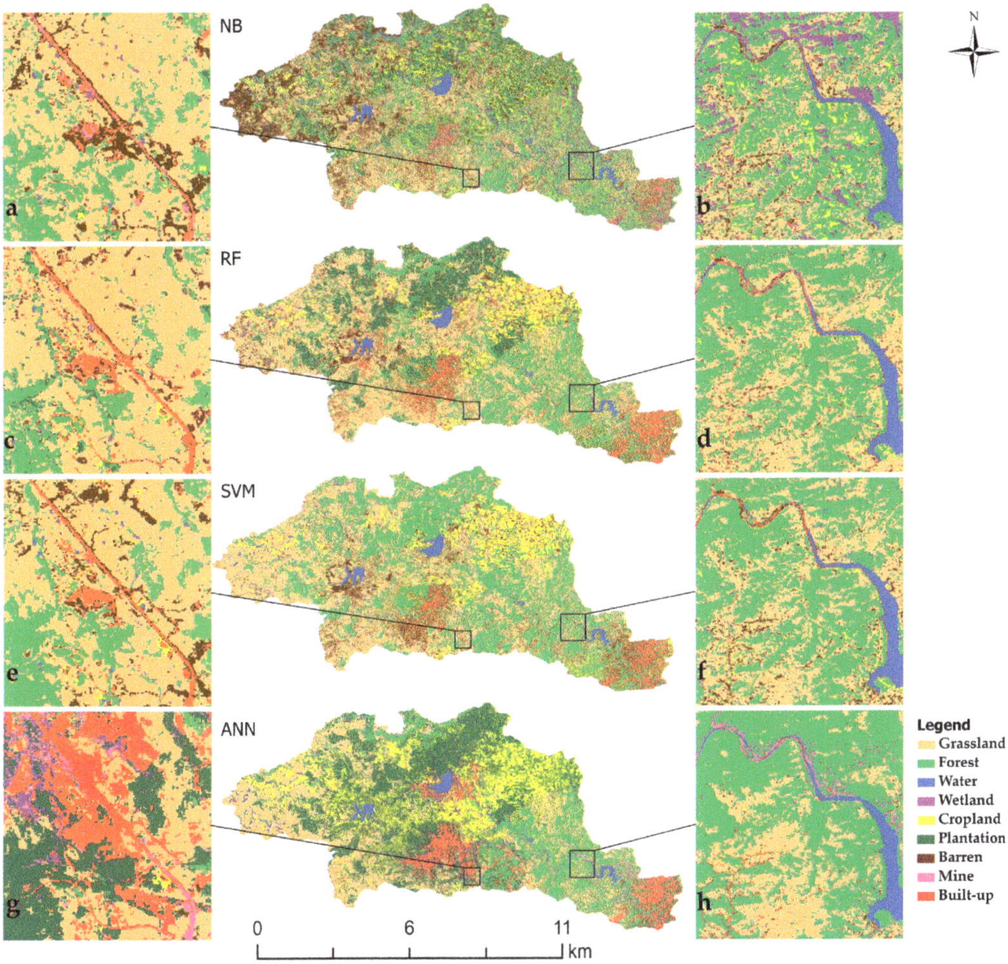

Figure 5. UMngeni River catchment LULC maps produced by NB, RF, SVM, and ANN models with detailed small portions in (**a**–**h**).

The geographic positions of thematic land use classes are consistent across four MLAs (Figure 5). In the grassland matrix, which dominates the catchment by 40.99%, cropland and barren land are observed in the proximal distance from the water bodies and in the periphery of the upper parts of the catchment. Meanwhile, natural forest land covers a large portion of the middle reaches and extends to the lower regions, with about 0.08% of mining classes on the forest banks. The plantation is scattered around the upper region, with higher density in the far north of the upper region. The distribution of 10.27% (Table 6) of the built-up class is observed mainly in the lower region of the study area, with dense areas towards the uMngeni river estuary (Figure 6). The wetland class covered tiny patches (0.12%) in the dam shorelines and was visible at the tip of the upper region.

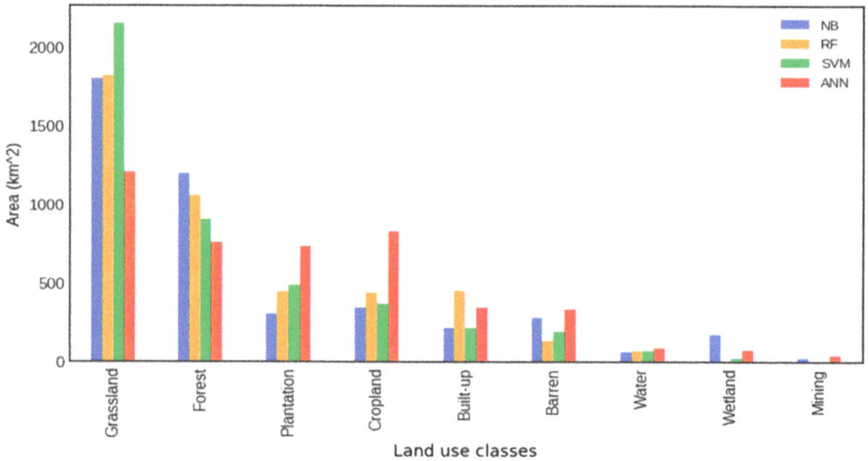

Figure 6. The spatial distribution of land use classes produced by each MLA.

Table 6. Cover Areas of Thematic Land Use Classes as Reported by RF [1].

LULC Class	Area in km^2	Area in %
Grassland	1819.31	40.94
Forest	1058.55	23.82
Built-up	455.69	10.25
Cropland	446.49	10.05
Plantation	440.29	9.91
Barren	134.68	3.03
Water	74.81	1.68
Wetland	5.29	0.12
Mining	3.35	0.08
Total	4444.00	100.00

[1] These are the results (km^2 and percentage) for class area cover that were used in the Discussion section since RF produced more accurate maps than SVM, NB, and ANN.

3.2. Comparison of the Mapping Accuracy of Machine Learning Algorithms

The accuracy matrices of the models were generated using 30% of the sample size (Table 4). The matrices of the MLAs (Figure 7) provide the statistical report on the agreement between reality on the ground and the maps produced. The OA, Kappa, UA, and PA algorithms were used to assess the accuracy of the models. According to the confusion matrix reports, the OA and Kappa were 64.38% and 0.58 for NB, 97.02% and 0.96 for RF, 89.74% and 0.88 for SVM, and 87% and 0.86 for ANN, respectively. The overall performance of the models was generated from the accuracy of the individual classes.

The classification accuracy for the land use classes varies significantly across the four MLAs, as presented in Figure 7. RF and SVM have effectively classified the built-up class at thresholds of >85.60% for both UA and PA. The NB model struggled to separate the built-up and mining classes. Hence, there is observable swapping between the two classes. Similarly, the ANN model showed a high commission of mining pixels to the built-up class. Thus, NB only achieved 48.54% and 49.39% for UA and PA for the built-up class, respectively. NB tends to confuse cropland with plantations. In contrast, RF, SVM, and ANN successfully extracted the cropland pixels with high metric percentages (Figure 7). All the MLAs delineated the surface water, plantation, forest, and barren classes with fairly high confidence, except for the NB model that commissioned some natural forest pixels to the plantations class.

It is worth reporting that only the RF model provided meaningful results for mapping uMngeni River catchment classes with a small area extent, which includes the barren, mining, and wetland classes. The SVM model experienced an error in omission regarding mining and wetland pixels. In contrast, the ANN model mapped the mining and wetlands classes with observable overfitting, which highly correlates with the decline in the built-up, grassland, and forest classes (Figure 5). NB significantly overfitted the barren and wetland classes at the expense of cropland and plantation land cover classes. Consequently, these accuracy metrics form the basis of the overall classification performance of each model.

NB

	Reference data										
	1	2	3	4	5	6	7	8	9	Sum row	PA%
Predicted classes 1	1797	140	0	14	50	0	328	97	157	2583	69.57
2	50	1132	0	409	582	61	0	0	0	2234	50.67
3	0	0	694	2	0	1	0	0	0	697	99.57
4	8	9	1	103	0	0	2	8	0	131	78.63
5	93	342	0	6	2429	400	11	1	2	3284	73.96
6	18	629	0	3	595	1585	27	0	2	2859	55.44
7	244	11	0	0	0	0	1171	3	239	1668	70.20
8	5	0	3	73	0	0	9	71	76	237	29.96
9	56	3	2	85	0	0	62	252	449	909	49.39
Sum column	2271	2266	700	695	3656	2047	1610	432	925		
UA%	79.13	49.96	99.14	14.82	66.44	77.43	72.73	16.44	48.54		
OA%	64.10										
Kappa	0.58										

RF

	Reference data										
	1	2	3	4	5	6	7	8	9	Sum row	PA%
Predicted classes 1	2554	9	0	0	7	0	13	0	0	2583	98.88
2	22	2093	0	0	98	21	0	0	0	2234	93.69
3	0	0	696	0	1	0	0	0	0	697	99.86
4	8	1	0	121	1	0	0	0	0	131	92.37
5	60	10	0	0	3210	0	4	0	0	3284	97.75
6	11	37	1	0	1	2801	8	0	0	2859	97.97
7	64	0	0	0	1	0	1597	0	6	1662	96.09
8	5	0	3	4	0	0	1	133	91	237	56.12
9	4	0	0	0	0	0	7	0	898	909	98.79
Sum column	2728	2150	700	125	3319	2822	1630	133	995		
UA %	93.62	97.35	99.43	96.80	96.72	99.26	97.98	100.00	90.25		
OA%	97.02										
Kappa	0.96										

SVM

	Reference data										
	1	2	3	4	5	6	7	8	9	Sum row	PA%
Predicted classes 1	2330	54	0	3	24	3	156	11	2	2583	90.21
2	75	1932	0	0	186	38	2	0	1	2234	86.48
3	0	0	697	0	0	0	0	0	0	697	100.00
4	19	1	3	96	1	0	0	11	0	131	73.28
5	134	52	0	6	3080	0	11	0	1	3284	93.79
6	16	86	0	1	8	2737	11	0	0	2859	95.73
7	297	0	0	0	4	0	1289	0	78	1668	77.28
8	17	2	3	18	0	0	24	47	126	237	19.83
9	24	1	0	5	0	0	60	27	792	909	87.13
Sum column	2912	2128	703	129	3303	2778	1553	96	1000		
UA%	80.01	90.79	99.15	74.42	93.25	98.52	83.00	48.96	79.20		
OA%	89.18										
Kappa	0.87										

ANN

	Reference data										
	1	2	3	4	5	6	7	8	9	Sum row	PA%
Predicted classes 1	2100	111	0	8	285	2	2	68	7	2583	81.30
2	119	190	0	2	37	166	0	10	0	524	36.26
3	0	0	685	0	0	0	0	11	1	697	98.28
4	24	6	0	1011	0	0	0	0	0	1041	97.12
5	184	13	0	0	3051	0	36	0	0	3284	92.90
6	0	24	0	0	6	2820	9	0	0	2859	98.64
7	167	0	0	0	38	0	1258	0	205	1668	75.42
8	17	0	0	0	0	0	0	74	146	237	31.22
9	45	0	0	0	2	0	1	73	788	909	86.69
Sum column	2656	344	685	1021	3419	2988	1306	236	1147		
UA %	79.07	55.23	100.00	99.02	89.24	94.38	96.32	31.36	68.70		
OA%	87.50										
Kappa	0.86										

Figure 7. Confusion matrix and accuracy metrics for NB, RF, SVM, and ANN.

3.3. Model Feature Importance

Land cover classes were mapped through the contribution of L8-OLI multispectral bands and derived indices (Figures 2 and 3). The impact of each input feature on the labeling of the classes was analysed based on the average SHAP absolute value. Figure 8 shows the feature importance bar in the degressive arrangement. Each bar represents the sum of the average SHAP values of classes per feature. In the y-axis, the longer the bar, the higher the importance of the feature, and the x-axis represents the average SHAP values for each feature.

In Figure 8, the model's interpretation graphs depict the two common patterns across four models. The top four contributors include SWIR (B6 and B7), Blue (B2), and Green (B3) bands, and the second half shows the gradual decreasing mean SHAP value trend from BSI, NDWI, Red (B4), and NDVI. Given that our study area is dominated by vegetation land use classes, this explains why the highly ranked features are sensitive to vegetation. Thus, SWIR bands were ranked as the most important feature in overall. They are centered at

1.56–2.29 µm, allowing the detection of water content among vegetation class members [67]. Similar results have been reported in studies based on assessing Landsat features for mapping vegetation foliage, such as [68,69]. This has also been observed in other studies that used different sensors (Sentinel-2) and found that SWIR has dominance over the other features [70–72].

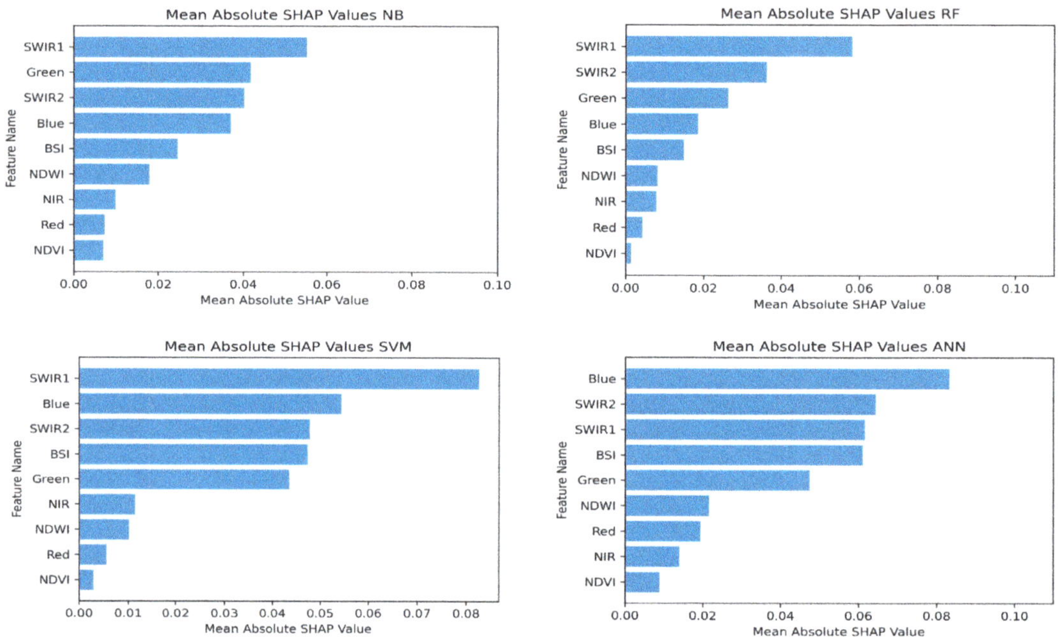

Figure 8. Model feature importance based on the mean absolute SHAP value.

4. Discussion

The mapping of the nine land use classes in GEE was accomplished by complementing the four non-parametric MLAs (NB, RF, SVM, and ANN) to examine the LULC's current state and the geographic distribution of each class in the uMngeni River basin. The results of this analysis will be useful in determining which planning instruments are best for protecting freshwater resources. It also sheds light on how MLAs are applied and how well they may explain the LULC in the region of undulating terrain.

Thematic land use classes were classified based on the use of the multi-spectral bands from L8-OLI imagery. It is imperative to underscore the variations in area coverage for land cover classes amongst the MLAs (Figure 6); these results align with comparative research [17,18,73] that examined machine learning classifiers in mapping LULC classes. This reflects the variability in the inherent design of the machine learning models [73,74] and the dimensionality of the data to be processed [46]. Thus, results from the RF model were used in the discussion of this work as it was the best-performing classifier.

The grassland class was identified as the most dominant class. It is widely distributed, with an area extent of 40.99% of the total catchment area. Namugize et al. [8] reported similar findings. The grassland class holds the topsoil together, increasing the water infiltration rate [75]. Therefore, it has a significant impact on groundwater recharge and water quality.

The forest provides similar ecological functions as the grassland; with an average coverage of 23.85% of the catchment area, the forest is the second largest class after grassland. It is found in abundance in the central sections and in somewhat dispersed patches in the upper and lower regions. Most notably, it mitigates climate change and acts as a carbon sink, hence preventing the deterioration of water quality [6,76]. As a result, efforts to preserve the current natural plant cover should be undertaken, and as suggested by [6,77,78], actions aimed at increasing the area covered by these classes may be encouraged.

The cropland is the fifth largest class, accounting for 10.06% of the study area. Although it is evenly dispersed throughout the catchment, homogeneous patches are observed in the northeast and stretch towards the tip of the western regions (Figure 5). Adjacent to cropland in the further north, about 9.92% of the plantation area cover was observed. The areas dominated by croplands and plantations are known for having low water quality [8,79] due to the inputs used to enhance plant life. The residuals remain in the soil and are eventually channelled to the waterbodies, impacting the freshwater ecosystem [78,79]. In light of these environmental risks, water resource managers must ensure that control measures are in place to promote sustainable agriculture and forestry.

Among the classes that account for the bigger share of the catchment area, the built-up class is the last at 10.27%. It is mostly concentrated in the southern and lower areas of the catchment. Driven by societal needs [7,8,76], it comprises a range of infrastructure types, as explained in Table 1. The built-up class creates impermeable surfaces with low water infiltration, raising water run-off [80], hiking the non-point pollutants to the rivers [8,63,79,81] and ultimately increasing the stream rate flow [82] with implications for water quality and hydrological structures. For such reasons, practical measures to control urban sprawling must be developed, as numerous studies have suggested [76,83].

Similar to built-up areas, barren areas and mining pose a hazard to freshwater resources; they occupy comparatively minor portions of the watershed, at 3.03% and 0.08%, respectively (Table 6). Water turbidity is caused by barren areas, which permit dirt particles to splash into waterbodies [84]. The mining spots are somewhat randomly dispersed along the catchment's forest banks. Their by-products endanger human health as well as aquatic biodiversity [63]. In order to minimize the influence on waterbodies, it is advised that mining developments and barren areas within the catchment be constantly monitored and subject to management measures.

Water and wetland classes cover 1.69% and 0.12% of the study area, respectively. The wetlands are found in the north-west periphery and similar findings were also reported by [85]. Both water and wetlands are the major role players in ensuring the health of the ecosystem in general [86]. Therefore, to guarantee the continuity of their service, initiatives aimed at water harvesting and natural restoration systems should be established, as recommended by [80].

The quantitative analysis of the LULC in the catchment unpacked the underlying data regarding the LULC distribution status. These data will serve as a reference for freshwater resource managers and environmental health specialists in developing management plans and policies for freshwater resources based on informed decisions.

Since various models yield different results in different classes, it is imperative to evaluate the MLAs' performance at the class level [29]. In light of this, only the ANN model confidently delineated the river paths (Figure 5h). However, it could not detect small dams and ponds. In contrast, RF, SVM, and NB struggled to distinguish between river trails and built-up areas (Figure 5b,d,f). This has been reported in prior studies that used pixel-based methods in LULC mapping [7,17,87] and linked them to spectral overlap between built-up areas and riverbeds or river rocks. A strong ability to map small waterbodies was demonstrated by the RF and SVM models. Similar results have been reported by [65], highlighting SVM's efficiency in mapping small waterbodies. However, it performed poorly in defining mining pits. On the other hand, NB and ANN showed bidirectional misclassification between the built-up and mining classes. These findings are in line with [46,56]. However, RF succeeded in separating these classes with a high user accuracy score. Similar results have been reported [28,54,56], claiming that RF possesses a strong generalization technique for LULC analysis, making it less sensitive to the small variations in input vector data.

It is also worth reporting that all MLAs except the RF model struggled to delineate the wetland areas. NB and ANN tended to overfit the wetland class, while SVM commissioned some of the wetland pixels to grassland. These findings were observed in other studies [19], suggesting that this can be associated with composite signatures from the mixed pixels caused by the L8-OLI's 30 m resolution [88], with small samples being allocated to the wetland class due to their small size. The ability to compile the MLAs' overall performance is made possible by the precise evaluation at the class level.

Overall, all the models performed well and produced more realistic LULC maps that resemble the reality on the ground. The findings of this study indicate that RF outperformed all the MLAs with slightly higher OA and Kappa values than those recorded by the SVM model as the second-best classifier, yielding more accurate results than ANN and NB. Previous studies [9,16,17,73] have debated the comparison of RF and SVM and concluded that RF is resilient in classifying complex LULC classes. Its success is credited to its insensitivity to sample size and the advantage of the random multi-decision trees in generalization which are immune to outliers. In contrast, other studies found SVM to be superior to RF; for example, [6] found that SVM outperformed the RF model in mapping natural forest cover classes, indicating that the model's success can be attributed to its default settings. Woldemariam et al. [18] evaluated MLAs for dynamic land use and found SVM to be superior to RF, crediting the model's performance to the usage of the kernel functions to optimize the hyperplanes to find the optimum margin.

An emerging MLA in the remote sensing field, NB, was found unfit to analyse complex land use landscapes. Other researchers [28,29,46] also reported similar findings. This highlights the limitations in the inherent design of the model in dealing with high-dimensional data. The performance can also be associated with the sensitivity of the model to sample size. The NB classifier can be effective in large areas with more linear land cover patterns.

The results of our study proved the excellence of the RF model, which produced a more accurate LULC analysis of the uMngeni River catchment. These results will fill the gap in the literature by identifying the most suitable MLA for quantitative analysis of dimensional LULC in undulating watershed areas. In addition, this can be a useful tool for water resource managers and environmental health stakeholders to facilitate the development of freshwater resource management strategies by unpacking visual and qualitative LULC information for watershed areas.

L8-OLI multi-spectral bands with integrated indices provided good results. SWIR (1 and 2) emerged as the most influential band in the delineation of LULC classes, as discussed in Section 3.3. For such reasons, L8-OLI data were found to be reliable for analysis of the complex LULC in the undulating landscape of the catchment areas. However, it was found to be challenging to classify the pixels at the edges of the big parcels of classes and built-up class members in the core of barren and/or grassland classes or areas partially masked by trees. These findings are consistent with [19], who found that pixels in the transition zone between two parcels have been misclassified due to mixed-pixel composite signatures, as explained by [88,89]. It is challenging to discriminate smaller classes from others in L8-OLI 30 m resolution images due to mixed pixels [17]. Other studies factored in the limitations of pixel-based algorithms [24,87], associating mixed-pixel effects with the restrictions of pixel-based models to go beyond the pixel level.

5. Conclusions

The production of LULC thematic maps based on moderate-resolution satellite data is commonly used in various land cover mapping studies. Using such data gives environmental managers the timely quantitative spatial information they need to implement best practices for natural resource conservation. In this study, four pixel-based non-parametric MLAs, NB, RF, SVM, and ANN, were attested in the classification of thematic LULC classes in the uMngeni River Catchment landscape, South Africa. Moderate-resolution (30 m) data from the L8-OLI sensor were harnessed for mapping nine identified LULC classes. The referenced data acquired from field surveys and Google Earth Pro high-resolution imagery were used for both training models and accuracy assessment. The findings of this study demonstrated the disparities in the MLAs' performances at the class level, resulting in variations in LULC class area coverage across four MLAs. Overall, the RF model outperformed all other models, producing a map that closely resembles the reality on the ground with 97.02% overall accuracy and a kappa value of 0.96, which are in line with a number of studies on mapping LULC in the catchment areas. We discovered that the ANN model performed well in defining the river trails compared to other models, while RF proved to be efficient in mapping classes with tiny area sizes. When viewed from the viewpoint of other models, SVM tends to experience the commission of mining class members to built-up classes, hence producing low accuracies in the mining class. As evidenced by their low accuracies, NB and ANN either overfitted or underestimated the LULC classes covering a small area. These errors stem from two directions: first, the variations in the designs of the models hinder them from analysing the data at the subpixel level; second, the 30 m resolution data used makes it difficult for MLA models to classify the impure Landsat pixels due to low resolution. The findings of this work will demonstrate the practical application of MLAs in mapping LULC and serve as an anchor for water resource managers and other relevant environmental health professionals to update or establish the policies and management plans aimed at conserving water resources based on informed decisions. In addition, using RF for classification, with balanced hyperparameter tuning and Landsat data as used in this study, proved to be efficient in mapping LULC in the Umngeni River Catchment. As a result, the method used in this study can be adapted to other research in similar geographic settings.

Author Contributions: Conceptualization, O.B., A.R. and M.G.; Methodology, O.B., A.R. and M.G.; Formal analysis, O.B. and M.G.; Supervision, M.G.; Writing—original draft, O.B.; Writing—review & editing, O.B., A.R. and M.G. All authors have read and agreed to the published version of the manuscript.

Funding: This study was funded by the South African National Space Agency (SANSA), postgraduate bursaries 2023, accessed on 3 August 2022 (www.sansa.org.za).

Data Availability Statement: The surface reflectance Landsat 8 data can be accessed in Earth Engine data catalog at https://developers.google.com/earth-engine/datasets/catalog/LANDSAT_LC08_C02_T1_L2, accessed on the 15 May 2023.

Conflicts of Interest: The authors declare no conflicts of interest.

References

1. Mekuria, W.; Gedle, A.; Tesfaye, Y.; Phimister, E. Implications of Changes in Land Use for Ecosystem Service Values of Two Highly Eroded Watersheds in Lake Abaya Chamo Sub-Basin, Ethiopia. *Ecosyst. Serv.* **2023**, *64*, 101564. [CrossRef]
2. Aneseyee, A.B.; Soromessa, T.; Elias, E. The Effect of Land Use/Land Cover Changes on Ecosystem Services Valuation of Winike Watershed, Omo Gibe Basin, Ethiopia. *Hum. Ecol. Risk Assess.* **2020**, *26*, 2608–2627. [CrossRef]
3. Pullanikkatil, D.; Palamuleni, L.G.; Ruhiiga, T.M. Land Use/Land Cover Change and Implications for Ecosystems Services in the Likangala River Catchment, Malawi. *Phys. Chem. Earth* **2016**, *93*, 96–103. [CrossRef]
4. Lynch, A.J.; Cooke, S.J.; Arthington, A.H.; Baigun, C.; Bossenbroek, L.; Dickens, C.; Harrison, I.; Kimirei, I.; Langhans, S.D.; Murchie, K.J.; et al. People Need Freshwater Biodiversity. *Wiley Interdiscip. Rev. Water* **2023**, *10*, e1633. [CrossRef]
5. Nyathi, N.A.; Zhao, W.; Musakwa, W. Land Use Land Cover Changes and Their Impacts on Ecosystem Services in the Nzhelele River Catchment, South Africa. *ISPRS Ann. Photogramm. Remote Sens. Spat. Inf. Sci.* **2020**, *5*, 809–816. [CrossRef]
6. Gyamfi-Ampadu, E.; Gebreslasie, M.; Mendoza-Ponce, A. Mapping Natural Forest Cover Using Satellite Imagery of Nkandla Forest Reserve, KwaZulu-Natal, South Africa. *Remote Sens. Appl.* **2020**, *18*, 100302. [CrossRef]
7. Obaid, A.; Adam, E.; Ali, K.A. Land Use and Land Cover Change in the Vaal Dam Catchment, South Africa: A Study Based on Remote Sensing and Time Series Analysis. *Geomatics* **2023**, *3*, 205–220. [CrossRef]
8. Namugize, J.N.; Jewitt, G.; Graham, M. Effects of Land Use and Land Cover Changes on Water Quality in the UMngeni River Catchment, South Africa. *Phys. Chem. Earth* **2018**, *105*, 247–264. [CrossRef]
9. Ouchra, H.; Belangour, A.; Erraissi, A. Comparison of Machine Learning Methods for Satellite Image Classification: A Case Study of Casablanca Using Landsat Imagery and Google Earth Engine. *J. Environ. Earth Sci.* **2023**, *5*, 118–134. [CrossRef]
10. Akar, O.; Tunc Gormus, E. Land Use/Land Cover Mapping from Airborne Hyperspectral Images with Machine Learning Algorithms and Contextual Information. *Geocarto Int.* **2022**, *37*, 3963–3990. [CrossRef]

11. Lekka, C.; Petropoulos, G.P.; Detsikas, S.E. Appraisal of EnMAP Hyperspectral Imagery Use in LULC Mapping When Combined with Machine Learning Pixel-Based Classifiers. *Environ. Model. Softw.* **2024**, *173*, 105956. [CrossRef]
12. Delogu, G.; Caputi, E.; Perretta, M.; Ripa, M.N.; Boccia, L. Using PRISMA Hyperspectral Data for Land Cover Classification with Artificial Intelligence Support. *Sustainability* **2023**, *15*, 13786. [CrossRef]
13. James, D.; Collin, A.; Mury, A.; Costa, S. Very High Resolution Land Use and Land Cover Mapping Using Pleiades-1 Stereo Imagery and Machine Learning. *Int. Arch. Photogramm. Remote Sens. Spat. Inf. Sci.* **2020**, *XLIII-B2-2020*, 675–682. [CrossRef]
14. Mahmoud, R.; Hassanin, M.; Al Feel, H.; Badry, R.M. Machine Learning-Based Land Use and Land Cover Mapping Using Multi-Spectral Satellite Imagery: A Case Study in Egypt. *Sustainability* **2023**, *15*, 9467. [CrossRef]
15. Bayas, S.; Sawant, S.; Dhondge, I.; Kankal, P.; Joshi, A. Land Use Land Cover Classification Using Different ML Algorithms on Sentinel-2 Imagery. In *Advanced Machine Intelligence and Signal Processing*; Springer: Singapore, 2022; pp. 761–777.
16. Parashar, D.; Kumar, A.; Palni, S.; Pandey, A.; Singh, A.; Singh, A.P. Use of Machine Learning-Based Classification Algorithms in the Monitoring of Land Use and Land Cover Practices in a Hilly Terrain. *Environ. Monit. Assess.* **2024**, *196*, 8. [CrossRef]
17. Loukika, K.N.; Keesara, V.R.; Sridhar, V. Analysis of Land Use and Land Cover Using Machine Learning Algorithms on Google Earth Engine for Munneru River Basin, India. *Sustainability* **2021**, *13*, 13758. [CrossRef]
18. Woldemariam, G.W.; Tibebe, D.; Mengesha, T.E.; Gelete, T.B. Machine-Learning Algorithms for Land Use Dynamics in Lake Haramaya Watershed, Ethiopia. *Model Earth Syst. Environ.* **2022**, *8*, 3719–3736. [CrossRef]
19. Dash, P.; Sanders, S.L.; Parajuli, P.; Ouyang, Y. Improving the Accuracy of Land Use and Land Cover Classification of Landsat Data in an Agricultural Watershed. *Remote Sens.* **2023**, *15*, 4020. [CrossRef]
20. Hughes, C.J.; De Winnaar, G.; Schulze, R.E.; Mander, M.; Jewitt, G.P.W. Mapping of Water-Related Ecosystem Services in the UMngeni Catchment Using a Daily Time-Step Hydrological Model for Prioritisation of Ecological Infrastructure Investment—Part 2: Outputs. *Water SA* **2018**, *44*, 590–600. [CrossRef]
21. Kusangaya, S.; Warburton, M.; Archer van Garderen, E. Use of ACRU, a Distributed Hydrological Model, to Evaluate How Errors from Downscaled Rainfall Are Propagated in Simulated Runoff in UMngeni Catchment, South Africa. *Hydrol. Sci. J.* **2017**, *62*, 1995–2011. [CrossRef]
22. Kaur, H.; Tyagi, S.; Mehta, M.; Singh, D. Time Series (2001/2002–2021) Analysis of Earth Observation Data Using Google Earth Engine (GEE) for Detecting Changes in Land Use Land Cover (LULC) with Specific Reference to Forest Cover in East Godavari Region, Andhra Pradesh, India. *J. Earth Syst. Sci.* **2023**, *132*, 86. [CrossRef]
23. Mandal, S.; Bandyopadhyay, A.; Bhadra, A. Dynamics and Future Prediction of LULC on Pare River Basin of Arunachal Pradesh Using Machine Learning Techniques. *Environ. Monit Assess* **2023**, *195*, 709. [CrossRef]
24. Maviza, A.; Ahmed, F. Analysis of Past and Future Multi-Temporal Land Use and Land Cover Changes in the Semi-Arid Upper-Mzingwane Sub-Catchment in the Matabeleland South Province of Zimbabwe. *Int. J. Remote Sens.* **2020**, *41*, 5206–5227. [CrossRef]
25. Foga, S.; Scaramuzza, P.L.; Guo, S.; Zhu, Z.; Dilley, R.D.; Beckmann, T.; Schmidt, G.L.; Dwyer, J.L.; Joseph Hughes, M.; Laue, B. Cloud Detection Algorithm Comparison and Validation for Operational Landsat Data Products. *Remote Sens. Environ.* **2017**, *194*, 379–390. [CrossRef]
26. Mashala, M.J.; Dube, T.; Ayisi, K.K.; Ramudzuli, M.R. Using the Google Earth Engine Cloud-Computing Platform to Assess the Long-Term Spatial Temporal Dynamics of Land Use and Land Cover within the Letaba Watershed, South Africa. *Geocarto Int.* **2023**, *38*, 2252781. [CrossRef]
27. Floreano, I.X.; de Moraes, L.A.F. Land Use/Land Cover (LULC) Analysis (2009–2019) with Google Earth Engine and 2030 Prediction Using Markov-CA in the Rondônia State, Brazil. *Environ. Monit. Assess.* **2021**, *193*, 239. [CrossRef]
28. Orieschnig, C.A.; Belaud, G.; Venot, J.-P.; Massuel, S.; Ogilvie, A. Input Imagery, Classifiers, and Cloud Computing: Insights from Multi-Temporal LULC Mapping in the Cambodian Mekong Delta. *Eur. J. Remote Sens.* **2021**, *54*, 398–416. [CrossRef]
29. Palanisamy, P.A.; Jain, K.; Bonafoni, S. Machine Learning Classifier Evaluation for Different Input Combinations: A Case Study with Landsat 9 and Sentinel-2 Data. *Remote Sens.* **2023**, *15*, 3241. [CrossRef]
30. Viana, C.M.; Girão, I.; Rocha, J. Long-Term Satellite Image Time-Series for Land Use/Land Cover Change Detection Using Refined Open Source Data in a Rural Region. *Remote Sens.* **2019**, *11*, 1104. [CrossRef]
31. Gao, B.C. NDWI—A Normalized Difference Water Index for Remote Sensing of Vegetation Liquid Water from Space. *Remote Sens. Environ.* **1996**, *58*, 257–266. [CrossRef]
32. Nguyen, C.T.; Chidthaisong, A.; Kieu Diem, P.; Huo, L.-Z. A Modified Bare Soil Index to Identify Bare Land Features during Agricultural Fallow-Period in Southeast Asia Using Landsat 8. *Land* **2021**, *10*, 231. [CrossRef]
33. Rouse, J.W.; Haas, R.H.; Schell, J.A.; Deering, D.W. Monitoring vegetation systems in the great plains with ERTS. In Proceedings of the Third NASA Earth Resources Technology Satellite Symposium, Washington, DC, USA, 10–14 December 1973; pp. 309–317.
34. Rikimaru, A.; Roy, P.S.; Miyatake, S. Tropical Forest Cover Density Mapping. *Trop. Ecol.* **2002**, *43*, 39–47.
35. Carlotto, M.J. Effect of Errors in Ground Truth on Classification Accuracy. *Int. J. Remote Sens.* **2009**, *30*, 4831–4849. [CrossRef]
36. Foody, G.M. Assessing the Accuracy of Land Cover Change with Imperfect Ground Reference Data. *Remote Sens. Environ.* **2010**, *114*, 2271–2285. [CrossRef]
37. Olofsson, P.; Foody, G.M.; Herold, M.; Stehman, S.V.; Woodcock, C.E.; Wulder, M.A. Good Practices for Estimating Area and Assessing Accuracy of Land Change. *Remote Sens. Environ.* **2014**, *148*, 42–57. [CrossRef]

38. Gandhi, Ujaval, 2021. End-to-End Google Earth Engine Course. Spatial Thoughts. Available online: https://courses.spatialthoughts.com/end-to-end-gee.html (accessed on 2 February 2024).
39. Gyamfi-Ampadu, E.; Gebreslasie, M.; Mendoza-Ponce, A. Evaluating Multi-sensors Spectral and Spatial Resolutions for Tree Species Diversity Prediction. *Remote Sens.* **2021**, *13*, 1033. [CrossRef]
40. Costache, R.; Tin, T.T.; Arabameri, A.; Crăciun, A.; Costache, I.; Islam, A.R.M.T.; Sahana, M.; Pham, B.T. Stacking State-of-the-Art Ensemble for Flash-Flood Potential Assessment. *Geocarto Int.* **2022**, *37*, 13812–13838. [CrossRef]
41. Breiman, L.E.O. Random Forests. *Mach. Learn.* **2001**, *45*, 5–32. [CrossRef]
42. Zhang, Q.; Liu, M.; Zhang, Y.; Mao, D.; Li, F.; Wu, F.; Song, J.; Li, X.; Kou, C.; Li, C.; et al. Comparison of Machine Learning Methods for Predicting Soil Total Nitrogen Content Using Landsat-8, Sentinel-1, and Sentinel-2 Images. *Remote Sens.* **2023**, *15*, 2907. [CrossRef]
43. Mather, P.M.; Koch, M. *Computer Processing of Remotely-Sensed Images: An Introduction*; John Wiley & Sons: Hoboken, NJ, USA, 2011; ISBN 1119956404.
44. Cuypers, S.; Nascetti, A.; Vergauwen, M. Land Use and Land Cover Mapping with VHR and Multi-Temporal Sentinel-2 Imagery. *Remote Sens.* **2023**, *15*, 2501. [CrossRef]
45. Kim, Y.H.; Im, J.; Ha, H.K.; Choi, J.K.; Ha, S. Machine Learning Approaches to Coastal Water Quality Monitoring Using GOCI Satellite Data. *GIsci Remote Sens.* **2014**, *51*, 158–174. [CrossRef]
46. Camargo, F.F.; Sano, E.E.; Almeida, C.M.; Mura, J.C.; Almeida, T. A Comparative Assessment of Machine-Learning Techniques for Land Use and Land Cover Classification of the Brazilian Tropical Savanna Using ALOS-2/PALSAR-2 Polarimetric Images. *Remote Sens.* **2019**, *11*, 1600. [CrossRef]
47. Yulianto, F.; Raharjo, P.D.; Pramono, I.B.; Setiawan, M.A.; Chulafak, G.A.; Nugroho, G.; Sakti, A.D.; Nugroho, S.; Budhiman, S. Prediction and Mapping of Land Degradation in the Batanghari Watershed, Sumatra, Indonesia: Utilizing Multi-Source Geospatial Data and Machine Learning Modeling Techniques. *Model Earth Syst. Environ.* **2023**, *9*, 4383–4404. [CrossRef]
48. Huang, C.; Davis, L.S.; Townshend, J.R.G. An Assessment of Support Vector Machines for Land Cover Classification. *Int. J. Remote Sens.* **2002**, *23*, 725–749. [CrossRef]
49. Lantzanakis, G.; Mitraka, Z.; Chrysoulakis, N. X-SVM: An Extension of C-SVM Algorithm for Classification of High-Resolution Satellite Imagery. *IEEE Trans. Geosci. Remote Sens.* **2021**, *59*, 3805–3815. [CrossRef]
50. Srivastava, P.K.; Han, D.; Rico-Ramirez, M.A.; Bray, M.; Islam, T. Selection of Classification Techniques for Land Use/Land Cover Change Investigation. *Adv. Space Res.* **2012**, *50*, 1250–1265. [CrossRef]
51. Martins, S.; Bernardo, N.; Ogashawara, I.; Alcantara, E. Support Vector Machine Algorithm Optimal Parameterization for Change Detection Mapping in Funil Hydroelectric Reservoir (Rio de Janeiro State, Brazil). *Model Earth Syst. Environ.* **2016**, *2*, 138. [CrossRef]
52. Roushangar, K.; Aalami, M.T.; Golmohammadi, H.; Shahnazi, S. Monitoring and Prediction of Land Use/Land Cover Changes and Water Requirements in the Basin of the Urmia Lake, Iran. *Water Supply* **2023**, *23*, 2299–2312. [CrossRef]
53. Saraf, N.M.; Lokman, M.F.; Abdul Rasam, A.R.; Hashim, N. Assessment of Urban Growth Changes in Klang District Using Support Vector Machine by Different Kernel. *IOP Conf Ser Earth Environ. Sci* **2022**, *1051*, 012023. [CrossRef]
54. Dahiya, N.; Gupta, S.; Singh, S. Qualitative and Quantitative Analysis of Artificial Neural Network-Based Post-Classification Comparison to Detect the Earth Surface Variations Using Hyperspectral and Multispectral Datasets. *J. Appl. Remote Sens.* **2023**, *17*, 032403. [CrossRef]
55. Wu, J.L.; Ho, C.R.; Huang, C.C.; Srivastav, A.L.; Tzeng, J.H.; Lin, Y.T. Hyperspectral Sensing for Turbid Water Quality Monitoring in Freshwater Rivers: Empirical Relationship between Reflectance and Turbidity and Total Solids. *Sensors* **2014**, *14*, 22670–22688. [CrossRef]
56. Chen, H.; Lin, X.; Sun, Y.; Wen, J.; Wu, X.; You, D.; Cheng, J.; Zhang, Z.; Zhang, Z.; Wu, C.; et al. Performance Assessment of Four Data-Driven Machine Learning Models: A Case to Generate Sentinel-2 Albedo at 10 Meters. *Remote Sens.* **2023**, *15*, 2684. [CrossRef]
57. Deshpande, A.R.; Emmanuel, M. Context Based Recommendation Methods: A Brief Review. Presented at Conference on Cognitive Knowledge Engineering, Aurangabad, Maharashtra, India, 2016; pp. 14–19. Available online: https://www.researchgate.net/profile/Ratnadeep-Deshmukh-2/publication/329059854_2nd_International_Conference_on_Knowledge_Engineering/links/5bf3dcdba6fdcc3a8de38181/2nd-International-Conference-on-Knowledge-Engineering.pdf#page=58 (accessed on 30 March 2024).
58. Hafeez, S.; Wong, M.S.; Ho, H.C.; Nazeer, M.; Nichol, J.; Abbas, S.; Tang, D.; Lee, K.H.; Pun, L. Comparison of Machine Learning Algorithms for Retrieval of Water Quality Indicators in Case-Ii Waters: A Case Study of Hong Kong. *Remote Sens.* **2019**, *11*, 617. [CrossRef]
59. Lundberg, S.M.; Allen, P.G.; Lee, S.-I. A Unified Approach to Interpreting Model Predictions. *Adv. Neural Inf. Process. Syst.* **2017**, *30*. Available online: https://github.com/slundberg/shap (accessed on 23 May 2024).
60. Wei, P.; Lu, Z.; Song, J. Variable importance analysis: A comprehensive review. *Reliab. Eng. Syst. Saf.* **2015**, *142*, 399–432. [CrossRef]
61. Hosseiny, B.; Abdi, A.M.; Jamali, S. Urban land use and land cover classification with interpretable machine learning—A case study using Sentinel-2 and auxiliary data. *Remote Sens. Appl. Soc. Environ.* **2022**, *28*, 100843. [CrossRef]

62. Yilmaz, E.O.; Kavzoglu, T. Classification of jilin-1 gp01 hyperspectral image using machine learning techniques with explainable artificial intelligence. *Intercont. Geoinf. Days* **2022**, *5*, 145–148.
63. Hua, A.K. Land Use Land Cover Changes in Detection of Water Quality: A Study Based on Remote Sensing and Multivariate Statistics. *J. Environ. Public Health* **2017**, *2017*, 5–7. [CrossRef]
64. Congalton, R.G. A Review of Assessing the Accuracy of Classifications of Remotely Sensed Data. *Remote Sens. Environ.* **1991**, *37*, 35–46. [CrossRef]
65. Kang, C.S.; Kanniah, K.D.; Mohd Najib, N.E. Google Earth Engine for Landsat Image Processing and Monitoring Land Use/Land Cover Changes in the Johor River Basin, Malaysia. In Proceedings of the International Geoscience and Remote Sensing Symposium (IGARSS), Brussels, Belgium, 11–16 July 2021; pp. 4236–4239.
66. Story, M.; Congalton, R.G. Accuracy Assessment: A User's Perspective. *Photogramm. Eng. Remote Sens.* **1986**, *52*, 397–399.
67. Lukeš, P.; Stenberg, P.; Rautiainen, M.; Mõttus, M.; Vanhatalo, K.M. Optical Properties of Leaves and Needles for Boreal Tree Species in Europe. *Remote Sens. Lett.* **2013**, *4*, 667–676. [CrossRef]
68. Schultz, B.; Immitzer, M.; Formaggio, A.; Sanches, I.; Luiz, A.; Atzberger, C. Self-Guided Segmentation and Classification of Multi-Temporal Landsat 8 Images for Crop Type Mapping in Southeastern Brazil. *Remote Sens.* **2015**, *7*, 14482–14508. [CrossRef]
69. Sothe, C.; Almeida, C.; Liesenberg, V.; Schimalski, M. Evaluating Sentinel-2 and Landsat-8 Data to Map Sucessional Forest Stages in a Subtropical Forest in Southern Brazil. *Remote Sens.* **2017**, *9*, 838. [CrossRef]
70. Chen, H.; Yang, L.; Wu, Q. Enhancing Land Cover Mapping and Monitoring: An Interactive and Explainable Machine Learning Approach Using Google Earth Engine. *Remote Sens.* **2023**, *15*, 4585. [CrossRef]
71. McCarty, D.A.; Kim, H.W.; Lee, H.K. Evaluation of Light Gradient Boosted Machine Learning Technique in Large Scale Land Use and Land Cover Classification. *Environments* **2020**, *7*, 84. [CrossRef]
72. Abdi, A.M. Land Cover and Land Use Classification Performance of Machine Learning Algorithms in a Boreal Landscape Using Sentinel-2 Data. *GIsci Remote Sens.* **2020**, *57*, 1–20. [CrossRef]
73. Ashiagbor, G.; Asare-Ansah, A.O.; Amoah, E.B.; Asante, W.A.; Mensah, Y.A. Assessment of Machine Learning Classifiers in Mapping the Cocoa-Forest Mosaic Landscape of Ghana. *Sci. Afr.* **2023**, *20*, e01718. [CrossRef]
74. Foody, G.M. Valuing Map Validation: The Need for Rigorous Land Cover Map Accuracy Assessment in Economic Valuations of Ecosystem Services. *Ecol. Econ.* **2015**, *111*, 23–28. [CrossRef]
75. Machado, R.A.S.; Oliveira, A.G.; Lois-González, R.C. Urban Ecological Infrastructure: The Importance of Vegetation Cover in the Control of Floods and Landslides in Salvador/Bahia, Brazil. *Land Use Policy* **2019**, *89*, 104180. [CrossRef]
76. Abbas, H.; Tao, W.; Khan, G.; Alrefaei, A.F.; Iqbal, J.; Albeshr, M.F.; Kulsoom, I. Multilayer Perceptron and Markov Chain Analysis Based Hybrid-Approach for Predicting Land Use Land Cover Change Dynamics with Sentinel-2 Imagery. *Geocarto Int.* **2023**, *38*, 2256297. [CrossRef]
77. Gyamfi-Ampadu, E.; Gebreslasie, M.; Mendoza-Ponce, A. Multi-Decadal Spatial and Temporal Forest Cover Change Analysis of Nkandla Natural Reserve, South Africa. *J. Sustain. For.* **2022**, *41*, 959–982. [CrossRef]
78. du Plessis, A.; Harmse, T.; Ahmed, F. Quantifying and Predicting the Water Quality Associated with Land Cover Change: A Case Study of the Blesbok Spruit Catchment, South Africa. *Electron. J. Theor. Phys.* **2014**, *11*, 2946–2968. [CrossRef]
79. Mararakanye, N.; Le Roux, J.J.; Franke, A.C. Long-Term Water Quality Assessments under Changing Land Use in a Large Semi-Arid Catchment in South Africa. *Sci. Total Environ.* **2022**, *818*, 151670. [CrossRef]
80. Schütte, S.; Schulze, R.E. Projected Impacts of Urbanisation on Hydrological Resource Flows: A Case Study within the UMngeni Catchment, South Africa. *J. Environ. Manag.* **2017**, *196*, 527–543. [CrossRef]
81. Kibena, J.; Nhapi, I.; Gumindoga, W. Assessing the Relationship between Water Quality Parameters and Changes in Landuse Patterns in the Upper Manyame River, Zimbabwe. *Phys. Chem. Earth* **2014**, *67–69*, 153–163. [CrossRef]
82. Tena, T.M.; Mwaanga, P.; Nguvulu, A. Impact of Land Use/Land Cover Change on Hydrological Components in Chongwe River Catchment. *Sustainability* **2019**, *11*, 6415. [CrossRef]
83. Jewitt, D.; Goodman, P.S.; Erasmus, B.F.N.; O'Connor, T.G.; Witkowski, E.T.F. Systematic Land-Cover Change in KwaZulu-Natal, South Africa: Implications for Biodiversity. *S. Afr. J. Sci.* **2015**, *111*, 9. [CrossRef]
84. Twesigye, C.K. The Impact of Land Use Activities on Vegetation Cover and Water Quality in the Lake Victoria Watershed. *Open Environ. Eng. J.* **2011**, *4*, 66–77. [CrossRef]
85. Moodley, K.; Toucher, M.L.; Lottering, R.T. Simulating Future Land-Use within the UThukela and UMngeni Catchments in KwaZulu-Natal. *Sci. Afr.* **2023**, *20*, e01666. [CrossRef]
86. Balha, A.; Mallick, J.; Pandey, S.; Gupta, S.; Singh, C.K. A Comparative Analysis of Different Pixel and Object-Based Classification Algorithms Using Multi-Source High Spatial Resolution Satellite Data for LULC Mapping. *Earth Sci. Inform.* **2021**, *14*, 2231–2247. [CrossRef]
87. Nayak, A.; Bhushan, B. Wetland Ecosystems and Their Relevance to the Environment. In *Handbook of Research on Monitoring and Evaluating the Ecological Health of Wetlands*; IGI Global: Hershey, PA, USA, 2022; pp. 1–16.

88. Campbell, J.B.; Wynne, R.H. *Introduction to Remote Sensing*; Guilford Press: New York, NY, USA, 2011; ISBN 1609181778.
89. Lillesand, T.; Kiefer, R.W.; Chipman, J. *Remote Sensing and Image Interpretation*; John Wiley & Sons: Hoboken, NJ, USA, 2015; ISBN 111834328X.

remote sensing

MDPI

Article

Spatiotemporal Analysis of Rainfall and Droughts in a Semiarid Basin of Brazil: Land Use and Land Cover Dynamics

Lizandra de Barros de Sousa [1,*], Abelardo Antônio de Assunção Montenegro [1], Marcos Vinícius da Silva [1], Thayná Alice Brito Almeida [1], Ailton Alves de Carvalho [2], Thieres George Freire da Silva [1,3] and João Luis Mendes Pedroso de Lima [4]

1 Department of Agriculture Engineering, Federal Rural University of Pernambuco, Rua Dom Manoel de Medeiros, Dois Irmãos, Recife 52171-900, PE, Brazil; montenegro.ufrpe@gmail.com (A.A.d.A.M.); marcos.viniciussilva@ufrpe.br (M.V.d.S.); thaynaalicee@gmail.com (T.A.B.A.); thieres.silva@ufrpe.br (T.G.F.d.S.)
2 National Institute of Semiarid, Avenida Francisco Lopes de Almeida, Serrotão, Campina Grande 58434-700, PB, Brazil; ailton.carvalho@insa.gov.br
3 Academic Unit of Serra Talhada, Federal Rural University of Pernambuco, Avenida Gregório Ferraz Nogueira, Serra Talhada 56909-535, PE, Brazil
4 MARE—Marine and Environmental Sciences Centre, ARNET—Aquatic Research Network, Department of Civil Engineering, Faculty of Science and Technology, University of Coimbra, Rua Luís Reis Santos, Pólo II, 3030-788 Coimbra, Portugal; plima@dec.uc.pt
* Correspondence: lizandradebarros@gmail.com

Abstract: Precipitation estimation is a challenging task, especially in regions where its spatial distribution is irregular and highly variable. This study evaluated the spatial distribution of annual rainfall in a semiarid Brazilian basin under different regimes and its impact on land use and land cover dynamics. Climate Hazards Group InfraRed Precipitation with Stations (CHIRPS) records and observed data from 40 weather stations over a time series of 55 years were used, in addition to the Standardized Precipitation Index. Spatiotemporal analysis was carried out based on geostatistics. Remote sensing images were also interpreted for different rainfall regimes using the Normalized Difference Vegetation Index and Modified Normalized Difference Water Index. The Gaussian semivariogram model best represented the rainfall spatial structure, showing strong spatial dependence. Results indicated that rainfall amount in the basin significantly increases with elevation, as expected. There is high variation in the dynamics of water storage that can threaten water security in the region. Our findings point out that the application of geostatistics for mapping both the annual precipitation and the Standardized Precipitation Index provides a powerful framework to support hydrological analysis, as well as land use and land cover management in semiarid regions.

Keywords: spatial dependence; temporal variability; sequential Gaussian simulation; kriging; cokriging; CHIRPS

1. Introduction

Precipitation is an essential factor in hydrological processes (e.g., erosion, flooding, and aquifer recharge) [1–4]. In that regard, the spatiotemporal distribution of rainfall has a major impact on the hydrological cycle, water security, agriculture, water resource management, and on several economic and productive activities [5–7].

Despite the relevance of the rainfall distribution in water availability, reliable and high-quality long-term rainfall time series remain scarce in South America, particularly in northeastern Brazil (NEB) [8]. Moreover, precipitation estimation is particularly challenging in the Brazilian semiarid region due to its highly variable spatial distribution and irregular periodicity, which has caused frequent droughts over the years [9,10]. Thus, applying traditional and widely adopted meteorological indices for monitoring droughts in such areas, such as the Standardized Precipitation Index (SPI), is usually a hard task that requires data from several rainfall stations [11–14]. The usefulness of SPI by analyzing long-term rainfall data from several drainage basins in the eastern Mediterranean region was evaluated by Tsesmelis et al. [15], emphasizing the importance of SPI as a tool for assessing changes in drought and wetness episodes, enabling the identification of areas requiring water management strategies to improve water security and to reduce the impact of hydrological extremes.

Remote sensing is an important tool for hydrological research, allowing for the estimation of rainfall using satellite images with high spatiotemporal resolution. The Tropical Rainfall Measuring Mission (TRMM), Climate Hazards Group InfraRed Precipitation with Stations (CHIRPS), and Precipitation Estimation from Remotely Sensed Information using Artificial Neural Networks (PERSIANN) have been widely adopted for rainfall analysis on a basin scale [8,10,16,17]. The CHIRPS product offers a high spatial resolution of approximately 5.3 km and a long temporal period (from 1981) for validation using in situ rain gauge measurements [8,16,18]. In addition, remote sensing can also be used to monitor landscape and biophysical changes in semiarid regions through the evaluation of land use and land cover (LULC) map spatial association with vegetation indices, such as the Normalized Difference Vegetation Index (NDVI) and the Modified Normalized Difference Water Index (MNDWI). These indexes allow the estimation of land use and cover, enabling the characterization of spatiotemporal dynamics and quantification of vegetation conditions and water features [5,19–24]. Additionally, the NDVI has proved to be a valuable tool in the analysis of irrigation in agricultural areas, allowing farmers and researchers to assess soil salinity and monitor crop growth and water stress to optimize yield and conserve resources [25,26]. Due to the high resolution and wide availability of remote sensing products for hydrological studies, there has been a surge of interest from the scientific community to validate such indirect information for different basins around the world [18,24,27]. Moreover, remote sensing products are highly valuable for climate change investigations, and especially for devising adaptive strategies to promote sustainability, particularly in semiarid regions.

The semiarid region of Brazil is known for its highly unpredictable weather patterns, featuring short wet periods and long dry spells. This region is considered the wettest semiarid region worldwide, as evidenced by Marengo et al. [28]. Furthermore, Montenegro and Ragab [29] demonstrated that rivers in this region predominantly have non-perennial hydrological regimes. One of the most prominent rivers in the region is the São Francisco River, with its tributary, the Brígida River, being the focus of this study. The Caatinga vegetation biome dominates the region, which is characterized by deciduous cover that sheds part of its canopy during the dry season.

Typically, rainfall data estimates are obtained by using regression techniques based on time series analysis [30]. The application of geostatistical modeling through kriging takes into account the structure of spatial variation and usually provides high-precision maps, which are important for the study of extreme events [27,31–33]. This method is especially important for the NEB, where the density of rainfall networks is usually low, and proper characterization of rainfall spatial variability is limited due to the lack of observational data [34]. Among the geostatistical estimation methods, kriging is often used in the literature [32,35–37].

The spatiotemporal variability of monthly rainfall data at 269 rainfall stations from 1994 to 2014 in the Paraíba State, located in the NEB, was analyzed by Medeiros et al. [37], who found that 85% of the total rainfall variability could be well represented using kriging, providing a reliable spatial distribution of regions where rainfall conditions varied from scarcity to excess. Regarding the spatiotemporal variability of annual rainfall, Silva et al. [35] found substantial annual spatiotemporal variability and a negative trend (decrease) for most rainy seasons in the Epitácio Pessoa reservoir watershed, Paraíba State, during the study period (1963–1991) using the kriging technique. Additionally, for the Brazilian semiarid region, Araújo et al. [27] investigated the spatiotemporal distribution of precipitation in the Brígida River basin, Pernambuco State, using kriging, providing an adequate precipitation estimate and classification of the aridity condition; however, the availability of observational data was a limiting factor. Furthermore, due to the high spatial variability of precipitation patterns related to topographic and climatic factors [38,39], a study conducted by Adhikary et al. [39] compared three geostatistical methods, finding that the cokriging method with elevation as a secondary variable was significant in Australian watersheds with complex mountainous terrain. Similarly, relationships between elevation and rainfall have been poorly studied in the semiarid regions of Brazil.

Despite the effectiveness of kriging in reducing the variance of the estimation error, it is strongly dependent on the quantity and quality of data, as well as their spatial distribution. However, it may not always accurately reproduce the existing spatial correlation, which can result in unrealistic spatial distributions [40,41]. An alternative estimation method is based on a stochastic simulation approach, which, by definition, reproduces the statistical behavior of the phenomenon. The most widely used simulation technique for forecasting and uncertainty assessment is the sequential Gaussian simulation (SGS) method [42]. The main advantage of the SGS is to produce alternative and equally probable stochastic realizations that reproduce model statistics rather than reduce local forecast variation [43]. In a study of the spatial variability of extreme rainfall in Mexico, Méndez-Venegas et al. [41] used stochastic simulation techniques with a geostatistical approach, highlighting the Gaussian simulation as a powerful tool for studying rainfall variability and modeling its spatial distribution. Another method that seeks to improve the accuracy of spatial estimates is cokriging, which typically reduces the forecast error variance and outperforms kriging when the chosen secondary variable is highly correlated with the primary variable [44,45].

Although a considerable number of studies have addressed the spatial variability of rainfall patterns in the literature, there is still a lack of research focusing on the spatial dependence structures for various hydrological regimes (from dry to wet years) based on the Standardized Precipitation Index (SPI) and drought indexes, especially in the Brazilian semiarid region. Furthermore, there is a need for a joint analysis of the relationship between water and vegetation in the region. Indeed, the SPI values for a 12-month period are strongly linked to streamflow regimes, reservoir levels, water security, and long-term management [12,13,34,46].

Therefore, the aim of this research was to analyze the distribution of rainfall in a basin area of the Brazilian semiarid region, using both observational and remote sensing data. The study seeks to provide a methodological framework for characterizing the spatial scales of rainfall dependence in long-term semiarid regions. This study applies both deterministic and stochastic geostatistical procedures to map and examine the correlation of rainfall with elevation, vegetation cover, and water availability in order to assist decision-making related to water security, vegetation–water nexus studies, reforestation, climate change initiatives, and public policies. Additionally, it assesses the impacts of rainfall variability on land use and land cover (LULC) dynamics using remote sensing data and vegetation indices.

2. Materials and Methods

2.1. Characterization of the Study Site

The study was conducted in the Brígida River basin (Figure 1), located in the High Sertão mesoregion of the Pernambuco State, Brazil (07°19′02″ to 08°36′32″S and 39°17′33″ to 40°43′06″W), with an average elevation of 557 m. The basin area belongs to the semi-arid region of Pernambuco, covering 13,495.73 km^2, which corresponds to 13.73% of the total surface of the state. It comprises 15 municipalities and is part of the São Francisco River basin area [47]. The study basin is one of the tributaries of the São Francisco River Integration Project (PISF), which plays a strategic role in ensuring future water availability in the semiarid northeast region. The PISF main objective is to provide water supply for the socioeconomic development of the most drought-vulnerable northeast states of Brazil (Ceará, Paraíba, Pernambuco, and Rio Grande do Norte) [48].

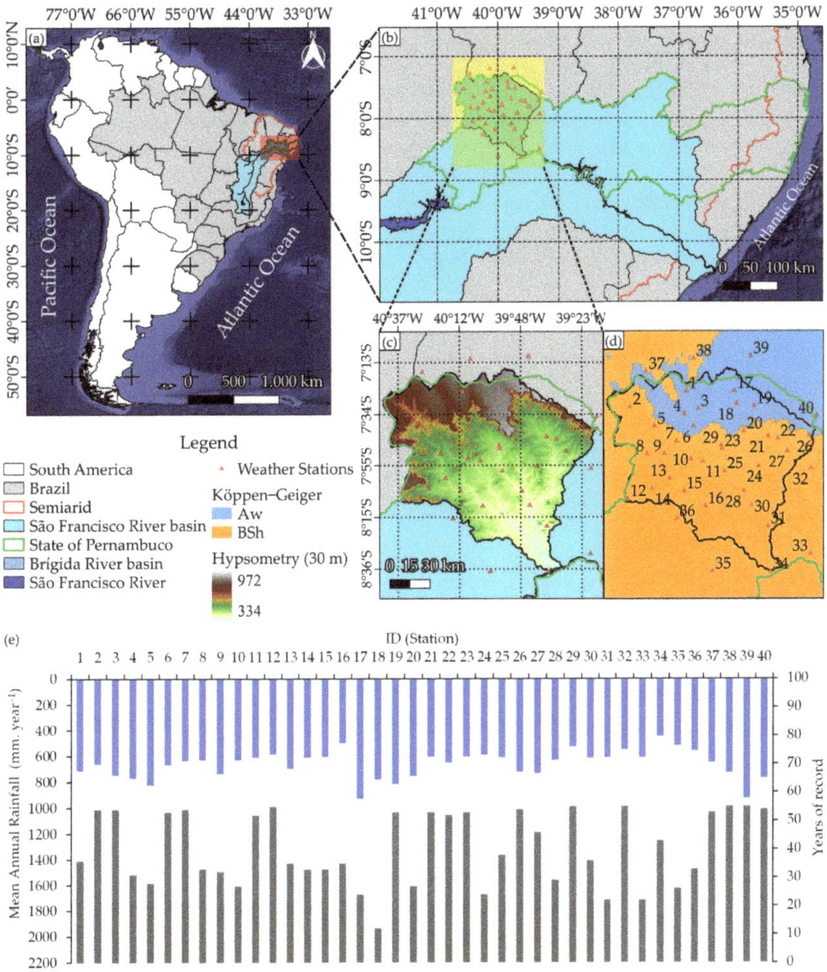

Figure 1. Location of the São Francisco Basin, Brazil, South America (**a**); Brígida River basin, Pernambuco State (**b**); Shuttle Radar Topography Mission (SRTM) map (**c**); Köppen–Geiger climate classification and the spatial distribution of rainfall stations (ID) (**d**); and mean annual rainfall and years of record of the 40 analyzed rainfall stations (**e**).

According to the Köppen–Geiger climate classification, the region is classified as type BSh (hot semiarid climates) and Aw (tropical with dry winter) [49,50], as shown in Figure 1. The monthly mean temperature is 26.1 °C, indicating an average potential evapotranspiration of 175.1 mm per month and an annual cumulative value greater than 2000 mm. The mean total annual precipitation is 608.0 mm (29 years; from 1991 to 2020), with the wet season occurring from December to April (with an average total rainfall of 558.5 mm) [51]. The predominant vegetation in the basin is the hyperxerophilous Caatinga, and the predominant soils are Oxisols, Ultisols, Alfisols, Entisols, and Aridisols [52].

2.2. Rainfall Data

Data from the Climate Hazards Group InfraRed Precipitation with Station (CHIRPS) was used from 1981 to 2017 (https://developers.google.com/earth-engine/datasets/catalog/UCSB-CHG_CHIRPS_DAILY, accessed on 9 May 2023), which includes data from 50°S to

50°N latitudes. CHIRPS combines in situ climatology and satellite imagery with a spatial resolution of 0.05° (about 5.3 km) spanning 35 years [8–14,16,53], and has already been validated for the Brazilian semiarid region [8]. The CHIRPS dataset was validated for the selected basin using observed data from 40 rainfall stations, with an annual time series of 55 years (1962 to 2017), provided by the Pernambuco State Water and Climate Agency (APAC). These data were tested for homogeneity and missing data by APAC [54].

To evaluate the performance of CHIRPS against the observed precipitation data, a pixel-by-pixel comparison was conducted based on the location of the rainfall stations, and various statistical indices were used, including the coefficient of determination (R^2), Pearson's correlation (r), Nash–Sutcliffe Efficiency Coefficient (NSE), and percentage of bias (PBIAS). CHIRPS data, with its high resolution, has the potential to provide a better representation of local-scale spatial variation of rainfall in the basin.

2.3. Standard Precipitation Index (SPI)

The Standard Precipitation Index (SPI) [11] was employed to evaluate the temporal variability of cumulative annual precipitation (SPI-12), commonly adopted to monitor drought and excessive rainfall conditions [12–14]. We used the time series of precipitation observed at rainfall stations from 1962 to 2017, which was later supplemented with CHIRPS data from 1981 to 2017. The SPI is given in thresholds, which correspond to the number of standard deviations that the observed cumulative precipitation deviates from the climatological mean. The precipitation classification used was described by McKee et al. [11] and grouped according to Table 1, which ranges from the most extreme deficit to excess precipitation situations. We selected years for each precipitation class from 1981 onwards, based on their classification using the SPI-12, to carry out the statistical analysis.

Table 1. SPI classification according to McKee et al. [11].

SPI	Classifications
≥2.00	Extremely Wet (EW)
1.00 to 1.99	Severely Wet (SW)
0.50 to 0.99	Moderately Wet (MW)
0.49 to −0.49	Normal (N)
−0.50 to −0.99	Moderately Dry (MD)
−1.00 to −1.99	Severely Dry (SD)
≤−2.00	Extremely Dry (ED)

2.4. Statistical Analysis

The CHIRPS data used, covering the entire basin area, consisted of 930 pixels, and were collected from 1981 to 2017. These data were subjected to classical statistics to verify their behavior in terms of statistical measures of position and variability. According to the coefficient of variation (CV) values, variability was classified as low (CV $\leq$ 12%), intermediate (12% < CV $\leq$ 60%), and high variability (CV > 60%), following Warrick and Nielsen [55]. Additionally, the data distribution was evaluated in relation to the normal distribution using the Kolmogorov–Smirnov test at a 5% significance level. The data were then used to estimate the average semivariogram of the study area. Trend analysis was performed by estimating the trend surface using a quadratic polynomial function and examining the coefficients of determination obtained. Trends were considered to exist for R^2 values greater than or equal to 0.7. In such cases, residual values were adopted for the analysis.

Geostatistical analysis was conducted using GS+ software. The sampling locations corresponded to the center point of the CHIRPS pixels and were supplemented with observed values from stations. Spatial dependence was evaluated using the classic semivariogram built from the estimate of classical semivariance, as given by Equation (1) [56].

$$\gamma(h) = \frac{1}{2N(h)} \sum_{i=1}^{N(h)} [Z(X_i) - Z(X_i+h)]^2 \quad (1)$$

where $\gamma(h)$ is the estimated value of the semivariance of the experimental data; $Z(X_i + h)$ and $Z(X_i)$ are the observed values of the regionalized variable; $N(h)$ denotes the number of pairs of measured values separated by distance h; and X_i and $X_i + h$ refer to two data points (locations where the variable Z is measured) that are a lag distance, h, apart [57].

According to Landim [57], the sums required to calculate $\gamma(h)$ must include a sufficient number of pairs to produce consistent results, with a minimum number of 30 pairs per distance class adopted [58]. After evaluating the experimental semivariogram, exponential, spherical, and Gaussian theoretical models were tested, and the parameters C_0 (nugget effect), $C_0 + C_1$ (sill), and A (range of spatial dependence) were estimated. The three theoretical models considered are presented below in Equations (2)–(4) [40]:

- Exponential Model:

$$\gamma(h) = C_0 + C_1\left\{1 - \exp\left[-\left(\frac{h}{A}\right)\right]\right\},\ h \neq 0 \quad (2)$$

- Spherical Model:

$$\begin{aligned} \gamma(h) &= C_0 + C_1,\ h \geq A \\ \gamma(h) &= C_0 + C_1\left[1.5\left(\tfrac{h}{A}\right) - 0.5\left(\tfrac{h}{A}\right)^3\right],\ 0 < h < A \end{aligned} \quad (3)$$

- Gaussian Model:

$$\gamma(h) = C_0 + C_1\left\{1 - \exp\left[-\left(\frac{h}{A}\right)^2\right]\right\},\ h \neq 0 \quad (4)$$

where C_0 is the nugget effect; $C = C_0 + C_1$ is the sill; A is the range; and h denotes the distance between points.

The degree of spatial dependence (DSD) was assessed based on the methodology proposed by Cambardella et al. [59], which uses the relationship between the nugget effect and the sill of the adjusted semivariogram, classifying the spatial dependence as strong, moderate, or weak, as observed in Equation (5). Values below 25% are characterized as strong spatial dependence, values between 25% and 75% are considered moderate, while values above 75% indicate weak spatial dependence.

$$DSD(\%) = \frac{C_0}{C_0 + C_1} \quad (5)$$

The best models were selected by analyzing the coefficient of determination (R^2) and performing cross-validation using the Jack-Knifing criterion [60], assuming a mean error close to zero and a standard deviation close to one.

2.4.1. Universal Kriging (UK)

After modeling the semivariograms, the values were interpolated at non-sampled sites using Universal Kriging (UK). This technique uses both correlations with auxiliary maps and spatial correlation, where the drift (or trend) is modeled as a function of coordinates [61], and then kriging is performed on the local average residuals [39].

2.4.2. Sequential Gaussian Simulation (SGS)

The sequential Gaussian simulation (SGS) was applied to different classifications of SPI. At each location, the SGS algorithm randomly selects a simulated value from an estimated conditional cumulative distribution function. The distribution function is determined by the mean and variance of the kriging calculated from neighborhood information. Running the SGS requires transforming the original data into a Gaussian distribution, which is accomplished by transforming the normal score. The cumulative density function is characterized by the mean and covariance [62], assuming a random Gaussian field. By repeating these sequential steps with different random paths, the SGS algorithm can provide various realizations of the spatial distribution of the simulated values. In this study, to obtain an accurate probability calculation, the SGS algorithm was run with ten, one hundred, and one thousand runs using GS+ software.

2.4.3. Ordinary Cokriging

After modeling the semivariograms, the values were interpolated at unsampled locations to estimate the primary variable using the ordinary cokriging (CKO) procedure, which is a modification of the simple kriging method. CKO offers the advantage of using more than one variable in the estimation process, thereby improving the estimation of the primary variable by assuming correlation between variables [44]. In this study, rainfall and elevation were considered the primary and secondary variables, respectively, in the CKO method. Thus, CKO aimed to establish a suitable model for cross-continuity and spatial dependence between the primary (rainfall) and secondary (elevation) variables. This correlation between variables is known as interregionalization [43], which can be quantified by cross semivariogram or cross-covariance, as seen in Equation (6).

$$\gamma(h) = \frac{1}{2N(h)} \sum_{i=1}^{N(h)} [Z_1(X_i) - Z_1(X_i+h)][Z_2(X_i) - Z_2(X_i+h)] \tag{6}$$

where $\gamma(h)$ is the estimated value of the cross semivariogram of the experimental data; $Z(X_i + h)$ and $Z(X_i)$ are the observed values of the regionalized variable; Z_1 and Z_2 are spatially correlated variables; N(h) is the number of pairs of measured values separated by a distance, h; and X_i and $X_i + h$ refer to two data points (locations where the variable Z is measured) that are a lag distance, h, apart [57].

The elevation was obtained through a Digital Elevation Model (DEM) using the Shuttle Radar Topography Mission (SRTM) product, which was acquired from the Google Earth Engine (GEE) with a spatial resolution of 30 m.

2.5. Assessment of Geostatistics Methods

In this study, the performance of all geostatistical methods for rainfall estimation is compared based on the values of mean bias error (MBE), root mean squared error (RMSE), and coefficient of determination (R^2) between observed and estimated rainfall values, according to Adhikary et al. and Hengl et al. [39,61].

The geostatistical method with the lowest MBE and RMSE values and the highest R^2 value is selected as the best geostatistical method.

2.6. Analysis of Biophysical Indexes and Land Use and Land Cover (LULC)

To characterize vegetation cover in the basin and highlight water features in a semiarid region for different years, the Normalized Difference Vegetation Index (NDVI) and Modified Normalized Difference Water Index (MNDWI) were calculated. Images from Landsat 5 TM and Landsat 8 OLI sensors were used, capturing different rainfall regimes (MD, N, MW, and EW) during selected years after the month of September, as detailed in Table 2. Images with less than 10% cloud coverage were processed on the Google Earth Engine (GEE) platform (https://developers.google.com/earth-engine/datasets/catalog/landsat, accessed on 9 May 2023), provided by the National Aeronautics and Space Administration (NASA). The images were then transferred to Google Earth Engine (GEE) where they were atmospherically corrected for surface reflectance. All images were processed using GEE, classified, and mapped using QGIS software version 3.22.

Table 2. Satellite images used in the study with hour and date.

Year	Satellite	Date	Hour (UTM)	Orbit	Point
ED (1993)		No Landsat images after September and no clouds			
SD (1983)		No Landsat images			
MD (2017)	Landsat 8	30 September 2017	12:47:34	217	65
		30 September 2017	12:47:58	217	66
		7 October 2017	12:53:47	218	65
N (2014)	Landsat 8	22 September 2014	12:47:26	217	65
		22 September 2014	12:47:50	217	66
		13 September 2014	12:53:40	218	65
MW (2004)	Landsat 5	13 November 2004	12:32:25	217	65
		13 November 2004	12:32:49	217	66
		4 November 2004	12:38:28	218	65
EW (1985)	Landsat 5	24 October 1985	12:16:14	217	65
		24 October 1985	12:16:38	217	66
		15 October 1985	12:22:31	218	65

Source: [63].

The Normalized Difference Vegetation Index (NDVI) is commonly used and disseminated in the literature. It is calculated as the ratio between the difference of the near-infrared and red bands' reflectance (ρ_{NIR} and ρ_R, respectively) and their sum, according to Equation (7) [19].

$$NDVI = \frac{\rho_{NIR} - \rho_R}{\rho_{NIR} + \rho_R} \tag{7}$$

In turn, the Modified Normalized Water Difference Index (MNDWI) proposed by Xu [64] aims to capture water bodies on the earth's surface and identify water courses. The MNDWI can be obtained by calculating the ratio between the difference in reflectance of the green (ρ_G) and mid-infrared (ρ_{MIR}) bands, divided by their sum (Equation (8)).

$$MNDWI = \frac{\rho_G - \rho_{MIR}}{\rho_G + \rho_{MIR}} \tag{8}$$

This index minimizes information about vegetation and accurately highlights water bodies such as rivers and reservoirs, where positive MNDWI values indicate pixels filled with water [64]. An estimate was made of the volume of water stored in the surface reservoirs of the basin using Equation (9), proposed by Molle [65], who studied more than 400 reservoirs in the Brazilian semiarid region and proposed a model that considers only the area of formed water bodies. The proposed model is highly accurate, with a coefficient of determination of 0.946.

$$V_R = 1.98\, S_x^{1.001} \tag{9}$$

where V_R is the volume of the reservoir (m^3), and S_x is the area of the water bodies of the reservoir (m^2), as adopted by Silva et al. [24].

The land use and land cover (LULC) maps were obtained using the MapBiomas Brazil platform (https://mapbiomas.org/, accessed on 9 May 2023) for the same years analyzed in the biophysical indices (Table 2) to characterize the study area and investigate the impacts of land use changes and climate variability. The extracted data were from the latest available collection, collection 7.1, with a spatial resolution of 30 m and a catalog with 27 classes, in which the following major classes were prioritized: Forest, Non-forest natural formation, Farming, Non-vegetated area, and Water [66]. Additionally, the area values (km^2) corresponding to each of the major classes studied were extracted to quantify the evolution of these areas over the different rainfall regimes.

2.7. Flowchart of the Data Analysis and Processing Steps

In order to represent the stages of statistical analysis, geostatistical modeling, development of kriging, sequential Gaussian simulation and cokriging maps, and analysis of biophysical indices, Figure 2 shows the flowchart of the processing steps.

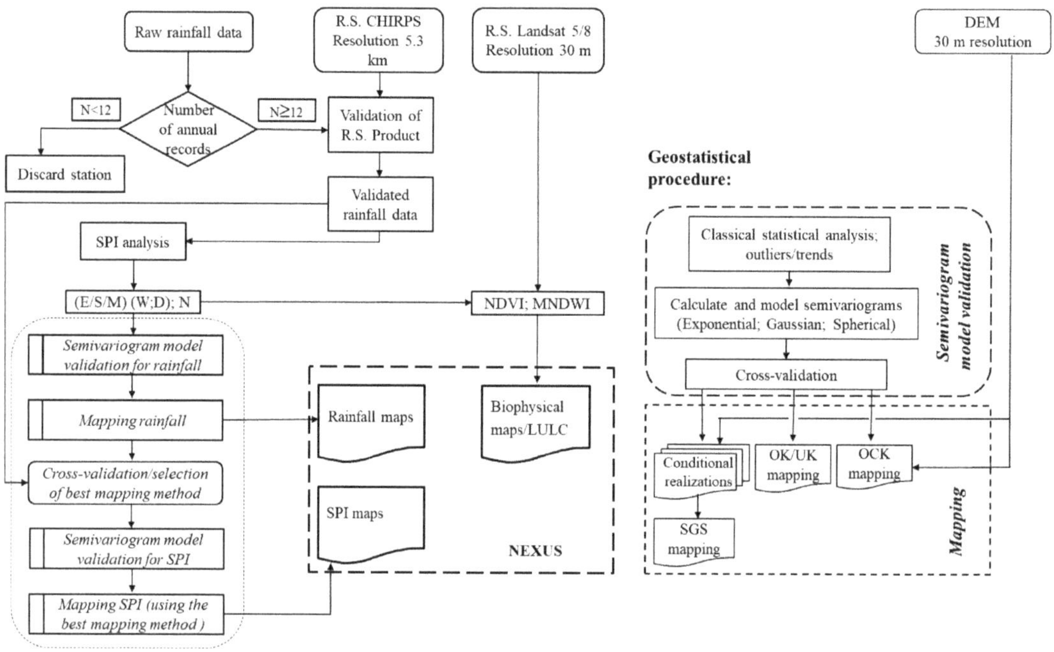

Figure 2. Flowchart of the data analysis, processing steps, and procedures.

3. Results and Discussion

3.1. Rainfall via CHIRPS and SPI

Figure 3 depicts the correlation and accumulated density curve of observed rainfall data and CHIRPS precipitation estimates. The percentage of trend (PBIAS) was 9.2%, indicating a "very good" performance (<10%). The coefficient of determination exhibited a high-quality fit (R^2 = 0.938). The Nash–Sutcliffe Efficiency Coefficient (NSE) was 0.63, classified as "satisfactory" (>0.50), and the Pearson correlation coefficient (r) was 0.82, further demonstrating the correlation of CHIRPS with weather stations. Additionally, the angular parameter of the linear regression model between observed and estimated rainfall data is very close to one, meaning no severe underestimation or overestimation when utilizing the estimated data to represent rainfall dynamics in the basin. These results align with those obtained by Paredes-Trejo et al. [8], where they validated satellite rainfall estimates based on CHIRPS for northeast Brazil via automatic meteorological stations and achieved a Pearson correlation coefficient (r) of 0.941, PBIAS = −3.58%, and NSE = 0.886.

Based on the observed rainfall data from stations before 1981 and after 1981, along with CHIRPS data on total annual rainfall, the Standard Precipitation Index (SPI-12) was calculated. The SPI-12 classifies years into Extremely Dry (ED), Severely Dry (SD), Moderately Dry (MD), Normal (N), Moderately Wet (MW), Severely Wet (SW), and Extremely Wet (EW) categories (Table 3). The majority of the years classified as N years; followed by MD years; MW years; ED years; SW years; and, with only one occurrence each, SD years (1983) and EW years (1985).

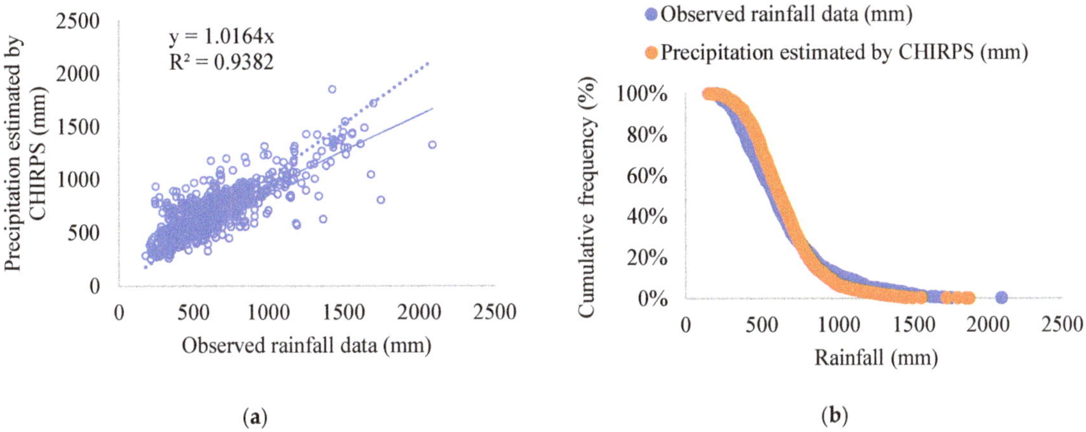

Figure 3. Correlation between observed annual rainfall at meteorological stations and simulated CHIRPS data (**a**) and cumulative density curve (**b**) for the period 1981–2017.

Table 3. SPI classification for the total annual rainfall measured at the station locations.

Year	SPI	Classification	Year	SPI	Classification	Year	SPI	Classification	Year	SPI	Classification
1962	0.47	N	1976	−0.33	N	1990	−1.27	MD	2004	1.32	MW
1963	0.55	N	1977	0.42	N	1991	−0.81	N	2005	0.29	N
1964	1.72	SW	1978	−0.04	N	1992	−0.78	N	2006	−0.32	N
1965	0.19	N	1979	0.70	N	1993	−2.21	ED	2007	−0.62	N
1966	0.11	N	1980	0.49	N	1994	0.01	N	2008	0.59	N
1967	1.09	MW	1981	−0.19	N	1995	0.29	N	2009	0.94	N
1968	0.29	N	1982	−1.01	MD	1996	0.68	N	2010	−0.15	N
1969	−0.13	N	1983	−1.98	SD	1997	0.54	N	2011	0.41	N
1970	−0.48	N	1984	0.70	N	1998	−1.26	MD	2012	−2.47	ED
1971	0.50	N	1985	3.30	EW	1999	0.05	N	2013	−0.78	N
1972	−0.65	N	1986	0.35	N	2000	0.02	N	2014	−0.27	N
1973	0.67	N	1987	−0.80	N	2001	−0.86	N	2015	−0.37	N
1974	1.84	SW	1988	1.31	MW	2002	−0.37	N	2016	−1.32	MD
1975	0.12	N	1989	1.25	MW	2003	−0.70	N	2017	−1.02	MD

Where ED = Extremely Dry, SD = Severely Dry, MD = Moderately Dry, N = Normal, MW = Moderately Wet, SW = Severely Wet, EW = Extremely Wet.

3.2. Statistical Analysis of Annual Rainfall

Table 4 presents the descriptive statistics of the annual rainfall data estimated by CHIRPS for the years 1993 (ED), 1983 (SD), 2017 (MD), 2014 (N), 2004 (MW), and 1985 (EW) in 930 pixels. The coefficient of variation (CV) for these years was intermediate ($12\% < CV \leq 60\%$), based on the SPI classification in Table 4, according to Warrick and Nielsen [55]. Additionally, Medauar et al. [36] evaluated the spatial behavior and temporal stability of the mean monthly rainfall in the Bahia State, Brazil, which is part of the NEB, from 1975 to 2011, and found intermediate CV for all analyzed months, which supports the findings of this study.

Figure 4 presents boxplots for annual rainfall estimated by CHIRPS for the years 1993 (ED), 1983 (SD), 2017 (MD), 2014 (N), 2004 (MW), and 1985 (EW). Rainy years (1985 and 2004) showed greater data dispersion, unlike dry years such as 1983, which presented lower dispersion, corroborating the findings of Medeiros et al. [37]. These authors analyzed monthly mean rainfall and observed that months with the lowest rainfall, from June to November, in the Sertão mesoregion in Paraíba State (also inserted in NEB), had the lowest variability. It is also worth noting the overestimation of estimated values in 2014 (N) and

2017 (MD) by the CHIRPS product. According to Rodrigues et al. [8], the CHIRPS product showed a moderate overestimation of monthly rainfall ranging between 0 and 100 mm while tending to underestimate values above 100 mm in the NEB.

Table 4. Descriptive statistics of annual rainfall estimated by CHIRPS for the Brígida basin.

Year	Mean	Median	Maximum	Minimum	n	SD	CV (%)	*p*-Value (KS)
Extremely Dry (1993)	329.48	280.82	843.10	138.21	930	137.17	41.63%	0.14
Severely Dry (1983)	361.05	359.68	658.09	175.30	930	76.19	21.10%	0.05
Moderately Dry (2017)	475.55	445.60	1001.36	212.67	930	145.69	30.64%	0.10
Normal (2014)	641.66	543.09	1444.71	333.15	930	248.55	38.74%	0.16
Moderately Wet (2004)	854.19	824.72	1566.14	355.03	930	213.5	24.99%	0.09
Extremely Wet (1985)	1292.89	1237.86	2103.26	692.42	930	243.65	18.88%	0.09

Where SD = standard deviation, CV = coefficient of variation, n = number of sampling points.

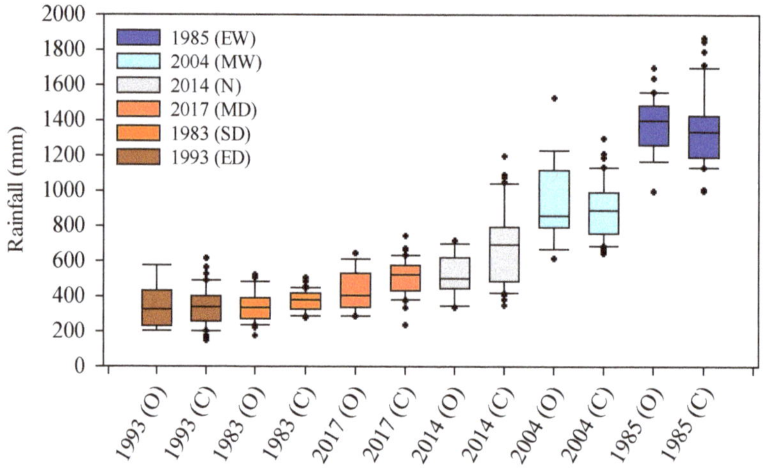

Figure 4. Box plots of CHIRPS annual rainfall data (C) and observed rainfall data (O) for Extremely Dry (1993), Severely Dry (1983), Moderately Dry (2017), Normal (2014), Moderately Wet (2004), and Extremely Wet (1985) years.

3.3. *Universal Kriging*

Trend occurrence was detected using linear and quadratic polynomials, and for such cases, geostatistical analysis was carried out based on the residuals. Well-defined experimental semivariograms with high spatial dependence were produced based on the autocorrelation between the residual values for the years 1993, 1983, 2017, 2014, 2004, and 1985, as shown in Figure 5. From 1981 onwards, which is the initial date of availability of CHIRPS data, there was no SW year based on CHIRPS data (1981–2017). The semivariograms parameters and validation results are presented in Table 5. Figure 5 displays a set of modeled semivariograms for the six classes of SPI under study. The semivariances increase with the increase in the rainfall index, and therefore, the sill ($C_0 + C_1$) is directly related to the rainfall amount for the Brígida River basin, as also shown in Table 5.

The semivariogram models that best fit the data were the spherical (for ED and SD), followed by the Gaussian (for N, MW, and EW) and exponential (for MD) models. The relationship between the nugget effect and the sill ranged from 0.04% (MD) to 10.88% (EW), indicating a strong spatial dependence for all analyzed years, according to the classification of Cambardella et al. [59]. For monthly mean rainfall, [36] also found strong spatial dependence for most months.

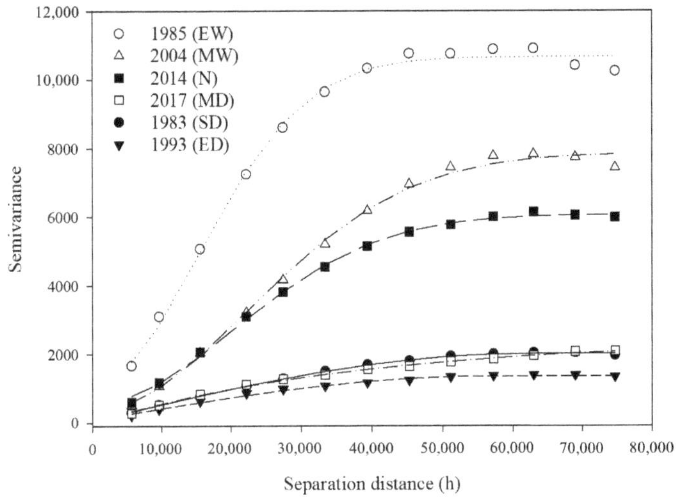

Figure 5. Semivariogram models for CHIRPS residual annual rainfall data for Extremely Dry (1993), Severely Dry (1983), Moderately Dry (2017), Normal (2014), Moderately Wet (2004), and Extremely Wet (1985) years. The separation distance is given in meters (m).

Table 5. Semivariogram parameters and models for annual rainfall residual data; statistics of Jack-Knifing procedure.

SPI Class	Model	Nugget (C_0)	Sill ($C_0 + C_1$)	Range (A, km)	R^2	DSD (%)	Jack-Knifing	
							M	SD
Extremely Dry (1993)	Spherical	80	1348	55.9	0.992	5.93%	0.01	1.04
Severely Dry (1983)	Spherical	59	1998	61.0	0.999	2.95%	0.01	1.03
Moderately Dry (2017)	Exponential	1	2405	114.0	0.998	0.04%	0.01	1.02
Normal (2014)	Gaussian	580	6059	50.2	0.998	9.57%	0.01	1.06
Moderately Wet (2004)	Gaussian	370	7852	55.4	0.996	4.71%	0.01	1.04
Extremely Wet (1985)	Gaussian	1160	10,660	37.6	0.996	10.88%	0.01	1.04

Where DSD = degree of spatial dependence, M = mean, SD = standard deviation.

Annual rainfall in the study area exhibited a spatial dependence structure, with ranges varying from 37.6 km (for EW) to 114.0 km (for MD), indicating a comprehensive structure of spatial dependence. Furthermore, Araújo et al. [27] also fitted semivariogram models for annual observed rainfall in the Brígida River basin, obtaining ranges from 20 to 45 km and a degree of spatial dependence ranging from moderate to strong. The coefficients of determination (R^2) for the fitted models were moderate to high, with a minimum of 0.992 (for ED) and a maximum of 0.999 (for SD). Moreover, the semivariogram models were validated using the Jack-Knifing method. However, experimental semivariances presented gaps and oscillations due to uneven spatial distribution of rain gauges.

3.4. Cross-Semivariogram Models of Annual Rainfall

Well-constructed experimental cross-semivariograms were generated, exhibiting high spatial dependence based on the correlation between annual rainfall and elevation data for the years 1993, 1983, 2017, 2014, 2004, and 1985 (Figure 6). Parameters and validation results of the cross-semivariograms are shown in Table 6. The sill is generally related to the amount of rainfall, except for the year 1983 (SD), which can be explained by the lower value of the parameter C_0 (nugget effect).

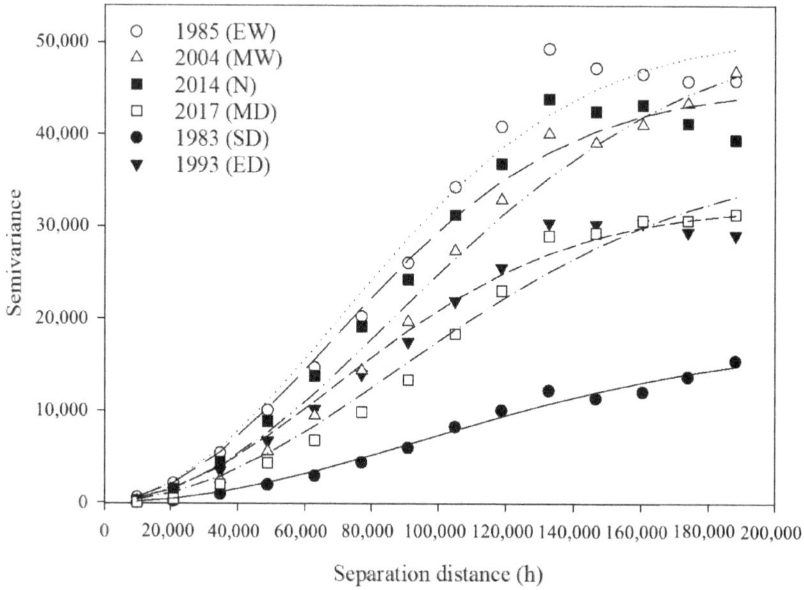

Figure 6. Cross-semivariogram models for CHIRPS annual rainfall and elevation data for Extremely Dry (1993), Severely Dry (1983), Moderately Dry (2017), Normal (2014), Moderately Wet (2004), and Extremely Wet (1985) years. Separation distance is in meters (m).

Table 6. Cross-semivariogram parameters and models for annual rainfall and elevation; statistics of Jack-Knifing procedure.

SPI Class	Model	Nugget (C_0)	Sill ($C_0 + C_1$)	Range (A, km)	R^2	DSD (%)	Jack-Knifing	
							M	SD
Extremely Dry (1993)	Gaussian	100	32,040	168.5	0.987	0.31%	0.00	1.00
Severely Dry (1983)	Gaussian	10	17,200	232.3	0.986	0.06%	0.01	1.01
Moderately Dry (2017)	Gaussian	100	37,160	217.0	0.985	0.27%	0.00	1.00
Normal (2014)	Gaussian	100	44,920	167.5	0.979	0.22%	0.00	1.00
Moderately Wet (2004)	Gaussian	100	51,300	212.9	0.990	0.19%	0.00	1.00
Extremely Wet (1985)	Gaussian	100	50,690	171.9	0.979	0.20%	0.00	1.00

Where DSD = degree of spatial dependence, M = mean, SD = standard deviation.

3.5. Spatiotemporal Analysis (Rainfall UK, SGS and Cokriging)

The adjusted models were used to produce CHIRPS annual rainfall isoline maps for the years 1993 (ED), 1983 (SD), 2017 (MD), 2014 (N), 2004 (MW), and 1985 (EW) using universal kriging (UK), sequential Gaussian simulation (SGS), and ordinary cokriging (OCK) techniques (Figure 7). The fitted models were also used to produce maps of annual rainfall with different runs (10, 100, and 1000) using the sequential Gaussian simulation (SGS) technique.

The northern region of the basin, which presents the highest elevation, encompasses the "Chapada do Araripe" highland, an area with considerable spatial variability of rainfall and a higher vegetation density. As expected, our study found that the amount of rainfall in the basin increased with elevation. Supporting the findings of this study, Mutiga et al. [67] calculated the water balance in the upper part of the northern Ewaso Ng'iro basin in Kenya, with altitudes ranging from 800 to 5200 m, and observed that with increasing elevation, the Normalized Difference Vegetation Index (NDVI) tended to increase due to greater rainfall occurrence, and actual evapotranspiration decreased. These results reinforce the well-known relationship between rainfall and elevation in hydrological systems.

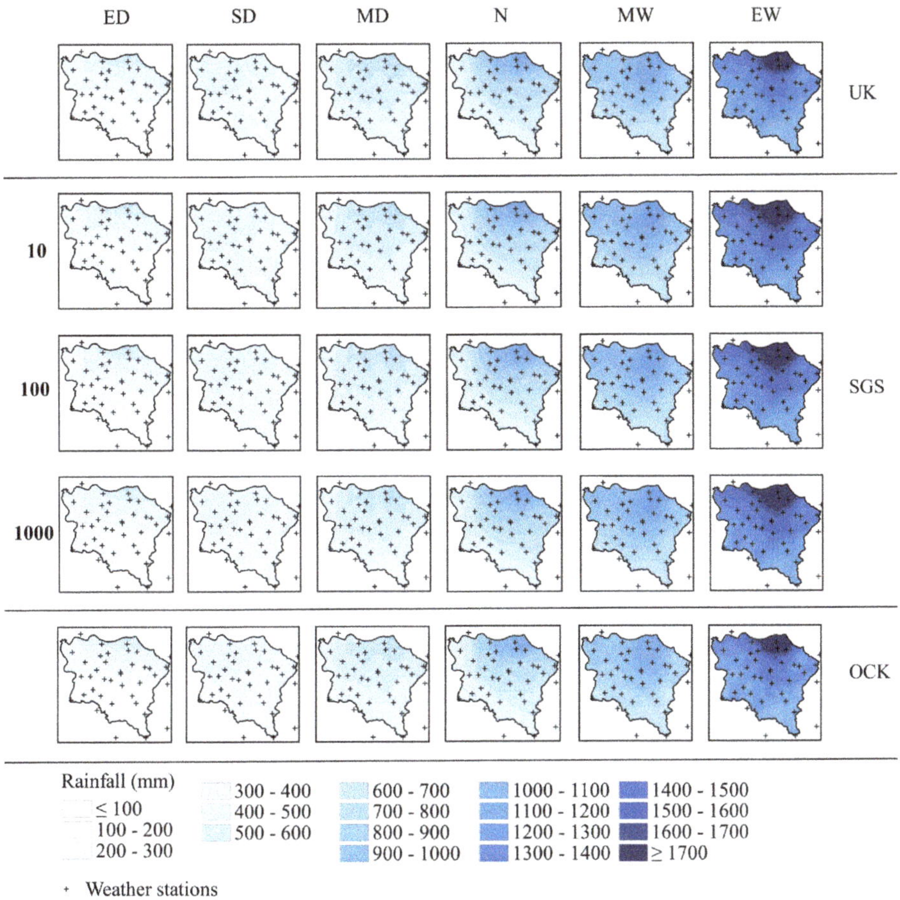

Figure 7. Universal kriging (UK), sequential Gaussian simulation (SGS), and ordinary cokriging (OCK) maps for CHIRPS annual rainfall data for Extremely Dry (1993), Severely Dry (1983), Moderately Dry (2017), Normal (2014), Moderately Wet (2004), and Extremely Wet (1985) years, indicating the location of weather stations used in the study.

The southern region of the Brígida River basin exhibited the lowest annual rainfall values for all years analyzed. This region is part of the Cabrobó desertification area located in the southern of the Pernambuco State and is characterized by a lack of vegetation cover and high degrees of soil erosion. Among the primary causes of the desertification process, low precipitation and deforestation should be noted [17]. The municipality of Exu (located at an altitude above 972 m, as shown in Figure 1b) consistently had the highest rainfall amounts in the basin for all maps.

The lowest rainfall occurred, in general, throughout the basin in the ED and SD years, while in the MD and N years, critical areas were located in the south and west. In the MW and EW years, the critical areas were in the southern part of the basin. The maps for the MD, N, MW, and EW years show the area of the Brígida basin that corresponds to the Köppen–Geiger climate classification of Aw (tropical climate with dry winter), as shown in Figure 1b.

Increasing the number of runs in the SGS maps often results in stability, with low or no improvement, and a convergence criterion can be achieved with 100 runs, reducing the computational effort, as observed in Bai and Tahmasebi [68] and Pereira et al. [69]. In addition, there was lower statistical error comparing the map of SGS-1000 with SGS-100. Furthermore, the SGS maps produced consistent results with low uncertainties and showed significant similarities to the maps of universal kriging (UK).

To evaluate the behavior of precipitation from the mouth to a high-altitude region in the basin, a transect analysis was performed on the maps of CHIRPS annual precipitation data produced using universal kriging (UK), sequential Gaussian simulation (SGS), and ordinary cokriging (OCK), as shown in Figure 8.

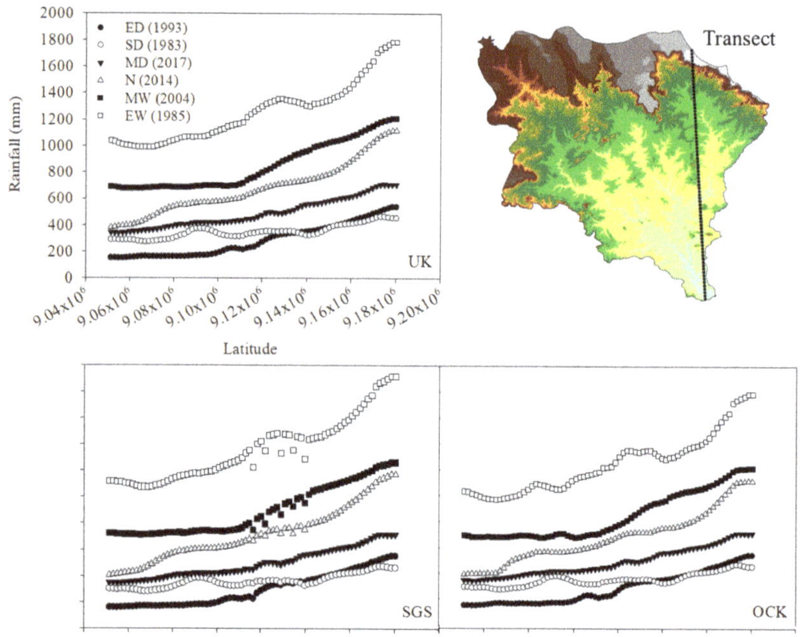

Figure 8. Transect analysis of CHIRPS annual precipitation data using universal kriging (UK), sequential Gaussian simulation (SGS), and ordinary cokriging (OCK) maps for Extremely Dry (1993), Severely Dry (1983), Moderately Dry (2017), Normal (2014), Moderately Wet (2004), and Extremely Wet (1985) years.

3.6. *Performance of Different Interpolation (UK, SGS, OCK) Methods for Annual Rainfall Estimation in the Study Area*

The performance of the different interpolations (UK, SGS, OCK; Table 7) shows that SGS has the best performance, with a coefficient of determination (R^2) of 100% and MBE and RMSE values close to zero. There is a slight advantage of OCK over UK with lower values of RMSE. This result is consistent with Adhikary et al. [39], who evaluated three geostatistical interpolation methods (ordinary kriging [OK], ordinary cokriging [OCK], external drift kriging [KED]) for mapping monthly rainfall from two basins in Australia and found that OCK interpolator yielded the best performance. In addition, maps of the standard deviations of universal kriging (UK), sequential Gaussian simulation (SGS), and ordinary cokriging (OCK) are presented (Figure 9). It can be verified that uncertainties associated with the mapping are smaller when using the SGS, and the errors observed are more pronounced for the wetter years.

Table 7. Performance of different interpolation methods (UK, SGS, OCK) for annual rainfall estimation in the study area.

SPI Class	MBE (mm)			RMSE (mm)			R^2		
	UK	SGS	OCK	UK	SGS	OCK	UK	SGS	OCK
Extremely Dry (1993)	−0.1415	−0.0003	−0.1615	55.877	0.0029	30.632	0.9972	1.0000	0.9992
Severely Dry (1983)	0.0010	−0.0001	0.0205	57.187	0.0028	36.781	0.9895	1.0000	0.9955
Moderately Dry (2017)	0.6475	0.0000	0.6481	35.520	0.0029	35.248	0.9984	1.0000	0.9984
Normal (2014)	0.1050	−0.0001	0.3873	18.392	0.0029	54.153	0.9910	1.0000	0.9992
Moderately Wet (2004)	0.3519	0.0000	−0.0886	160.142	0.0029	47.650	0.9896	1.0000	0.9991
Extremely Wet (1985)	0.4379	−0.0003	0.7608	309.794	0.0028	87.278	0.9770	1.0000	0.9982

Where MBE = mean bias error, RMSE = root mean squared error, and R^2 = coefficient of determination.

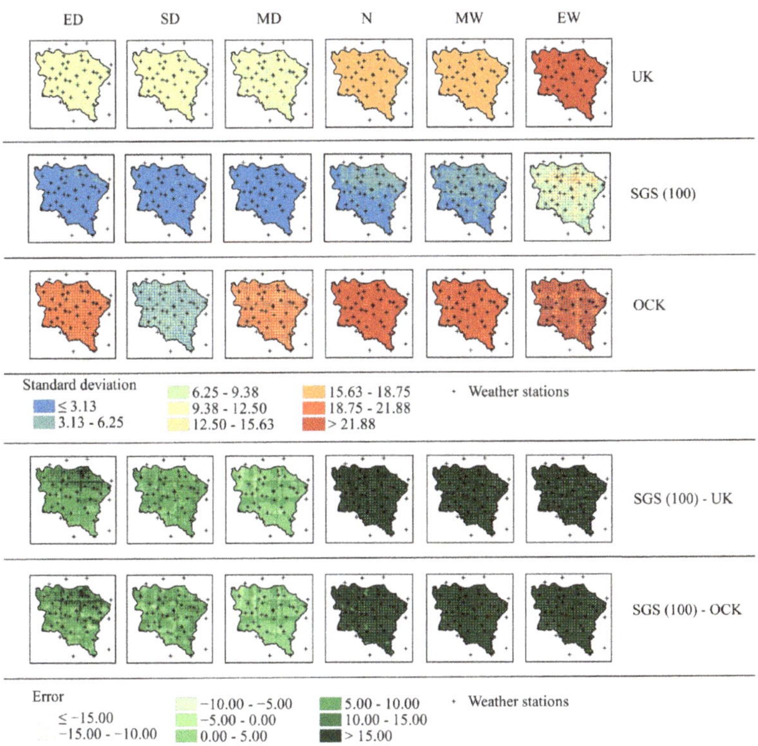

Figure 9. Standard deviation maps for universal kriging (UK), sequential Gaussian simulation (SGS), and ordinary cokriging (OCK) of CHIRPS annual rainfall data. In addition, errors (SGS-UK) and (SGS-OCK) for Extremely Dry (1993), Severely Dry (1983), Moderately Dry (2017), Normal (2014), Moderately Wet (2004) and Extremely Wet (1985) years.

3.7. SPI Semivariogram Models

Rainfall data for each CHIRPS pixel was transformed into SPI data to investigate the spatial and temporal variation in historical rainfall patterns and drought severity. Since there is a spatial trend in the data, semivariograms were constructed with the residual values obtained for the years 1993, 1983, 2017, 2014, 2004, and 1985 (Figure 10), along with the adjustment parameters presented in Table 8.

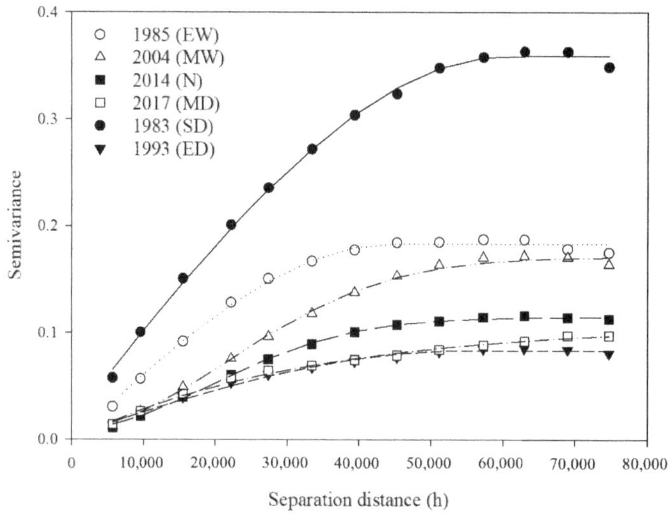

Figure 10. Semivariogram models fitted for residual SPI data during Extremely Dry (1993), Severely Dry (1983), Moderately Dry (2017), Normal (2014), Moderately Wet (2004), and Extremely Wet (1985) years. Separation distance is in meters (m).

Table 8. Semivariogram adjustment parameters of SPI residual data; Jack-Knifing statistics.

SPI Class	Model	Nugget (C_0)	Sill ($C_0 + C_1$)	Range (A, km)	R^2	DSD (%)	Jack-Knifing	
							M	SD
Extremely Dry (1993)	Spherical	0.0044	0.0833	54.2	0.992	5.28%	0.01	1.04
Severely Dry (1983)	Spherical	0.0180	0.3590	60.4	0.998	5.01%	0.01	1.04
Moderately Dry (2017)	Exponential	0.0001	0.1078	97.2	0.995	0.09%	0.01	1.03
Normal (2014)	Gaussian	0.0099	0.1148	47.8	0.999	8.62%	0.01	1.06
Moderately Wet (2004)	Gaussian	0.0110	0.1710	53.7	0.997	6.43%	0.01	1.05
Extremely Wet (1985)	Spherical	0.0001	0.1832	44.4	0.995	0.05%	0.01	1.03

Where DSD = degree of spatial dependence, M = mean, SD = standard deviation.

The SPI showed a spatial dependence structure with ranges from 44.4 km (EW year) to 97.2 km (MD year), similar to the previous analysis with rainfall data. The semivariogram models that best fit the data were spherical (ED, SD, and EW years), Gaussian (N and MW), and exponential (MD). The relationship between the nugget effect and the sill of semivariograms ranged from 0.05% (EW) to 8.62% (N), indicating a strong spatial dependence for all analyzed years, according to the classification by Cambardella et al. [59]. The coefficients of determination (R^2) for the model fitting were high, with a minimum of 0.992 (ED) and a maximum of 0.999 (N), similar to the results obtained previously with CHIRPS rainfall data. Furthermore, the semivariogram models were validated using the Jack-Knifing method.

3.8. SPI SGS

The fitted semivariogram models were used with the sequential Gaussian simulation (SGS) technique to produce maps of the SPI classifications with 100 runs (Figure 11). The municipalities of Ouricuri (ID-11), Parnamirim (ID-29), and Orocó (ID-34), as well as the northern regions of the municipalities of Santa Maria da Boa Vista (ID-35), Santa Cruz (ID-36), and Santa Filomena (ID-12), were identified as experiencing the most severe water deficits for all analyzed years, as classified by the SPI.

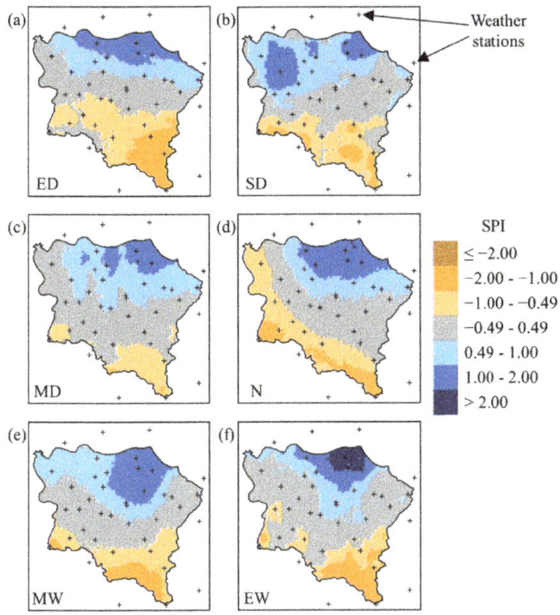

Figure 11. Sequential Gaussian simulation (SGS) maps for the SPI with 100 runs for Extremely Dry (1993) (**a**), Severely Dry (1983) (**b**), Moderately Dry (2017) (**c**), Normal (2014) (**d**), Moderately Wet (2004) (**e**), and Extremely Wet (1985) (**f**) years, indicating the location of weather stations used in the study.

Even during Extremely Dry, Severely Dry, and Moderately Dry years, high (positive) SPI values are observed in the north of the hydrographic basin, which is a higher altitude area with an Aw (tropical climate with dry winter). This is consistent with a study conducted by Santos et al. [21], which estimated actual evapotranspiration (ETa) in the Brígida River basin and compared the dry year of 1990 and the wet year of 2009. The study found that areas with higher altitudes in the northern part of the basin had greater vegetation cover; lower surface temperature and albedo values; and higher NDVI (Normalized Difference Vegetation Index), SAVI (Soil-adjusted Vegetation Index), LAI (Leaf Area Index), and ETa (Actual evapotranspiration) values for both years analyzed. Therefore, a well-defined spatial pattern has been identified for these variables in the basin.

The anthropization of vegetated areas in the Caatinga biome is an extremely important component causing the reduction of precipitation, as reported by Silva et al. [70]. The authors found a reduction in precipitation for a deforested area compared to a region with native vegetation in the non-deforested Caatinga in the Pajeú watershed, Brazil. According to Panday et al. [71], the simultaneous occurrence of deforestation associated with climate variability has promoted significant impacts on the water balance in the Xingu watershed, Brazil, where abstraction of native vegetation significantly reduced rainfall in the basin. In the Brígida Basin, it was found that 57.32% of the area is anthropized and that the spatial distribution of rainfall is strongly influenced by the relief [20].

3.9. Analysis of Biophysical Indexes and Land Use and Land Cover (LULC)

Figure 12 shows the maps of biophysical indexes (Figure 12a,b) and major spatial classes of LULC (Figure 12c). In the maps for the MW years, it is possible to identify two reservoirs located in the central-west and central-south regions of the basin, which are the Entremontes Reservoir (capacity of 339.33 hm^3, built in 1982) and Chapéu Reservoir (capacity of 188 hm^3, built in 1936). These reservoirs are essential for domestic water supply, agricultural activities, and animal watering, according to the Brazilian National Water Authority [72,73]. Furthermore, the estimated area and stored volume of the reservoirs is consistent with the rainfall regime of a given year, assessed through the SPI-12 index (Table 9).

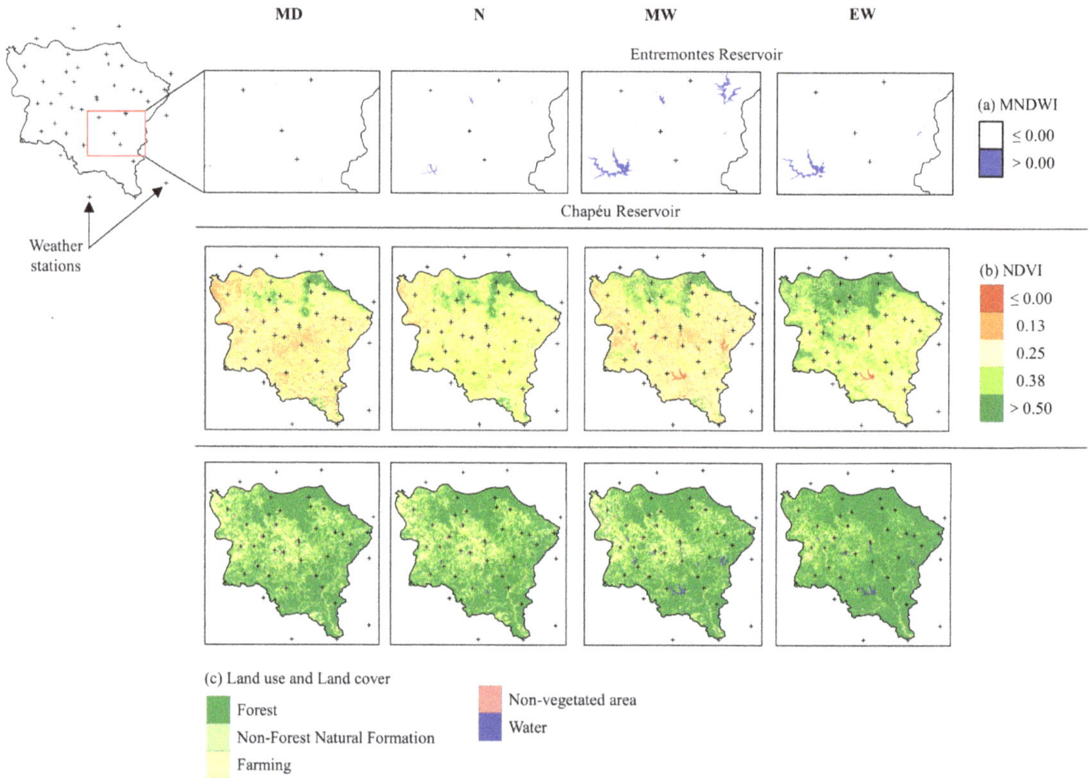

Figure 12. Modified Normalized Difference Water Index (MNDWI) maps (**a**), Normalized Difference Vegetation Index (NDVI) maps (**b**), and annual spatiotemporal dynamics of land use and land cover (LULC) maps (**c**) for Moderately Dry (2017), Normal (2014), Moderately Wet (2004), and Extremely Wet (1985) years, indicating the location of the weather stations used in the study.

Table 9. Water mirror area and volume of water stored in the water reservoirs of the Brígida River basin, semiarid region of Pernambuco, Brazil.

Year	Total Area of Entremontes Reservoir ($m^2 \times 10^5$)	Total Area of Chapéu Reservoir ($m^2 \times 10^5$)	Total Volume of Water Stored in Entremontes Reservoir ($m^3 \times 10^5$)	Total Volume of Water Stored in Chapéu Reservoir ($m^3 \times 10^5$)
Moderately Dry (2017)	3.35	0	6.71	0
Normal (2014)	38.0	8.28	76.4	16.6
Moderately Wet (2004)	327	191	659	386
Extremely Wet (1985)	259	1.47	522	2.94

There are no images for 1983 and 1993.

There is an increase in the area of water bodies and stored volume for the wetter rainfall regimes (from MD to EW). However, for the year 1985 (EW), there is a smaller area and volume than in 2004 (MW). This can be explained by the greater rainfall that occurred during the initial months of the wet season in 1985 (January and February), while in 2004, the highest rainfall occurred in the last month of the wet season (April). Additionally, this behavior is also associated with an increase in the use of stored water for irrigation. For a nearby basin, Silva et al. [24] estimated the total volume of water stored in small reservoirs to be 8.8×10^5 m^3, also using the MNDWI index. As expected, SPI-12 Index reflects long-term precipitation patterns and is linked to stored volumes in reservoirs due to the associated longer time scale.

The number of water bodies (blue pixels of the MNDWI) in the maps of the ED, MD, and N years is low (Figure 12a), as reservoir storage is strongly dependent of rainfall regime and water use allocation for different multiple uses. It should be highlighted that the hydrological river flow regime is typically temporary, and hence in periods of droughts, the river flow is interrupted until the following rainy season [74]. According to a study by Marengo et al. [28], which evaluated the climate patterns of drought in the NEB from 2010 to 2016, there was a sequence of dry years that affected the levels of reservoirs in the region and led to a water crisis. In addition, Silva et al. [23] evaluated the spatiotemporal dynamics of water bodies in the micro-region of Vale do Ipojuca (included in the NEB) from 2014 to 2021 and found a marked decrease in the area of water bodies between 2014 (1226 ha, the year of greatest water availability) and 2017 (683 ha), mainly in 2016 (558 ha). According to the authors, this dynamic is associated with the occurrence of severe droughts in the NEB, which caused significant reductions in water bodies volumes that lasted until mid-2019.

In relation to the total basin area (13,495.73 km^2), 9.0%, 13.2%, 11.8%, and 31.2% were covered by arboreal Caatinga vegetation, considering a NDVI (Figure 12b) greater than or equal to 0.375, for the years MD (2017), N (2014), MW (2004), and EW (1985), respectively. These results support the findings of Marengo et al. [28], who evaluated the climatological aspects of the droughts that occurred in the semiarid region of northeast Brazil, and also concluded that the conditions of the La Niña event in 2013 were not enough to attenuate the drought conditions established years before, and the El Niño in 2015 worsened drought conditions in later years.

Similar results of NDVI were also obtained in studies by Brito et al. [13], who quantitatively evaluated drought events that occurred in the semiarid region of the NEB between 1981 and 2016 and found that droughts from 2011 to 2016 were more frequent, severe, and affected a larger area with significant impacts on the population and economic activities. Additionally, these results also support Cunha et al. [34], who reported that unlike other drought events that occurred in the NEB, the 2012–2013 drought persisted for 6 years (until 2017). One possible reason for the smaller areas corresponding to agriculture and livestock in the 2004 (MW) maps compared to the 2014 (N) maps may be due to the abandonment of agricultural areas from 2000 to 2009 in the semiarid region of Pernambuco State, influencing an expansion of exposed soil areas with low and stunted vegetation, as found by [22] while studying the changes caused by agricultural activities in the municipality of Capoeiras, also in the semiarid region of the Pernambuco State. A reduction in vegetation cover was also observed when comparing the years 1990 and 2009, according to a study by Santos et al. [21]. Furthermore, Silva et al. [23] evaluated the spatiotemporal dynamics of vegetation cover in the micro-region of Vale do Ipojuca (included in the NEB) from 2014 to 2021 and observed, based on the NDVI maps, that the year 2014 had denser vegetation and the resilience of the vegetation is observed in native vegetation (Caatinga) compared to 2017.

To the north of the basin, where Chapada do Araripe is located at higher elevations, there is a region with higher spatial variability of rainfall, as evidenced by the UK rainfall maps. This region also has greater vegetation density, as verified by the NDVI maps, which in turn influences the distribution of rainfall. It is noteworthy to compare Figure 11a–c (for SPI-12) to Figure 12b (MD year) in order to verify the rainfall–vegetation nexus in the basin, particularly in the higher areas. Even during dry years, the Caatinga canopy of arboreal species is present, highlighting the need to preserve and regulate land use in those areas. Undoubtedly, higher zones provide ecological services for the region, even during dry periods.

In their studies in the Brígida River basin, Galvíncio et al. [20] also found that the spatial distribution of rainfall is influenced by elevation. NDVI values between 0 and 0.250 are assumed to correspond to areas more susceptible to degradation, as they present areas with exposed soil and herbaceous-graminoid vegetation due to the low resilience of native vegetation in drier periods. Similar results were observed by Tomasella et al. [75], who monitored degraded areas in Northeast Brazil from 2000 to 2016 and found that NDVI values ranging from 0 to 0.30 were mostly associated with exposed soil and herbaceous-graminoid vegetation, as well as a very low density of subshrubs. Similar results were found by Silva et al. [22] when identifying this type of vegetation and exposed soil at NDVI values between 0.001 and 0.264. According to the authors, NDVI values lower than 0.264 were associated with areas occupied by agriculture or livestock.

In terms of the total area of the basin (13,495.73 km^2), the Forest class (Caatinga vegetation) accounts for 66.6%, 66.6%, 69.6%, and 82.5% of it, according to LULC (Figure 12c), for the MD (2017), N (2014), MW (2004), and EW (1985) years, respectively. Meanwhile, the Farming class comprises 31.6%, 31.7%, 27.9%, and 15.7% of the basin for the same years. Indeed, there exists an inverse relationship between the Forest and Farming classes, where one class increases while the other decreases, mainly between the MD and EW years. For instance, the Forest class (Caatinga biome) had an area of 9155 and 11,335 km^2 for the MD and EW years, respectively, while the agricultural activity area accounted for 4339 and 2159 km^2 for the MD and EW years, respectively. During the transition period, there was a net increase of 2180 km^2 in the Forest class and a reduction of 2181 km^2 in the Farming class. This was due to the deforestation of Caatinga forest areas that were cleared for agricultural activities. These findings align with previous studies, where Silva et al. [22,23] evaluated degradation processes in the semiarid region of northeastern Brazil (NEB) using vegetation indices and LULC classes, highlighting the strong influence of agricultural activity on the Caatinga biome and vice versa.

The results of this study demonstrate the effectiveness of using spatial interpolation techniques and geostatistical analysis of rainfall data and provide valuable information for the management of public policies related to river basins at both national and international levels. These findings can aid in the characterization of river basins and inform the development of management and water management techniques.

4. Conclusions

We investigated rainfall variability patterns in a Brazilian semiarid basin and properly validated the CHIRPS rainfall product, which performed well in capturing rainfall and drought patterns for contrasting rainfall regimes. Among the semivariograms established and validated, the Gaussian model produced the best fit for most of the datasets that were used. Higher areas in the basin systematically present positive values for SPI-12, even in dry years, and exhibited high NDVI, which are directly related to canopy cover, ecological services, and streamflow regimes at the longer time scales. By using the sequential Gaussian simulation (SGS) technique, we identified low uncertainties in the CHIRPS data due to its proper reproduction of the spatial variability of annual rainfall. We suggest the adoption of 100 runs as a stopping criterion to reduce computational effort for simulations.

The use of sequential Gaussian simulation (SGS) for mapping both the annual precipitation and the Standard Precipitation Index (SPI-12) proved to be an efficient procedure, being able to provide essential data for hydrological analysis in semiarid regions. The developed framework successfully identified the vegetation cover dynamics and water features in the basin for different rainfall regimes. The Normalized Difference Vegetation Index (NDVI) properly detected the environmental degradation occurred during the 2012–2013 severe drought. Additionally, it is worth noting that the maps of generated SPI classifications provided valuable information on the spatial distribution of severe rainfall deficits in the basin. Through a spatiotemporal analysis of the indices, it was possible to identify the most vulnerable regions to intense droughts, thus emphasizing the need for effective management measures to be implemented for monitoring and mitigating their effects.

Moreover, the use of the Modified Normalized Difference Water Index (MNDWI) provided additional insights into water body dynamics in the basin and also indicated that, even in normal years, the number of water bodies is low, threatening water security in the region. Additionally, our study also analyzed the land use and land cover (LULC) in the basin, which revealed a significant change in the landscape over the years, with the expansion of cropland and livestock (Farming class) areas. The conversion of natural vegetation to Farming may have an impact on the water cycle and ecosystem services in the region, affecting both the water availability and quality. Effective management measures need to be implemented to maintain the water bodies and their associated ecological services. In conclusion, our study provides important insights into rainfall variability and hydrological processes in semiarid regions and presents a valuable framework for future studies in similar environments.

Author Contributions: Conceptualization, A.A.d.A.M., M.V.d.S. and L.d.B.d.S.; methodology, A.A.d.A.M., L.d.B.d.S., M.V.d.S., T.A.B.A. and A.A.d.C.; software, L.d.B.d.S., M.V.d.S., T.A.B.A. and A.A.d.C.; validation, L.d.B.d.S.; investigation, L.d.B.d.S. and M.V.d.S.; resources, L.d.B.d.S. and M.V.d.S.; data curation, L.d.B.d.S.; writing—original draft preparation, L.d.B.d.S.; writing—review and editing, A.A.d.A.M., J.L.M.P.d.L., M.V.d.S. and L.d.B.d.S.; visualization, A.A.d.A.M., T.G.F.d.S. and J.L.M.P.d.L.; supervision, A.A.d.A.M.; project administration, A.A.d.A.M. and L.d.B.d.S.; funding acquisition, A.A.d.A.M. and J.L.M.P.d.L. All authors have read and agreed to the published version of the manuscript.

Funding: The Coordination for the Improvement of Higher Education Personnel (CAPES), the Foundation for the Support of Science and Technology of the State of Pernambuco (FACEPE—APQ-0300-5.03/17 and IBPG-0855-5.03/20), the National Council for Scientific and Technological Development (CNPq—308890/2018-3; 140281/2022-3), and the CAPES-PrInt/UFRPE financed and provided scholarships. This study also had the support of Portuguese funds through the Foundation for Science and Technology, I. P (FCT), under the projects UIDB/04292/2020 and UIDP/04292/2020 granted to MARE, and LA/P/0069/2020 granted to the Associate Laboratory ARNET.

Data Availability Statement: The data presented in this study are available upon request from the corresponding author.

Acknowledgments: We thank the Postgraduate Program in Agricultural Engineering (PGEA) of the Federal Rural University of Pernambuco (UFRPE) for supporting the development of this research.

Conflicts of Interest: The authors declare no conflict of interest.

References

1. Cristiano, E.; ten Veldhuis, M.C.; van de Giesen, N. Spatial and temporal variability of rainfall and their effects on hydrological response in urban areas—A review. *Hydrol. Earth Syst. Sci.* **2017**, *21*, 3859–3878. [CrossRef]
2. Peña-Angulo, D.; Nadal-Romero, E.; González-Hidalgo, J.C.; Albaladejo, J.; Andreu, V.; Bagarello, V.; Barhi, H.; Batalla, R.J.; Bernal, S.; Bienes, R.; et al. Spatial variability of the relationships of runoff and sediment yield with weather types throughout the Mediterranean basin. *J. Hydrol.* **2019**, *571*, 390–405. [CrossRef]
3. Kim, J.; Lee, J.; Kim, D.; Kang, B. The role of rainfall spatial variability in estimating areal reduction factors. *J. Hydrol.* **2019**, *568*, 416–426. [CrossRef]
4. Peña-Angulo, D.; Nadal-Romero, E.; González-Hidalgo, J.C.; Albaladejo, J.; Andreu, V.; Barhi, H.; Bernal, S.; Biddoccu, M.; Bienes, R.; Campo, J.; et al. Relationship of weather types on the seasonal and spatial variability of rainfall, runoff, and sediment yield in the western Mediterranean basin. *Atmosphere* **2020**, *11*, 609. [CrossRef]
5. Da Silva, J.L.B.; Moura, G.B.D.A.; da Silva, M.V.; Lopes, P.M.O.; Guedes, R.V.D.S.; Silva, F.D.F.E.; Ortiz, P.F.S.; Rodrigues, J.A.D.M. Changes in the water resources, soil use and spatial dynamics of Caatinga vegetation cover over semiarid region of the Brazilian Northeast. *Remote Sens. Appl. Soc. Environ.* **2020**, *20*, 100372. [CrossRef]
6. Santos, J.F.S.; Naval, L.P. Spatial and temporal dynamics of water footprint for soybean production in areas of recent agricultural expansion of the Brazilian savannah (Cerrado). *J. Clean. Prod.* **2020**, *251*, 119482. [CrossRef]
7. Singh, V.P. *Handbook of Applied Hydrology*, 2nd ed.; McGraw-Hill Education—Europe: London, UK, 2016; ISBN 0071835091.
8. Paredes-Trejo, F.J.; Barbosa, H.A.; Kumar, T.V.L. Validating CHIRPS-based satellite precipitation estimates in Northeast Brazil. *J. Arid Environ.* **2017**, *139*, 26–40. [CrossRef]

9. Rodrigues, D.T.; Gonçalves, W.A.; Spyrides, M.H.C.; e Silva, C.M.S. Spatial and temporal assessment of the extreme and daily precipitation of the Tropical Rainfall Measuring Mission satellite in Northeast Brazil. *Int. J. Remote Sens.* **2019**, *41*, 549–572. [CrossRef]
10. Neto, R.M.B.; Santos, C.A.G.; Silva, J.F.C.B.D.C.; da Silva, R.M.; dos Santos, C.A.C.; Mishra, M. Evaluation of the TRMM product for monitoring drought over Paraíba State, northeastern Brazil: A trend analysis. *Sci. Rep.* **2021**, *11*, 1097. [CrossRef]
11. McKee, T.B.; Doesken, N.J.; Kleist, J. The Relationship of Drought Frequency and Duration to Time Scales. In Proceedings of the 8th Conference on Applied Climatology, Anaheim, CA, USA, 17–22 January 1993; pp. 179–184.
12. Juliani, B.H.T.; Okawa, C.M.P. Application of a standardized precipitation index for meteorological drought analysis of the semi-arid climate influence in Minas Gerais, Brazil. *Hydrology* **2017**, *4*, 26. [CrossRef]
13. Brito, S.S.B.; Cunha, A.P.M.A.; Cunningham, C.C.; Alvalá, R.C.; Marengo, J.A.; Carvalho, M.A. Frequency, duration and severity of drought in the Semiarid Northeast Brazil region. *Int. J. Climatol.* **2018**, *38*, 517–529. [CrossRef]
14. Santos, C.A.G.; Brasil Neto, R.M.; de Araújo Passos, J.S.; da Silva, R.M. Drought assessment using a TRMM-derived standardized precipitation index for the upper São Francisco River basin, Brazil. *Environ. Monit. Assess.* **2017**, *189*, 250. [CrossRef]
15. Tsesmelis, D.E.; Leveidioti, I.; Karavitis, C.A.; Kalogeropoulos, K.; Vasilakou, C.G.; Tsatsaris, A.; Zervas, E. Spatiotemporal Application of the Standardized Precipitation Index (SPI) in the Eastern Mediterranean. *Climate* **2023**, *11*, 95. [CrossRef]
16. Toté, C.; Patricio, D.; Boogaard, H.; van der Wijngaart, R.; Tarnavsky, E.; Funk, C. Evaluation of satellite rainfall estimates for drought and flood monitoring in Mozambique. *Remote Sens.* **2015**, *7*, 1758–1776. [CrossRef]
17. Mariano, D.A.; dos Santos, C.A.; Wardlow, B.D.; Anderson, M.C.; Schiltmeyer, A.V.; Tadesse, T.; Svoboda, M.D. Use of remote sensing indicators to assess effects of drought and human-induced land degradation on ecosystem health in Northeastern Brazil. *Remote Sens. Environ.* **2018**, *213*, 129–143. [CrossRef]
18. Paca, V.H.d.M.; Espinoza-Dávalos, G.E.; Moreira, D.M.; Comair, G. Variability of Trends in Precipitation across the Amazon River Basin Determined from the CHIRPS Precipitation Product and from Station Records. *Water* **2020**, *12*, 1244. [CrossRef]
19. Rouse, J.W.; Haas, R.H.; Schell, J.A.; Deering, D.W.; Harlan, J.C. *Monitoring the Vernal Advancement of Retrogradation of Natural Vegetation*; NASA: Washington, DC, USA, 1973.
20. Galvíncio, J.D.; Sá, I.I.S.; de Moura, M.S.B.; Ribeiro, J.G. Determinação das características físicas, climáticas e da paisagem da bacia hidrográfica do rio Brigida com o auxilio de técnicas de geoprocessamento e sensoriamento remoto. *Rev. Geogr.* **2007**, *24*, 83–96.
21. Santos, C.A.G.; da Silva, R.M.; Silva, A.M.; Neto, R.M.B. Estimation of evapotranspiration for different land covers in a Brazilian semi-arid region: A case study of the Brígida River basin, Brazil. *J. S. Am. Earth Sci.* **2017**, *74*, 54–66. [CrossRef]
22. Da Silva, M.V.; Pandorfi, H.; Lopes, P.M.O.; da Silva, J.L.B.; de Almeida, G.L.P.; Silva, D.A.D.O.; dos Santos, A.; Rodrigues, J.A.D.M.; Batista, P.H.D.; Jardim, A.M.D.R.F. Pilot monitoring of caatinga spatial-temporal dynamics through the action of agriculture and livestock in the brazilian semiarid. *Remote Sens. Appl. Soc. Environ.* **2020**, *19*, 100353. [CrossRef]
23. Da Silva, M.V.; Pandorfi, H.; de Oliveira-Júnior, J.F.; da Silva, J.L.B.; de Almeida, G.L.P.; Montenegro, A.A.d.A.; Mesquita, M.; Ferreira, M.B.; Santana, T.C.; Marinho, G.T.B.; et al. Remote sensing techniques via Google Earth Engine for land degradation assessment in the Brazilian semiarid region, Brazil. *J. S. Am. Earth Sci.* **2022**, *120*, 104061. [CrossRef]
24. Silva, J.R.I.; Montenegro, A.A.D.A.; Farias, C.W.L.D.A.; Jardim, A.M.D.R.F.; da Silva, T.G.F.; Montenegro, S.M.G.L. Morphometric characterization and land use of the Pajeú river basin in the Brazilian semi-arid region. *J. S. Am. Earth Sci.* **2022**, *118*, 103939. [CrossRef]
25. De Lima, I.P.; Jorge, R.G.; de Lima, J.L.M.P. Remote Sensing Monitoring of Rice Fields: Towards Assessing Water Saving Irrigation Management Practices. *Front. Remote Sens.* **2021**, *2*, 762093. [CrossRef]
26. Silva, A.P.; de Lima, I.P.; Santo, M.F.E.S.; Pires, V.C. Assessing changes in drought and wetness episodes in drainage basins using the Standardized Precipitation Index. *Die Bodenkult.* **2014**, *65*, 31–37.
27. Araújo, H.L.; Montenegro, A.A.d.A.; Lopes, I.; de Carvalho, A.A.; e Silva, E.C.; Gonçalves, G.E. Precipitation spatialization in the Brígida river basin in Pernambuco semi-arid. *Rev. Bras. Geogr. Fis.* **2020**, *13*, 391–405. [CrossRef]
28. Marengo, J.A.; Alves, L.M.; Alvala, R.C.S.; Cunha, A.P.; Brito, S.; Moraes, O.L.L. Climatic characteristics of the 2010-2016 drought in the semiarid northeast Brazil region. *An. Acad. Bras. Cienc.* **2018**, *90*, 1973–1985. [CrossRef]
29. Montenegro, A.A.A.; Ragab, R. Hydrological response of a Brazilian semi-arid catchment to different land use and climate change scenarios: A modelling study. *Hydrol. Process.* **2010**, *24*, 2705–2723. [CrossRef]
30. Wanderley, H.S.; de Amorim, R.F.C.; de Carvalho, F.O. Variabilidade espacial e preenchimento de falhas de dados pluviométricos para o estado de Alagoas. *Rev. Bras. Meteorol.* **2012**, *27*, 347–354. [CrossRef]
31. Hengl, T.; Walsh, M.G.; Sanderman, J.; Wheeler, I.; Harrison, S.P.; Prentice, I.C. Global mapping of potential natural vegetation: An assessment of machine learning algorithms for estimating land potential. *PeerJ* **2018**, *2018*, e5457. [CrossRef]
32. Wolff, W.; Duarte, S.N.; Fernandes, R.D.M. Spatialization of the annual and seasonal average precipitations in the state of Santa Catarina, Brazil. *J. S. Am. Earth Sci.* **2020**, *103*, 102735. [CrossRef]
33. Da Silva, M.V.; Pandorfi, H.; Jardim, A.M.D.R.F.; de Oliveira-Júnior, J.F.; da Divincula, J.S.; Giongo, P.R.; da Silva, T.G.F.; de Almeida, G.L.P.; Moura, G.B.D.A.; Lopes, P.M.O. Spatial modeling of rainfall patterns and groundwater on the coast of northeastern Brazil. *Urban Clim.* **2021**, *38*, 100911. [CrossRef]
34. Cunha, A.P.M.A.; Tomasella, J.; Ribeiro-Neto, G.G.; Brown, M.; Garcia, S.R.; Brito, S.B.; Carvalho, M.A. Changes in the spatial–temporal patterns of droughts in the Brazilian Northeast. *Atmos. Sci. Lett.* **2018**, *19*, e855. [CrossRef]

35. Da Silva, R.M.; Santos, C.A.G.; Silva, J.F.C.B.D.C.; Silva, A.M.; Neto, R.M.B. Spatial distribution and estimation of rainfall trends and erosivity in the Epitácio Pessoa reservoir catchment, Paraíba, Brazil. *Nat. Hazards* **2020**, *102*, 829–849. [CrossRef]
36. Medauar, C.C.; Silva, S.A.; Carvalho, L.C.C.; Galvão, Í.M.; Macêdo, P.V. Spatial-temporal variability of rainfall and mean air temperature for the state of Bahia, Brazil. *An. Acad. Bras. Cienc.* **2020**, *92*, e20181283. [CrossRef]
37. De Medeiros, E.S.; de Lima, R.R.; de Olinda, R.A.; dos Santos, C.A.C. Modeling spatiotemporal rainfall variability in Paraíba, Brazil. *Water* **2019**, *11*, 1843. [CrossRef]
38. Page, T.; Beven, K.J.; Hankin, B.; Chappell, N.A. Interpolation of rainfall observations during extreme rainfall events in complex mountainous terrain. *Hydrol. Process.* **2022**, *36*, 14758. [CrossRef]
39. Adhikary, S.K.; Muttil, N.; Yilmaz, A.G. Cokriging for enhanced spatial interpolation of rainfall in two Australian catchments. *Hydrol. Process.* **2017**, *31*, 2143–2161. [CrossRef]
40. Deutsch, C.V.; Journel, A.G. *GSLIB: Geostatistical Software Library and User's Guide*, 2nd ed.; Oxford University Press: New York, NY, USA, 1998; ISBN 0195073924.
41. Méndez-Venegas, J.; Díaz-Viera, M.A.; Herrera, G.S.; Valdés-Manzanilla, A. Geostatistical simulation of spatial variability of convective storms in Mexico City Valley. *Geofis. Int.* **2013**, *52*, 111–120. [CrossRef]
42. Wang, G.; Gertner, G.; Singh, V.; Shinkareva, S.; Parysow, P.; Anderson, A. Spatial and Temporal Prediction and Uncertainty Analysis of Rainfall Erosivity for the Revised Universal Soil Loss Equation. *Ecol. Modell.* **2002**, *153*, 143–155. [CrossRef]
43. Goovaerts, P. *Geostatistics for Natural Resources Evaluation*; Oxford University Press: New York, NY, USA, 1997; ISBN 0195115384.
44. Isaaks, E.H.; Srivastava, R.M. *An Introduction to Applied Geostatistics*; Oxford University Press: New York, NY, USA, 1989; ISBN 0195050134.
45. Goovaerts, P. Geostatistical approaches for incorporating elevation into the spatial interpolation of rainfall. *J. Hydrol.* **2000**, *228*, 113–129. [CrossRef]
46. Cuartas, L.A.; Cunha, A.P.M.D.A.; Alves, J.A.; Parra, L.M.P.; Deusdará-Leal, K.; Costa, L.C.O.; Molina, R.D.; Amore, D.; Broedel, E.; Seluchi, M.E.; et al. Recent Hydrological Droughts in Brazil and Their Impact on Hydropower Generation. *Water* **2022**, *14*, 601. [CrossRef]
47. Agência Pernambucana de Águas e Clima (APAC). Relatório de Situação de Recursos Hídricos do Estado de Pernambuco 2011/2012. Available online: https://www.lai.pe.gov.br/apac/wp-content (accessed on 16 June 2022).
48. Instituto de Pesquisa Econômica Aplicada (IPEA). *Transposição do Rio São Francisco: Análise de Oportunidade do Projeto*; IPEA: Rio de Janeiro, Brazil, 2011; ISBN 1415-4765.
49. Alvares, C.A.; Stape, J.L.; Sentelhas, P.C.; Gonçalves, J.L.D.M.; Sparovek, G. Köppen's climate classification map for Brazil. *Meteorol. Z.* **2013**, *22*, 711–728. [CrossRef] [PubMed]
50. Beck, H.E.; Zimmermann, N.E.; McVicar, T.R.; Vergopolan, N.; Berg, A.; Wood, E.F. Present and future köppen-geiger climate classification maps at 1-km resolution. *Sci. Data* **2018**, *5*, 180214. [CrossRef] [PubMed]
51. Instituto Nacional de Meteorologia (INMET). Normais Climatológicas. Available online: https://clima.inmet.gov.br/GraficosClimatologicos/ (accessed on 10 June 2022).
52. Soil Survey Staff. *Keys to Soil Taxonomy*, 9th ed.; US Department of Agriculture: Washington, DC, USA, 2006.
53. Gerardo, R.; de Lima, I.P. Sentinel-2 Satellite Imagery-Based Assessment of Soil Salinity in Irrigated Rice Fields in Portugal. *Agriculture* **2022**, *12*, 1490. [CrossRef]
54. Agência Pernambucana de Águas e Clima (APAC). Coordenadas das estações pluviométricas do Estado de Pernambuco. Available online: http://old.apac.pe.gov.br/meteorologia/coordenadas/index.html (accessed on 16 June 2022).
55. Warrick, A.W.; Nielsen, D.R. Spatial variability of soil physical properties in the field. In *Applications of Soil Physics*; Academic Press: New York, NY, USA, 1980.
56. Journel, A.G. *Fundamentals of Geostatistics in Five Lessons*; American Geophysical Union: Washington, DC, USA, 1989; ISBN 0-87590-708-3.
57. Landim, P.M.B. *Análise Estatística de Dados Geológicos*, 2nd ed.; Edunesp: Rio Claro, Brazil, 2003; ISBN 9788571395046.
58. Journel, A.G.; Huijbregts, C.J. *Mining Geostatistics*; Oxford University Press: London, UK, 1978; ISBN 0123910501.
59. Cambardella, C.A.; Moorman, T.B.; Novak, J.M.; Parkin, T.B.; Karlen, D.L.; Turco, R.F.; Konopka, A.E. Field-Scale Variability of Soil Properties in Central Iowa Soils. *Soil Sci. Soc. Am. J.* **1994**, *58*, 1501–1511. [CrossRef]
60. Vauclin, M.; Vieira, S.R.; Vachaud, G.; Nielsen, D.R. The Use of Cokriging with Limited Field Soil Observations. *Soil Sci. Soc. Am. J.* **1983**, *47*, 175–184. [CrossRef]
61. Hengl, T.; Heuvelink, G.B.M.; Stein, A. *Comparison of Kriging with External Drift and Regression-Kriging*; ITC: Geneve, Switzerland, 2003; p. 17.
62. Goovaerts, P. Geostatistical modelling of uncertainty in soil science. *Geoderma* **2001**, *103*, 3–26. [CrossRef]
63. Google Earth Engine. Landsat Collections. Available online: https://developers.google.com/earth-engine/datasets/catalog/landsat (accessed on 10 May 2022).
64. Xu, H. Modification of normalised difference water index (NDWI) to enhance open water features in remotely sensed imagery. *Int. J. Remote Sens.* **2006**, *27*, 3025–3033. [CrossRef]
65. Molle, F. *Geometria de Pequenos Açudes*; SUDENE: Recife, Brazil, 1994; Volume 1, ISBN 55655812813.
66. MapBiomas Brazil. Coleção 7.1 da Série Anual de Mapas de Cobertura e Uso de Solo do Brasil. Available online: https://mapbiomas.org/colecoes-mapbiomas-1?cama_set_language=pt-BR (accessed on 11 March 2023).

67. Mutiga, J.K.; Su, Z.; Woldai, T. Using satellite remote sensing to assess evapotranspiration: Case study of the upper Ewaso Ng'iro North Basin, Kenya. *Int. J. Appl. Earth Obs. Geoinf.* **2010**, *12*, 100–108. [CrossRef]
68. Bai, T.; Tahmasebi, P. Sequential Gaussian simulation for geosystems modeling: A machine learning approach. *Geosci. Front.* **2022**, *13*, 101258. [CrossRef]
69. Pereira, G.T.; Teixeira, D.D.B.; de Souza, Z.M.; de Oliveira, I.R.; Marques Júnior, J. Stochastic simulations of calcium contents in sugarcane area. *Rev. Bras. Eng. Agric. Ambient.* **2015**, *19*, 767–772. [CrossRef]
70. Da Silva, T.G.F.; de Queiroz, M.G.; Zolnier, S.; de Souza, L.S.B.; de Souza, C.A.A.; de Moura, M.S.B.; de Araújo, G.G.L.; Neto, A.J.S.; dos Santos, T.S.; de Melo, A.L.; et al. Soil properties and microclimate of two predominant landscapes in the Brazilian semiarid region: Comparison between a seasonally dry tropical forest and a deforested area. *Soil Tillage Res.* **2021**, *207*, 104852. [CrossRef]
71. Panday, P.K.; Coe, M.T.; Macedo, M.N.; Lefebvre, P.; de Castanho, A.D. Deforestation offsets water balance changes due to climate variability in the Xingu River in eastern Amazonia. *J. Hydrol.* **2015**, *523*, 822–829. [CrossRef]
72. Agência Nacional de Águas e Saneamento Básico (ANA). Reservatórios do Semiárido Brasileiro: Hidrologia, Balanço Hídrico e Operação—Reservatório Entremontes. Available online: http://portal1.snirh.gov.br/arquivos/semiarido/204res/SF_Entremontes.pdf (accessed on 1 February 2022).
73. Agência Nacional de Águas e Saneamento Básico (ANA). Reservatórios do Semiárido Brasileiro: Hidrologia, Balanço Hídrico e Operação—Reservatório Chapéu. Available online: http://portal1.snirh.gov.br/arquivos/semiarido/204res/SF_Chapéu.pdf (accessed on 1 February 2022).
74. De Araújo, J.C.; Mamede, G.L.; De Lima, B.P. Hydrological guidelines for reservoir operation to enhance water governance: Application to the Brazilian Semiarid region. *Water* **2018**, *10*, 1628. [CrossRef]
75. Tomasella, J.; Silva Pinto Vieira, R.M.; Barbosa, A.A.; Rodriguez, D.A.; de Oliveira Santana, M.; Sestini, M.F. Desertification trends in the Northeast of Brazil over the period 2000–2016. *Int. J. Appl. Earth Obs. Geoinf.* **2018**, *73*, 197–206. [CrossRef]

remote sensing

MDPI

Article

Assessing the Impact of Land Use and Land Cover Change on Environmental Parameters in Khyber Pakhtunkhwa, Pakistan: A Comprehensive Study and Future Projections

Mehjabeen Khan [1] and Ruishan Chen [2,*]

[1] School of Geographic Sciences, East China Normal University, Shanghai 200241, China; mehjabeenkhanECNU@outlook.com
[2] School of Design, Shanghai Jiaotong University, Shanghai 200240, China
* Correspondence: rschen@sjtu.edu.cn

Abstract: Land use and land cover (LULC) change, driven by environmental and human activities, significantly impacts ecosystems, climate, biodiversity, and socio-economic systems. This study focuses on Khyber Pakhtunkhwa (KPK), Pakistan, a region with sensitive ecosystems and diverse landscapes, to analyze LULC dynamics and their environmental consequences. Based on Landsat imagery from 2000, 2010, and 2020, we used the Random Forest algorithm on Google Earth Engine (GEE) to classify LULC, and the CA-ANN model to project future scenarios for 2030, 2050, and 2100. Additional simulations were conducted using the MOLUSCE Plugin in QGIS. The results revealed a 138.02% (4071.98 km^2) increase in urban areas from 2000 to 2020, marking urbanization as a major driver of LULC change. Urban expansion strongly correlated with land surface temperature (LST) (R^2 = 0.89), amplifying the urban heat island effect. Rising LST showed negative correlations with the key environmental indices NDVI (−0.88), MNDWI (−0.49), and NDMI (−0.62), signaling declining vegetation cover, water resources, and soil moisture, respectively. Projections for 2100 predict LST rising to 55.3 °C, with NDVI, MNDWI, and NDMI dropping to 0.36, 0.17, and 0.21, respectively. Vegetation health, as indicated by the Leaf Area Index (LAI), also declined, with maximum and minimum values falling from 4.66 and −5.75 in 2000 to 2.16 and −2.55 in 2020, reflecting increased barren land and reduced greenness. The spatial analysis highlights significant transitions from vegetated to barren or urban land, leading to declining moisture levels, water stress, soil erosion, and biodiversity. Projections show continued reductions in forests, vegetation, and agricultural lands, replaced by barren and built-up areas. Declines in key indices such as NDVI, MNDWI, and NDMI indicate deteriorating vegetation, water resources, and soil moisture levels. These findings emphasize the need for sustainable urban planning and environmental management. Expanding urban green spaces, using reflective materials, and preserving vegetation and water resources are vital to mitigating heat island effects and maintaining ecological balance. Anticipated declines in LST, NDVI, MNDWI, NDMI, and LAI stress the urgency for climate adaptation strategies to protect human health, ecosystem services, and economic stability in KPK.

Keywords: land use land cover changes; Google Earth Engine; environmental impact; future projection; remote sensing

1. Introduction

Land use and land cover (LULC) change refers to the modification of natural or semi-natural landscapes, typically into artificial environments. It is a complicated process triggered by various factors, such as population growth, urban sprawl, changes in local and regional environmental characteristics [1,2], loss of agricultural land, and consumption of other environmental resources. Some of the environmental impacts include effects on water resources, climate change, and biodiversity [3–5].

Land use and land cover decisions are made in different parts of the world depending on socio-economic and bio-physical factors. Changes in LULC affect the hydrological cycle, population density, urbanization, and agricultural activities [6–9]. Alterations in the use of resources and their cover are clearly in line with activities that affect the climatic conditions of a particular region. Such circumstances include land use change encompassing urbanization and deforestation that can impact hydrologic systems' infiltration, evaporation, and runoff rates [10,11]. Information on LULC changes can be helpful to watershed managers in developing and implementing interventions to protect water resources. For instance, they may uncover areas that experience frequent erosion and sedimentation and then respond by planting trees or facilitating the installation of riparian strips to counter such negative impacts. They can also mark areas that are flood-prone and put up barriers or other measures to avoid floods [8].

Remote sensing (RS) and geographical information systems (GISs) are useful approaches for surveying and tracking LULC changes. Researchers can also merge land use data, climate data, satellite images, cartographic information, and census data into GIS [12–14]. This information could be used to map how the LULC has evolved over the years to arrive at its current state. RS can help researchers obtain LULC data for large-scale areas at specific timelines. This information enables them to track the changes in LULC and patterns that may occur [3,15,16].

Khyber Pakhtunkhwa (KPK) is a province vulnerable to LULC change, given that it is a region dominated by mountains and has a sensitive landscape. These transformations could drastically change critical environmental parameters like land surface temperature (LST), modified normalized difference water index (MNDWI), normalized difference vegetation index (NDVI), leaf area index (LAI), and normalized difference moisture index (NDMI). LULC changes have previously been influenced by population, urbanism, and industrialization, among other factors in KPK [17].

While social factors such as population growth and urban expansion are significant drivers of LULC dynamics, the primary focus of this study is on quantifying the environmental consequences of these changes [18,19]. By leveraging spatial and remote sensing data, the study investigates key indicators such as LST, NDVI, MNDWI, LAI, and NDMI. The emphasis on environmental impacts reflects the nature of the available data, which are well-suited for analyzing physical changes but less effective in capturing social dynamics directly. Social factors are considered contextually but remain outside the scope of this research. This focused approach provides valuable insights into the physical consequences of LULC changes while laying the groundwork for future interdisciplinary studies [20].

The reduced forest and agricultural land and increased urbanization that KPK has experienced bring significant shifts in environmental stability, improved natural resource management, and sustainable livelihood adaptation [21]. It was also observed that LULC change exerts a relative and prominent influence on the environment in KPK. This has brought about higher LST, lower NDVI, and lower MNDWI [22]. Some of these changes include increased heat stress, reduced availability of water, and reduced biodiversity. Other studies also note that LULC change has a relative and more prominent influence on KPK's environment, resulting in higher LST, lower NDVI, and lower MNDWI [23]. Other changes include heat stress, the quantity and quality of water, and the levels of diversity [24]. Hence, based on the findings in the existing literature concerning the effects of change in LULC on the environmental parameters of the province of KPK, it is possible to pinpoint the following limitations. Most of these studies were based on analysis by focusing on urbanization, deforestation, and LULC modeling techniques [21,25,26].

The primary purpose of this research paper is to explore the impact of land use and land cover transformation on the environmental attributes of Khyber Pakhtunkhwa, Pakistan. We analyzed spatio-temporal trends in land use and land cover (LULC) for 2000, 2010, and 2020 using Landsat images and the Random Forest (RF) algorithm within Google Earth Engine (GEE). Future changes for 2030, 2050, and 2100 were predicted using the Cellular Automata-Artificial Neural Network (CA-ANN) method and the MOLUSCE plugin in QGIS [23,27]. We also calculated other critical environmental parameters such as LST, MNDWI, NDVI, LAI, and NDMI. We used Landsat 5 and 8 satellite imagery from 2000, 2010, and 2020 to classify land use and land cover (LULC) for each year independently using the Random Forest (RF) model. This enables the modeling and forecasting of these changes and their environment up to 2100. Such an approach provides a conceptual perspective on the multi-faceted interactions between LULC changes and environmental factors within the local spatial scale. The predictive perceptions that emerge from this study provide policy-relevant knowledge on land use planning and enhancing environmental management in KPK These findings support sustainable strategies for addressing emerging environmental challenges, offering long-term solutions to manage future land use and ecological impacts.

This study highlights the need for integrating remote sensing and geospatial sciences to assess the LULC change and its relationship with critical environmental factors. Consequently, the study shall provide policymakers, land users, and conservationists with enough information to assist them in improving sustainable land management and conservation of the environment, as well as enhancement of resilience within the territorial landscape of KPK, Pakistan. The findings and the method employed in this research can be helpful for other similar environments across the world experiencing LULC change. Based on our analysis, we find that the provincial administration should adopt sustainable land use methods and encourage farmers to cultivate cultivars that thrive even during droughts. Introducing soil conservation measures is an option for coping with the expected reduction in LAI that will help build environmental resilience in KPK's territories. The study results will significantly aid KPK in managing a fast-changing environment and urban planning.

2. Study Area

Khyber Pakhtunkhwa (KPK), which lies in northwestern Pakistan, is a vast and geographically diverse region, ranging from its most northern point (at relatively 35.3213°N) to its southernmost point (at about 30.1575°N), and from its easternmost to westernmost points (at about 77.8359°E and 69.3216°E, respectively), and covering an area of around 101,741 square kilometers [28–31].

KPK is one of Pakistan's most densely populated provinces, containing 36 districts. Extending from the Indus River plains to the Hindu Kush and the Himalayas, the region exhibits a diversity of climates, ranging from the hot, arid south to the soaring humidity of the alpine north. Soaked in water through the Indus River and its system of tributaries, KPK supports irrigated agriculture on a significant scale [10,31]. The province also has various climate zones ranging from subtropical to temperate to subalpine [32]. These different climatic zones and the various topographical features of the province exert a significant influence on land-use decisions, agricultural practices, and the kind of vegetation found in the province However, Khyber Pakhtunkhwa faces many serious environmental hazards, including floods and earthquakes, which necessitate serious planning effort for resilience [17,33,34]. The driving forces of KPK's dynamics are its increasing population, burgeoning urbanization, and intensified pressures on agricultural lands [35–37]. Growing numbers of people and the rapid spread of urbanization are most notable in the fast-expanding city of Peshawar. Both lead directly to land use changes: a loss of vegetation, a build-out of infrastructure, and rising surface temperatures. These two federally administered areas of KPK seem to have an almost inevitable momentum toward urbanization and sprawling land use, forcing competition for land between human settlers and agriculture Spatio-temporal Analysis of Urban Expansion on Farmland [14,38]. According to the 2017 census by the Pakistan Bureau of Statistics, the total number of people living in the KPK province was around 30.5 million. The Bureau has projected that this figure will swell to nearly 40.9 million by 2023. The PBS also estimates the density of the population at the provincial level. As per the PBS, the average population density in KPK was about 409.40 persons per square kilometer as of 2017. The agricultural sector has a significant position in the economy of KPK, as there is a vast area for agriculture and its products, including crops, meat, etc. [39,40].

Alterations in land use for agriculture have effects on indices such as the NDVI and may lead to changes in cropping patterns and water management. The province regularly experiences floods and landslide disasters that require sustainable land use management programs. Knowledge of LULC changes helps address issues related to disaster risk reduction and make communities more resilient [31]. The geographical variations in KPK, ranging from mountainous areas to near-sea level regions, offer a perfect laboratory to understand the impact of altitude and climate on environmental variables such as LST, NDVI, MNDWI, LAI, and NDMI. Lastly, the rivers and water courses of KPK, comprising the Indus River, have provided a favorable location for analyzing the correlation of water sector indices, such as MNDWI, with land use. The red circle in (Figure 1) designates our study area [14,39].

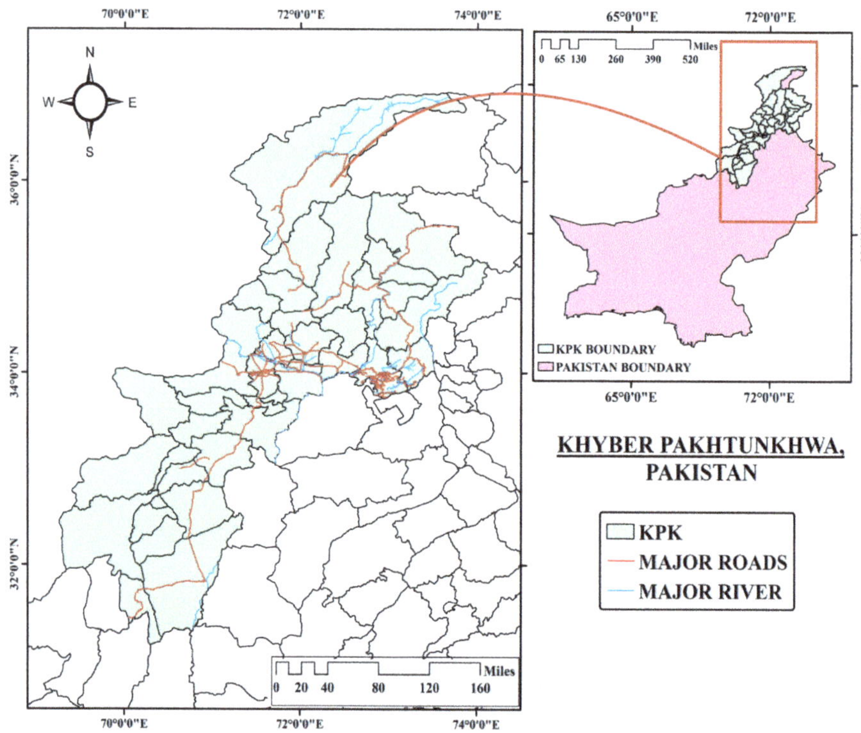

Figure 1. Study area map of Khyber Pakht-unkhwa showing major road networks and major rivers.

3. Data and Methodology

3.1. Accuracy Assessment of LULC Classification

By analyzing Landsat imagery and several environmental parameters of the KPK region for 2000, 2010, and 2020, LULC changes were determined using Landsat 5 and Landsat 8 satellites. These satellite images were sourced from the United States Geological Survey (USGS). The primary geospatial data for analyzing the study area's vital environmental parameters and land use changes involved Landsat 5 Thematic Mapper TOA. Since Landsat satellite images are already geometrically and radiometrically corrected, no additional image-to-image georeferencing or correction was required. A maximum cloud threshold of 5% was applied during image selection to obtain images with minimal cloud cover. The downloaded images were time-divided into 10-year intervals (2000, 2010, and 2020) based on the availability of satellite data. To maintain consistency, Landsat images of the same or similar dates, specifically December and January, were used to minimize seasonal variability and ensure comparable temporal coverage [39,40]. Training samples for MLSC were selected exclusively from the Landsat imagery by analyzing spectral reflectance patterns and utilizing true-color composites (TCCs). This process involved systematically selecting representative pixels from each LULC class across diverse geographic locations within the study area. Band combinations were optimized to enhance spectral separability, particularly for similar classes such as barren land and built-up areas.

Our LULC classification methodology was based on the maximum likelihood supervised classification (MLSC) technique, a widely used approach in remote sensing that assumes each class follows a Gaussian distribution. The MLSC algorithm assigns pixels to the class with the highest likelihood based on statistical probability, ensuring robust classification. To achieve this, sufficient training samples were carefully selected to capture the variability within each class, with samples manually identified from Landsat imagery for seven LULC categories: forest cover, vegetation, agricultural land, built-up land, snow cover, waterbodies, and barren land. The spectral diversity of categories like vegetation and agricultural land was accounted for by selecting samples across varying topographies and vegetation densities. Training data were validated by cross-referencing with high-resolution, satellite-derived composites and available ancillary information, while spectral separability analysis was conducted to minimize overlap among classes and enhance classification accuracy.

The seven LULC categories were defined to reflect the specific landscape features of KPK. Forest cover includes areas dominated by dense tree canopies, encompassing both natural and managed forests. Vegetation represents non-forest natural vegetation, such as grasslands and shrubs, primarily in low-density areas. Agricultural land is defined as cultivated fields and other areas dedicated to crop production, distinctly separate from natural vegetation. This differentiation between forest cover, vegetation land, and agricultural land was essential for accurately capturing the diverse land use patterns in KPK. Although tailored to local conditions, the classification approach adheres to principles of internationally recognized systems, such as the FAO land cover classification system (LCCS), ensuring compatibility with global studies while accurately representing KPK's unique environmental and agricultural diversity. Consistency checks during the classification process reduced ambiguity and ensured that the LULC categories accurately reflect the region's land cover characteristics, providing a reliable and meaningful representation of KPK's landscape (Figure 2) [33,41]. Accuracy assessment of the LULC maps was performed by comparing the classified outputs with independent verification samples derived from satellite imagery. Accuracy metrics were calculated to quantify the model's performance, including overall accuracy, producer accuracy, user accuracy, and the kappa coefficient. This accuracy assessment process ensured that the LULC classifications were reliable and consistent, enabling robust analysis of spatiotemporal LULC changes and their implications for environmental dynamics.

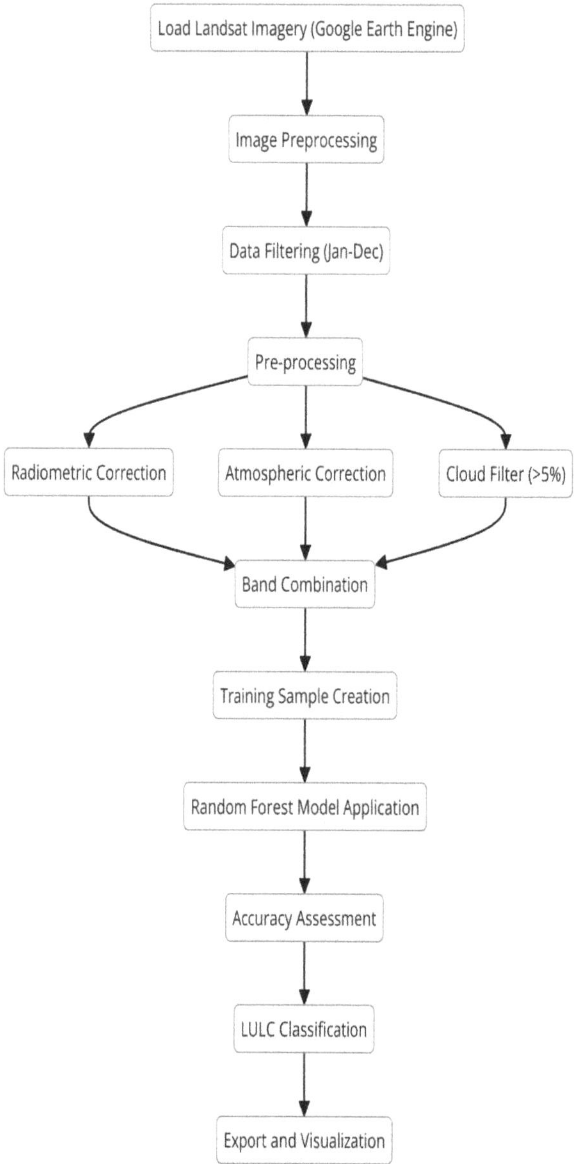

Figure 2. LULC calculate framework diagram.

The classification accuracy for the years 2000, 2010, and 2020 was rigorously assessed using key performance metrics: overall accuracy, producer's accuracy, users' accuracy, and the kappa coefficient [42,43]. Overall accuracy, representing the proportion of correctly classified pixels relative to the total number of pixels analyzed, consistently demonstrated high performance, averaging approximately 90% across the study years. This result underscores the robustness of the classification models employed. Producer accuracy, which quantifies the probability that a reference land cover type is correctly identified, maintained averages ranging from 88.6% to 90.2%, highlighting the model's reliability in comprehensive classification [44]. User accuracy, indicating the probability that a pixel classified as a certain type accurately corresponds to that type in reality, showed stable and dependable values between 89.3% and 91.1%. The kappa coefficient, a measure that evaluates classification agreement beyond random chance, further confirmed the strong performance of the models, with values of 0.81 in 2000, 0.83 in 2010, and 0.89 in 2020 [45]. These kappa values signify substantial to near-perfect agreement, reinforcing the validity and reliability of the classification results. Collectively, these metrics confirm that the classification approach was effective, ensuring the credibility of subsequent analyses on LULC changes and their implications for environmental dynamics. Table 1 lists the detailed classification accuracy of each LULC type.

Table 1. Users' accuracy, producer's accuracy, overall accuracy, and kappa coefficient for LULC classification (unit %).

LULC Classes	Users' Accuracy 2000	Producer's Accuracy 2000	Users' Accuracy 2010	Producer's Accuracy 2010	Users' Accuracy 2020	Producer's Accuracy 2020
Forest Cover	90	89.6	89.3	88.5	88.6	87.8
Vegetation	90.7	83.4	91.2	85.2	92	87
Agriculture Land	98.2	100	96.7	91.5	95.4	92.3
Built-up Land	86.7	84.9	88.4	89.1	89.7	90.4
Barren Land	89.3	88.1	90.1	90.2	91.4	91.3
Snow Cover	87.5	91.2	86.9	87.6	87.8	88.5
Waterbodies	91.6	92.4	92.5	93.7	93.8	94.1
Over all Accuracy	90.05		89.75		90.4	
Kappa Coefficient	0.81		0.83		0.89	

3.2. *Future Land Use Projections*

To establish the prediction model and calculate the LULC changes in 2030, 2050, and 2100, the transition matrices for 2000, 2010, and 2020 were first created. Then the combined CA-ANN method was used to model the future changes (Figure 3). The study on establishing the prediction model and calculating the LULC changes in 2030, 2050, and 2100 was conducted using the MOLUSCE plugin of QGIS software version 2.9. ArcGIS 8.1 was used to map LULC changes and other environmental parameter maps. Our methodology was systematically extended to Landsat 8 imagery, ensuring consistency in parameter computation across both datasets, thus enabling practical comparative analysis. Synchronization between the Landsat 5 and 8 datasets provided a one-stop comparison of the land use and land cover changes within the study area. Landsat 5 and Landsat 8 satellite imagery to predict land use and land cover (LULC) changes for 2030, 2050, and 2100, with historical data from 2000, 2010, and 2020, were used to establish baseline conditions. Preprocessing was conducted in Google Earth Engine (GEE), including geometric and radiometric corrections, cloud masking using bit-masking algorithms, and extraction of surface reflectance bands at a 30-m spatial resolution to ensure precision. Future projections were generated using the cellular automata-artificial neural network (CA-ANN) model, implemented via the MOLUSCE plugin in QGIS. This model incorporates transition matrices derived from historical LULC data and spatial constraints such as proximity to urban centers, elevation, slope, and road networks to simulate realistic LULC dynamics. Predictions followed a "present-change trend scenario", extrapolating historical patterns without assuming policy interventions or major environmental disruptions. Raster outputs from the CA-ANN model were processed in ArcGIS Pro, where raster data were converted to vector polygons to delineate LULC boundaries accurately. The total area for each LULC class was calculated by summing polygon attributes, ensuring robust quantification. Validation was performed through hindcasting, a process in which the model was trained on historical data from

2000 and 2010 to predict LULC for 2020. The predicted outputs were compared against observed 2020 LULC data. Statistical metrics such as overall accuracy, producer's accuracy, users' accuracy, and the kappa coefficient demonstrated high reliability. While providing a robust baseline, these predictions are limited to the continuation of observed historical trends and do not consider potential future interventions or unforeseen socio-economic changes. Validation employed a hindcasting approach, comparing predicted outputs against observed 2020 data. Statistical accuracy metrics, including the kappa coefficient, producer accuracy, and user accuracy, demonstrated the model's high reliability (Table 2). The kappa coefficient measured agreement beyond chance; producer accuracy quantified the proportion of correctly classified reference pixels for each land cover type; and user accuracy evaluated the reliability of the classifications.

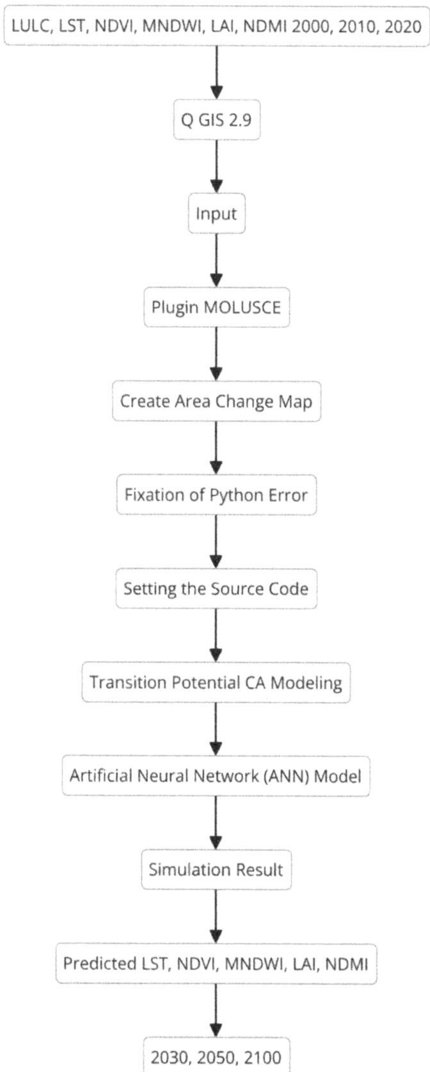

Figure 3. Methodology for predicting LULC, NDVI, MNDWI, LST, LAI, and NDMI and for the years 2030, 2050, and 2100.

Table 2. Users' accuracy, producer's accuracy, overall accuracy, and kappa coefficient for LULC classification (unit %).

LULC Classes	Users' Accuracy (2030)	Producer's Accuracy (2030)	Users' Accuracy (2050)	Producer's Accuracy (2050)	Users' Accuracy (2100)	Producer's Accuracy (2100)
Forest Cover	89.6	90	88.5	89.3	87.8	88.6
Vegetation	83.4	90.7	85.2	91.2	87	92
Agriculture Land	100	98.2	91.5	96.7	92.3	95.4
Built-up Land	84.9	86.7	89.1	88.4	90.4	89.7
Barren Land	88.1	89.3	90.2	90.1	91.3	91.4
Snow Cover	91.2	87.5	87.6	86.9	88.5	87.8
Waterbodies	92.4	91.6	93.7	92.5	94.1	93.8
Over all Accuracy	90.05		89.75		90.04	
Kappa Coefficient	0.85		0.82		0.8	

The model was further calibrated to ensure alignment with historical patterns, refining transition probabilities and incorporating spatial constraints. While direct validation for 2030, 2050, and 2100 is not possible, the high accuracy achieved during hindcasting strengthens confidence in the model's ability to predict future scenarios. Based on the "present-change trend scenario", these projections offer a scientifically grounded baseline for understanding long-term land use dynamics.

3.3. Calculation and Predictions of Environmental Parameters

To ensure a comprehensive representation of the study area's environmental parameters, waterbodies were intentionally included in the calculations of LST, NDVI, MNDWI, NDMI, and LAI (Figure 4). The process began by collecting Landsat 5 and Landsat 8 data for 2000, 2010, and 2020, which provided the basis for these calculations. These data, with a spatial resolution of 30 m, enabled consistent measurement of environmental parameters and land-use changes across the study area. Including waterbodies ensures the natural variability of the landscape is preserved, capturing lower NDVI values and cooler LST measurements that reflect the region's mixed land cover, including lakes, rivers, and snow-covered areas. This approach aligns with its design for LST to capture surface temperatures across all features, not just dry land. Similarly, NDVI and other indices benefit from including waterbodies, as their unique spectral properties provide critical contrast and serve as reference points for distinguishing land cover types. The identical spatial resolution of 30 m for all parameters facilitates accurate comparisons and integration of results across indices. Given the relatively small proportion of waterbodies in the study area, their inclusion does not significantly affect aggregate values but ensures continuity and a holistic environmental analysis. This approach provides a more accurate depiction of surface conditions and avoids artificial dataset segmentation, enhancing the interpretability and reliability of the results. Table 3 below are the key data sources used in our study and their resolutions.

Our analysis began with the computation of LST. To estimate LST from Landsat 5 imagery, we initiated the process by converting radiance values to top-of-atmosphere brightness temperature (BT) utilizing the thermal infrared Band 6 of Landsat 5, characterized by a central wavelength of approximately 10.4 μm. Furthermore, normalizations of LST from Kelvin (K) to Celsius (°C) or Fahrenheit (°F) were also performed where necessary for further evaluation [34,45–48].

Once the emissivity coefficient (W), the spectral conversion factor (ρ), and the radiance emitted by the Earth in the thermal band (Et) had been calculated, (LST) was determined using Equation (1):

$$LST = \frac{Bt}{\frac{1+W}{\rho \times \ln(Et)}} \quad (1)$$

where (Bt) can be calculated directly from the thermal band radiance (L) using Equation (2):

$$Bt = \frac{L}{(\rho \times 10^6)} \tag{2}$$

where LST stands for land surface temperature in Kelvin (K), (Bt) for brightness temperature in Kelvin (K), (W) for emissivity coefficient, (ρ) for spectral conversion factor, (Et) for radiance emitted by the Earth in the thermal band, L is the thermal band radiance in watts per square meter per steradian per micrometer ($W/m^2/sr/\mu m$), and the emissivity coefficient (W) and spectral conversion factor (ρ) are specific to the Landsat sensor and the land cover type. For Landsat 5 and 8, the emissivity coefficient for the thermal band (Band 10) can be estimated using Equation (3) [49–53]:

$$W = 0.986 + 0.004 \times NDVI \tag{3}$$

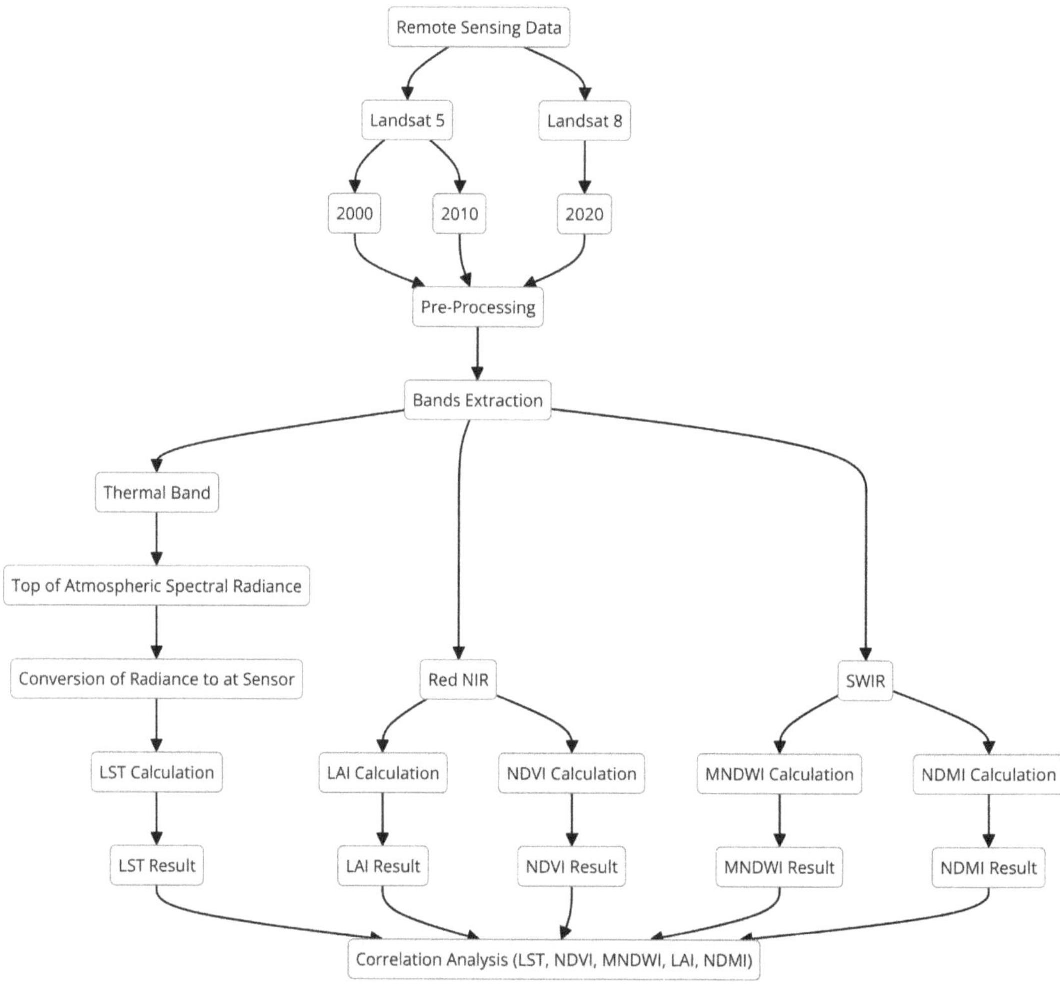

Figure 4. Methodology for calculating environmental parameters (LST, NDVI, MNDWI, LAI, and NDMI) in KPK, Pakistan.

Table 3. Data sources and resolutions for the initial parameters.

Data	Source	Resolution
Satellite Imagery of Landsat 5 and 8	United States Geological Survey	30 m
Digital Elevation Model	Shuttle Radar Topography Mission	30 m
Normalized Difference Vegetation Index	RED Band and Near-Infrared Band of Landsat Imagery	30 m
Modified Normalized Difference Water Index	Landsat 5: Green Band (SR_B2), Shortwave Infrared Band (SR_B5); Landsat 8: Green Band (SR_B3), Shortwave Infrared Band (SR_B6)	30 m
Leaf Area Index (LAI)	Near-Infrared Band of Landsat Imagery	30 m
Land Surface Temperature	Thermal Band, Brightness Temperature in Kelvin	30 m
Normalized Difference Moisture Index	Landsat 5: Near-Infrared Band (SR_B4) and Shortwave Infrared Band (SR_B5) Landsat 8: Near-Infrared Band (SR_B5) and Shortwave Infrared Band (SR_B6)	30 m

ρ is a calibration parameter that converts the thermal band radiance values to brightness temperature. Its value depends on the sensor specifications of Landsat 5 and 8, as provided in the calibration metadata $\rho = 1.44 \times 10^{-4}$ μm/K [54]. The radiance emitted by the Earth in the thermal band (Et) can be calculated using Equation (4):

$$Et = \frac{h \times (Bt)^3}{\lambda} \tag{4}$$

where (h) is the Planck constant (6.63×10^{-34} J/s) and λ is the wavelength of the thermal band in micrometers (10.78 μm for Landsat 5 and 8).

Concurrently, MNDWI is calculated using Equation (5), utilizing the green (GREEN) and SWIR bands, which enhance detection of water features while suppressing built-up land noise. For Landsat 5, the green band (SR_B2) and the shortwave infrared (SWIR) band (SR_B5) were used, while for Landsat 8, the green band (SR_B3) and SWIR band (SR_B6) were utilized. Calculations of MNDWI for a given pixel result in values ranging from minus one (−1) to plus one (+1) [2,42].

$$MNDWI = \frac{GREEN - SWIR}{GREEN + SWIR} \tag{5}$$

NDVI is calculated using near-infrared (NIR) and red (RED) reflectance values. These indicators provide crucial insight into vegetation health and water content in the landscape. NDVI was calculated using Equation (6):

$$NDVI = \frac{NIR - Red}{NIR + Red} \tag{6}$$

LAI is defined in terms of area units and refers to the ratio between total leaf area and ground area in square meters per square meter (m^2/m^2) units. This factor can easily determine vegetation health and productiveness. In our study, LAI is an important factor in controlling an ecosystem's energy and water balance in the KPK region [12,55], LAI was calculated using Equation (7):

$$LAI = 0.97\left(e^{2.23 \times NDVI}\right) \tag{7}$$

The normalized difference moisture index (NDMI) is calculated using the Equation (8), which utilizes the near-infrared (NIR) and shortwave infrared (SWIR) bands. These bands provide critical information for detecting vegetation moisture content and differentiating between wet and dry vegetation conditions. For Landsat 5, the NIR band (SR_B4) and the SWIR band (SR_B5) were used, while for Landsat 8, the NIR band (SR_B5) and the SWIR band (SR_B6) were applied. NDMI calculations for a given pixel result in values ranging from −1 to +1, where higher values indicate higher vegetation moisture levels.

$$NDMI = \frac{NIR - SWIR}{NIR + SWIR} \tag{8}$$

NDMI provides an effective means of assessing vegetation moisture status in the study area. It is particularly valuable for understanding the impact of climate change, land use changes, and hydrological variability on vegetation health.

Processed Landsat 5 and 8 satellite data was used to estimate the environmental factors, such as the LST, NDVI, MNDWI, LAI, and NDMI, in 2000, 2010, and 2020. Using these calculated values, we developed an Artificial Neural Network (ANN) model for the MOLUSCE plugin in QGIS software. These parameters were used to train the ANN model, and, once validated against the historical data, the model was used to predict these parameters in the 2030, 2050 and 2100 scenarios. The subsequent positions were then visualized in the QGIS analysis platform to illustrate future trends in vegetation, water content, and land surface temperature. Another significant factor is the dynamic changes observed in the environment and future environmental conditions over time. This methodological consistency ensures reliable assessments of water-related changes over the years, providing valuable information for understanding the impact on land use and environmental parameters. A strong correlation exists between these factors (LST, NDVI, MNDWI, LAI and NDMI), which we will cover in the "correlation" section.

4. Results and Discussion

4.1. Evaluation of LULC Change Through Mapping and Accuracy Assessment

After producing an LULC change map, an accuracy assessment is needed to show that the map is accurate. An accuracy assessment compares a classified cover map with a reference cover map. A reference land cover map collects onsite data, including aerial photographs and field surveys. There are many ways to make an accurate assessment. We used a confusion matrix, which is a table that presents the number of correctly and incorrectly classified pixels for each land cover class in 2000, 2010, and 2020.

Trend changes are evident in (Figures 5 and 6). Significant changes were observed across various categories by analyzing LULC changes from 2000 to 2020 using Landsat 5 and 8 data. The area of waterbodies initially expanded from 8.13% (9099.29 km^2) in 2000 to 16.74% (18,732.76 km^2) in 2010, but dropped sharply to 4.96% (5549.78 km^2) by 2020. The forest area showed an increasing trend from 18.16% (20,322.02 km^2) in 2000 to 22.77% (25,482.21 km^2) in 2010, followed by a slight reduction to 20.16% (22,556.68 km^2) in 2020. The vegetation area experienced a decline from 29.28% (32,762.39 km^2) in 2000 to 22.51% (25,191.63 km^2) in 2010, but rebounded to 28.63% (32,030.81 km^2) in 2020. In contrast, barren land area decreased substantially from 28.28% (31,638.85 km^2) in 2000 to 14.78% (16,537.83 km^2) in 2010 and subsequently increased to 19.16% (21,435.77 km^2) in 2020. Agricultural land exhibited significant fluctuations, dropping from 10.98% (12,283.53 km^2) in 2000 to 9.04% (10,117.69 km^2) in 2010 and remaining stable at 9.04% (10,119.95 km^2) in 2020. Built-up land experienced substantial growth during this period, expanding from 2.63% (2950.28 km^2) in 2000 to 7.21% (8072.39 km^2) in 2010 and reaching 9.28% (10,383.14 km^2) in 2020. Similarly, snow cover increased steadily from 2.50% (2807.54 km^2) in 2000 to 6.90% (7729.40 km^2) in 2010 and further to 8.74% (9787.77 km^2) in 2020.

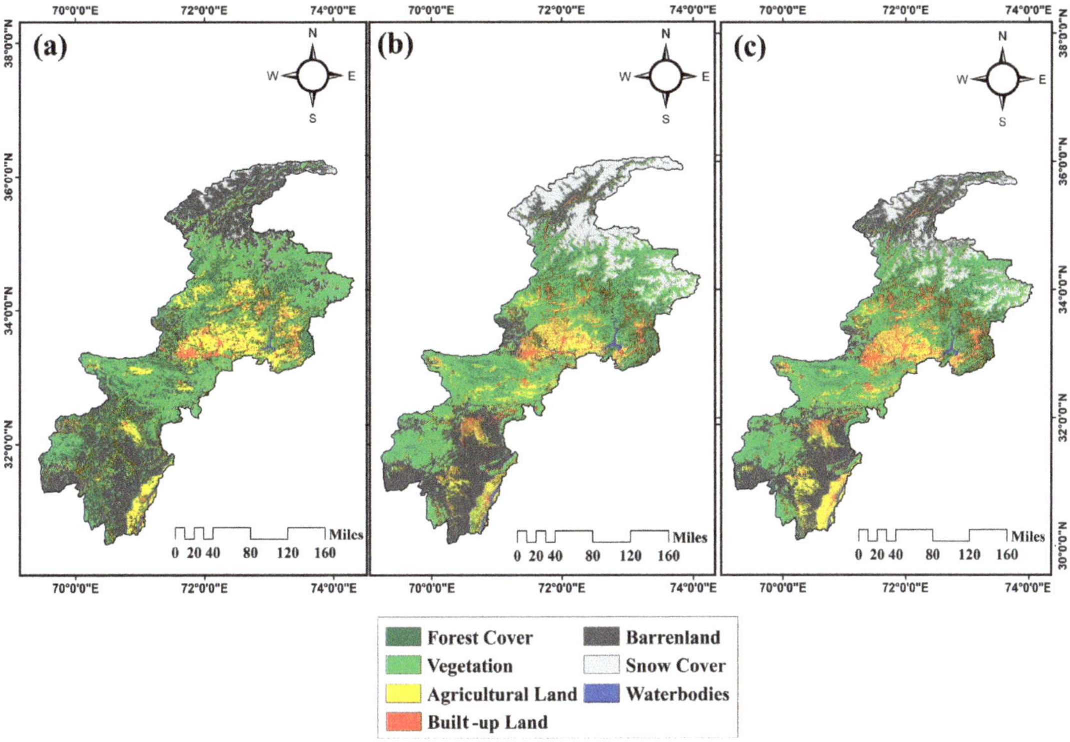

Figure 5. Land use and land cover (LULC) classification in Khyber Pakhtunkhwa, Pakistan for the years (**a**) 2000, (**b**) 2010, (**c**) 2020.

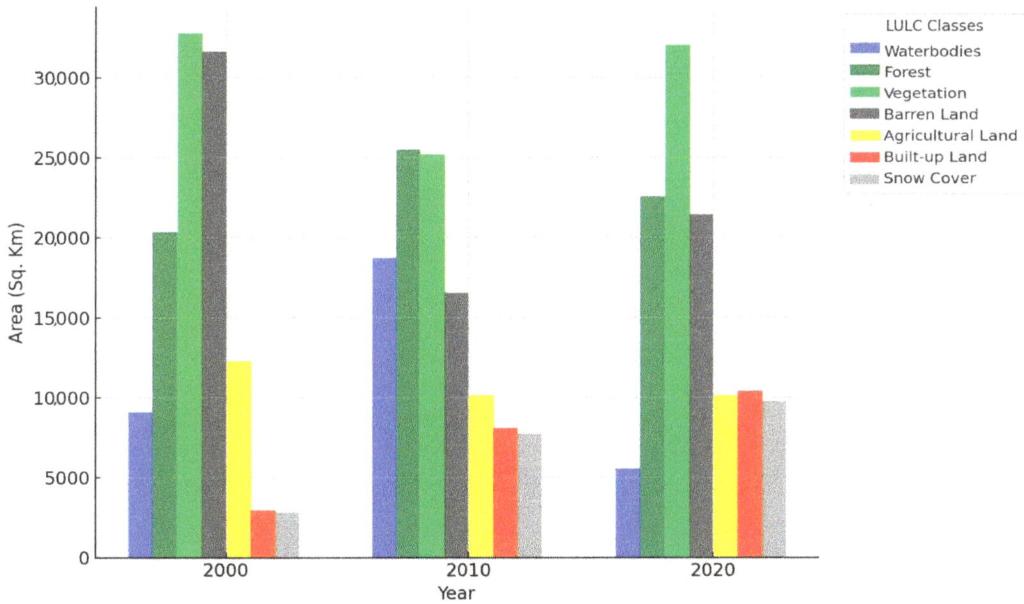

Figure 6. Area of each land cover class for the years 2000, 2010, and 2020.

These LULC trends highlight complex dynamics influenced by anthropogenic and environmental factors. Expanding built-up areas and barren land, alongside reductions in waterbodies and vegetation, emphasize the need for sustainable land management strategies to mitigate adverse impacts on biodiversity, agricultural productivity, and water availability. The increase in snow cover has several negative environmental impacts. Excessive snow accumulation can disrupt regional hydrological systems by delaying snowmelt, which reduces immediate water runoff critical for agriculture and ecosystems during dry seasons. This delay exacerbates water scarcity and impacts water resource management. Furthermore, increased snow cover raises the risk of glacial surges and avalanches, posing significant hazards to infrastructure and ecosystems. It can also alter regional energy balances by increasing surface albedo, leading to reduced solar radiation absorption, which may impact vegetation growth cycles and soil processes. These changes indicate dynamic changes in the landscape, significant urban expansion, and changes in natural land cover.

The changes calculated above (Table 4) are alarming, as the decrease in the area of waterbodies, forests, vegetation, and agricultural land could lead to water scarcity, loss of biodiversity, more extreme weather events, air pollution, soil degradation, and food insecurity. The increase in the area of built-up land and snow cover could also lead to negative environmental impacts affecting quality of life in the region.

Table 4. Area statistics for each land cover class for the years 2000, 2010, and 2020.

Class	Area 2000 (Sq. Km)	Area in % (Year 2000)	Area 2010 (Sq. Km)	Area in % (Year 2010)	Area 2020 (Sq. Km)	Area in % (Year 2020)
Waterbodies	9099.29	8.13	18,732.76	16.74	5549.78	4.96
Forest	20,322.02	18.16	25,482.21	22.77	22,556.68	20.16
Vegetation	32,762.39	29.28	25,191.63	22.51	32,030.81	28.63
Barren Land	31,638.85	28.28	16,537.83	14.78	21,435.77	19.16
Agricultural Land	12,283.53	10.98	10,117.69	9.04	10,119.95	9.04
Built-up Land	2950.28	2.63	8072.39	7.21	10,383.14	9.28
Snow Cover	2807.54	2.5	7729.4	6.9	9787.77	8.74

4.2. *Temporal and Spatial Dynamics of LST, NDVI, MNDWI, LAI, and NDMI*

This section examines the temporal and spatial dynamics of LST, NDVI, MNDWI, LAI, and NDMI in KPK, Pakistan, using satellite imagery from 2000, 2010, and 2020.

4.2.1. Land Surface Temperature (LST)

LST measures the temperature of the land surface, calculated using satellite imagery in the thermal infrared band. The spatial and temporal distribution of LST was determined using the Landsat 5 TM and Landsat 8 OLI thermal bands for 2000, 2010, and 2020. LST data were synthesized using a median composite method to minimize outliers, such as cloud cover and atmospheric anomalies, ensuring stable and representative conditions across diverse environmental settings. LST values were divided into five ranges (≤20 °C, 20–25 °C, 25–30 °C, 30–35 °C, and >35 °C) for consistency in interpretation across the study area. This fixed-interval classification is a widely accepted method in remote sensing and environmental studies, providing clarity and comparability across geographically diverse regions like KPK. While the classification does not assume a simple linear relationship between LST and LULC, it enables the identification of broad temperature patterns, which are critical for analyzing long-term trends and regional impacts (Table 5). The relationship between LULC and LST in KPK reveals the significant influence of land use changes on regional thermal patterns. Between 2000 and 2020, forest cover decreased from 31.49% to 23.54%, and vegetation cover dropped from 50.95% to 33.44%. These reductions contributed to an increase in areas with temperatures above 35 °C, rising from 21.57% in 2000 to 23.03% in 2020. Forests and vegetation mitigate surface heating through evapotranspiration and albedo effects; their loss exacerbates urban heat island effects, particularly as built-up land expanded from 4.56% to 10.89% over the same period. This trend underscores the direct role of LULC changes in driving LST increases (Figure 7). Additional analysis was conducted to evaluate the influence of topography on LST patterns, as the relationship between LST and LULC is inherently complex and influenced by various factors. A digital elevation model (DEM) was incorporated into the study to assess how elevation variations impact temperature distribution across the study area. The DEM (Figure 8) analysis revealed a strong negative correlation between elevation and LST (R^2 values: 0.98 in 2000, 0.95 in 2010, and 0.975 in 2020). This finding confirms that higher elevations, particularly in the northern mountainous regions with permanent snow cover, consistently experience lower temperatures compared to the southern regions. These results emphasize the dominant role of topography in shaping LST patterns, alongside the effects of LULC changes. The relationship between LULC and LST is primarily unidirectional, where changes in LULC influence LST patterns. However, over longer timescales, the relationship may become reciprocal, as sustained LST increases could degrade vegetation and soil, altering future LULC patterns. Previous studies provide evidence of urbanization and deforestation elevating LST. These findings align with the IPCC (2021) report and historical temperature data from the Pakistan Meteorological Department (PMD), which confirm regional warming trends exacerbated by greenhouse gas emissions [56].

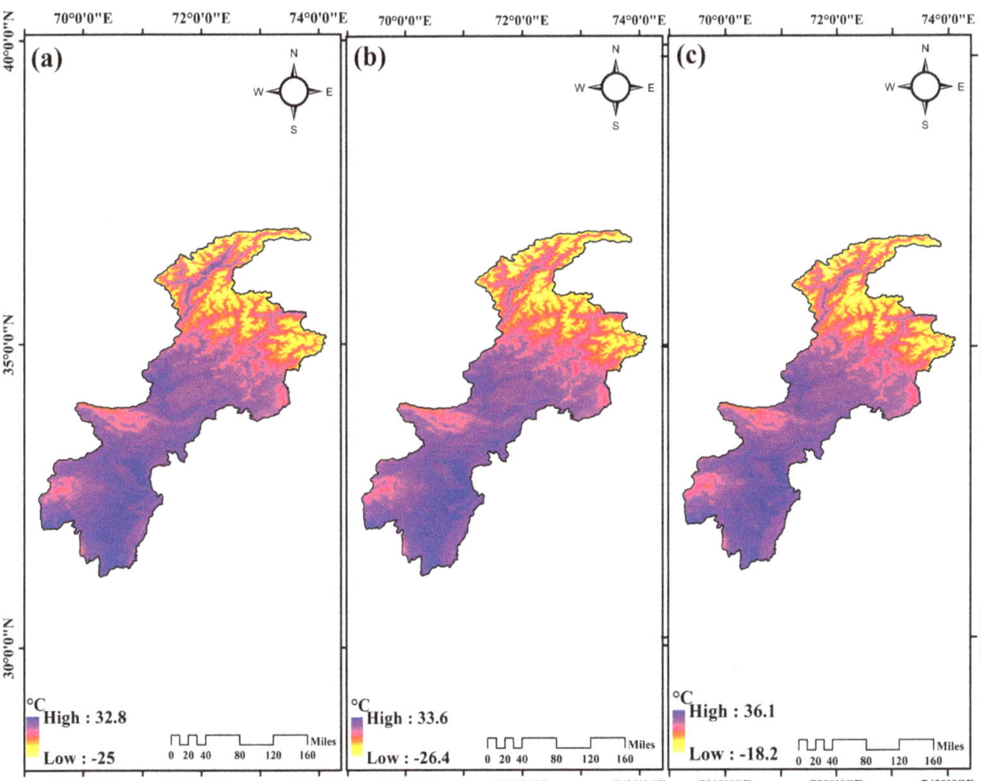

Figure 7. Land surface temperature (LST) distribution in Khyber Pakht-unkhwa, Pakistan in (**a**) 2000, (**b**) 2010, (**c**) 2020.

Table 5. Area statistics of the actual land surface temperature for the years 2000, 2010, and 2020.

Year	2000		2010		2020	
LST (°C)	Area (km^2)	Area (%)	Area (km^2)	Area (%)	Area (km^2)	Area (%)
≤20	240.48	6.36	361.19	9.56	18.92	0.48
20–25	923.32	24.43	779.25	20.61	514.34	13.18
25–30	841.16	19.85	875.17	23.15	629.6	16.06
30–35	1014.31	26.05	1019.34	26.97	607.75	15.49
≥35	760.77	21.57	745.07	19.71	894.48	23.03

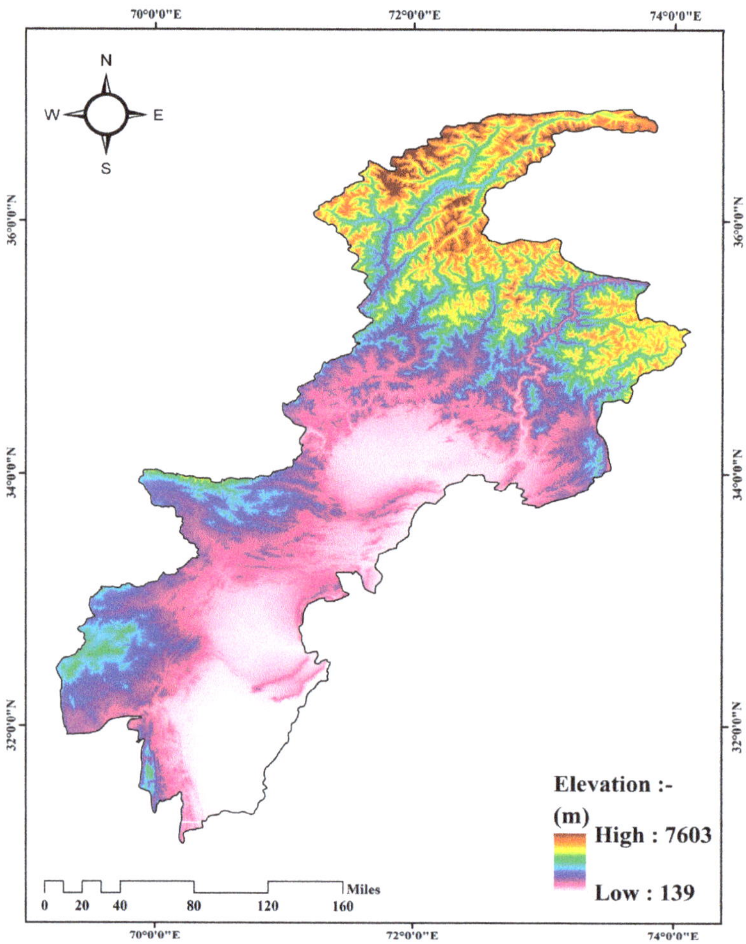

Figure 8. Digital elevation model (DEM) of Khyber Pakht-unkhwa (KPK) region highlighting topographical variation.

The rise in LST from 32.8 °C in 2000 to 36.1 °C in 2020 stems from both anthropogenic LULC changes and climate change. These elevated temperatures have significant implications for KPK, including increased heat stress, higher water demand, reduced crop yields, and greater energy consumption. Urban areas are particularly vulnerable due to heat-retaining impervious surfaces. These findings highlight the need for sustainable urban planning and land management strategies to mitigate the combined effects of LULC changes and climate change.

4.2.2. Modified Normalized Difference Water Index (MNDWI)

The modified normalized difference water index (MNDWI) measures the presence and dynamics of surface waterbodies using satellite imagery from the green and shortwave infrared (SWIR) bands. The green band enhances water feature detection, while the SWIR band suppresses noise from built-up areas and vegetation. MNDWI is calculated as the difference between reflectance in the green and SWIR bands. For this study, the spatial and temporal distribution of MNDWI was determined using Landsat 5 TM (2000 and 2010) and Landsat 8 OLI (2020) sensors.

Over the study period, MNDWI values showed a clear decline, indicating a reduction in surface water availability (Figure 9). The maximum MNDWI dropped from 0.78 in 2000 to 0.71 in 2020, while the average MNDWI decreased sharply from 0.27 to 0.03 over the same period (Table 6). The southern region consistently exhibited the lowest MNDWI values, reflecting significant water loss and increased aridification. These results highlight the fragmentation of waterbodies over time, with continuous water features in 2000 breaking into smaller, isolated patches by 2020. Northern regions retained relatively stable MNDWI values due to the presence of perennial rivers and reservoirs, while southern regions were more vulnerable to water scarcity. These changes are strongly associated with the expansion of barren land, urbanization, and agricultural intensification.

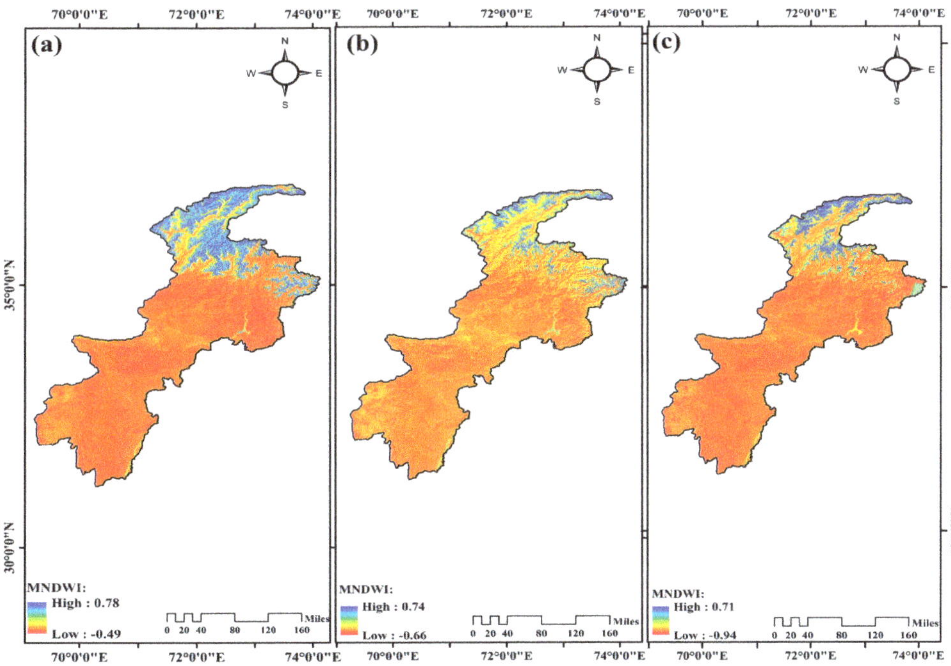

Figure 9. Modified normalized difference water index (MNDWI) in Khyber Pakht-unkhwa, Pakistan for the years (**a**) 2000, (**b**) 2010, (**c**) 2020.

Population growth in KPK has exacerbated surface water stress. Urban expansion, driven by increasing population, has led to the proliferation of impervious surfaces, such as roads and buildings, which reduce infiltration, heighten runoff, and elevate flooding risks during intense precipitation events. This relationship is well-documented in studies linking urbanization to increased hydrological disruptions [57]. The reduced infiltration capacity of impervious surfaces amplifies surface water stress by limiting groundwater recharge, while heightened runoff exacerbates the likelihood of urban flooding [58,59]. This underscores the complex relationship between population dynamics and water availability, emphasizing the need for sustainable urban planning to mitigate these risks.

Table 6. Actual normalized difference water index (MNDWI) values for the years 2000, 2010, and 2020.

Year	Min MNDWI	Max MNDWI	Mean MNDWI	StdDev
2000	−0.49	0.78	0.2655	0.859
2010	−0.66	0.74	0.256	0.8287
2020	−0.94	0.71	0.026	0.7778

4.2.3. Normalized Difference Vegetation Index (NDVI)

The calculation of the NDVI serves as a quantitative indicator to assess vegetation coverage within a given region. This index is derived from satellite imagery, specifically utilizing data from the red and near-infrared bands. Through this process, valuable information regarding chlorophyll content in vegetation can be obtained by analyzing the red band [20]. Simultaneously, by examining reflected radiation levels in the near-infrared band, insights into vegetative conditions become accessible [60]. Consequently, NDVI values are established based on disparities between reflectance measurements taken from these two distinct bands. For comprehensive analysis throughout multiple years, including 2000, 2010, and 2020, Landsat 5 TM and 8 OLI sensors were employed to determine spatial and temporal distribution patterns associated with NDVI assessment (Figure 10).

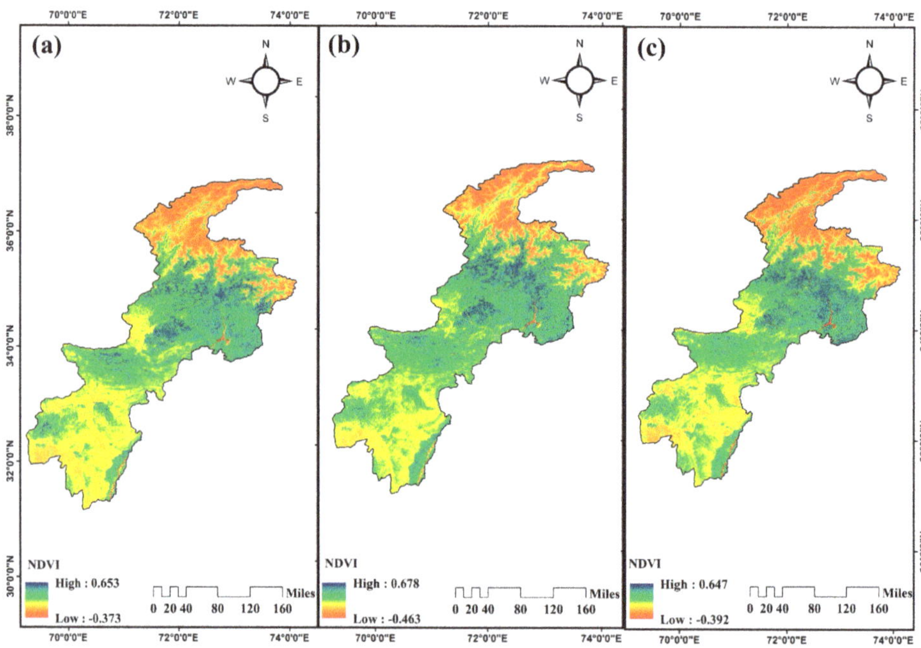

Figure 10. Normalized difference vegetation index (NDVI) in Khyber Pakht-unkhwa, Pakistan for the years (**a**) 2000, (**b**) 2010, (**c**) 2020.

As is evident from (Table 7), the NDVI values have declined overall, with the area covered by low NDVI values increasing between 2000 and 2020. The mean NDVI dropped from 0.14 in 2000 to 0.11 in 2010 and further decreased to 0.52 in 2020, while the standard deviation sharply declined from 0.73 to 0.18 over the same period. This decline highlights significant vegetation loss, particularly in areas experiencing deforestation and land degradation [47,61,62].

Table 7. Actual normalized difference vegetation index (NDVI) values for 2000, 2010, and 2020.

Year	Min NDVI	Max NDVI	Mean NDVI	StdDev
2000	−0.37	0.65	0.14	0.73
2010	−0.46	0.68	0.11	0.81
2020	0.392	0.647	0.5195	0.1803

Several factors explain the decline in NDVI in KPK. One significant factor is deforestation due to agricultural expansion, timber extraction, and the use of firewood. The expansion of agricultural land is a major driver of forest and vegetation loss in the region [56]. Additionally, climate change has played a crucial role, with rising temperatures and decreased rainfall levels creating conditions unfavorable for vegetation growth (IPCC, 2021). These climatic shifts have reduced plant water availability, negatively affecting transpiration processes and resulting in prolonged drought periods.

The consequences of declining NDVI are far-reaching. Reduced vegetation limits the ability of plants to sequester carbon dioxide, exacerbating climate change impacts. Moreover, water scarcity associated with declining NDVI reduces habitat quality and food resources for wildlife, increasing the vulnerability of species dependent on vegetation for survival [60]. These findings highlight the critical need for sustainable land management practices to mitigate further vegetation loss and promote regional ecological resilience.

4.2.4. LAI (Leaf Area Index)

The LAI refers to the leaf area per unit area of the ground surface: the leaf area is that part of the plant that directly intercepts radiation. It is an important factor when determining plant growth, photosynthesis, and evapotranspiration. Most users of LAI calculate LAI from satellites where the data can be obtained in the red and one of the near-infrared bands. The last five bands are divided into two parts: the red band decides the amount of chlorophyll in the vegetation, and the near-infrared band calculates the amount of reflected radiation. LAI is the total green leaf area per ground area beneath the canopy [28,54,63,64]. The spatial and temporal distribution of LAI was determined using the Landsat 5 TM and Landsat 8 OLI sensors for the entirety of the study years (2000, 2010, and 2020).

LAI data from 2000 to 2020 (Figure 11) show evident fluctuations in vegetation density, consistent with observed changes in land use and moisture content. Maximum LAI values decreased from 4.66 in 2000 to 2.16 in 2020, reflecting a reduction in the area of dense vegetation, particularly forest cover. This trend is consistent with agricultural and built-up land expansion observed in the LULC maps, indicating a shift from natural vegetation to more anthropogenic land use.

The mean LAI value in 2010 (−1.48) decreased significantly compared to 2000 (−0.55), indicating a decline in overall vegetation health and cover. However, by 2020, the mean LAI improved slightly to −0.20, which may be due to localized reforestation or changes in agricultural practices but is still lower than in 2000. The standard deviation (StdDev) of LAI values has also decreased over time, from 7.36 in 2000 to 3.33 in 2020 (Table 8), indicating a decrease in variability in vegetation density across the region. However, the decrease in LAI variability and the decrease in MNDWI indicate a homogenization of the landscape towards drier, less vegetated conditions, especially in the southern region where barren lands are expanding. The correlation between the decrease in LAI and the increase in the built-up area suggests that urbanization is leading to a loss of vegetation cover, affecting the region's microclimate and water balance. In contrast, the decrease in LAI indicates that the region is increasingly vulnerable to desertification and land degradation. Also, the availability of waterbodies has reduced, and barren land has increased, which makes them a probable cause of increased environmental pressures. Supposing the incidence and extent of the present patterns of land use and land cover persist, a further reduction in LAI and MNDWI will occur, causing greater effects on the current levels of biodiversity, food production, and regional climate. Also, variations in LAI in KPK could be attributed to factors such as deforestation, climate change, and use of land for agriculture. Loss of trees and plants will, therefore, reduce the LAI as a result of deforestation. Climate change can also reduce LAI, because it hinders plant growth in regions with higher temperatures and less rainfall. Another way that agricultural practices depress the LAI is through activities such as land conversion from natural vegetation to agricultural lands. As a result of reduced LAI, there are many implications in KPK, such as climate change, drought, and reduction in food and habitat needed for wildlife. Hence, it is high time we focused on cutting down LAI through tree planting, forest rehabilitation, and developing agricultural practices that are relatively resilient to climate change.

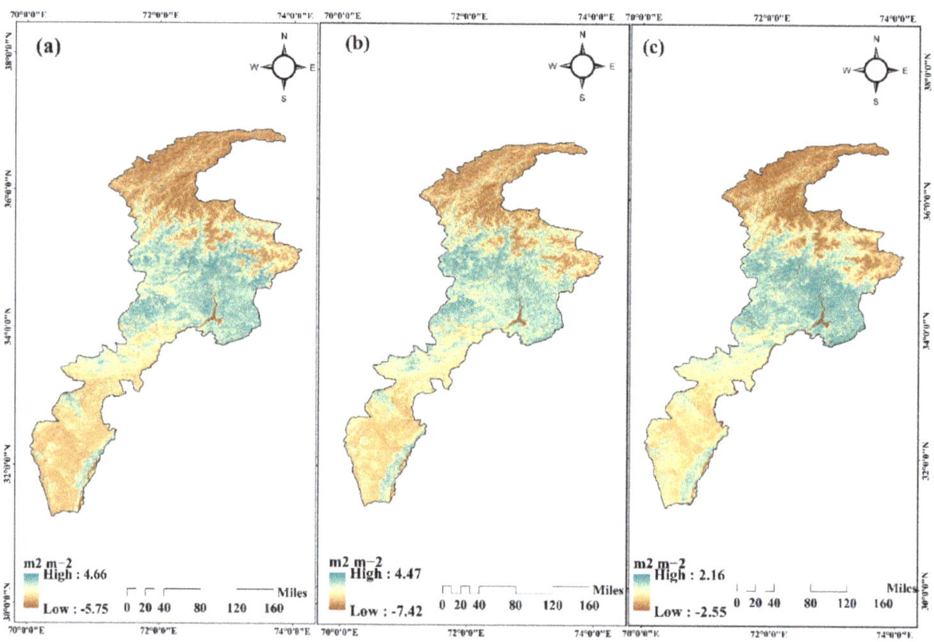

Figure 11. Leaf area index (LAI) analysis in Khyber Pakh-tunkhwa, Pakistan for the years (**a**) 2000, (**b**) 2010, (**c**) 2020.

Table 8. Actual leaf area index (LAI) values for the years 2000, 2010, and 2020.

Year	Min LAI	Max LAI	Mean LAI	StdDev
2000	−5.75	4.66	−0.545	7.3609
2010	−7.42	4.47	−1.475	8.4074
2020	−2.55	2.16	−0.195	3.3304

4.2.5. Normalized Differences Moisture Index (NDMI)

The NDMI is a key indicator used to measure vegetation moisture by utilizing reflectance data from the near-infrared (NIR) and shortwave infrared (SWIR) bands. The NIR band is sensitive to vegetation structure, while the SWIR band is responsive to vegetation water content, making NDMI an effective tool for monitoring water stress in vegetated areas. NDMI is calculated as the normalized difference between the NIR and SWIR bands, providing insights into the water retention capacity of vegetation. This study used Landsat 5 TM and Landsat 8 OLI sensors to calculate NDMI values for the years 2000, 2010, and 2020. Specifically, the NIR (SR_B4) and SWIR (SR_B5) bands of Landsat 5 and the NIR (SR_B5) and SWIR (SR_B6) bands of Landsat 8 were employed. The resulting NDMI values were aggregated using a median composite method to reduce noise and capture the typical vegetation moisture conditions across the study period.

The results, summarized in (Table 9) and visualized in (Figure 12), reveal a significant decline in vegetation water content over time. In 2000, NDMI values ranged from −0.39 to 0.76, with a mean of 0.35 and a standard deviation of 0.475, reflecting relatively healthy vegetation water content. By 2010, NDMI values declined, ranging from −0.54 to 0.69, with a mean of 0.32 and a slightly higher standard deviation of 0.48. This declining trend continued in 2020, with NDMI values ranging from −0.41 to 0.69, with a mean of 0.33 and a standard deviation of 0.48. These changes highlight a reduction in the vegetation's ability to retain moisture, particularly in areas undergoing land use and land cover changes. The reduction in NDMI values aligns with the observed increase in barren land and urbanization in KPK. Deforestation and the conversion of forests to agricultural land or built-up areas reduce vegetation cover and soil water retention, exacerbating water stress in the region. Additionally, climate change-induced factors such as rising temperatures and altered precipitation patterns have further contributed to the loss of vegetation water content, particularly in southern and low elevation regions.

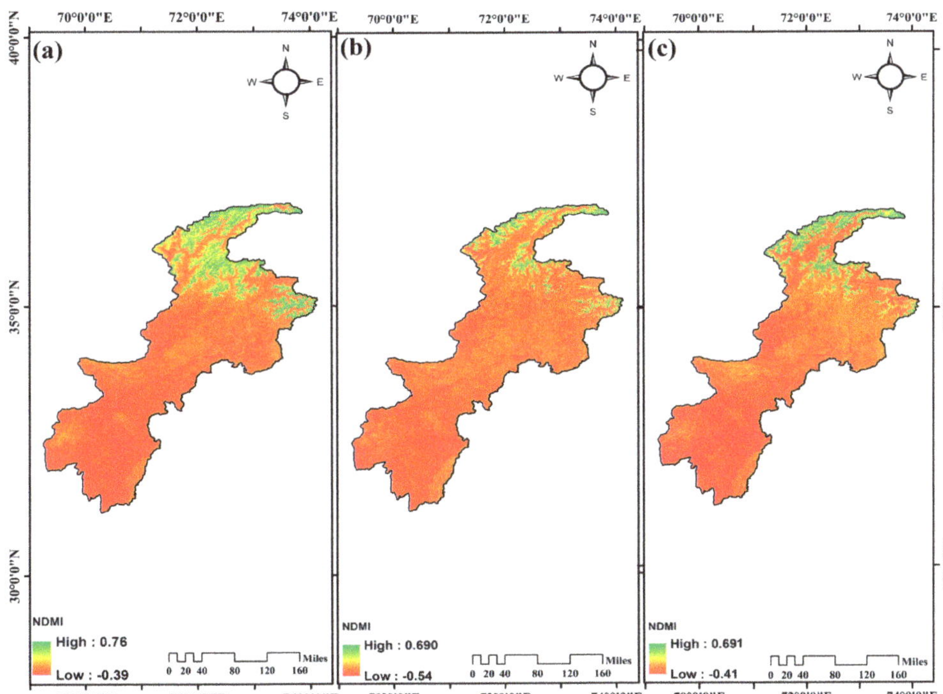

Figure 12. Normalized difference moisture index (NDMI) analysis in Khyber Pakh-tunkhwa, Pakistan for the years (**a**) 2000, (**b**) 2010, (**c**) 2020.

Table 9. Normalized Difference Moisture Indes (NDMI) values for the years 2000, 2010, and 2020.

Year	Min NDMI	Max NDMI	Mean NDMI	StdDev
2000	−0.39	0.76	0.35	0.48
2010	−0.54	0.69	0.32	0.48
2020	−0.41	0.69	0.33	0.48

These findings underscore the urgent need to implement sustainable land management practices to mitigate the adverse impacts of declining vegetation water content. Strategies such as reforestation, afforestation, and the adoption of water-efficient agricultural practices are essential to restoring vegetation health and improving the region's resilience to environmental changes. By monitoring NDMI trends, this study provides valuable insights into vegetation water dynamics, enabling better-informed decision-making for ecosystem management and climate adaptation.

4.3. Predictions of LULC for 2030, 2050 and 2100

For predictions of the LULC map for the years 2030, 2050, and 2100, we used the 2000–2010, 2010–2020, and 2020–2030 simulated LULC maps as the base maps. Figures 13 and 14 show the predicted land use and land cover (LULC) changes in KPK, Pakistan, for the years 2030, 2050, and 2100. Analyzing predicted LULC changes will identify potential environmental impacts, assess vulnerabilities, and inform adaptation and strategies. The LULC predictions were generated using the cellular automata-artificial neural network (CA-ANN) model, which integrates historical LULC trends with key influencing factors such as climate change, population growth, and urbanization. While specific census or economic datasets were not directly utilized, these factors were modeled based on observed patterns of urbanization, agricultural intensification, and deforestation between 2000 and 2020. Population growth and economic development effects were incorporated indirectly by extrapolating land cover changes driven by urban expansion and land demand trends. Climate change impacts were represented through temperature and precipitation trends guided by regional climatic models, ensuring a comprehensive approach to future scenario prediction. This methodology assumes a continuation of the observed trajectory of LULC changes, with the proportions of each class evolving based on historical dynamics and projected influences.

Table 10 presents the predicted land use and land cover (LULC) changes in KPK, Pakistan, for the years 2030, 2050, and 2100. These predictions are based on various factors, including climate change, population growth, and economic development. The most significant predicted change is the increase in barren land. Barren land is expected to rise from 21.14% (15,947.03 km^2) in 2030 to 33.96% (25,616.86 km^2) in 2100. This dramatic increase is attributed to various factors, including climate change, deforestation, and overgrazing. The second most significant change is observed in built-up land, which is projected to increase from 10.24% (7724.48 km^2) in 2030 to 16.44% (12,408.39 km^2) in 2100. This growth is likely due to population growth and urbanization, highlighting the challenges of expanding urban areas. Forest cover is predicted to decline steadily, decreasing from 19.21% (14,492.62 km^2) in 2030 to 13.22% (9977.37 km^2) in 2100. Similarly, vegetation cover is expected to drop from 27.28% (20,579.72 km^2) in 2030 to 18.78% (14,168.00 km^2) in 2100. These reductions reflect the impacts of land use changes, deforestation, and declining vegetation health due to climatic stress. Agricultural land is projected to shrink from 8.61% (6502.04 km^2) in 2030 to 5.93% (4476.30 km^2) in 2100. This decrease is a significant concern, underscoring the pressure on agricultural productivity caused by land degradation and urban expansion. Waterbodies are expected to diminish from 4.72% (3565.70 km^2) in 2030 to 3.25% (2454.79 km^2) in 2100, further highlighting the region's increasing vulnerability to water shortages. In contrast, snow cover is predicted to remain relatively stable, with a slight fluctuation of around 8.39% (6329.49 km^2) by 2100.

Table 10. Projected land use and land cover in KPK, Pakistan, for the years 2030, 2050, and 2100, with area in square kilometers and percentage of the total area for each class.

Class	Area 2030 (Sq. Km)	Area in % (Year 2030)	Area 2050 (Sq. Km)	Area in % (Year 2050)	Area 2100 (Sq. Km)	Area in % (Year 2100)
Waterbodies	3565.7	4.72	3199.79	4.24	2454.79	3.25
Forest	14,492.62	19.21	13,005.38	17.24	9977.37	13.22
Vegetation	20,579.72	27.28	18,467.81	24.48	14,168	18.78
Barren Land	15,947.03	21.14	19,080.71	25.29	25,616.86	33.96
Built-up Land	7724.48	10.24	9242.38	12.25	12,408.39	16.44
Snow Cover	6619.6	8.77	6600.32	8.75	6329.49	8.39
Agricultural Land	6502.04	8.61	5834.8	7.735	4476.3	5.93

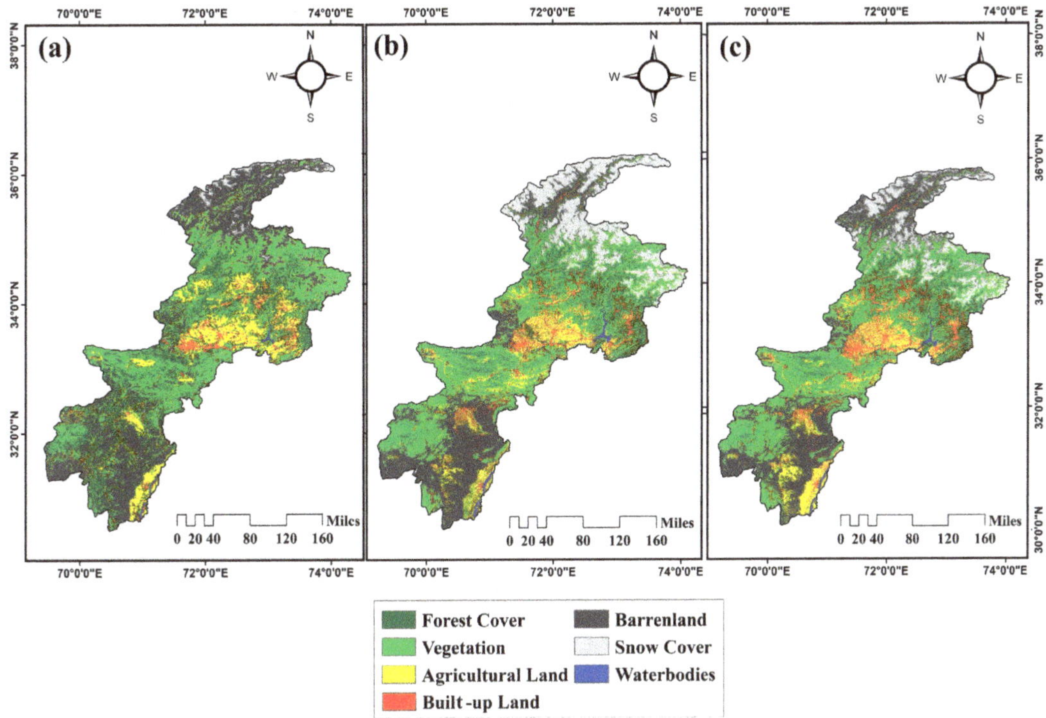

Figure 13. Land use and land cover (LULC) classification in Khyber Pakh-tunkhwa, Pakistan in (**a**) 2030, (**b**) 2050, (**c**) 2100.

The predicted changes in LULC have many implications for KPK. The increase in barren land is likely to lead to a decrease in agricultural productivity, an increase in soil erosion, and a decline in biodiversity. The increased built-up land will likely lead to increased traffic congestion, air pollution, and heat island effects. Decreased forest cover will likely lead to increased soil erosion, decreased water availability, and reduced carbon storage. Decreased vegetation cover will likely lead to decreased water availability, reduced carbon storage, and increased soil erosion. The decrease in agricultural land will likely lead to a decrease in food production and an increase in food prices. Decreased waterbodies will likely lead to water shortages and a decline in aquatic ecosystems.

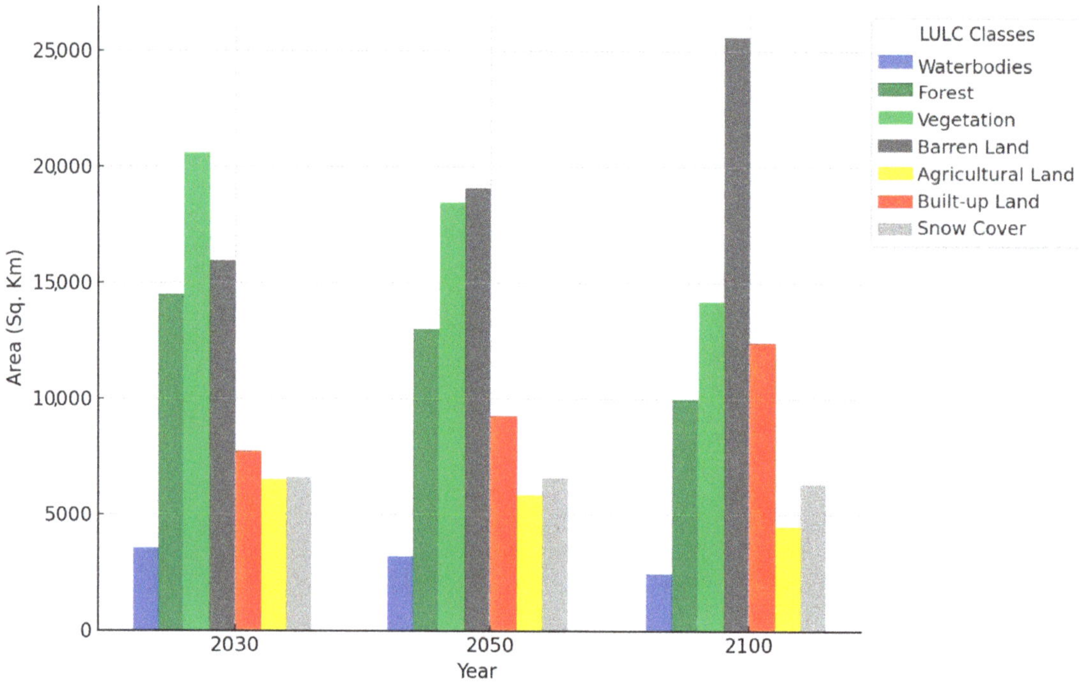

Figure 14. Areas in square kilometers for each land cover class for the years 2030, 2050, and 2100.

4.4. Predictions of LST, NDVI, MNDWI, LAI, and NDMI

4.4.1. Land Surface Temperature (LST)

Our analysis extends to future years, including 2030, 2050, and 2100, ensuring a comprehensive understanding of the temporal dynamics. The spatial and temporal distribution of LST was systematically examined, allowing the classification of LST values into distinct ranges (≤20 °C, 20–25 °C, 25–30 °C, 30–35 °C, and >35 °C) based on the observed lower and higher LST values across different timeframes. This categorization enables computation of the proportion of areas falling within each specified temperature range, contributing to a nuanced portrayal of thermal variations over the projected years (Figure 15).

By 2030, the area with temperatures ≤20 °C will cover 36.64 km^2, but by 2100, this figure will decrease significantly to only 13.42 km^2, highlighting the sharp decline in colder climate areas. At the same time, the most extreme temperature range (≥35 °C) will not only become more common but also will nearly double in area, expanding from 34.21% in 2030 to 39.8% in 2100. This trend is accompanied by a decrease in areas that may currently be considered to have "moderate" temperatures (20–25 °C), which shrink as a percentage of the total observed area, indicating that milder climates are retreating.

The area with temperatures between 30–35 °C will increase modestly by 2050 and decrease slightly by 2100, which may indicate that climate change will continue through the middle of this century before entering more consistently extreme conditions.

The percentage of areas with the highest temperatures will continue to increase (Table 11), highlighting a trend of increasing extreme heat conditions over time. Long-term temperatures will also increase over 70 years, likely significantly impacting ecosystems, agriculture and urban planning.

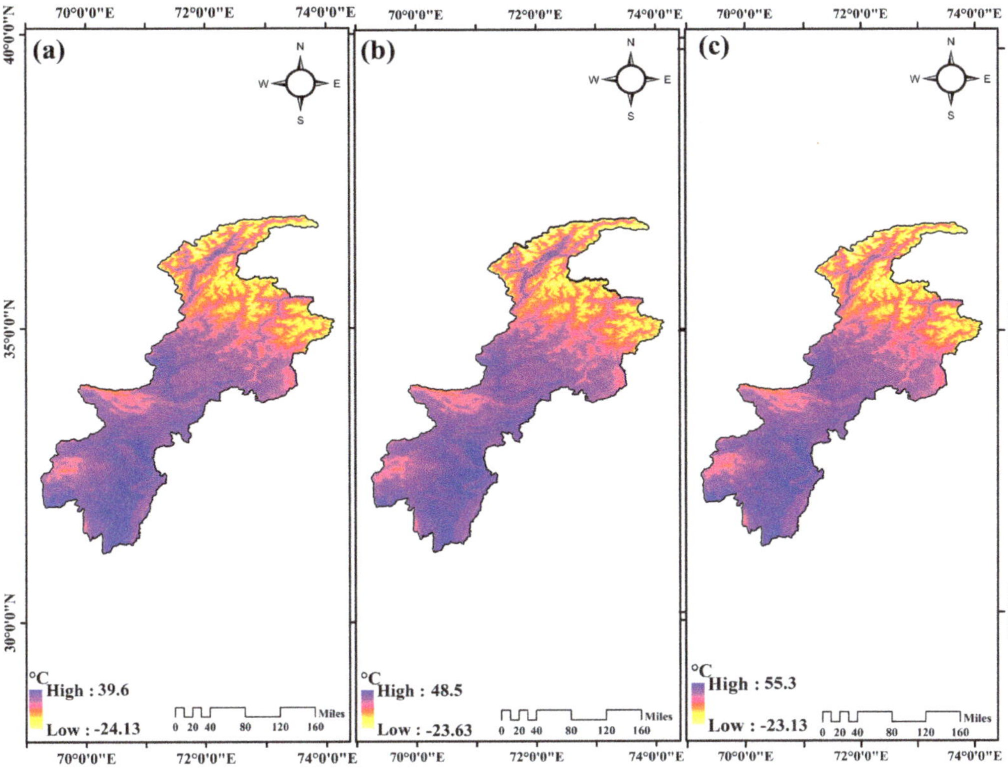

Figure 15. Land surface temperature (LST) distribution in Khyber Pakh-tunkhwa, Pakistan in (**a**) 2030, (**b**) 2050, (**c**) 2100.

Table 11. Prediction of the area statistics of the land surface temperature for the years 2030, 2050, and 2100.

Year	LST (°C)	Area (km^2)	Area (%)
2030	≤20	36.64	0.94
	20–25	582.1	15.01
	25–30	644.94	16.52
	30–35	990.78	25.33
	≥35	1332.52	34.21
2050	≤20	22.54	0.58
	20–25	571.71	14.64
	25–30	617.92	15.89
	30–35	1116.73	28.25
	≥35	1228.58	31.59
2100	≤20	13.42	0.34
	20–25	381.74	9.71
	25–30	485.91	12.39
	30–35	1102.15	27.97
	≥35	1566.79	39.8

4.4.2. Normalized Difference Vegetation Index (NDVI)

To anticipate changes in the years 2030, 2050, and 2100 (Figure 16), our focus extends to the NDVI, a key indicator of vegetation health. Employing satellite imagery from the Landsat 5 TM and Landsat 8 OLI sensors, NDVI calculations were performed for 2000, 2010, and 2020. The resulting NDVI values in our analysis provide insights into the temporal evolution of vegetation health, shedding light on trends and potential future variations. In 2030, the NDVI values will range from a minimum of −0.473 and a maximum of 0.644, with an average of 0.0855, revealing relatively healthy vegetation. However, by 2050, minimum and maximum values will decrease, indicating a decline in vegetation health, and the average will become negative (−0.071). This trend will worsen by the year 2100, with the minimum NDVI increasing slightly to −0.546, but the maximum value will drop further to 0.362, and the average will further decrease to −0.092. Furthermore, the standard deviation will decrease from 0.7898 in 2030 to 0.6420 in 2100 (Table 12). This means that vegetation and water content in the area are expected to decline. Model adjustments were made to correct initial inconsistencies, such as negative average NDVI values for 2050 and 2100, which do not represent real-world conditions. Constraints were introduced to maintain NDVI within logical positive ranges, ensuring the predictions reflect expected vegetation trends while avoiding unrealistic outcomes. The decline in NDVI can be due to various factors, including climate change, deforestation, and urbanization. Climate change leads to higher temperatures and drier weather, which can stress vegetation and reduce water availability [21]. Deforestation is also a major factor in the decline in NDVI, as trees are replaced by land with less vegetation. Urbanization can also cause the NDVI to decline, as impervious surfaces such as roads and buildings reflect more sunlight than vegetation.

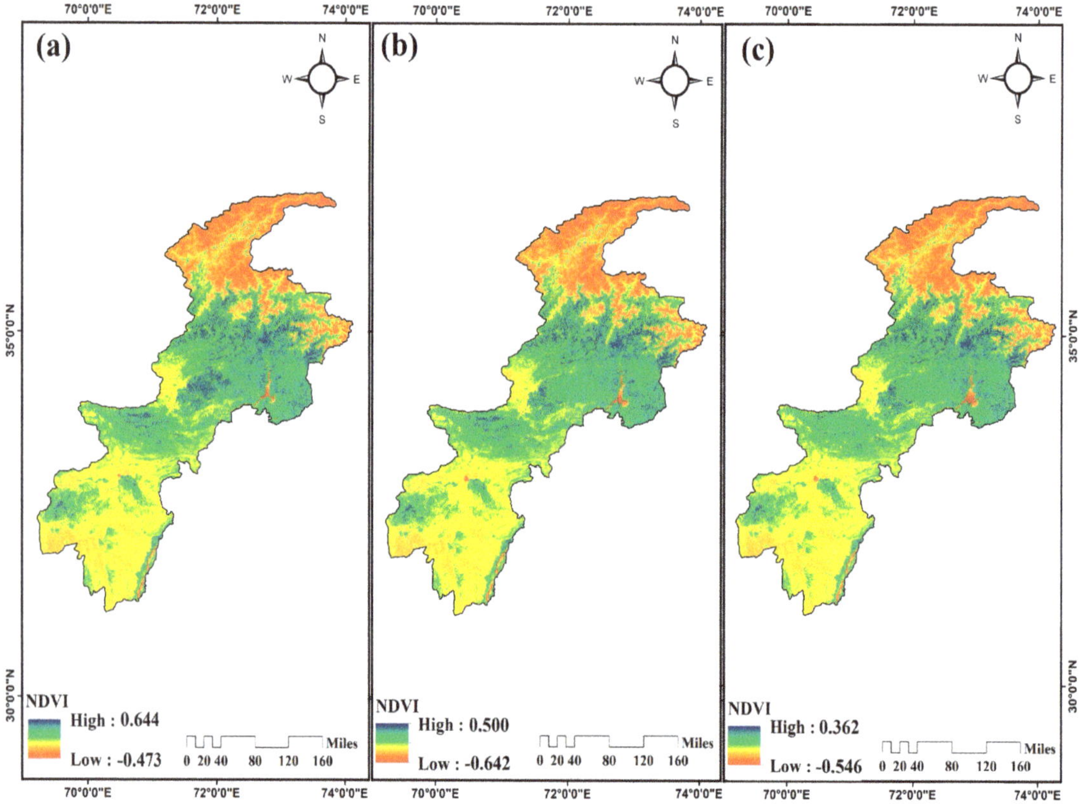

Figure 16. Normalized difference vegetation index (NDVI) in Khyber Pakh-tunkhwa, Pakistan in (**a**) 2030, (**b**) 2050, (**c**) 2100.

Table 12. Prediction of normalized difference vegetation index (NDVI) values for the years 2030, 2050, and 2100.

Year	Min NDVI	Max NDVI	Mean NDVI	StdDev
2030	−0.473	0.644	0.0855	0.7898
2050	−0.642	0.500	−0.071	0.8075
2100	−0.546	0.362	−0.092	0.6420

4.4.3. Modified Normalized Difference Water Index (MNDWI)

The predicted MNDWI changes for the years 2030, 2050, and 2100 indicate a continuous decline in surface water content and increasing environmental stress (Figure 17). Our analysis encompasses the spatial and temporal distribution of MNDWI, utilizing data from the Landsat 5 TM and Landsat 8 OLI sensors. The minimum MNDWI values will drop sharply from −0.6 in 2030 to −0.9 in 2050, followed by a partial recovery to −0.45 in 2100. This fluctuation suggests worsening water stress in the mid-term, with a slight stabilization in the long-term, potentially due to land use changes or water management interventions. Similarly, the maximum MNDWI values will decline steadily, from 0.69 in 2030 to 0.61 in 2100, reflecting a reduction in areas with higher water content.

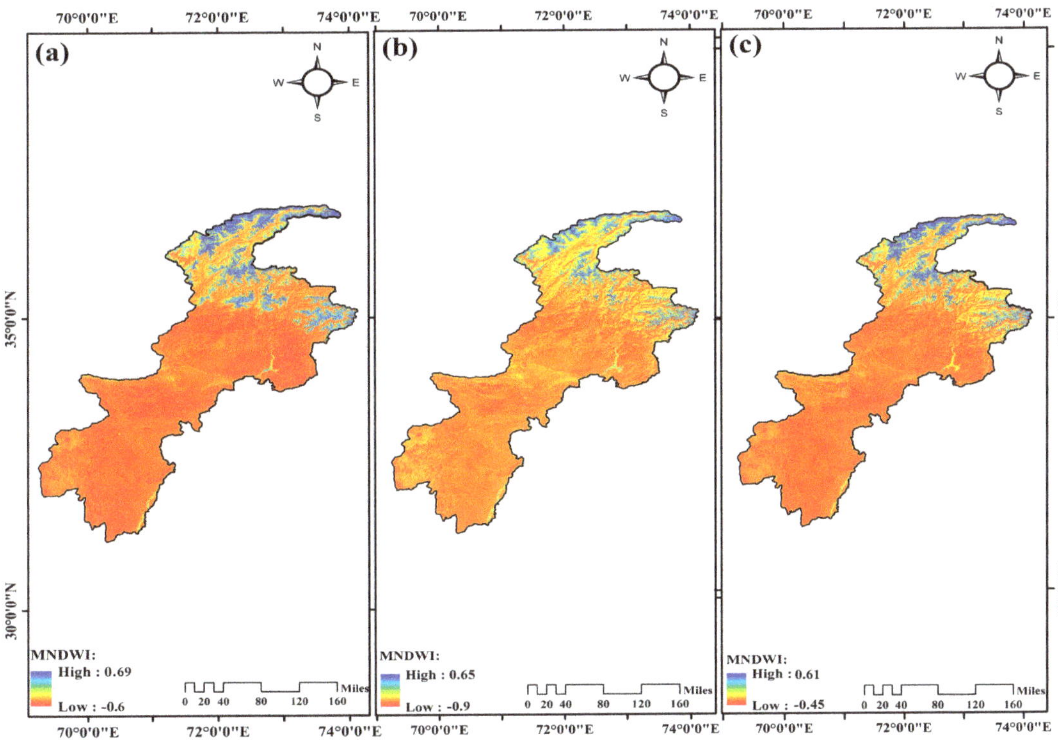

Figure 17. Modified normalized difference water index (MNDWI) in Khyber Pakh-tunkhwa, Pakistan in (**a**) 2030, (**b**) 2050, and (**c**) 2100.

The mean MNDWI values will decrease significantly, dropping from 0.25 in 2030 to 0.19 in 2100. This trend highlights the overall degradation of water resources across the region, likely caused by climate change-induced shifts in precipitation patterns, increased evapotranspiration due to rising temperatures, and human-induced factors such as urbanization and agricultural expansion. The increasing standard deviation, from 0.789 in 2030 to 0.85 in 2100, is presented in (Table 13), indicating growing spatial disparities in water availability, with some areas experiencing severe water depletion while others may retain relatively stable water levels. The declining MNDWI values have far-reaching implications for the region's hydrology and ecology. Reduced surface water availability can disrupt aquatic ecosystems, limit water supplies for human and agricultural use, and exacerbate the impacts of droughts and heat waves. Additionally, areas with persistently low MNDWI values may face challenges such as soil degradation, reduced agricultural productivity, and biodiversity loss. The continued decline in MNDWI underscores the urgent need for sustainable water management strategies, including rainwater harvesting, groundwater recharge programs, and the restoration of degraded wetlands and natural waterbodies.

Table 13. Prediction of modified normalized difference water index (MNDWI) values for the years 2030, 2050, and 2100.

Year	Min MNDWI	Max MNDWII	Mean MNDWI	StdDev
2030	−0.6	0.69	0.25	0.789
2050	−0.9	0.65	0.22	0.823
2100	−0.45	0.61	0.19	0.85

To mitigate these challenges, policymakers must adopt a multi-faceted approach integrating land use planning, water conservation, and climate adaptation measures. Encouraging the use of water-efficient agricultural practices, controlling urban sprawl, and promoting afforestation efforts could help stabilize MNDWI values and ensure long-term water security for the region

4.4.4. Leaf Area Index (LAI)

To anticipate the impact of the LAI in 2030, 2050, and 2100, we delved into this pivotal metric representing the total green leaf area per unit of ground area. Employing the Landsat 5 TM and Landsat 8 OLI sensors, we scrutinized LAI's spatial and temporal distribution across the years 2000, 2010, and 2020 in (Figure 18).

Table 14 shows that predictions of the LAI for the future years (2030, 2050, and 2100), which are the essential factors in the vegetation density. For the year 2030, the variation in LAI is quite significant: the lowest figure is −5.99, and the highest figure is 1.62. This variation suggests places where vegetation cover can be extremely low or even non-existent, perhaps because of human activities, such as deforestation, or unfavorable climatic conditions. The large standard deviation of 5.38 indicates high variability; therefore, the environmental conditions could be highly dynamic. By 2050, there is moderate improvement, where the reproduced average of the LAI is −1.76, and the variation decreases to 4.23. This shift may point to a modest improvement in vegetation cover due to enhanced environmental stewardship practices or favorable climate conditions. However, matters seem to get worse by 2100, with LAI values varying between a low of −6.32 and a high of only 1.01. This is relative to a volatility level shown by the corresponding standard deviation of 5.18, which shows a relative shift to what could be described as more volatile status in terms of vegetation density.

Table 14. Prediction of leaf area index (LAI) values for the years 2030, 2050, and 2100.

Year	Min LAI	Max LAI	Mean LAI	StdDev
2030	−5.99	1.62	−2.185	5.3810
2050	−4.75	1.23	−1.76	4.2284
2100	−6.32	1.01	−2.655	5.1830

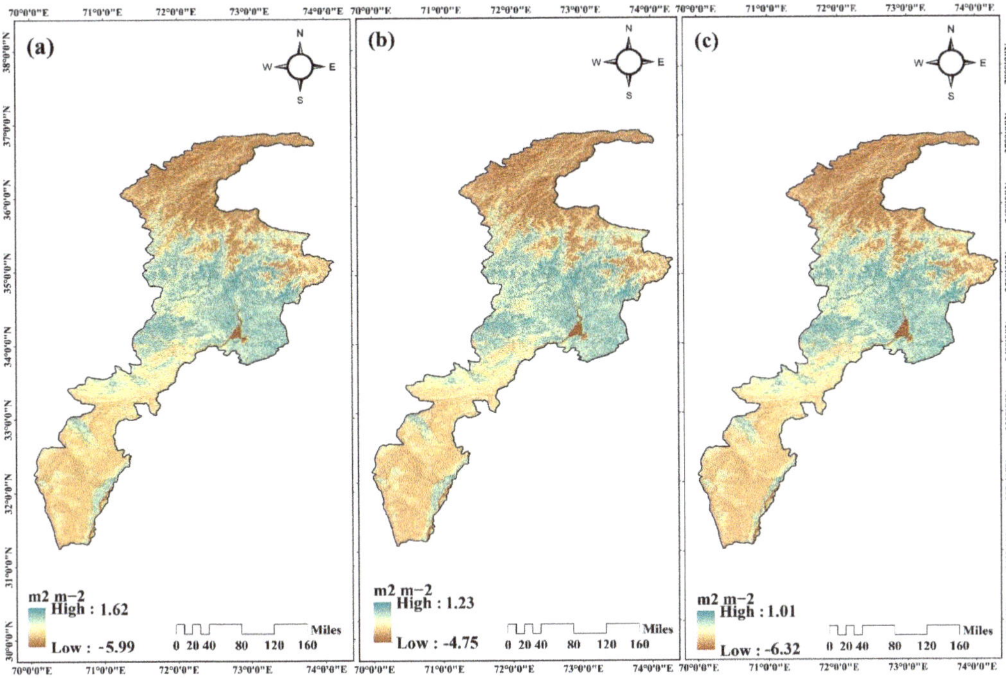

Figure 18. Leaf area index (LAI) analysis in Khyber Pakh-tunkhwa, Pakistan for the years (**a**) 2030, (**b**) 2050, and (**c**) 2100.

4.4.5. Normalized Differences Moisture Index (NDMI)

The predicted normalized difference moisture index (NDMI) values for the years 2030, 2050, and 2100 offer a detailed representation of the changing moisture conditions in KPK (Figure 19). These projections are based on variations in the region's land use and climatic conditions, with NDMI providing a reliable index to evaluate vegetation water content. In 2030, the NDMI values range from a minimum of −0.44 to a maximum of 0.68, with an average value of 0.27 and a standard deviation of 0.48. This indicates moderate vegetation moisture content across the region, with relatively consistent distribution patterns. However, by 2050, there is a significant decline in both maximum and mean NDMI values. The range shifts to a minimum of −0.51 and a maximum of 0.66, with the mean value decreasing to 0.23 and the standard deviation slightly dropping to 0.475 (Table 15). This suggests a reduction in vegetation moisture and highlights the region's increasing vulnerability to water stress.

By 2100, the NDMI values will demonstrate further decline, with the minimum value improving slightly to −0.11, but the maximum dropping to 0.60. The average NDMI value will decrease to 0.21, and the standard deviation will reduce further to 0.47. These changes reflect severe degradation in vegetation moisture content, likely influenced by expanding barren land and urbanization, as observed in the corresponding land use and land cover (LULC) predictions. Spatial analysis of the NDMI reveals that areas with dense vegetation, such as forests and agricultural lands, show a progressive reduction in moisture content over the years. In contrast, regions transitioning to barren or built-up land will demonstrate pronounced declines in NDMI, particularly in the southern and central parts of the study area. These trends align with the LULC predictions, which indicate significant increases in barren land and built-up areas, alongside reductions in forests, vegetation, and agricultural lands.

The implications of these findings are critical. Declining vegetation moisture will likely exacerbate soil erosion, reduce agricultural productivity, and impact biodiversity. Moreover, the expansion of impervious surfaces in urban areas may further degrade the hydrological balance of the region. Strategies such as sustainable water management, afforestation, and land use planning must be prioritized to mitigate these impacts.

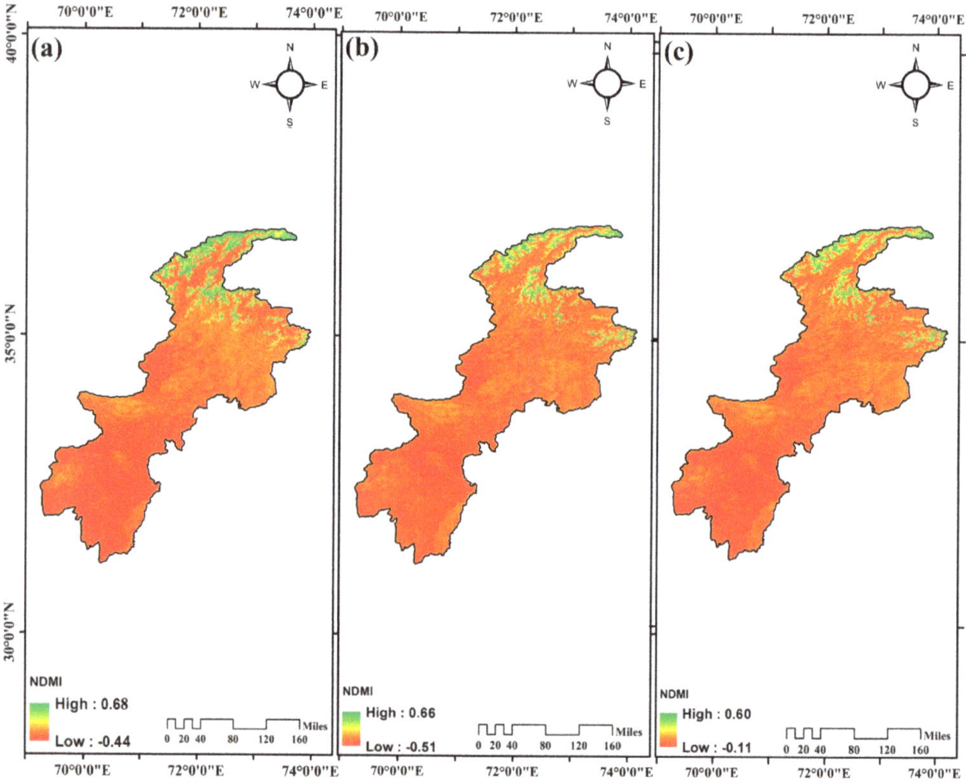

Figure 19. Prediction of the normalized difference moisture index (NDMI) values in Khyber Pakhtunkhwa, Pakistan for the years (**a**) 2030, (**b**) 2050, and (**c**) 2100.

Table 15. Prediction of normalized difference moisture index (NDMI) values for the years 2030, 2050, and 2100.

Year	Min NDMI	Max NDMI	Mean NDMI	StdDev
2030	−0.44	0.68	0.27	0.482
2050	−0.51	0.66	0.23	0.475
2100	−0.11	0.6	0.21	0.468

4.5. Assessment and Predictions of Multiple Correlation Matrices for LST, NDVI, MNDWI, LAI, and NMDI

This section explores the relationship between critical environmental factors, including LAI and LST, LAI and NDVI, LST and NDVI, MNDWI and NDVI, MNDWI and LST, NDMI and LAI, LST and NDMI, and NDVI and NDMI. These indices provide insights into vegetation health, moisture content, and temperature variations, and their interrelationships help explain the dynamics of land use and land cover (LULC) changes over time. The inclusion of MNDWI allows for a more accurate assessment of water-related dynamics in the region, particularly vegetation water content and the influence of waterbodies on surface temperature.

The correlation analysis, enhanced with statistical significance testing, highlights these interactions to better understand the drivers behind LULC changes and their implications for environmental sustainability in the study area.

The relationship between LST and LAI (Figure 20) shows a strong negative correlation ($R^2 = 0.89$, $p < 0.01$), suggesting that, as surface temperatures increase, vegetation density and health (represented by LAI) decrease. This indicates heat stress reducing plant productivity, particularly in areas experiencing deforestation or land degradation. Such trends are evident in the predicted LULC changes, where reductions in forest and vegetation cover are accompanied by increased barren land and built-up areas. This dynamic exacerbates the vulnerability of ecosystems by intensifying surface temperatures and reducing vegetation's cooling effect through shading and evapotranspiration.

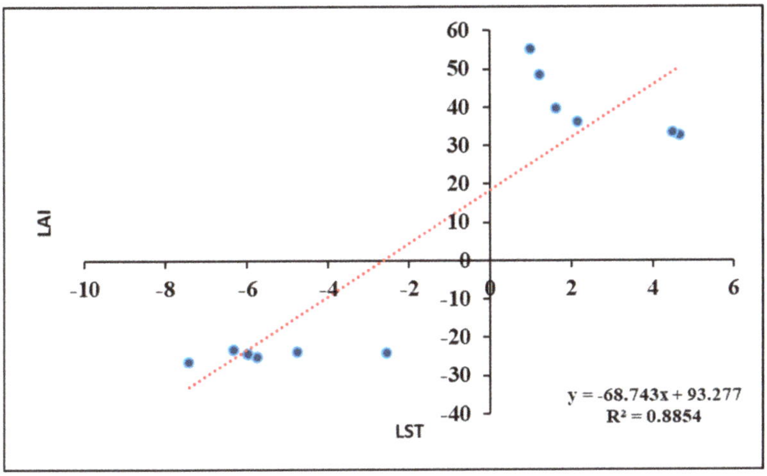

Figure 20. Correlation between LAI-LST.

A strong positive correlation between LAI and NDVI ($R^2 = 0.89$, $p < 0.01$) (Figure 21) underscores the close relationship between vegetation density and overall vegetation health. Higher NDVI values in areas with dense vegetation reflect robust photosynthetic activity and better ecological stability. However, LULC predictions indicate declines in forest and vegetation cover, which are likely to result in lower NDVI and LAI values. This reduction signifies deteriorating vegetation health, decreasing carbon sequestration and ecosystem stability.

The correlation between LST and NDVI is moderately negative ($R^2 = 0.89$, $p < 0.01$) (Figure 22), revealing that vegetation plays a significant role in mitigating surface temperature. As vegetation cover decreases, barren and urbanized areas expand, contributing to higher LST due to the absence of vegetative cooling mechanisms. This effect is most pronounced in regions transitioning from vegetation or forest cover to barren land or built-up areas, as evident in the LULC predictions for 2050 and 2100.

The analysis of MNDWI and NDVI demonstrates a strong positive correlation ($R^2 = 0.99$, $p < 0.01$), with (Figure 23) indicating that areas with dense vegetation also retain higher water content. However, the projected reduction in forest and vegetation cover will lead to declining NDVI and MNDWI values, particularly in areas transitioning to barren or built-up land. This suggests a loss of water retention capacity and heightened vulnerability to drought and land degradation, particularly in the southern and central regions.

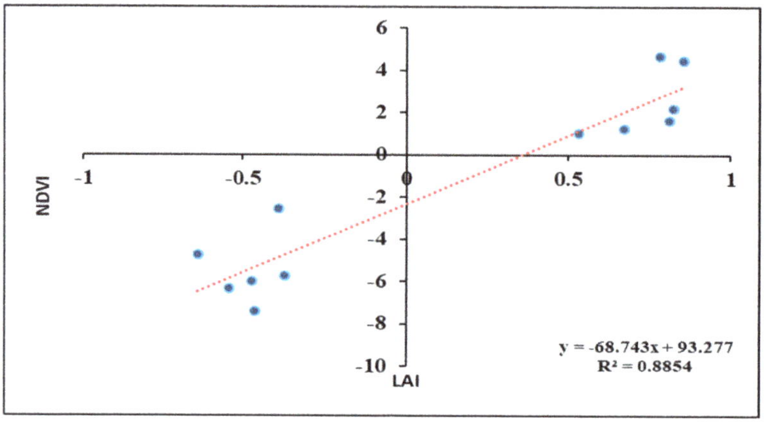

Figure 21. Correlation between LAI-NDVI.

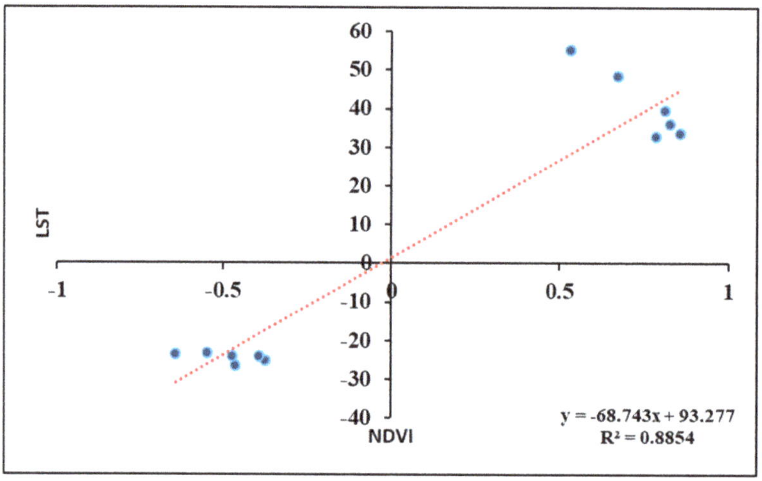

Figure 22. Correlation between LST-NDVI.

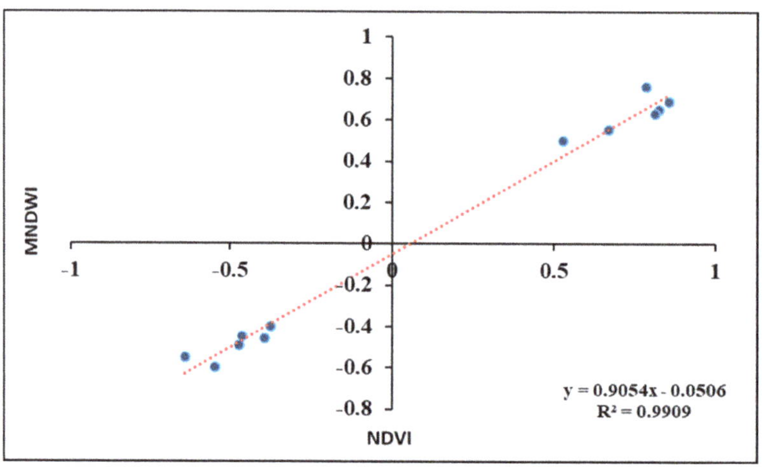

Figure 23. Correlation between MNDWI-NDVI.

The relationship between MNDWI and LST shows a positive correlation ($R^2 = 0.91$, $p < 0.01$) (Figure 24), highlighting the interplay between water availability and surface temperature. While vegetation and water content help moderate LST, reductions in these resources due to land use changes will likely result in increased surface temperatures and more severe heat stress, compounding environmental challenges.

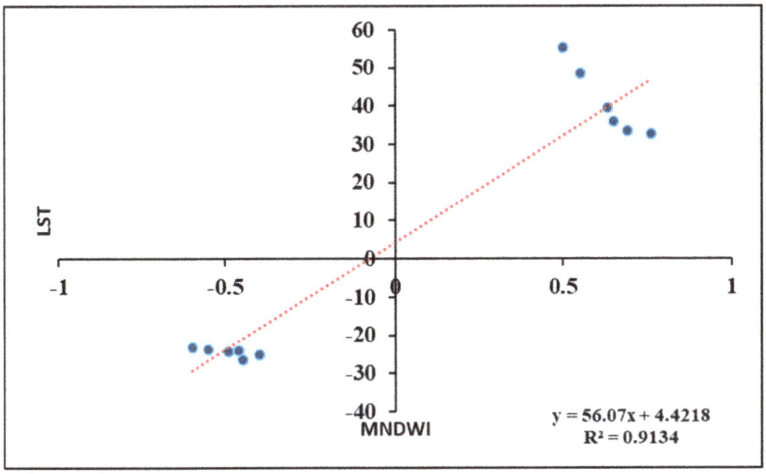

Figure 24. Correlation between MNDWI-LST.

A moderate positive correlation between NDMI and LAI ($R^2 = 0.79$, $p < 0.01$) (Figure 25) reflects the role of healthy vegetation in retaining moisture. However, as forest and vegetation areas decrease, NDMI values decline, further indicating reduced vegetation moisture content. Similarly, the weak positive correlation between LST and NDMI ($R^2 = 0.62$, $p < 0.05$) (Figure 26) suggests that vegetation moisture plays a role in temperature moderation. However, other factors, such as land use type and soil properties, also contribute.

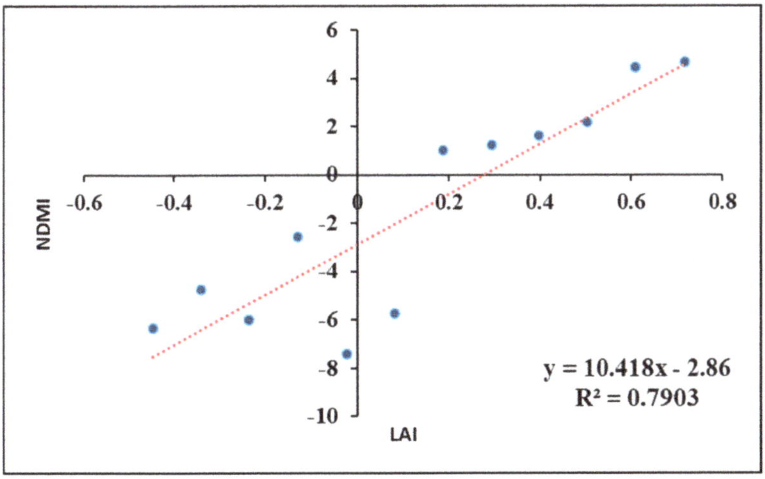

Figure 25. Correlation between NDMI-LAI.

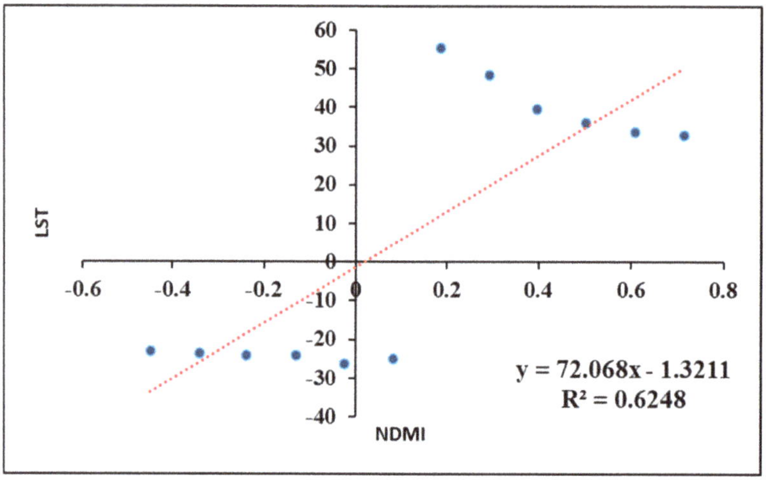

Figure 26. Correlation between LST-NDMI.

Finally, the strong positive correlation between NDVI and NDMI ($R^2 = 0.85$, $p < 0.01$) (Figure 27) highlights the interdependence of vegetation health and moisture content. Areas with high NDVI values, representing healthier vegetation, correspond to higher NDMI values, emphasizing the importance of vegetation in maintaining ecological and hydrological balance.

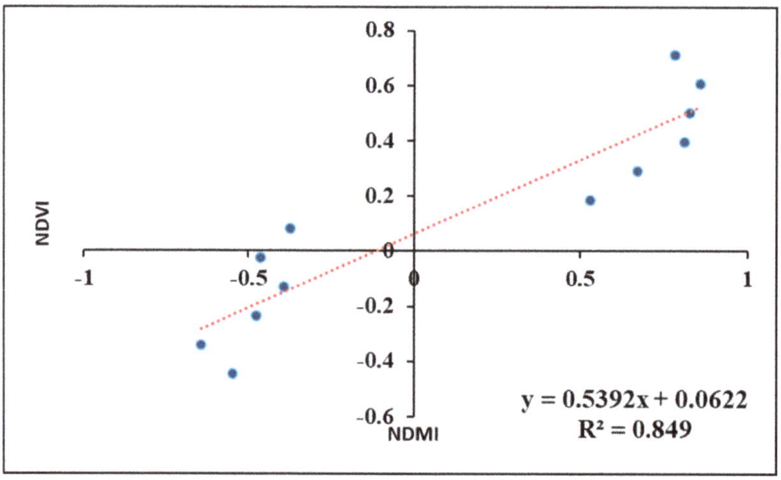

Figure 27. Correlation between NDVI-NDMI.

These results demonstrate the cascading effects of LULC changes on temperature, vegetation health, and water availability. The projected increase in barren land and urbanization, coupled with decreases in forest and vegetation cover, will likely amplify environmental challenges such as soil erosion, biodiversity loss, and reduced agricultural productivity. Strategies such as afforestation, sustainable water management, and climate-resilient urban planning are essential to mitigate these impacts and promote ecological stability in the region.

5. Conclusions

This study comprehensively assessed the dynamic interactions between land use and land cover (LULC) changes and environmental parameters, including Land Surface Temperature (LST), vegetation indices such as NDVI and LAI, and moisture indices such as MNDWI and NDMI, in KPK, Pakistan. The research highlighted significant patterns and predictions in LULC transformations and their environmental implications from 2000 to 2100 using multi-temporal Landsat imagery and advanced modeling techniques like the Random Forest algorithm and the CA-ANN model. The findings underline the pressing need for integrated land management strategies and climate adaptation measures in the region. Over the study period, urban expansion emerged as the most pronounced LULC transformation, with urban areas increasing by over 138% between 2000 and 2020. This urban growth is directly correlated with rising LST, as evidenced by a positive correlation coefficient of 0.89, suggesting that impervious surfaces, such as roads and buildings, significantly contribute to localized heating effects. Simultaneously, LST negatively correlated with NDVI (−0.88) and MNDWI (−0.49), indicating that increasing temperatures reduce vegetation health and moisture content. Projections for 2030, 2050, and 2100 further confirm this trend, with barren land expanding from 21.41% in 2030 to 36.00% in 2100, accompanied by decreases in forests, vegetation, and agricultural lands.

The decline in vegetation greenness, reflected in decreasing NDVI and LAI values, points to escalating environmental stress. In 2020, the mean NDVI was 0.14, while projections indicate a drop to negative values by 2050 and 2100, highlighting severe vegetation degradation. Similarly, LAI values demonstrated fluctuating yet declining patterns, with the lowest recorded in 2020 and further reductions predicted by 2100. These changes emphasize the compounded effects of climate change, deforestation, and land-use practices on vegetation health and ecosystem services. Moisture indices further revealed the region's vulnerability. NDMI values steadily declined, from 0.35 in 2000 to 0.21 in 2100, underscoring diminishing vegetation moisture content. MNDWI patterns reflected similar trends, with reductions in moisture availability correlating with LULC changes such as deforestation and agricultural land conversion.

The interplay between environmental parameters and LULC highlights a bidirectional relationship. Rising LST and declining moisture indices exacerbate vegetation loss, while urbanization and deforestation amplify regional heating effects and disrupt water balance. These findings underscore the urgent need for sustainable urban planning, afforestation initiatives, and adaptive land management to mitigate these impacts. The study underscores the urgent need for sustainable land management practices. Urban expansion should be accompanied by green infrastructure and afforestation to counteract the urban heat island effect. Conservation of waterbodies and vegetation restoration are essential to mitigate the adverse impacts of LULC changes. By providing robust correlations and future projections, this research offers valuable insights for policymakers to devise strategies that balance development with environmental sustainability in KPK.

Author Contributions: M.K.: conceptualization, data collection, methodology, software, investigation, visualization, formal analysis, and writing an original draft. R.C.: supervision, review and editing, investigation, and project administration. All authors have read and agreed to the published version of the manuscript.

Funding: This research received no external funding.

Data Availability Statement: The raw data supporting the conclusions of this article will be made available by the authors on request.

Acknowledgments: The first author would like to extend heartfelt thanks to Annah Zhu for her invaluable assistance with English editing. Her comments and suggestions have been instrumental in enhancing the clarity and quality of this research work.

Conflicts of Interest: The authors declare that they have no other financial or non-financial interests or personal relationships that could have influenced the work reported in this paper.

References

1. Wang, Y.; He, Z.; Zhai, W.; Wang, S.; Zhao, C. How do the 3D urban morphological characteristics spatiotemporally affect the urban thermal environment? A case study of San Antonio. *Build. Environ.* **2024**, *261*, 111738. [CrossRef]
2. McDonald, R.; Guneralp, B.; Zipperer, W.; Marcotullio, P. The future of global urbanization and the environment. *Solutions* **2014**, *5*, 60–69.
3. Pal, S.; Ziaul, S.K. Detection of land use and land cover change and land surface temperature in English Bazar urban centre. *Egypt. J. Remote Sens. Space Sci.* **2017**, *20*, 125–145. [CrossRef]
4. Bastawesy, M.A.; Khalaf, F.I.; Arafat, S.M. The use of remote sensing and GIS for the estimation of water loss from Tushka lakes, southwestern desert, Egypt. *J. Afr. Earth Sci.* **2008**, *52*, 73–80. [CrossRef]
5. Zheng, G.; Moskal, L.M. Retrieving leaf area index (LAI) using remote sensing: Theories, methods and sensors. *Sensors* **2009**, *9*, 2719–2745. [CrossRef] [PubMed]
6. UN. Population Division. *World Urbanization Prospects*; no. 237; United Nations, Department of International Economic and Social Affairs: New York, NY, USA, 2003.
7. Svirejeva-Hopkins, A.; Schellnhuber, H.J.; Pomaz, V.L. Urbanised territories as a specific component of the Global Carbon Cycle. *Ecol. Model.* **2004**, *173*, 295–312. [CrossRef]
8. Parveen, S.; Basheer, J.; Praveen, B. A literature review on land use land cover changes. *Int. J. Adv. Res.* **2018**, *6*, 1–6. [CrossRef]
9. Xu, L.; Li, B.; Yuan, Y.; Gao, X.; Zhang, T.; Sun, Q. Detecting different types of directional land cover changes using MODIS NDVI time series dataset. *Remote Sens.* **2016**, *8*, 495. [CrossRef]
10. Bibi, T.; Nawaz, F.; Rahman, A.A.; Razak, K.A.; Latif, A. Flood risk assessment of river Kabul and Swat catchment area: District Charsadda, Pakistan. *Int. Arch. Photogramm. Remote Sens. Spat. Inf. Sci.* **2018**, *42*, 105–113. [CrossRef]
11. Tewabe, D.; Fentahun, T. Assessing land use and land cover change detection using remote sensing in the Lake Tana Basin, Northwest Ethiopia. *Cogent Environ. Sci.* **2020**, *6*, 1778998. [CrossRef]
12. Clarke, K.C.; Gaydos, L.J. Loose-coupling a cellular automaton model and GIS: Long-term urban growth prediction for San Francisco and Washington/Baltimore. *Int. J. Geogr. Inf. Sci.* **1998**, *12*, 699–714. [CrossRef]
13. Naz, A.; Rasheed, H. Modeling the rice land suitability using GIS and multi-criteria decision analysis approach in Sindh, Pakistan. *J. Basic Appl. Sci.* **2017**, *13*, 26–33.
14. Tayyab, M.; Zhang, J.; Hussain, M.; Ullah, S.; Liu, X.; Khan, S.N.; Baig, M.A.; Hassan, W.; Al-Shaibah, B. Gis-based urban flood resilience assessment using urban flood resilience model: A case study of peshawar city, khyber pakhtunkhwa, pakistan. *Remote Sens.* **2021**, *13*, 1864. [CrossRef]
15. Ali, L.; Mi, J.; Shah, M.; Shah, S.J.; Khan, S.; Ullah, R.; Bibi, K. Local residents' attitude towards road and transport infrastructure (a case of China Pakistan economic corridor). *J. Chin. Econ. Foreign Trade Stud.* **2018**, *11*, 104–120. [CrossRef]
16. Deng, J.S.; Wang, K.; Hong, Y.; Qi, J.G. Spatio-temporal dynamics and evolution of land use change and landscape pattern in response to rapid urbanization. *Landsc. Urban Plan.* **2009**, *92*, 187–198. [CrossRef]
17. Rahman, G.; Rahman, A.; Ullah, S.; Dawood, M.; Moazzam, M.F.U.; Lee, B.G. Spatio-temporal characteristics of meteorological drought in Khyber Pakhtunkhwa, Pakistan. *PLoS ONE* **2021**, *16*, e0249718. [CrossRef]
18. Hamid, I.; Dar, L.A.; Akintug, B. Assessing land use land cover change using remote sensing and GIS techniques: A case study of Kashmir Valley. *J. Earth Syst. Sci.* **2024**, *133*, 168. [CrossRef]
19. Abebe, G.; Getachew, D.; Ewunetu, A. Analysing land use/land cover changes and its dynamics using remote sensing and GIS in Gubalafito district, Northeastern Ethiopia. *SN Appl. Sci.* **2021**, *4*, 30. [CrossRef]
20. Tucker, C.J. Red and photographic infrared linear combinations for monitoring vegetation. *Remote Sens. Environ.* **1979**, *8*, 127–150. [CrossRef]
21. Zoungrana, B.J.B.; Conrad, C.; Thiel, M.; Amekudzi, L.K.; Da, E.D. MODIS NDVI trends and fractional land cover change for improved assessments of vegetation degradation in Burkina Faso, West Africa. *J. Arid. Environ.* **2018**, *153*, 66–75. [CrossRef]
22. Pandi, D.; Kothandaraman, S.; Kumarasamy, M.V.; Kuppusamy, M. Assessment of Land Use and Land Cover Dynamics Using Geospatial Techniques. *Pol. J. Environ. Stud.* **2022**, *31*. [CrossRef]
23. Chen, L.; Li, M.; Huang, F.; Xu, S. Relationships of LST to NDBI and NDVI in Wuhan City based on Landsat ETM+ image. In Proceedings of the 2013 6th International Congress on Image and Signal Processing, CISP, Hangzhou, China, 16–18 December 2013; Volume 2, pp. 840–845. [CrossRef]

24. Verma, P.; Raghubanshi, A.; Srivastava, P.K.; Raghubanshi, A.S. Appraisal of kappa-based metrics and disagreement indices of accuracy assessment for parametric and nonparametric techniques used in LULC classification and change detection. *Model Earth Syst. Environ.* **2020**, *6*, 1045–1059. [CrossRef]
25. Nega, W.; Balew, A. The relationship between land use land cover and land surface temperature using remote sensing: Systematic reviews of studies globally over the past 5 years. *Environ. Sci. Pollut. Res.* **2022**, *29*, 42493–42508. [CrossRef]
26. Karimian, R.; Rangzan, K.; Karimi, D.; Einali, G. Spatiotemporal Monitoring of Land Use-Land Cover and Its Relationship with Land Surface Temperature Changes Based on Remote Sensing, GIS, and Deep Learning. *J. Indian Soc. Remote Sens.* **2024**, *52*, 2461–2481. [CrossRef]
27. Parker, D.; Manson, S.; Janssen, M.; Hoffmann, M.; Deadman, P. Multi-Agent Systems for the Simulation of Land-Use and Land-Cover Change: A Review. *Ann. Assoc. Am. Geogr.* **2003**, *93*, 314–337. [CrossRef]
28. Alexander, C. Normalised difference spectral indices and urban land cover as indicators of land surface temperature (LST). *Int. J. Appl. Earth Obs. Geoinf.* **2020**, *86*, 102013. [CrossRef]
29. Veldkamp, A.; Lambin, E.F. *Predicting Land-Use Change*; Elsevier: Amsterdam, The Netherlands, 2001.
30. Zhang, W.; Jia, Z.; Li, B.; Liu, Q.; Zhu, X.; Jia, C.; Gu, X.; Yu, T. Research on Landsat 8 land surface temperature retrieval and spatial–temporal migration capabilities based on random forest model. *Adv. Space Res.* **2024**, *74*, 610–627. [CrossRef]
31. Beroho, M.; Briak, H.; Cherif, E.K.; Boulahfa, I.; Ouallali, A.; Mrabet, R.; Kebede, F.; Bernardino, A.; Aboumaria, K. Future Scenarios of Land Use/Land Cover (LULC) Based on a CA-Markov Simulation Model: Case of a Mediterranean Watershed in Morocco. *Remote Sens.* **2023**, *15*, 1162. [CrossRef]
32. Rauf, T.; Khan, N.; Shah, S.J.; Zada, M.; Malik, S.Y.; Yukun, C.; Sadique, A. Poverty and Prosperity: Impact on Livelihood Assets of Billion Trees Afforestation Program in Khyber Pakhtunkhwa (KPK), Pakistan. *Forests* **2019**, *10*, 916. [CrossRef]
33. Yasir, M.; Hui, S.; Rahman, S.U.; Ilyas, M.; Zafar, A.; Mehmood, A. Estimation of land surface temperature using LANDSAT-8 data-A case study of district Malakand, Khyber Pakhtunkhwa, Pakistan. *J. Lib. Arts Humanit.* **2020**, *1*, 140–148.
34. Nazir, N.; Farooq, A.; Jan, S.A.; Ahmad, A. A system dynamics model for billion trees tsunami afforestation project of Khyber Pakhtunkhwa in Pakistan: Model application to afforestation activities. *J. Mt. Sci.* **2019**, *16*, 2640–2653. [CrossRef]
35. Waqas, H.; Lu, L.; Tariq, A.; Li, Q.; Baqa, M.F.; Xing, J.; Sajjad, A. Flash flood susceptibility assessment and zonation using an integrating analytic hierarchy process and frequency ratio model for the Chitral District, Khyber Pakhtunkhwa, Pakistan. *Water* **2021**, *13*, 1650. [CrossRef]
36. Khan, S.U.; Hussain, I. Inequalities in health and health-related indicators: A spatial geographic analysis of Pakistan. *BMC Public Health* **2020**, *20*, 1800. [CrossRef] [PubMed]
37. Sabir, M.; Ali, Y.; Khan, I.; Salman, A. Plants species selection for afforestation: A case study of the Billion Tree Tsunami Project of Pakistan. *J. Sustain. For.* **2022**, *41*, 537–549. [CrossRef]
38. Yar, P.; Khan, M.A.; Shah, S.A.A. Spatio-temporal Analysis of Urban Expansion on Farmland and Its Impact on the Agricultural Land Use of Mardan City, Pakistan: Spatio-temporal Analysis of Urban Expansion on Farmland. *Proc. Pak. Acad. Sci. B Life Environ. Sci.* **2016**, *53*, 35–46.
39. Fang, C.; Liu, H.; Li, G. International progress and evaluation on interactive coupling effects between urbanization and the eco-environment. *J. Geogr. Sci.* **2016**, *26*, 1081–1116. [CrossRef]
40. Yaqoob, U. *Population Distribution and Water Resources in Pakistan*; Università Ca' Foscari Venezia: Venezia, Italy, 2018.
41. Ezimand, K.; Kakroodi, A.A.; Kiavarz, M. The development of spectral indices for detecting built-up land areas and their relationship with land-surface temperature. *Int. J. Remote Sens.* **2018**, *39*, 8428–8449. [CrossRef]
42. Yang, J.; Ren, J.; Sun, D.; Xiao, X.; Xia, J.C.; Jin, C.; Li, X. Understanding land surface temperature impact factors based on local climate zones. *Sustain. Cities Soc.* **2021**, *69*, 102818. [CrossRef]
43. Yang, Y.Z.; Cai, W.H.; Yang, J. Evaluation of MODIS land surface temperature data to estimate near-surface air temperature in Northeast China. *Remote Sens.* **2017**, *9*, 410. [CrossRef]
44. Olofsson, P.; Foody, G.M.; Herold, M.; Stehman, S.V.; Woodcock, C.E.; Wulder, M.A. Good practices for estimating area and assessing accuracy of land change. *Remote Sens. Environ.* **2014**, *148*, 42–57. [CrossRef]
45. Foody, G.M. Explaining the unsuitability of the kappa coefficient in the assessment and comparison of the accuracy of thematic maps obtained by image classification. *Remote Sens. Environ.* **2020**, *239*, 111630. [CrossRef]
46. Bharatkar, P.S.; Patel, R. Approach to accuracy assessment tor RS image classification techniques. *Int. J. Sci. Eng. Res.* **2013**, *4*, 79–86.
47. Khan, M.S.; Ullah, S.; Sun, T.; Rehman, A.U.R.; Chen, L. Land-use/land-cover changes and its contribution to urban heat Island: A case study of Islamabad, Pakistan. *Sustainability* **2020**, *12*, 3861. [CrossRef]
48. Congalton, R. Putting the map back in map accuracy assessment. *Remote Sens. GIS Accuracy Assess.* **2004**, 1–11.
49. Lambin, E.F.; Geist, H.J.; Lepers, E. Dynamics of land-use and land-cover change in tropical regions. *Annu. Rev. Environ. Resour.* **2003**, *28*, 205–241. [CrossRef]

50. Pathan, S.A.; Ashwini, K.; Sil, B.S. Spatio-temporal variation in land use/land cover pattern and channel migration in Majuli River Island, India. *Environ. Monit. Assess.* **2021**, *193*, 811. [CrossRef] [PubMed]
51. Somayajula, V.K.A.; Ghai, D.; Kumar, S.; Tripathi, S.L.; Verma, C.; Safirescu, C.O.; Mihaltan, T.C. Classification and Validation of Spatio-Temporal Changes in Land Use/Land Cover and Land Surface Temperature of Multitemporal Images. *Sustainability* **2022**, *14*, 5677. [CrossRef]
52. Nogueira, M.; Albergel, C.; Boussetta, S.; Johannsen, F.; Trigo, I.F.; Ermida, S.L.; Martins, J.P.A.; Dutra, E. Role of vegetation in representing land surface temperature in the CHTESSEL (CY45R1) and SURFEX-ISBA (v8. 1) land surface models: A case study over Iberia. *Geosci. Model Dev.* **2020**, *13*, 3975–3993. [CrossRef]
53. Omran, E.-S.E. Detection of land-use and surface temperature change at different resolutions. *J. Geogr. Inf. Syst.* **2012**, *4*, 189–203. [CrossRef]
54. Mutiibwa, D.; Strachan, S.; Albright, T. Land surface temperature and surface air temperature in complex terrain. *IEEE J. Sel. Top. Appl. Earth Obs. Remote Sens.* **2015**, *8*, 4762–4774. [CrossRef]
55. Wang, F.; Qin, Z.; Song, C.; Tu, L.; Karnieli, A.; Zhao, S. An improved mono-window algorithm for land surface temperature retrieval from landsat 8 thermal infrared sensor data. *Remote Sens.* **2015**, *7*, 4268–4289. [CrossRef]
56. Weng, Q. A remote sensing?GIS evaluation of urban expansion and its impact on surface temperature in the Zhujiang Delta, China. *Int. J. Remote Sens.* **2001**, *22*, 1999–2014. [CrossRef]
57. Shuster, W.D.; Bonta, J.; Thurston, H.; Warnemuende, E.; Smith, D.R. Impacts of impervious surface on watershed hydrology: A review. *Urban Water J.* **2005**, *2*, 263–275. [CrossRef]
58. Chen, J.; Theller, L.; Gitau, M.W.; Engel, B.A.; Harbor, J.M. Urbanization impacts on surface runoff of the contiguous United States. *J. Environ. Manag.* **2017**, *187*, 470–481. [CrossRef] [PubMed]
59. Fletcher, T.D.; Andrieu, H.; Hamel, P. Understanding, management and modelling of urban hydrology and its consequences for receiving waters: A state of the art. *Adv. Water Resour.* **2013**, *51*, 261–279. [CrossRef]
60. Pettorelli, N.; Vik, J.O.; Mysterud, A.; Gaillard, J.-M.; Tucker, C.J.; Stenseth, N.C. Using the satellite-derived NDVI to assess ecological responses to environmental change. *Trends Ecol. Evol.* **2005**, *20*, 503–510. [CrossRef]
61. Sobrino, J.A.; Raissouni, N. Toward remote sensing methods for land cover dynamic monitoring: Application to Morocco. *Int J Remote Sens.* **2000**, *21*, 353–366. [CrossRef]
62. Li, Z.-L.; Tang, B.-H.; Wu, H.; Ren, H.; Yan, G.; Wan, Z.; Trigo, I.F.; Sobrino, J.A. Satellite-derived land surface temperature: Current status and perspectives. *Remote Sens. Environ.* **2013**, *131*, 14–37. [CrossRef]
63. Banerjee, R. Expansion or over growth of municipal area: A case study of Burdwan Municipality, Burdwan District, West Bengal, India. *Int. J. Curr. Res.* **2013**, *5*, 59–64.
64. Avdan, U.; Jovanovska, G. Algorithm for automated mapping of land surface temperature using LANDSAT 8 satellite data. *J. Sens.* **2016**, *2016*, 1480307. [CrossRef]

remote sensing

MDPI

Article

A "Region-Specific Model Adaptation (RSMA)"-Based Training Data Method for Large-Scale Land Cover Mapping

Congcong Li [1,*], George Xian [2] and Suming Jin [2]

[1] ASRC Federal Data Solutions, Contractor to the U.S. Geological Survey (USGS) Earth Resources Observation and Science (EROS) Center, Work Performed under USGS Contract 140G0124D0001, Sioux Falls, SD 57198, USA

[2] U.S. Geological Survey, Earth Resources and Observation Science (EROS) Center, 47914 252nd St., Sioux Falls, SD 57198, USA; xian@usgs.gov (G.X.); sjin@usgs.gov (S.J.)

* Correspondence: congcongli@contractor.usgs.gov

Abstract: An accurate and historical land cover monitoring dataset for Alaska could provide fundamental information for a range of studies, such as conservation habitats, biogeochemical cycles, and climate systems, in this distinctive region. This research addresses challenges associated with the extraction of training data for timely and accurate land cover classifications in Alaska over longer time periods (e.g., greater than 10 years). Specifically, we designed the "Region-Specific Model Adaptation (RSMA)" method for training data. The method integrates land cover information from the National Land Cover Database (NLCD), LANDFIRE's Existing Vegetation Type (EVT), and the National Wetlands Inventory (NWI) and machine learning techniques to generate robust training samples based on the Anderson Level II classification legend. The assumption of the method is that spectral signatures vary across regions because of diverse land surface compositions; however, despite these variations, there are consistent, collective land cover characteristics that span the entire region. Building upon this assumption, this research utilized the classification power of deep learning algorithms and the generalization ability of RSMA to construct a model for the RSMA method. Additionally, we interpreted existing vegetation plot information for land cover labels as validation data to reduce inconsistency in the human interpretation. Our validation results indicate that the RSMA method improved the quality of the training data derived solely from the NLCD by approximately 30% for the overall accuracy. The validation assessment also demonstrates that the RSMA method can generate reliable training data on large scales in regions that lack sufficient reliable data.

Keywords: fine-tuning; CNN; U-net; training data; validation data; Alaska

1. Introduction

Understanding land cover and land cover changes (LCCs) in Alaska is important for understanding the global biogeochemical cycle [1,2]. Unique land cover types like tundra and glaciers are found in Alaska [3,4]. Tundra and glaciers contribute to the environmental dynamics of the state and the world and provide vital services. Forests, tundra, wetlands, and permafrost store large amounts of carbon dioxide, functioning as carbon sinks and helping to control global temperatures [5]. Changes in land surface, like deforestation [6] and thawing permafrost [7], can affect hydrological patterns [8], alter wildlife habitats [9], release stored carbon [4], and exacerbate climate change [10] on different scales. As a result, precise land cover monitoring is needed to comprehend and support conservation initiatives, as well as climate change research, in Alaska.

Two land cover and vegetation disturbance products provide full coverage of Alaska, as follows: the National Land Cover Database (NLCD) [3] and LANDFIRE's Existing Vegetation Type (EVT) [11]. Additionally, approximately 63% of the nation's wetlands are located in Alaska [12]. Two notable wetland products were identified that provide additional wetland information. One is the National Wetlands Inventory (NWI) wetland extent [13], which only covers certain regions of Alaska and has not been updated for a long period of time [14]. The second is a comprehensive vegetated wetland map of Alaska [14] with a spatial resolution of 50 m. The map was produced using variables from the Phased-Array L-band SAR (PALSAR) from Advanced Land Orbiting Satellite (ALOS) data for 2007 and other ancillary data as classification features. Despite these resources, more timely and accurate land cover and change maps are needed for Alaska.

Training data, features, and classification algorithms are three important factors in land cover classification [15]. Over the past decades, classification algorithms evolved from conventional classification to deep learning algorithms, improving the classification accuracy [16,17]. Classification features, like texture [18], contextual [19,20], and time-series features [21], are used to further enhance the land cover classification process. Sufficient and accurate extraction of training data continues to be the bottleneck in the creation of timely and reliable land cover maps that facilitate progress in classification techniques and feature extraction [22]. Currently, extraction of accurate training data remains largely dependent on labor-intensive collection methods [23,24]. Human interpretation is often considered a reliable method but time-consuming and difficult to implement, especially in large and diverse geographic regions like Alaska. Nevertheless, identifying vegetated wetland without periodically occurring standing water is challenging when using human interpretation based on Landsat or other online map images. Moreover, the unique characteristics of the Alaskan landscape and the lack of expert knowledge and data (e.g., historical high-resolution images) [25] complicate efforts to collect accurate and reliable training data for the region. Additionally, research that focuses on the automatic extraction of training data is limited, and new, innovative approaches could help improve the classification process.

In previous studies, three training data methods were developed and used to collect training samples automatically. First, the method of collecting samples from existing reliable products and sources was adopted by LCMAP Collection 1.0 products [26,27]; however, the accuracy is limited by the quality of the sources used. Second is the method of generating consensus samples from various published products, which improves the quality of training data, or from different epochs of land cover products to obtain time-consistent samples. This method solves the problem of a lack of suitable land cover products (e.g., incompatible classification scheme and product mapping uncertainties) from which to extract training data for the mapping year. Li et al. [28] designed a series of rules to integrate several published products and collect consensus samples for land-cover mapping of Hawaii. However, the mapping accuracy is determined by the leveraged spatial coverage and quality of the training samples. Generating consensus samples provides training samples that partially cover the mapping region, which can reduce the representativeness of training samples for land cover types with high intraclass spectral variability or rare classes. Third, Li et al. [22] developed an automatic phenology learning (APL) method to generate reliable training data with full spatial coverage of the mapping region without human intervention. The APL method was developed based on the assumption that the time-series patterns for the same land cover type on a local scale are the same or similar. Thus, the APL method measures the time-series similarity between standard phenology patterns and unknown sample units using the time-weighted dynamic time warping (twDTW) method [29] and labels the unknown sample units automatically based on analysis and other land cover products. Nevertheless, the limited availability of Landsat observations due to the removal of clouds, ice, and snow and the short "leaf-on" season constrains the construction of reliable phenological curves in applying the APL method. Given these challenges, the APL method may not accurately represent the Alaska region.

Validating a training data method ensures the reliability of the method in real-world applications for end users. Traditionally, in large-scale land-cover mapping, the generation of validation samples has heavily relied on human interpretation to ensure their distributions and quantity without introducing subjectivity and potential biases [23–25,30]. The inconsistency among validation samples interpreted by different experts poses additional challenges in evaluating the quality of land cover maps. Therefore, we developed an alternative way of interpreting the Alaska Vegetation Plots Database (AKVEG) [31] to collect validation data that aims to minimize reliance on human interpretation and reduce subjectivity in the validation process.

In this study, we present a novel approach involving machine learning for training data to bridge the knowledge and information gaps across regions in Alaska. Our research objectives were the following:

(1) Design a framework that integrates machine learning with multiple scientific datasets to minimize errors and uncertainties in the training samples, which are often sourced from a single dataset source.
(2) Evaluate the efficacy of the training data method through experiments with two different sampling strategies and two different deep learning classifiers in the meta-learning framework.
(3) Convert the online vegetation plot information into distinct land cover types (including wetland) for validating the developed training data method.

2. Method

We designed a "Region-Specific Model Adaptation (RSMA)"-based training data method (Figure 1) to generate training data for land-cover mapping over a large geographical region. The assumption of the method is that despite variations in spectral signatures for the same vegetation type collected from different regions within a large area, there are still shared characteristics that remain consistent across the entire region. The method involves the integration of a meta-training process [32] and fine-tuning [33] and can be divided into the following several steps:

Meta-training process: The process constructed a basic deep-learning model using a set of consensus sample units obtained from large geographical regions within the same area. These consensus sample units were identified based on their agreement across two or more existing land cover products. Pixels with high confidence labels were used to build the machine learning model.

Fine-tuning process: The pretrained basic model was transferred to a specific target Landsat "Analysis Ready Data" (ARD) tile within the same geographical region. The model's parameters were updated with a different set of consensus sample units in the target tile. This adapted the model to the local characteristics of land cover in that region.

Prediction process: The adapted model was utilized to predict the land cover probabilities of undefined sample units (any location without an agreed land cover label) within the same geographical region. This enables the model to generate predictions for areas where consensus training data are not available.

Optimal label selection: For every unlabeled training sample unit that was taken out of the common region, the approach identified the best land cover label based on the probabilities using the predetermined criteria.

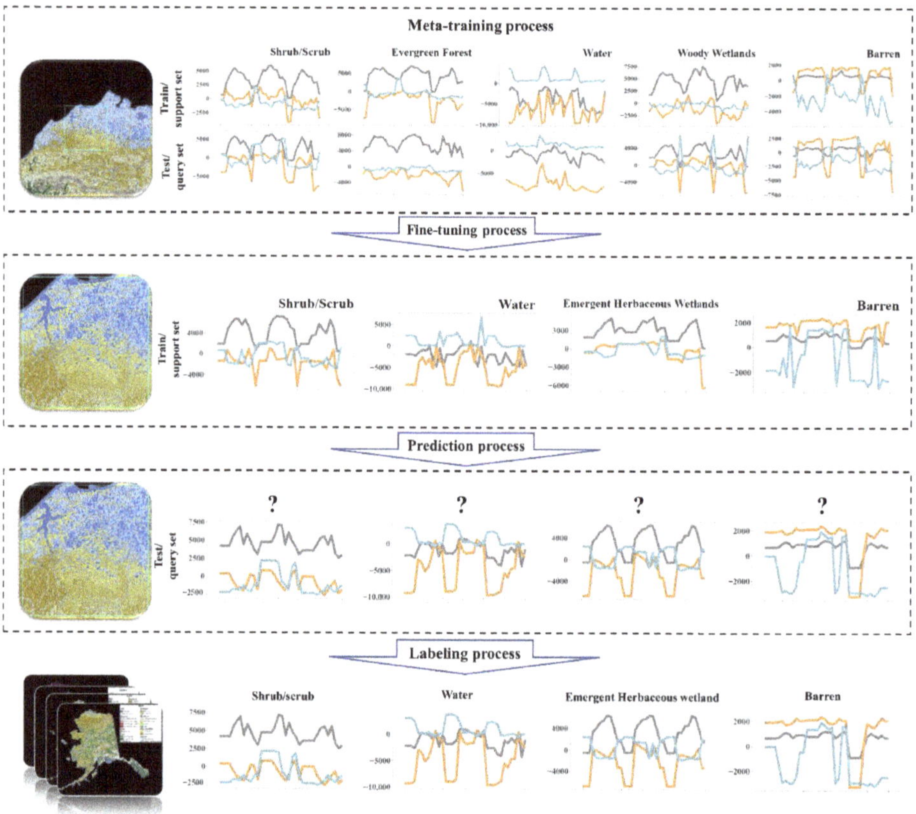

Figure 1. The framework of the "Region-Specific Model Adaptation (RSMA)"-based training data method. The land cover images are NLCD land cover maps. The different colored lines represent various input features.

2.1. Dataset

In the RSMA training data method, a common layer was developed to extract consensus sample units for building the model. The common layer was produced by integrating available land products in Alaska, including the NLCD for the year 2011, LF EVT for the year 2010, the PALSAR wetland product for the year 2007, and the NWI. Specifically for wetlands, regions classified as woody wetland or emergent herbaceous wetland in the NLCD, LF EVT, and PALSAR wetland products were preserved in the common layer. These regions were considered "reliable wetland areas". In contrast, the areas that were classified as wetland by at least one of the four products (NLCD, LF EVT, PARSAR, and NWI) were identified and grouped as "potential-wetland (pWet) masks". These areas showed a high probability of being wetlands and were considered "potential wetland regions" and used for further processing in the methodology.

For non-wetland types, which include water, forest, shrubland, herbaceous, and barren regions, common land cover types between the NLCD and LF EVT products based on the Anderson level I classification legend [34] were kept in the common layer. This common layer did not contain any potential wetland pixels from the pWet masks. The pixels in the common layer were labeled by the NLCD product according to the Anderson Level II classification legend [34]. Additionally, developed and planted/cultivated pixels from the NLCD remained in the common layer as training data for classification but were not involved in the RSMA training data method. The final "consensus" sample units were then collected from this common layer. As a result, the generated common layer and final training data should represent the year 2011.

The following procedure was designed to generate training data for a Landsat ARD tile. We randomly selected 20,000 sample units for each land cover category from the common layer as meta-training data from the surrounding 3 × 3 Landsat ARD Tiles (9 tiles in total, hereafter referred to as "the surrounding broad region"). Based on previous research, a 3 × 3 Landsat ARD region size is adequate for capturing the spectral variation in land cover categories while avoiding unnecessary complexity [26]. The number of sample units was determined to balance the various spectral signatures and computation times, after testing 10,000 and 40,000 sample units per class. We used these selected random sample units in the region to develop a comprehensive model for the meta-training processing. An equal-number sampling method can help reduce bias toward categories with larger sample sizes and improve the model's generalization abilities for other regions while ensuring the representativeness of each land cover category in the training data.

We examined two sampling methods for constructing the fine-tuning dataset in the target Landsat ARD tile. First, we used a random sampling approach to gather a total of 100,000 sample units. Second, we used an equal-number random sampling method for each category [35] to collect an equivalent of 100,000 sample units in total. The sample sets were used to refine the model from the surrounding broad region to the target region in the fine-tuning procedure.

After generating the sample locations, we obtained all cloud-free records from the Landsat Collection 2 [36] data spanning 2009–2012, along with the 25 m resolution PALSAR dataset for each sample location. By spanning four years that encompass 2011, the aim is to increase the available cloud-free and snow-free data for constructing machine learning models on the study area. In addition to the original band values, we calculated several indices to enhance the characterization of vegetation, water, and barren areas. These indices are listed in Table 1.

Table 1. List of indicators used in the RSMA-based training data method.

Index	Formula	Citation
Normalized Difference Vegetation Index (NDVI)	$NDVI = \frac{NIR-Red}{NIR+Red}$	[37]
Normalized Difference Built-Up Index (NDBI)	$NDBI = \frac{SWIR-NIR}{SWIR+NIR}$	[38]
Normalized Difference Water Index (NDWI)	$NDWI = \frac{Green-NIR}{Green+NIR}$	[39]
Modified Normalized Difference Water Index (MNDWI)	$MNDWI = \frac{Green-SWIR}{Green+SWIR}$	[40]
Automated Water Extraction Index (AWEI)	$AWEIsh = Blue + 2.5 \times Green - 1.5 \times (NIR + SWIR1) - 0.25 \times SWIR2$ $AWEInsh = 4 \times (Green - SWIR1) - (0.25 \times NIR + 2.75 \times SWIR1)$	[41]
Wetness	For Landsat 5, $0.0315 \times Blue + 0.2021 \times Red + 0.3102 \times Green + 0.1594 \times NIR + 0.6806 \times SWIR1 - 0.6109 \times SWIR2$ For Landsat 7, $0.2626 \times Blue + 0.2141 \times Red + 0.0926 \times Green + 0.0656 \times NIR - 0.7629 \times SWIR1 - 0.5388 \times SWIR2$	[42,43]
Brightness	For Landsat 5, $0.2043 \times Blue + 0.4158 \times Red + 0.5524 \times Green + 0.5741 \times NIR + 0.3124 \times SWIR1 + 0.2303 \times SWIR2$ For Landsat 7, $0.3561 \times Blue + 0.3972 \times Red + 0.3904 \times Green + 0.6966 \times NIR + 0.2286 \times SWIR1 + 0.1596 \times SWIR2$	[42,43]
Normalized Burn Ratio (NBR)	$NBR = \frac{NIR-SWIR2}{NIR+SWIR2}$	[3,44]
Bare Soil Index (BSI)	$BSI = ((Red + SWIR) - (NIR + Blue))/((Red + SWIR) + (NIR + Blue))$	[45]

Upon extracting the remotely sensed spectral features for the sample units, we observed variations in the numbers and intervals of the observations within the time series.

To standardize the data, we generated semi-monthly data composites using all available clear Landsat data from 2009 to 2012. Taking the near-infrared (NIR) reflectance values as an example (Figure 2), we calculated the maximum, median, and minimum values for the first half and second half of each month. The new time series spanned from the second half of April to the first half of October after removal of the pixels contaminated by clouds, cloud shadows, and snow/ice. To address missing values, we filled them with the mean values of neighboring data points. The resulting reconstructed time series represents 2011 to match the consensus layer.

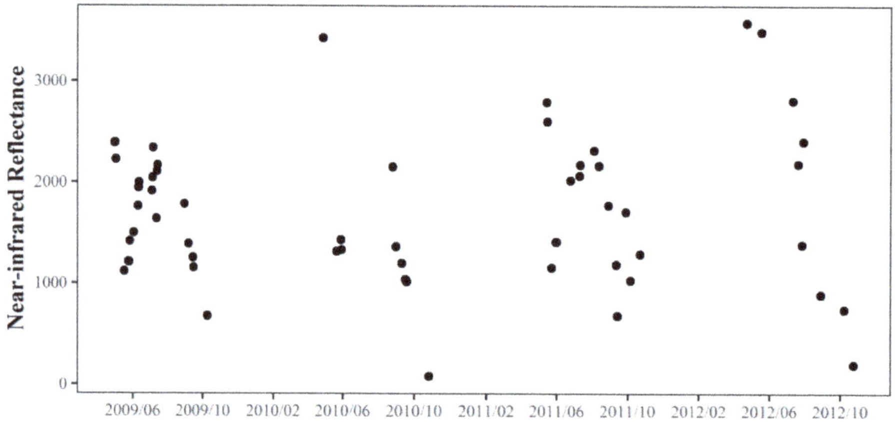

(**A**)

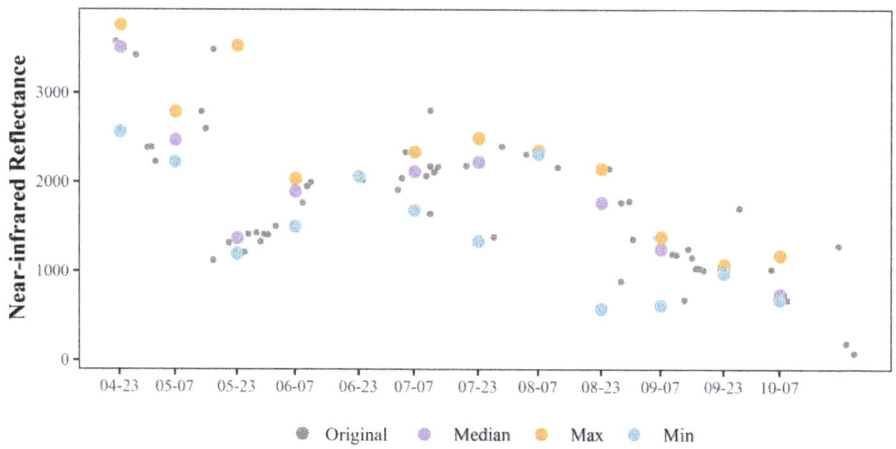

(**B**)

Figure 2. Temporal trends (**A**) and semi-monthly NIR composition (**B**) of the NIR reflectance values (2009–2012). The reflectance values were scaled by a factor of 10,000.

2.2. Basic Model of the RSMA-Based Training Data Method

In this research, the input features are multi-spectral temporal time series. When processing time series, long short-term memory (LSTM) is often deemed as a natural starting point, because it was initially proposed to analyze sequential data. Convolutional neural networks (CNNs) are network architectures of deep learning methods that are commonly used in spatial and spectral domains [46]. However, Zhong, Hu, and Zhou [16] suggested that a one-dimensional CNN model (1D-CNN) can deal with the temporal features and produce better results than LSTM. Therefore, we selected a 1D-CNN model as the basic model of the RSMA. For our CNN architecture, we used two convolutional blocks, each adapted to understand diverse patterns, especially in sequential data. Each convolutional block had a specific pattern, including the use of a 3 $\times$ 3 convolutional kernel for local feature capture, a max-pooling layer to further improve the network's capacity to extract important information, rectified linear activation function (ReLU) for enhanced nonlinearity enhancement [47], and batch normalization for stability [48]. Following the convolutional blocks, a fully connected layer was adjusted to the information provided, and a 50% dropout was applied after flattening, ensuring that the model remained adaptable [49]. The process is named the "1D-CNN-RSMA" in this research.

Another widely used neural network for land cover classification is UNet [50], which was originally designed for semantic segmentation and adapted for image classification. The classifier followed the encoder–decoder architecture, which is known for its capacity to preserve spatial information while capturing both local and global features [51]. In this study, we designed a UNet Classifier that was especially made for one-dimensional (1D) time-series classification applications. We applied skip connections with the encoder–decoder architecture to capture temporal patterns and hierarchically extract features at different temporal resolutions; the encoder module used three successive 1D convolutional layers with a kernel size of 3. The first layer had 64 filters, the second layer had 128 filters, and the third layer had 256 filters. Convolutional operations were used to link each layer of the encoder network. Batch normalization and ReLU activation functions were then added to reduce overfitting. The feature maps were downsampled using max-pooling layers with a stride of 2 and a convolutional kernel size of 2 $\times$ 2. Furthermore, a bottleneck layer with 512 filters was incorporated into the architecture to reduce the dimensionality of the feature maps and to facilitate more efficient learning of deep representations. Moreover, skip connections that store feature maps established direct connections between corresponding encoder and decoder layers, which maintain critical temporal information across network layers by preserving critical temporal details. In the decoder module, up-convolutional layers were used to upsample the feature maps back to the original temporal resolution. Each up-convolutional layer was then followed by batch normalization and ReLU activation functions to enhance the learning capabilities of the network. The process was named "UNet-RSMA" in this research.

2.3. Implementation of the RSMA-Based Training Data Method

The meta-training process is the same as in the Model-Agnostic Meta-Learning (MAML) method [52], in which the meta-training dataset is partitioned into 80% $T_{support}$ (train) and 20% T_{query} (test). As shown in Figure 3, each variable of the meta-training set was input as an individual channel of the model. The convolutional kernel moved from the beginning of the median values toward the minimum of the values, accommodating the following three distinct patterns: median, maximum, and minimum time series. PALSAR data were integrated as an individual channel in this process. The training procedure was divided into an inner loop and an outer loop. The neural network parameters, $\varnothing$, were iteratively optimized by gradient descent [53] after being randomly initialized. In the inner loop, the basic model (1D-CNN or UNet) was trained with a batch size of 512 sample units from the $T_{support}$ data. The parameters, $\varnothing_i$, of the model were updated to each task via gradient descent on the $T_{support}$ data. The outer loop then refined the parameters, $\varnothing_i$, based on the performance of model on T_{query} data with a batch size of 512 sample units.

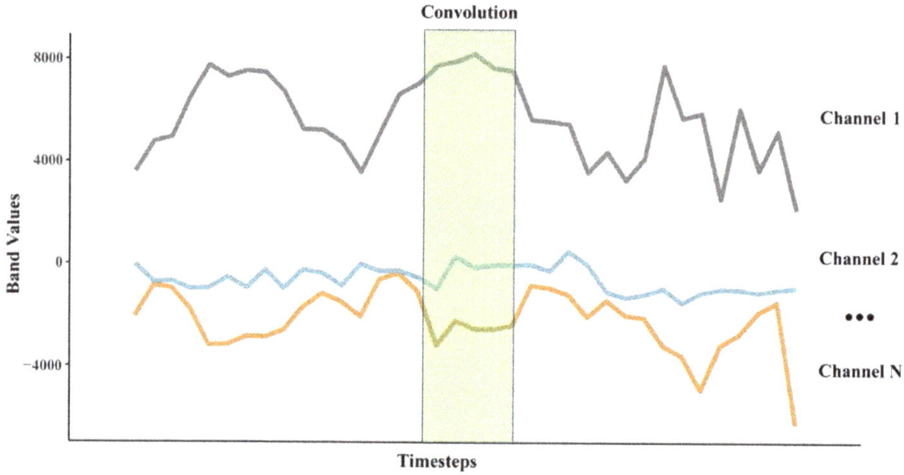

Figure 3. Diagram of the convolution along the time dimension.

During the fine-tuning stage, the pretrained basic model was designed to adjust the model's parameters by evaluating the fine-tuning data, which helps reduce differences among datasets from various areas. The pretrained basic model aligned the model with the characteristics of the local region to improve the model's performance in the target area.

For the prediction stage, the adapted model was applied to predict data, which comprise samples outside the common region. The model predicted the probabilities of each class for each sample location, serving as references in the selection of appropriate land cover labels.

Finally, the optimal land cover label from either the NLCD, LF EVT, or PALSAR wetland product was selected for each sample location in the order of the probabilities. We prioritized the labeling process based on confidence levels. Pixels were initially assigned labels corresponding to the Anderson Level II land cover types from the existing products based on whether the probability exceeded 0.1 to select the most detailed and specific land cover category first. Subsequently, labels were assigned based on the Anderson Level I classification legend in a similar manner, i.e., whether the probability exceeded 0.1 based on the order of probabilities. This means that if the probabilities indicate that the categories of the Anderson Level II classification legend are not confident enough, the method falls back on broader categories based on the Anderson Level I classification legend, ensuring all pixels are labeled with a reasonable category. Finally, the remaining unlabeled pixels are assigned the land cover type with the maximum probabilities, ensuring the entire dataset is labeled.

2.4. Method Assessment

The AKVEG is maintained by the Alaska Center for Conservation Science and includes over 14,000 vegetation plots from multiple projects conducted across Alaska. Most of the projects were conducted by the Alaska Center for Conservation Science, Bureau of Land Management, and the US Fish and Wildlife Service (USFWS). Macander et al. [54] also used the database to validate plant functional type cover.

For each plot, the survey date and the coverage of individual vegetation species were recorded. The survey date was considered as the reference date for the validation of land cover maps. Concerning the vegetation species, our initial step involved determining the life-form for each species by cross-referencing the accepted name with all names or accepted names in the checklist, which was supplied by the database provider.

Furthermore, we checked the wetland indicator status for each individual species by matching the accepted name with the scientific name on the national wetland plant list. This indicator status served as a measure of the likelihood of a particular species occurring in wetlands. Although facultative upland (FACU) and obligate upland (UPL) indicate high chances of being present in non-wetland areas, statuses such as obligatory wetland (OBL) and facultative wetland (FACW) might suggest a high possibility of occurring in wetlands.

Subsequently, we created a summary table for each plot that included the derived cover estimation for each life-form group and wetland status group. Additionally, we categorized life-form into deciduous forest, evergreen forest, mixed forest, graminoid, forb, dwarf scrub and shrub/scrub, and wetland status into confident wetland and confident non-wetland. Next, we calculated the total cover estimation for each life-form group and wetland status group. The created summary table is a repository of prior knowledge and provides percentage indicators for the plots. We established appropriate rules to correctly interpret each plot's class label based on the derived vegetation coverage and land cover definitions. The interpretation rules were applied sequentially, from top to bottom. We first verified whether the plots met the percentage requirements for wetland, followed by forest, then shrub, and, finally, grass based on the definitions outlined in the NLCD legend [55]. All interpretations underwent validation with remote sensing images to ensure geometric alignment.

The study area was chosen by taking the plot survey methods and considering the plot numbers of each project in the database. The following two projects were selected: North Slope Land Cover, conducted by North Slope Science Initiative [56], and Vegetation Monitoring in Interior Refuges, conducted by USFWS [54]. These projects were chosen because they covered a variety of Alaskan landscapes, such as the Interior Alaska region, and were mostly made up of forest and shrub areas, the northern coastal region, and wetland, water, and grass areas. These projects were also conducted with ground survey techniques and a semi-quantitative visual estimation method [54]. Based on the locations of the plot data, two Landsat ARD tiles, H04V01 and H06V04, were selected for testing the method. These two tiles capture the diversity of Alaskan land cover and represent the geographic conditions typical of the Alaskan landscape.

3. Results

3.1. Validation Samples

In this research, the AKVEG ground survey plots were used to assist in the interpretations. For example, 10 distinct vegetation types were identified within survey plot KAN226 (66.139274°N, 151.279178°W). As shown in Table 2, the plant life-forms and wetland indicator statuses were ascertained by cross-referencing the vegetation names with existing databases. The total vegetation coverage within the plot was calculated to be 97.1%. Among the identified vegetation types, deciduous trees constituted the highest canopy, composing around 36% of the plot area and 37% of the total vegetation cover. Therefore, deciduous trees were the dominant life-form vegetation cover within the plot. A coverage of above 20% matched the NLCD system's classification criteria for deciduous forests. Therefore, the plot was identified as a deciduous forest. In addition, the wetland indicator statuses indicated that area was mainly covered by facultative upland (FACU) vegetation, which composed 37% of the total plot, and facultative (FAC) vegetation, which constituted 58% of the total plot. Therefore, the plot was considered non-wetland. We interpreted 111 ground plots as validation sample units (Figure 4). The overall accuracy (OA), producer's accuracy (PA), and user's accuracy (UA) were calculated for evaluation.

Table 2. An example of the vegetation's interpretation into land cover types.

ID	Project	Site Code	Date	Accepted Name	Cover (%)	Plant Life-Form	Wetland Indicator Status
34027	USFWS Interior	KAN226	6/27/2013	*Betula neoalaskana*	36	deciduous tree	FACU
34024	USFWS Interior	KAN226	6/27/2013	*Alnus viridis*	23	shrub	FAC
34029	USFWS Interior	KAN226	6/27/2013	*Empetrum nigrum*	12	dwarf shrub	FAC
34031	USFWS Interior	KAN226	6/27/2013	*Vaccinium uliginosum*	12	dwarf shrub, shrub	FAC
34026	USFWS Interior	KAN226	6/27/2013	*Betula nana*	6	shrub	FAC
34030	USFWS Interior	KAN226	6/27/2013	*Rhododendron tomentosum*	4	dwarf shrub, shrub	FACW
34032	USFWS Interior	KAN226	6/27/2013	*Vaccinium vitis-idaea*	3	dwarf shrub	FAC
34025	USFWS Interior	KAN226	6/27/2013	*Arctous alpina*	0.5	dwarf shrub	FACU
34028	USFWS Interior	KAN226	6/27/2013	*Calamagrostis canadensis*	0.5	graminoid	FAC
34033	USFWS Interior	KAN226	6/27/2013	*Carex bigelowii*	0.1	graminoid	FAC

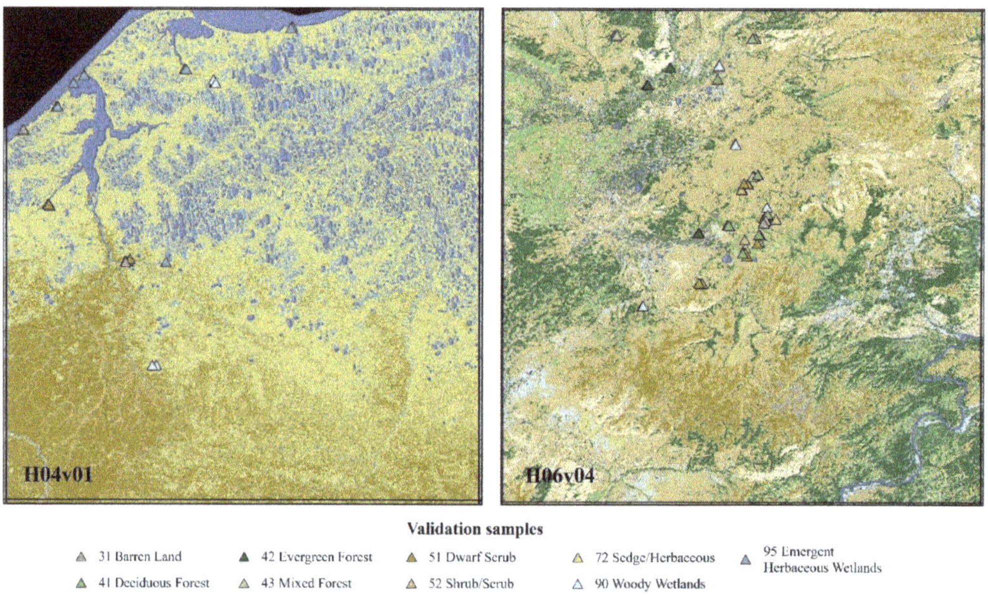

Figure 4. Interpreted validation sample units from plots in the North Slope Land Cover and USFWS. The background images are NLCD land cover maps.

3.2. Validation Results

Using the interpreted validation sample units, we completed an evaluation of the RSMA-based training data method. We compared the land cover labels from the NLCD product (Table 3) with labels produced by UNet-RSMA using a simple random sampling method in the fine-tuning phase (Table 4). Substantial increases in PA were observed in the woody wetlands (90) and emergent herbaceous wetlands (95) categories, which are predominant land cover types in the Alaskan region [12]. The increases in the PA accuracy for the forest classes were primarily found across the broader forest category based on the Anderson Level I classification legend, with only slight improvements for the Level II classification legend. The distinction between the barren and vegetation classes also increased. Although there were no increases in PA for the dwarf shrub (51) and sedge/herbaceous (72) categories, the commission errors (100 − UA) were greatly reduced, indicating a decrease in confusion with other categories. The accuracies of shrub/scrub (52) also showed improvements in the method. The OA increased from 29.7% with the NLCD to 60.4% with the new RSMA-based training data method based on the Anderson Level II classification legend.

Table 3. The agreement between the interpreted validation sample units (interpreted plot data from AKVEG) and the NLCD. The OA was 29.7%. (31: Barren; 41: Deciduous Forest; 42: Evergreen Forest; 43: Mixed Forest; 51: Dwarf Scrub; 52: Shrub/Scrub; 71: Grassland/Herbaceous; 72: Sedge/Herbaceous; 90: Woody Wetlands; 95: Emergent Herbaceous Wetlands). When the ground total is 0, the corresponding accuracy was not calculated, indicated by a dash (-). Each grey block highlights the group of classes under the same Anderson Level I legend. The bolded values along the diagonal are the correctly labeled sample units. The same applies to the following tables.

		NLCD 2011 Land Cover											
		31	**41**	**42**	**43**	**51**	**52**	**71**	**72**	**90**	**95**	**Grand Total**	**PA (%)**
Ground Truth	31	**3**				1			1			5	60.0
	41		**2**				4	2	1			9	22.2
	42			**2**	2	1						5	40.0
	43		1		**3**	1	1					6	50.0
	51	1				**8**	1		3			13	61.5
	52					5	**13**		1			19	68.4
	71							**0**				0	-
	72	1				2			**1**			4	25.0
	90			2		2	12		7	**0**		23	0.0
	95		1	1			2		22		**1**	27	3.7
Grand Total		5	4	5	5	20	33	2	36	-	1	111	29.7%
UA (%)		60.0	50.0	40.0	60.0	40.0	39.4	0.0	2.8	-	100.0		

Table 4. Agreement between the interpreted validation sample units (interpreted plot data from AKVEG) and land cover labels produced by UNet-RSMA with simple random sampling in the fine-tuning process. The OA was 60.4%. (31: Barren; 41: Deciduous Forest; 42: Evergreen Forest; 43: Mixed Forest; 51: Dwarf Scrub; 52: Shrub/Scrub; 72: Sedge/Herbaceous; 90: Woody Wetlands; 95: Emergent Herbaceous Wetlands). The grand total for 71: Grassland/Herbaceous in both validation sample units and predicted land cover labels was zero, so it was removed from the matrix. The same applies to the following tables.

		UNet-RSMA Based Training Data Method										
		31	**41**	**42**	**43**	**51**	**52**	**72**	**90**	**95**	**Grand Total**	**PA (%)**
Ground Truth	31	**4**				1					5	80.0
	41		**3**	1			3			2	9	33.3
	42			**5**							5	100.0
	43		3		**3**						6	50.0
	51	1				**8**	2	2			13	61.5
	52			2			**17**				19	89.5
	72	1				2		**1**			4	25.0
	90		1	1	1	1	6	2	**8**	3	23	34.8
	95		3	2			1	2	1	**18**	27	66.7
Grand Total		6	10	11	4	12	29	7	9	23	111	60.4%
UA (%)		66.7	30.0	45.5	75.0	66.7	58.6	14.3	88.9	78.3		

In the RSMA method, the pretrained global model must be adjusted to the target region to facilitate adaption of the model to local landscapes. For the UNet-RSMA, we calculated the PA, UA, and OA for the pretrained model after the meta-training process, as shown in Table 5. Comparing the results with those in Table 4, the increases in both PA and UA were found for various land cover categories including dwarf scrub (51), shrub/scrub (52), sedge/herbaceous (72), and emergent herbaceous wetlands (95), following the model transfer (i.e., fine-tuning) process. The OA increased by 18.1% through the model's fine-tuning. More accurate predictions were produced as a result of this adaptation process by incorporating the diversity and sharing characteristics of land cover across the large geographical region.

Table 5. Agreement between the interpreted validation sample units (interpreted plots data from AKVEG) and land cover labels produced by the pretrained model from the meta-training process in UNet-RSMA. The OA was 42.3%. (31: Barren; 41: Deciduous Forest; 42: Evergreen Forest; 43: Mixed Forest; 51: Dwarf Scrub; 52: Shrub/Scrub; 72: Sedge/Herbaceous; 90: Woody Wetlands; 95: Emergent Herbaceous Wetlands).

		Pretrained Model by Meta-Training Process in UNet-RSMA										
		31	**41**	**42**	**43**	**51**	**52**	**72**	**90**	**95**	**Grand Total**	**PA (%)**
Ground Truth	31	**4**								1	5	80.0
	41		**3**		1		2		1	2	9	33.3
	42		1	**3**	1						5	60.0
	43		3		**3**						6	50.0
	51	1				**5**	1	3	1	2	13	38.5
	52	1	2	2			**7**	2	4	1	19	36.8
	72	2				2		**0**			4	0.0
	90		1	1	1	2	5		**9**	4	23	39.1
	95		2		1	2	1	7	1	**13**	27	48.1
Grand Total		8	12	6	7	11	16	12	16	23	111	42.3
UA (%)		50.0	25.0	50.0	42.9	45.5	43.8	0.0	56.3	56.5		

We explored the sampling method used for fine-tuning the pretrained UNet model derived by the meta-training process for the UNet-RSMA. Our results revealed that land cover labels obtained through the simple random sampling method reported higher OAs than those acquired through the equal-number random sampling method for each category (refer to Tables 4 and 6). This observation indicates that the simple random sampling method accurately represented the distribution and characteristics of land cover classes at the given region. When comparing accuracies, the equal-number random sampling method for each category produced higher PA only for the woody wetlands (90) category, which was a rare class in both test regions. Using simple random sampling might arise from its ability to focus the model updates on major classes, thereby increasing the overall accuracy. Additionally, when comparing the results with the labels derived by the pre-trained model by the meta-training process in the UNet-RSMA (Table 5). Dwarf scrub (51), sedge/herbaceous (72), Woody Wetlands (90), and emergent herbaceous wetlands (95) show higher accuracy in both the PA and UA using equal-number random sampling method for each category in the fine-tuning process of the UNet-RSMA (Table 6). And the OA increased by 6.3% compared with Table 5. These results demonstrate that fine-tuning the model can improve accuracies of model predictions.

We conducted a comparative analysis for two basic models, the 1D-CNN and UNet classifier, in the meta-training process to assess their performances as a basic model in the RSMA based training data method. The overall accuracies achieved by the UNet-RSMA were 60.4% for the Anderson Level II classes and 69.4% for the Anderson Level I classes. On the other hand, the 1D-CNN- RSMA based training data method achieved an overall accuracy of 57.7% for the Anderson Level II classes (Table 7) and 73.0% for the Anderson Level I classes. Further analysis revealed the differences in the model performances based on the complexity of the land cover legend. For the land cover classifications with a simple legend with distinguishable differences, such as the Level I classification legend, simpler architecture models were found to be better. Conversely, in scenarios involving complex legends with intricate relationships in time-series data, more complex models were superior to others.

Table 6. The agreement between the interpreted validation sample units (interpreted plots data from AKVEG) and land cover labels produced by the equal-number random sampling method for each category in UNet-RSMA based training data method. The OA was 48.6% (31: Barren; 41: Deciduous Forest; 42: Evergreen Forest; 43: Mixed Forest; 51: Dwarf Scrub; 52: Shrub/Scrub; 71: Grassland/Herbaceous; 72: Sedge/Herbaceous; 90: Woody Wetlands; 95: Emergent Herbaceous Wetlands).

		Transferring Model Using the Equal-Number Random Sampling Method for Each Category											
		31	**41**	**42**	**43**	**51**	**52**	**71**	**72**	**90**	**95**	**Grand Total**	**PA (%)**
Ground Truth	31	**4**									1	5	80.0
	41		**3**		1		2	1			2	9	33.3
	42		1	**3**	1							5	60.0
	43		3		**3**							6	50.0
	51	1				**6**	1		2	1	2	13	46.2
	52	1	2	2			**7**		2	4	1	19	36.8
	71							**0**				0	-
	72	2				1			**1**			4	25.0
	90		1	1	1	2	5		1	**10**	2	23	43.5
	95		3			2	1		4		**17**	27	63.0
Grand Total		8	13	6	6	11	16	1	10	15	25	111	
UA (%)		50.0	23.1	50.0	50.0	54.5	43.8	0.0	10.0	66.7	68.0		48.6%

Table 7. Agreement between the interpreted validation sample units (interpreted plots data from AKVEG) and land cover labels produced by 1D-CNN-RSMA-based training data method. The OA was 57.7% (31: Barren; 41: Deciduous Forest; 42: Evergreen Forest; 43: Mixed Forest; 51: Dwarf Scrub; 52: Shrub/Scrub; 71: Grassland/Herbaceous; 72: Sedge/Herbaceous; 90: Woody Wetlands; 95: Emergent Herbaceous Wetlands).

		1D-CNN-RSMA										
		31	**41**	**42**	**43**	**51**	**52**	**72**	**90**	**95**	**Grand Total**	**PA (%)**
Ground Truth	31	**3**				1		1			5	60.0
	41		**4**				5				9	44.4
	42			**3**	2						5	60.0
	43		1		**4**		1				6	66.7
	51					**9**	2			2	13	69.2
	52			3		4	**12**				19	63.2
	72	1				2	1	**0**			4	0.0
	90		1	3	1		4		**6**	8	23	26.1
	95		2	1			1			**23**	27	85.2
Grand Total		4	8	10	7	16	26	1	6	33	111	
UA (%)		75.0	50.0	30.0	57.1	56.3	46.2	0.0	100.0	69.7		57.7%

4. Discussion

In large-scale land cover classification, training data obtained from full spatial coverage of a mapping region are beneficial for capturing the representative spectral characters to exhaust the spectral information for each land cover type. In this research, we developed an RSMA-based training data method (UNet-RSMA and 1D-CNN-RSMA) for the training data generation. The method combined the meta-training process of meta learning and fine-tuning and involves transferring a pretrained model from a larger region centered in the same geographic area and adapting it to perform on a smaller region within the same area. The method overcomes the challenge of consensus data scarcity in certain regions and produced accurate and consistent training samples at any given location that lacks consensus data. By determining optimal land cover labels for each generated training sample based on the probabilities associated with the model's predictions and existing products, we confirmed reliable training samples.

However, the RSMA-based training data method has certain limitations. The accuracy of the newly labeled training sample was determined by the existing land cover products because the final label was selected from the products based on the probabilities. Additionally, the accuracies of the predicted probabilities used as indicators were influenced by various factors, including the distribution, quantity, and quality of the consensus samples and model parameters in the model. Despite these limitations, the developed RSMA-based training data method generated training samples with improved land cover labels compared to the single land cover product for various applications. For instance, the method could be utilized to produce reliable and accurate training data for the USGS-led annual land cover product, which produces annual land cover maps with Anderson Level II classification legend for the contiguous United States and Alaska since the 1980s.

The framework proposed in this paper can be applied to regions outside the continental United States (CONUS), Alaska, and Hawaii, where datasets like the NLCD, LANDFIRE's EVT, and NWI are unavailable. In these regions, global land cover datasets can be used as substitutes. There are several global land cover products [57–59], forest [60], and water [61,62] products available at different scales and covering different time periods that provide a foundational basis for this approach.

For the validation data collection process, interpreting the AKVEG plots addresses another challenge of inconsistency among the validation samples and lack of knowledge for interpretation. One issue is the discrepancy between the spatial coverage of the database records and the remote sensing pixels. For example, the plots vary in size from 5 × 5 to 10 × 10 and 20 × 20 m, often smaller than the resolution of a single pixel or covering partial part of a pixel in a Landsat image. The vegetation coverage of the remaining pixel was often estimated when evaluating the vegetation coverage inside a pixel to identify land cover type. Furthermore, the surrounding pixels of the plot could exhibit spatial heterogeneity, The misalignment between the geolocation of the plot and the image pixel can impact the accurate determination of land cover types based solely on vegetation plot data. Therefore, the interpretation process requires cross-referencing with remote sensing images or high-resolution imagery.

5. Conclusions

In this research, we developed a novel method, the RSMA-based training data method, to automatically produce reliable training data at any location for mapping Alaska's land cover. The method integrated the meta-training and fine-tuning techniques to refine the land cover labels compared with the labels from the NLCD products. The RSMA-based training data method demonstrates the potential for using the produced training data to generate high-quality land cover maps especially for regions like Alaska. The RSMA-based training data method is tailored to address the scarcity of training data with limited prior knowledge. Furthermore, our research examined the effects of basic machine learning models (1D-CNN vs. UNet) on training the global model in the meta-training process and sampling strategies (equal-number strategy vs. simple random sampling) on updating model parameters in the specific region in the fine-tuning process. Based on the results, we have provided insights and guidelines for using the method in different scenarios. Our research results indicate that for less complex classes such as Anderson Level I classification legend, 1D-CNN as the basic model choice in the RMSA method derived a higher OA, whereas UNet as a basic model derived higher OA for more complex classes such as Anderson Level II classification legend. Additionally, the OA was highest using simple random sampling for transferring the pretrained model.

In addition to generating automatic training data, our study addressed challenges related to validation sample interpretation, especially for wetlands. Beyond using manually interpreted samples for the validation, our research interpreted existing vegetation plots from other vegetation survey projects as validation samples, which also could be used as prior landscape knowledge to guide the selection of random samples for validation purposes, thereby mitigating potential biases.

Overall, this research advances the field of land-cover mapping in Alaska by developing an automatic approach to generate training data. In addition to overcoming the lack of prior knowledge, the method provided consistent training data across regions.

Author Contributions: Conceptualization, C.L.; methodology, C.L.; validation, C.L.; formal analysis, C.L.; writing—original draft preparation, C.L.; writing—review and editing, C.L., G.X. and S.J.; supervision, G.X. and S.J. All authors have read and agreed to the published version of the manuscript.

Funding: This research was funded by the U.S. Geological Survey National Land Imaging (NLI) program under USGS Contract 140G0124D0001.

Data Availability Statement: Landsat-5 and Landsat-7 images courtesy of the U.S. Geological Survey, National Land Cover Dataset (NLCD; Jon, 2019; [63]), and the existing vegetation type LANDFIRE datasets (LANDFIRE, 2010; [64]).

Acknowledgments: We thank T. Adamson and F. Dwomoh for editing the manuscript. We also appreciate all anonymous reviewers and editors for their constructive comments. Any use of trade, firm, or product names is for descriptive purposes only and does not imply endorsement by the U.S. Government. This journal article has been peer reviewed and approved for publication consistent with USGS Fundamental Science Practices (https://pubs.usgs.gov/circ/1367/, accessed on 20 September 2024).

Conflicts of Interest: The authors declare no conflicts of interest.

References

1. McGuire, A.D.; Anderson, L.G.; Christensen, T.R.; Dallimore, S.; Guo, L.; Hayes, D.; Heimann, M.; Lorenson, T.; Macdonald, R.; Roulet, N. Sensitivity of the carbon cycle in the Arctic to climate change. *Ecol. Monogr.* **2009**, *79*, 523–555. [CrossRef]
2. McGuire, A.D.; Apps, M.; Chapin, F.S.; Dargaville, R.; Flannigan, M.D.; Kasischke, E.S.; Kicklighter, D.; Kimball, J.; Kurz, W.; McRae, D.J.; et al. Land Cover Disturbances and Feedbacks to the Climate System in Canada and Alaska. In *Land Change Science: Observing, Monitoring and Understanding Trajectories of Change on the Earth's Surface;* Gutman, G., Janetos, A.C., Justice, C.O., Moran, E.F., Mustard, J.F., Rindfuss, R.R., Skole, D., Turner, B.L., Cochrane, M.A., Eds.; Springer: Dordrecht, The Netherlands, 2004; pp. 139–161.
3. Jin, S.; Yang, L.; Zhu, Z.; Homer, C. A land cover change detection and classification protocol for updating Alaska NLCD 2001 to 2011. *Remote Sens. Environ.* **2017**, *195*, 44–55. [CrossRef]
4. Vynne, C.; Dovichin, E.; Fresco, N.; Dawson, N.; Joshi, A.; Law, B.E.; Lertzman, K.; Rupp, S.; Schmiegelow, F.; Trammell, E.J. The Importance of Alaska for Climate Stabilization, Resilience, and Biodiversity Conservation. *Front. For. Glob. Change* **2021**, *4*, 701277. [CrossRef]
5. Fisher, J.; Sikka, M.; Oechel, W.; Huntzinger, D.N.; Melton, J.R.; Koven, C.D.; Ahlström, A.; Arain, M.A.; Baker, I.; Chen, J.M.; et al. Carbon cycle uncertainty in the Alaskan Arctic. *Biogeosciences* **2014**, *11*, 4271–4288. [CrossRef]
6. Wang, J.A.; Sulla-Menashe, D.; Woodcock, C.E.; Sonnentag, O.; Keeling, R.F.; Friedl, M.A. Extensive land cover change across Arctic–Boreal Northwestern North America from disturbance and climate forcing. *Glob. Change Biol.* **2020**, *26*, 807–822. [CrossRef]
7. Douglas, T.; Hiemstra, C.; Anderson, J.; Barbato, R.; Bjella, K.; Deeb, E.; Gelvin, A.; Nelsen, P.E.; Newman, S.D.; Saari, S.P.; et al. Recent degradation of interior Alaska permafrost mapped with ground surveys, geophysics, deep drilling, and repeat airborne lidar. *Cryosphere* **2021**, *15*, 3555–3575. [CrossRef]
8. Crumley, R.; Hill, D.; Beamer, J.; Holzenthal, E. Hydrologic Diversity in Glacier Bay Alaska: Spatial Patterns and Temporal Change. In *The Cryosphere Discussions;* European Geosciences Union: Munich, Germany, 2019; pp. 1–31.
9. Marcot, B.G.; Jorgenson, M.T.; Lawler, J.P.; Handel, C.M.; DeGange, A.R. Projected changes in wildlife habitats in Arctic natural areas of northwest Alaska. *Clim. Change* **2015**, *130*, 145–154. [CrossRef]
10. Pielke Sr, R.A.; Pitman, A.; Niyogi, D.; Mahmood, R.; McAlpine, C.; Hossain, F.; Goldewijk, K.K.; Nair, U.; Betts, R.; Fall, S.; et al. Land use/land cover changes and climate: Modeling analysis and observational evidence. *WIREs Clim. Change* **2011**, *2*, 828–850. [CrossRef]
11. Nelson, K.J.; Long, D.G.; Connot, J.A. *LANDFIRE 2010—Updates to the National Dataset to Support Improved Fire and Natural Resource Management;* Report 2016-1010; USGS: Reston, VA, USA, 2016.
12. Hall, J.V.; Frayer, W.E.; Wilen, B.O. *Status of Alaska Wetlands;* US Fish & Wildlife Service: Washington, DC, USA, 1994.
13. U.S. Department of Interior (USDI); Fish and Wildlife Service (FWS). *National Wetlands Inventory Website;* U.S. Department of the Interior, Fish and Wildlife Service: Washington, DC, USA, 2018. Available online: http://www.fws.gov/wetlands/ (accessed on 2 November 2020).
14. Clewley, D.; Whitcomb, J.; Moghaddam, M.; McDonald, K.; Chapman, B.; Bunting, P. Evaluation of ALOS PALSAR Data for High-Resolution Mapping of Vegetated Wetlands in Alaska. *Remote Sens.* **2015**, *7*, 7272–7297. [CrossRef]

15. Li, C.C.; Wang, J.; Wang, L.; Hu, L.Y.; Pong, P. Comparison of classification algorithms and training sample sizes in urban land classification with Landsat Thematic Mapper imagery. *Remote Sens.* **2014**, *6*, 964–983. [CrossRef]
16. Zhong, L.; Hu, L.; Zhou, H. Deep learning based multi-temporal crop classification. *Remote Sens. Environ.* **2019**, *221*, 430–443. [CrossRef]
17. Zhang, G.; Roslan, S.N.A.b.; Wang, C.; Quan, L. Research on land cover classification of multi-source remote sensing data based on improved U-net network. *Sci. Rep.* **2023**, *13*, 16275. [CrossRef] [PubMed]
18. Shakya, A.K.; Ramola, A.; Vidyarthi, A. *Landcover Pattern Recognization through Texture Classification Using LANDSAT Data of Dallas*; Springer: Singapore, 2020; pp. 283–293.
19. Zhao, W.; Peng, S.; Chen, J.; Peng, R. Contextual-Aware Land Cover Classification With U-Shaped Object Graph Neural Network. *IEEE Geosci. Remote Sens. Lett.* **2022**, *19*, 6510705. [CrossRef]
20. Ghimire, B.; Rogan, J.; Miller, J. Contextual land-cover classification: Incorporating spatial dependence in land-cover classification models using random forests and the Getis statistic. *Remote Sens. Lett.* **2010**, *1*, 45–54. [CrossRef]
21. Zhu, Z.; Woodcock, C.E. Continuous change detection and classification of land cover using all available Landsat data. *Remote Sens. Environ.* **2014**, *144*, 152–171. [CrossRef]
22. Li, C.; Xian, G.; Zhou, Q.; Pengra, B. A novel automatic phenology learning (APL) method of training sample selection using multiple datasets for time-series land cover mapping. *Remote Sens. Environ.* **2021**, *266*, 112670. [CrossRef]
23. Li, C.; Gong, P.; Wang, J.; Yuan, C.; Hu, T.; Wang, Q.; Yu, L.; Clinton, N.; Li, M.; Guo, J.; et al. An all-season sample database for improving land-cover mapping of Africa with two classification schemes. *Int. J. Remote Sens.* **2016**, *37*, 4623–4647. [CrossRef]
24. Li, C.; Gong, P.; Wang, J.; Zhu, Z.; Biging, G.S.; Yuan, C.; Hu, T.; Zhang, H.; Wang, Q.; Li, X.; et al. The first all-season sample set for mapping global land cover with Landsat-8 data. *Sci. Bull.* **2017**, *62*, 508–515. [CrossRef]
25. Zhao, Y.; Gong, P.; Yu, L.; Hu, L.; Li, X.; Li, C.; Zhang, H.; Zheng, Y.; Wang, J.; Zhao, Y.; et al. Towards a common validation sample set for global land-cover mapping. *Int. J. Remote Sens.* **2014**, *35*, 4795–4814. [CrossRef]
26. Zhou, Q.; Tollerud, H.; Barber, C.; Smith, K.; Zelenak, D. Training data selection for annual land cover classification for the Land Change Monitoring, Assessment, and Projection (LCMAP) initiative. *Remote Sens.* **2020**, *12*, 699. [CrossRef]
27. Brown, J.F.; Tollerud, H.J.; Barber, C.P.; Zhou, Q.; Dwyer, J.L.; Vogelmann, J.E.; Loveland, T.R.; Woodcock, C.E.; Stehman, S.V.; Zhu, Z.; et al. Lessons learned implementing an operational continuous United States national land change monitoring capability: The Land Change Monitoring, Assessment, and Projection (LCMAP) approach. *Remote Sens. Environ.* **2020**, *238*, 111356. [CrossRef]
28. Li, C.; Xian, G.; Wellington, D.; Smith, K.; Horton, J.; Zhou, Q. Development of the LCMAP annual land cover product across Hawai'i. *Int. J. Appl. Earth Obs. Geoinf.* **2022**, *113*, 103015. [CrossRef]
29. Maus, V.; Camara, G.; Cartaxo, R.; Sanchez, A.; Ramos, F.M.; Queiroz, G.R. A Time-Weighted Dynamic Time Warping method for land-use and land-cover mapping. *IEEE J. Sel. Top. Appl. Earth Obs. Remote Sens.* **2016**, *9*, 3729–3739. [CrossRef]
30. Pengra, B.; Stehman, S.V.; Horton, J.A.; Wellington, D.F. *Land Change Monitoring, Assessment, and Projection (LCMAP) Version 1.0 Annual Land Cover and Land Cover Change Validation Tables*; U.S. Geological Survey Data Release: Reston, VA, USA, 2020.
31. Nawrocki, T.W. Alaska Vegetation Plots Database (AKVEG). Git Repository. 2021. Available online: https://github.com/accs-uaa/vegetation-plots-database (accessed on 20 April 2022).
32. Hospedales, T.; Antoniou, A.; Micaelli, P.; Storkey, A. Meta-Learning in Neural Networks: A Survey. *IEEE Trans. Pattern Anal. Mach. Intell.* **2022**, *44*, 5149–5169. [CrossRef] [PubMed]
33. Alshalali, T.; Josyula, D. Fine-Tuning of Pre-Trained Deep Learning Models with Extreme Learning Machine. In Proceedings of the 2018 International Conference on Computational Science and Computational Intelligence (CSCI), Las Vegas, NV, USA, 12–14 December 2018; pp. 469–473.
34. Anderson, J.R.; Hardy, E.E.; Roach, J.T.; Witmer, R.E. *A Land Use and Land Cover Classification System for Use with Remote Sensor Data, Report 964*; US Government Printing Office: Washington, DC, USA, 1976.
35. Cochran, W.G. *Sampling Techniques*, 3rd ed.; John Wiley & Sons: Hoboken, NJ, USA, 1977.
36. Crawford, C.J.; Roy, D.P.; Arab, S.; Barnes, C.; Vermote, E.; Hulley, G.; Gerace, A.; Choate, M.; Engebretson, C.; Micijevic, E.; et al. The 50-year Landsat collection 2 archive. *Sci. Remote Sens.* **2023**, *8*, 100103. [CrossRef]
37. Tucker, C.J.; Sellers, P.J. Satellite remote sensing of primary production. *Int. J. Remote Sens.* **1986**, *7*, 1395–1416. [CrossRef]
38. Zha, Y.; Gao, J.; Ni, S. Use of normalized difference built-up index in automatically mapping urban areas from TM imagery. *Int. J. Remote Sens.* **2003**, *24*, 583–594. [CrossRef]
39. McFeeters, S.K. The use of the Normalized Difference Water Index (NDWI) in the delineation of open water features. *Int. J. Remote Sens.* **1996**, *17*, 1425–1432. [CrossRef]
40. Xu, H. Modification of normalised difference water index (NDWI) to enhance open water features in remotely sensed imagery. *Int. J. Remote Sens.* **2006**, *27*, 3025–3033. [CrossRef]
41. Feyisa, G.L.; Meilby, H.; Fensholt, R.; Proud, S. Automated Water Extraction Index: A new technique for surface water mapping using Landsat imagery. *Remote Sens. Environ.* **2014**, *140*, 23–35. [CrossRef]
42. Crist, E.P. A TM Tasseled Cap equivalent transformation for reflectance factor data. *Remote Sens. Environ.* **1985**, *17*, 301–306. [CrossRef]
43. Huang, C.; Wylie, B.; Yang, L.; Homer, C.; Zylstra, G. Derivation of a Tasseled Cap Transformation Based On Landsat 7 At-Satellite Reflectance. *Int. J. Remote Sens.* **2002**, *23*, 1741–1748. [CrossRef]

44. Miller, J.D.; Yool, S.R. Mapping forest post-fire canopy consumption in several overstory types using multi-temporal Landsat TM and ETM data. *Remote Sens. Environ.* **2002**, *82*, 481–496. [CrossRef]
45. Diek, S.; Fornallaz, F.; Schaepman, M.E.; Rogier, D.J. Barest Pixel Composite for Agricultural Areas Using Landsat Time Series. *Remote Sens.* **2017**, *9*, 1245. [CrossRef]
46. Tegegne, A.M. Applications of Convolutional Neural Network for Classification of Land Cover and Groundwater Potentiality Zones. *J. Eng.* **2022**, *2022*, 6372089. [CrossRef]
47. Yamashita, R.; Nishio, M.; Do, R.K.G.; Togashi, K. Convolutional neural networks: An overview and application in radiology. *Insights Into Imaging* **2018**, *9*, 611–629. [CrossRef]
48. Ioffe, S.; Szegedy, C. Batch normalization: Accelerating deep network training by reducing internal covariate shift. In Proceedings of the 32nd International Conference on International Conference on Machine Learning, Lille, France, 6–11 July 2015; Volume 37, pp. 448–456.
49. Krizhevsky, A.; Sutskever, I.; Hinton, G.E. ImageNet classification with deep convolutional neural networks. *Commun. ACM* **2017**, *60*, 84–90. [CrossRef]
50. Ronneberger, O.; Fischer, P.; Brox, T. *U-Net: Convolutional Networks for Biomedical Image Segmentation*; Springer International Publishing: Cham, Switzerland, 2015; pp. 234–241.
51. Hesamian, M.H.; Jia, W.; He, X.; Kennedy, P. Deep Learning Techniques for Medical Image Segmentation: Achievements and Challenges. *J. Digit. Imaging* **2019**, *32*, 582–596. [CrossRef]
52. Finn, C.; Abbeel, P.; Levine, S. Model-Agnostic Meta-Learning for Fast Adaptation of Deep Networks. *ICML* **2017**, *70*, 1126–1135.
53. Bishop, C.M. *Pattern Recognition and Machine Learning*; Springer: Berlin/Heidelberg, Germany, 2006.
54. Macander, M.J.; Nelson, P.R.; Nawrocki, T.W.; Frost, G.V.; Orndahl, K.M.; Palm, E.C.; Wells, A.F.; Goetz, S.J. Time-series maps reveal widespread change in plant functional type cover across Arctic and boreal Alaska and Yukon. *Environ. Res. Lett.* **2022**, *17*, 054042. [CrossRef]
55. Selkowitz, D.J.; Stehman, S.V. Thematic accuracy of the National Land Cover Database (NLCD) 2001 land cover for Alaska. *Remote Sens. Environ.* **2011**, *115*, 1401–1407. [CrossRef]
56. Ducks Unlimited (DU). *North Slope Science Initiative Landcover Mapping Summary Report*; Ducks Unlimited: Rancho Cordova, CA, USA, 2013.
57. Gong, P.; Wang, J.; Yu, L.; Zhao, Y.; Zhao, Y.; Liang, L.; Niu, Z.; Huang, X.; Fu, H.; Liu, S.; et al. Finer resolution observation and monitoring of global land cover: First mapping results with Landsat TM and ETM+ data. *Int. J. Remote Sens.* **2013**, *34*, 2607–2654. [CrossRef]
58. Zanaga, D.; Van De Kerchove, R.; Daems, D.; De Keersmaecker, W.; Brockmann, C.; Kirches, G.; Wevers, J.; Cartus, O.; Santoro, M.; Fritz, S.; et al. *ESA WorldCover 10 m 2020 v100*; Zenodo: Genève, Switzerland, 2021. [CrossRef]
59. Friedl, M.A.; Woodcock, C.E.; Olofsson, P.; Zhu, Z.; Loveland, T.; Stanimirova, R.; Arevalo, P.; Bullock, E.; Hu, K.T.; Zhang, Y.; et al. Medium Spatial Resolution Mapping of Global Land Cover and Land Cover Change Across Multiple Decades From Landsat. *Front. Remote Sens.* **2022**, *3*, 894571. [CrossRef]
60. Hansen, M.C.; Potapov, P.V.; Moore, R.; Hancher, M.; Turubanova, S.A.; Tyukavina, A.; Thau, D.; Stehman, S.V.; Goetz, S.J.; Loveland, T.R.; et al. High-resolution global maps of 21st-century forest cover change. *Science* **2013**, *342*, 850–853. [CrossRef] [PubMed]
61. Pickens, A.H.; Hansen, M.C.; Hancher, M.; Stehman, S.V.; Tyukavina, A.; Potapov, P.; Marroquin, B.; Sherani, Z. Mapping and sampling to characterize global inland water dynamics from 1999 to 2018 with full Landsat time-series. *Remote Sens. Environ.* **2020**, *243*, 111792. [CrossRef]
62. Pekel, J.F.; Cottam, A.; Gorelick, N.; Belward, A.S. High-resolution mapping of global surface water and its long-term changes. *Nature* **2016**, *540*, 418–422. [CrossRef]
63. Dewitz, J. *National Land Cover Database (NLCD) 2016 Products (ver. 3.0, November 2023) [National Land Cover Database (NLCD) 2011 Land Cover—Alaska]*; U.S. Geological Survey: Reston, VA, USA, 2019. [CrossRef]
64. LANDFIRE. Existing Vegetation Type Layer, LANDFIRE 1.2.0, U.S. Department of the Interior, Geological Survey, and U.S. Department of Agriculture. 2010. Available online: http://www.landfire/viewer (accessed on 1 April 2021).

remote sensing

MDPI

Article

Explainability Feature Bands Adaptive Selection for Hyperspectral Image Classification

Jirui Liu [1], Jinhui Lan [1,*], Yiliang Zeng [1], Wei Luo [1], Zhixuan Zhuang [1] and Jinlin Zou [2]

[1] School of Automation and Electrical Engineering, China University of Science and Technology Beijing, Beijing 100083, China; b20200294@xs.ustb.edu.cn (J.L.); zengyl@ustb.edu.cn (Y.Z.); d202210336@xs.ustb.edu.cn (W.L.); d202310352@xs.ustb.edu.cn (Z.Z.)

[2] Equipment Management and Unmanned Aerial Vehicle Engineering School, Air Force Engineering University, Xi'an 710051, China; yuxinumberone@163.com

* Correspondence: lanjh@ustb.edu.cn

Abstract: Hyperspectral remote sensing images are widely used in resource exploration, urban planning, natural disaster assessment, and feature classification. Aiming at the problems of poor interpretability of feature classification algorithms for hyperspectral images, multiple feature dimensions, and difficulty in effectively improving classification accuracy, this paper proposes a feature band adaptive selection method for hyperspectral images. The proposed feature band adaptive selection model focuses on the joint salient feature regions of the hyperspectral image, visualizes the feature contribution of the bands, more intuitively reveals the selection basis of the feature bands of the hyperspectral features in the process of deep learning, and selects the feature bands with high contribution to carry out classification experiments for verification. Quantitative evaluations on four hyperspectral benchmarks (Pavia University/Centre, Washington DC, GF-5) demonstrate that EFBASN achieves state-of-the-art classification accuracy, with an overall accuracy (OA) of 97.68% on Pavia U, surpassing 12 recent methods including SSCFA (94.48%) and CNCMN (93.12%). Crucially, the attention weights of critical bands (e.g., Band 26 at 672 nm for iron oxide detection) are 3.2 times higher than redundant bands, providing physically interpretable selection criteria.

Keywords: hyperspectral image; hyperspectral image classification; deep learning

1. Introduction

The rapid advancement of remote sensing technology has positioned hyperspectral imaging as a cornerstone for Earth observation, offering unprecedented capabilities in resource exploration [1], urban planning [2], and disaster assessment [3]. Unlike conventional multispectral systems, hyperspectral sensors capture hundreds of contiguous spectral bands, enabling precise material discrimination through unique spectral signatures [4]. This richness in spectral dimensionality, combined with high-resolution spatial information [5], allows hyperspectral data to resolve ambiguities in feature classification that challenge traditional remote sensing methods. For instance, subtle spectral variations in the visible to short-wave infrared range (400–2500 nm) can distinguish mineral compositions in geological surveys [6] or detect early-stage vegetation stress in precision agriculture [7].

Despite these advantages, the complexity of hyperspectral data introduces significant challenges. Deep learning models, particularly convolutional neural networks (CNNs) and transformers, have become dominant tools for hyperspectral image classification due to their ability to extract hierarchical spatial–spectral features. Architectures such as

SSRN [8], which integrates 3D convolutions with residual learning, and SpectralFormer [9], a Transformer-based model for spectral sequence modeling, have pushed classification accuracy to new heights. However, these models inherently operate as "black boxes" [10], providing limited insight into how specific spectral bands or spatial regions contribute to classification decisions. This opacity hinders trust in critical applications such as environmental monitoring [11], where erroneous predictions could lead to flawed policy decisions.

The pursuit of interpretability in deep learning has spawned diverse approaches. Sensitivity analysis methods [12] quantify how input perturbations affect model outputs, while activation visualization techniques [13] highlight regions of interest in feature maps. Recent work by Samet Aksoy [14] demonstrates the potential of SHAP values for soil salinity estimation, and Shinnosuke Ishikawa's explainable AI framework [15] validates model generalizability across remote sensing domains. Despite these advances, most interpretability techniques remain decoupled from the feature alignment process, treating explanation as a post hoc analysis [16] rather than an integrated component of model training. Furthermore, existing methods predominantly target visible-light imagery [17], neglecting the unique spectral complexity of hyperspectral data. For example, band selection strategies [18] effective in RGB domains often fail to account for correlations between adjacent hyperspectral bands, leading to information loss [19].

Current hyperspectral classification frameworks thus face a dual challenge: achieving high accuracy while ensuring interpretability. While transformer-based models [20] excel at capturing long-range spectral dependencies, their self-attention mechanisms lack explicit links to physical material properties [21]. Conversely, CNN architectures leveraging 3D convolutions for joint spatial–spectral processing [22] struggle to quantify band-specific contributions. This gap becomes critical in scenarios requiring human-AI collaboration, such as disaster response [23], where domain experts demand transparent decision-making processes.

Addressing these limitations, this work introduces a unified framework that synergizes feature alignment with physics-aware interpretability. Our approach (EFBASN, Explainability Feature Bands Adaptive Selection Net) diverges from conventional post hoc explanation methods [24] by embedding interpretability directly into the learning process. Specifically, adaptive attention mechanisms prioritize spectral bands correlated with known material signatures—such as iron oxide absorption at 672 nm [25] or chlorophyll reflectance at 730 nm [26]—while a novel alignment loss function ensures spatial and spectral features converge into a coherent embedding space [27]. Extensive validation across four benchmark datasets demonstrates that the proposed method not only achieves state-of-the-art classification accuracy [28] but also provides actionable insights into band selection criteria [29], bridging the critical gap between model performance and interpretability in hyperspectral analysis.

2. Method

The main structure of the hyperspectral image feature self-attention interpretable classification method [30] mainly includes a feature alignment model and a feature self-attention interpretable model [31,32]. To design the feature alignment model, the multidimensional spatial features of hyperspectral remote sensing features are firstly extracted using deep neural networks, and the multifaceted feature alignment metric loss function is proposed to map the extracted multidimensional features into a common high-dimensional space for the alignment metric. The interstate multidimensional feature alignment loss function and the intrastate multidimensional feature alignment loss function are adopted so that the inter-class feature distance is farther for different kinds of features, which enhances the inter-class separability of features, and the intraclass feature distance is closer for the same kind of features, which improves the intraclass convergence of features. The self-attention and interpretable model of feature features is designed to extract the spatial and spectral features, focus on the spectral–space joint salient feature regions in the hyperspectral image features, and allow the model to calculate the contribution of each region to the final result in a quantitative form so as to analyze the correlation decision-making process of the model and realize the interpretability of the network. The feature features with high contribution are selected [33] to experimentally validate the proposed algorithmic model in a publicly available hyperspectral dataset. The overall framework of the proposed method in this paper is shown in Figure 1.

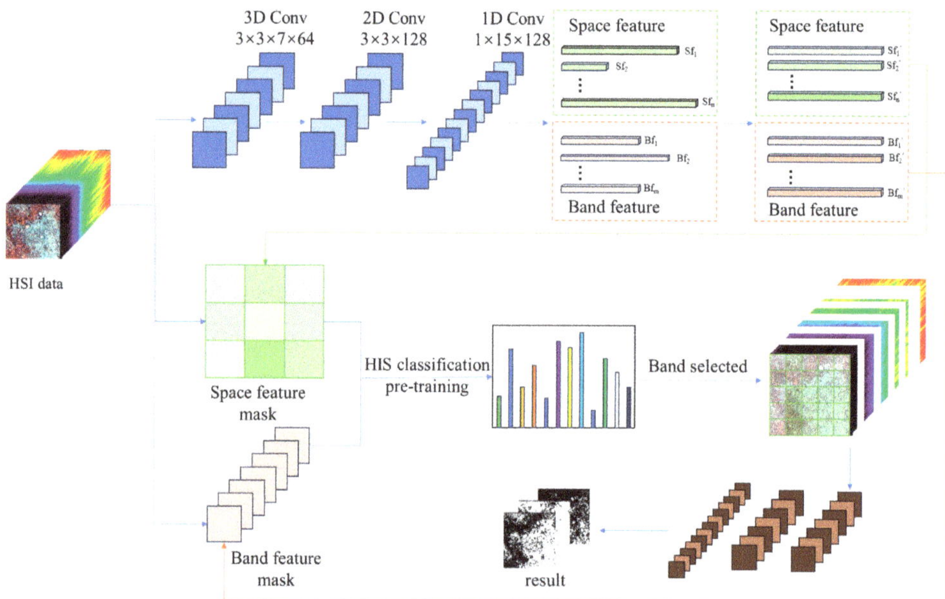

Figure 1. Overall framework. (The figure shows the feature extraction of spatial and spectral features in the hyperspectral image using a CNN structure, where the spatial dimensional features are represented by green vectors and the spectral dimensional features are represented by light red vectors).

2.1. Feature Alignment Modelling

The feature alignment model proposed in this paper is able to align the extracted multi-scale spatial features with the alignment metric. The hyperspectral remote sensing feature data are taken as input, and a convolutional neural network (CNN) containing multiple convolutional layers is used to extract features from the input hyperspectral remote sensing data [34,35]. The CNN model consists of several stages:

Input Layer: The input is a 3D hyperspectral image with dimensions $H \times W \times B$, where H and W represent the spatial dimensions (height and width) and B represents the spectral bands.

Convolutional Layers: Multiple convolutional layers are employed to capture spatial and spectral features. Three-dimensional convolutions (kernel $3 \times 3 \times 7$) extract joint spectral–spatial features $F_{joint} \in \mathbb{R}^{H \times W \times 64}$. Two-dimensional spatial (kernel 3×3) and 1D spectral (kernel 1×15) convolutions yield $F_{spa} \in \mathbb{R}^{H/4 \times W/4 \times 256}$ and $F_{spec} \in \mathbb{R}^{256 \times B}$. These layers use kernels that convolve over both the spatial dimensions and the spectral bands, enabling the network to learn multi-scale features at different levels of abstraction.

Each convolutional layer typically applies multiple filters (kernels) to extract different spatial and spectral patterns. The number of filters generally increases as the network depth increases, allowing the network to capture more complex features. Specifically, the CNN backbone consists of four sequentially structured blocks. The initial stage (Block 1) utilizes 3D convolutions with 3 × 3 × 7 kernels and 1 × 1 × 2 stride to jointly capture local spatial patterns and spectral correlations across seven adjacent bands, followed by max-pooling (2 × 2 × 2) to reduce spatial–spectral dimensions by half. This design preserves critical spectral resolution while mitigating computational complexity in early processing. Subsequent layers (Block 2) apply shallower 3 × 3 × 5 kernels with 1 × 1 × 1 stride to refine spatial features without further spectral downsampling, maintaining dimensionality at H/4 × W/4 × 64 through strategic pooling configurations.

Convolutional Layers: The architecture employs a hierarchical convolutional framework comprising four sequential blocks to progressively extract and integrate spatial–spectral features. Initial processing stages (Block 1–2) utilize 3D convolutions for joint feature learning, where Block 1 applies 3 × 3 × 7 kernels with 1 × 1 × 2 stride to capture local spatial patterns (3 × 3 window) and mid-range spectral correlations (7-band span), followed by 2 × 2 × 2 max-pooling to halve spatial–spectral dimensions. This strategic design preserves discriminative spectral signatures (e.g., iron oxide absorption at 672 nm) while reducing computational complexity early in the network. Block 2 refines these features using shallower 3 × 3 × 5 kernels with 1 × 1 × 1 stride, maintaining spectral resolution through restrained pooling (2 × 2 × 1), thereby outputting features at $\frac{H}{4} \times \frac{W}{4} \times 64$ dimensions.

To address the divergent nature of spatial textures and spectral responses, Block 3 implements dual-path processing: The spatial branch employs 2D convolutions (3 × 3 kernels, 128 channels) to analyze localized geometric patterns. The spectral branch leverages 1D convolutions (1 × 15 kernels, 128 channels) to model long-range band dependencies.

This decoupled architecture prevents cross-modality interference while enabling specialized feature learning. Channel dimensions systematically expand from 64 to 256 across blocks, balancing representational capacity with memory efficiency. The final stage (Block 4) introduces parametric channel weighting to dynamically fuse spatial ($F_{spa} \in R^{\frac{H}{4} \times \frac{W}{4} \times 256}$) and spectral ($F_{spec} \in R^{256 \times B}$) features through a learnable coefficient.

Pooling Layers: Following the convolutional layers, pooling layers (such as max-pooling) are used to reduce the spatial resolution while preserving important feature information. This helps in reducing the computational burden and extracting high-level spatial features.

Feature Fusion: After passing through several convolutional and pooling layers, the multi-scale spatial spectral features are merged and aligned into two distinct vectors: the spatial feature vector f^{spa} ($f^{spa} = F_{spa} \times K$) and the spectral feature vector f^{spec} ($f^{spec} = F_{spec} \times L$). Learnable transformation matrices $K \in \mathbb{R}^{d \times d}$ and $L \in \mathbb{R}^{d \times d}$ map the disentangled spatial and spectral features into a unified d-dimensional embedding space. These vectors capture the essential spatial and spectral information of the hyperspectral data.

By leveraging the integrated effects of pixels and bands, the model ensures that the extracted multi-dimensional features capture both local spatial information and global spectral relationships. The multi-scale feature extraction provided by the CNN allows the model to capture varying levels of abstraction, which is crucial for hyperspectral data where both spatial and spectral variations are important.

To enhance the classification performance of the algorithm, we propose a multidimensional feature alignment metric loss function. This loss function maps multidimensional spatial and spectral feature vectors into a high-dimensional space for transformation and alignment. Specifically, it incorporates an inter-state multidimensional feature alignment metric function to unify the dimensions of spatial and spectral features within the feature space. Additionally, two intra-state multidimensional feature alignment metric functions are introduced: one aligns spatial features of the same state to enforce homomorphic spatial representations, while the other aligns spectral features of the same state to ensure homomorphic spectral representations. By jointly optimizing these components, the proposed loss function enhances inter-class separability (distinguishing features across different classes) and improves intra-class convergence (clustering features within the same class). The multidimensional feature alignment loss function is as follows:

$$L_{alignment} = L_{\text{spatial}}(\text{f}^{\text{spa}}, \text{f}^{\text{spa}}) + L_{\text{spectral}}(\text{f}^{\text{spec}}, \text{f}^{\text{spec}}) + L_{\text{inter}}(\text{f}^{\text{spa}}, \text{f}^{\text{spec}}) \tag{1}$$

where $L_{\text{inter}}(\text{f}^{\text{spa}}, \text{f}^{\text{spec}})$ denotes the interstate multidimensional feature alignment metric function. Its purpose is to make the multi-dimensional spatial and spectral feature vectors become uniform in dimension in the feature space. The interstate multidimensional feature alignment metric function is as follows:

$$L_{\text{inter}}\ (\text{f}^{\text{spa}}, \text{f}^{\text{spec}}) = \frac{1}{N}\sum_{i,j} Sigmoid\left(\left|\text{f}_i^{\text{spa}} - \text{f}_j^{\text{spec}}\right|\right) \tag{2}$$

The sigmoid function is used to represent the disparity between spatial and spectral feature vectors in multiple dimensions in the same dimension and to judge whether the two still have distinguishable feature differences. The formula for the sigmoid function is as follows:

$$Sigmoid(x) = \frac{1}{1+e^{-x}} \tag{3}$$

Two in-state multidimensional feature alignment metric functions, $L_{\text{spatial}}(\text{f}^{\text{spa}}, \text{f}^{\text{spa}})$ and $L_{\text{spectral}}(\text{f}^{\text{spec}}, \text{f}^{\text{spec}})$, respectively, are used to make the spatial features of the homomorphic state and the spectral features of the homomorphic state in the feature space so that different classes of features are further apart within the feature space to enhance the interclass distinguishability of the features and to make the features of the same class more compactly distant in the feature space, enhancing the intra-class convergence of the features. Here, N denotes the total number of pixel pairs in the dataset. The calculation formula of the multidimensional feature alignment metric function within two states is as follows:

$$L_{\text{spatial}}\ (\text{f}^{\text{spa}}, \text{f}^{\text{spa}}) = -\sum_{i,j}^{N}\left(\log\left(1+e^{\tau_{ij}^{\text{spa-spa}}}\right) - s_{ij}^{\text{spa-spa}}\tau_{ij}^{\text{spa-spa}}\right) \tag{4}$$

$$\tau_{ij}^{\text{spa-spa}} = \frac{1}{2}\rho_{ij}^{\text{spa-spa}} = \frac{1}{2}\frac{\text{cov}\left(\text{f}_i^{\text{spa}}, \text{f}_j^{\text{spa}}\right)}{\sqrt{\text{var}\left(\text{f}_i^{\text{spa}}\right)}\sqrt{\text{var}\left(\text{f}_j^{\text{spa}}\right)}} \tag{5}$$

$$L_{\text{spectral}}(\text{f}^{\text{spec}}, \text{f}^{\text{spec}}) = -\sum_{i,j}^{N}\left(\log\left(1+e^{\tau_{ij}^{\text{spec-spec}}}\right) - s_{ij}^{\text{spec-spec}}\tau_{ij}^{\text{spec-spec}}\right) \tag{6}$$

$$\tau_{ij}^{\text{spec-spec}} = \frac{1}{2}\rho_{ij}^{\text{spec-spec}} = \frac{1}{2}\frac{\text{cov}\left(\text{f}_i^{\text{spec}}, \text{f}_j^{\text{spec}}\right)}{\sqrt{\text{var}\left(\text{f}_i^{\text{spec}}\right)}\sqrt{\text{var}\left(\text{f}_j^{\text{spec}}\right)}} \tag{7}$$

$\tau_{ij}^{\text{spa-spec}}$ denotes the similarity coefficient and ρ_{ij} denotes Pearson's correlation coefficient (PCC). When the spatial feature vector x_i^{spa} of pixel point x_i is of the same category as the spatial feature vector x_j^{spa} of pixel point x_j, $s_{ij}^{\text{spa-spa}} = 1$; and when the categories of x_i^{spa} and x_j^{spa} are different, $s_{ij}^{\text{spa-spa}} = 0$. $s_{ij}^{\text{spec-spec}}$ is taken in the same way as $s_{ij}^{\text{spa-spa}}$ is consistent.

The closer the feature vectors of local objects are in the high-dimensional feature space, the higher the probability that they are of the same category, and the further the distance, the lower the probability that they are of the same category. The formula for calculating the distance of the feature vector in the high-dimensional feature space is as follows:

$$d_{ij} = D\left(\mathrm{f}_i^{\mathrm{P}}, \mathrm{f}_j^{\mathrm{M}}\right) \tag{8}$$

where D denotes the distance metric, and, in this paper, we use the Euclidean distance metric. P and M represent the projected transformations for spatial and spectral features, respectively.

2.2. Self-Attention and Interpretable Model of Feature Features

The self-attention and interpretable model of feature features proposed in this paper is able to adaptively perceive and extract the spatial and spectral information of hyperspectral remote sensing features, distill and extract the data successively through the spatial and spectral feature self-attention modules, focus on the joint salient feature regions of hyperspectral images, and output the contribution of features to the features in the visual form to provide an interpretable basis for the selection of features in the "black box" process of deep learning. The contribution of the features to the features is output, which provides an interpretable basis for the selection of null-spectrum feature selection in the "black box" process of deep learning and realizes the interpretability. The schematic diagram of the feature self-attention interpretation model is shown in Figure 1. Hyperspectral remote sensing data are generally a cubic structure, which is denoted as $\boldsymbol{X} \in \mathbb{R}^{b \times r \times c}$, where b denotes the number of bands of hyperspectral remote sensing data, r denotes the number of rows of hyperspectral remote sensing data, and c denotes the number of columns of hyperspectral remote sensing data. The original hyperspectral remote sensing data and the spatial dimension feature vector generated after feature alignment are jointly inputted into the spatial dimension feature self-attention module, and the spatial self-attention feature map is obtained after the action of the softmax activation function, which is denoted as $\mathrm{F}^{\mathrm{spa}}$. The softmax activation function makes the generated element values between (0,1), and the sum of the elements is 1. The softmax activation function's expression is as follows:

$$Softmax(z_i) = \frac{e^{z_i}}{\sum_{k=1}^{K} e^{z_k}} \tag{9}$$

The spatial self-attention feature map is multiplied with the 2D transformed hyperspectral remote sensing data $\boldsymbol{X}'$ to obtain the spatial self-attention feature matrix $\mathrm{X}^{\mathrm{spa}}$. The hyperspectral original remote sensing data X, the spatial self-attention feature matrix $\mathrm{X}^{\mathrm{spa}}$ and the spectral dimension feature vector generated after feature alignment are used as inputs to the spectral dimension feature self-attention notation module, and after the action of the softmax activation function, a joint spectral–space attention feature map, denoted as $\mathrm{F}^{\text{spa-spec}}$, is obtained. Joint spectral–space attention feature map, denoted as $\mathrm{F}^{\text{spa-spec}}$.

The joint spectral–space attention feature map is further transformed into the joint spectral–space attention feature matrix $\mathrm{X}^{\text{spa-spec}}$. Each row of this matrix is the degree of contribution of the joint spectral–space features to the hyperspectral remotely sensed features obtained after one round of training to complete the self-perception analysis of the features.

The contribution of the joint spectral–space self-perception features is visualized and analyzed, and a spectral–space self-perception saliency map is obtained, which reveals the basis for the interpretation of the selection of spectral–space features in the process of the "black box" of deep learning, and based on the spectral–space self-perception saliency map, we obtain the features with high contributions selected by fusing spatial information and spectral information. Band subset. The original hyperspectral data are extracted from the salient feature information, reconstructed into new hyperspectral cube data based on the subset of feature bands, input into the classification algorithm, and then train the network to obtain the classification results.

3. Result

This segment can be organized through thematic subdivisions to systematically present research outcomes. It requires a clear articulation of empirical evidence coupled with critical analysis while maintaining academic rigor in demonstrating derived inferences. The narrative should strategically integrate data visualization with methodological validation to reinforce the validity of key findings.

To validate the effectiveness of the method proposed in this paper, experiments were conducted using four sets of publicly available Pavia University, Pavia Centre, Washington DC, and GF-5 hyperspectral datasets.

3.1. Experimental Process

By comparing and analyzing three typical hyperspectral image datasets, using the original data as well as the true value data, the spectral–space attention visualization results are obtained through the feature self-attention interpretable model. And according to the visualization results, a subset of feature bands is obtained, and the results are compared with the algorithms' composite neighbor-aware convolutional metric network (CNCMN.2023 [36]), dual-channel convolutional network (DCCN.2022 [37]), modified version of k-means with spectral similarity measures algorithms for HSI classification (IKMC.2023 [38]), novel spatial–spectral cross-fusion attention networks (SSCFA.2024 [39]) classification accuracy comparison, to verify the effectiveness and reliability of the proposed method.

(1) Pavia University and Pavia Centre datasets. The Pavia University and Pavia Centre datasets were acquired during flights over Pavia in northern Italy, both taken by the ROSIS sensor, in the spectral range 0.430–0.86 μm. The Pavia University is used for training with 103 bands, size 610 × 340, with 9 categories of features, including asphalt roads, metal plates, pastures, etc. Pavia Centre is 1096 × 715 in size and is used for training with 102 bands, also with 9 classifications of features, including trees, asphalt roads, and other features.

(2) Washington DC dataset. The Washington DC dataset is an aerial hyperspectral image over the Washington Mall acquired by the Hydice sensor, with a data size of 1280 × 307 and a spectral range of 191 bands from 0.4 to 2.4 μm. Feature classes include streets, grass, water, gravel paths, trees, shadows, and roofs.

(3) GF-5 dataset. The GF-5 dataset is an aerospace hyperspectral image of an area over Beijing acquired by the visible short-wave infrared hyperspectral camera on board the GF-5 satellite, with a data size of 301 × 301 and a spectral range of 180 bands from 0.4 to 2.5 μm. The feature categories include grassland, roads, etc. A comprehensive comparison of the basic information of the four hyperspectral datasets is shown in Table 1.

Table 1. Basic information of the four hyperspectral datasets.

Dataset	Pavia University	Pavia Centre	Washington DC	GF-5
Size	610 × 340	1096 × 715	1280 × 307	301 × 301
Band	103	102	191	180
Class	9	9	7	6
Number of Marker Pixels	42,776	7456	19,204	3649
spectral range (μm)	0.43–0.86	0.43–0.86	0.4–2.4	0.4–2.5

After a comprehensive comparison of the four hyperspectral datasets, it can be seen that each dataset has different sizes, numbers of bands, and spectral ranges and contains typical terrestrial and man-made feature types. Therefore, adopting the above datasets can more comprehensively verify the effectiveness of the algorithm proposed in this paper.

3.2. Comparison Experiment Setup

The size of the parameter batch is set to 16 in the training phase, the optimization method Adam is selected, the learning rate is set to 0.0001, and the training period of the model is set to 100. In order to further prove the advantages of the method proposed in this paper, it is compared with algorithms such as MVPCA, SpaBS, MOBS, and BS-Net-Conv. Support vector machine (SVM), a classification algorithm, is chosen to validate the comparative classification results on the set of bands selected by the above algorithms. Five percent of the labeled samples in each dataset are randomly selected as the training set, and the rest of the samples are used as the test set. In this paper, quantitative evaluation is performed by comparing the overall accuracy (OA), average accuracy (AA), and Kappa coefficient of the classification results. The overall classification accuracy is calculated by calculating the ratio of the number of correctly classified image elements for each kind of feature to the overall number, and its expression is shown below.

$$OA_i = \frac{x_{i,i}}{N} \tag{10}$$

where OA_i denotes the proportion of correctly classified features in the ith category. The average classification accuracy is first calculated for each class of features, and then the average of the classification proportion of all classes in the hyperspectral image is calculated; the expression of which is shown below.

$$y_i = \frac{x_{i,i}}{x_{+i}} \tag{11}$$

$$AA = \frac{1}{k}\sum_{i=1}^{k} y_i \tag{12}$$

where y_i is the classification accuracy of the ith feature. The Kappa coefficient can also be used to quantify the consistency or accuracy of the classification results of hyperspectral remote sensing images with the actual situation of the ground surface, thus reflecting the classification accuracy of the algorithm. The algorithm utilizes the discrete multivariate method, which effectively solves the dependence on the number of classifications and the number of samples. The expression of the Kappa coefficient is shown in the following equation.

$$\text{Kappa} = \frac{N\sum_{i=1}^{k} x_{i,i} - \sum_{i=1}^{k}(x_{i+} \times x_{+i})}{N^2 - \sum_{i=1}^{k}(x_{i+} \times x_{+i})} \tag{13}$$

4. Discussion

4.1. Comparative Analysis of Interpretable Results

In order to analyze and verify the interpretability of the feature self-attention interpretable model in the algorithm proposed in this paper, the spectral–space attention visualization fusing feature spatial and spectral information generated by the model training is visualized to reveal the average band weights normalized after iterations. The horizontal and vertical axes of the spectral–space attention visualization represent the number of bands and the number of training iterations of the hyperspectral data, respectively, and each column represents the change of the weight value of the band after iteration-by-iteration training from top to bottom. The changes in the weight values are shown in the greyscale, where the closer the color is to white means the higher the weight of the band, and the closer the color is to black means the lower the weight. For comparison, the weight values of the bands are normalized so that the weight values of the bands are in the range of (0,1).

Comparing the differences in spectral–space information focused on different hyperspectral datasets under the same network shows the trend of the weights of the bands after feature self-attention training and achieves the interpretability of the algorithm.

The results of the spectral–space attention visualization for different hyperspectral datasets are shown in Figures 2–5:

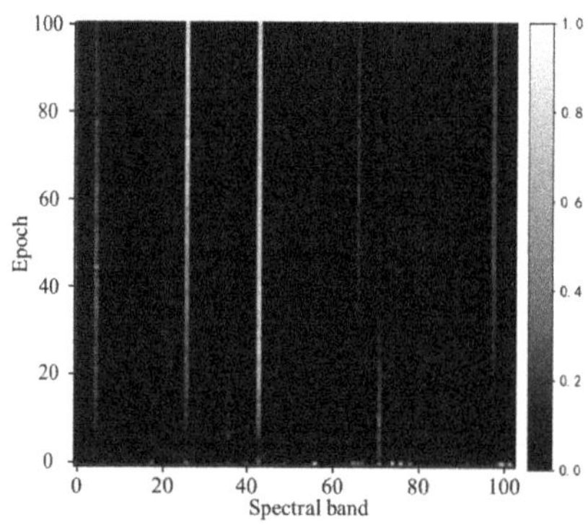

Figure 2. Visualization results of spectral–space attention for the Pavia University dataset.

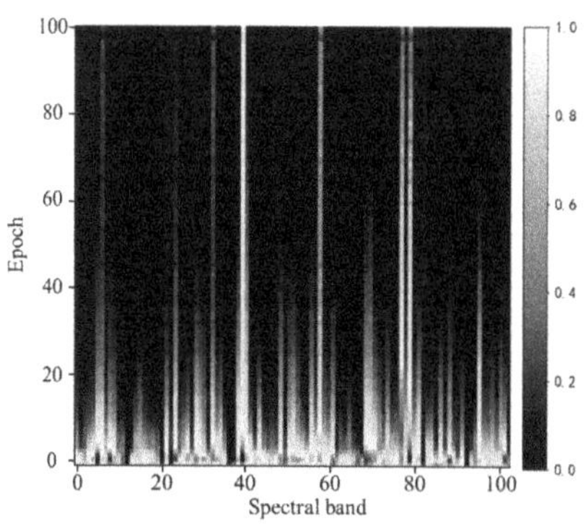

Figure 3. Visualization results of spectral–space attention for the Pavia Centre dataset.

Figure 4. Visualization results of spectral–space attention for the Washington DC dataset.

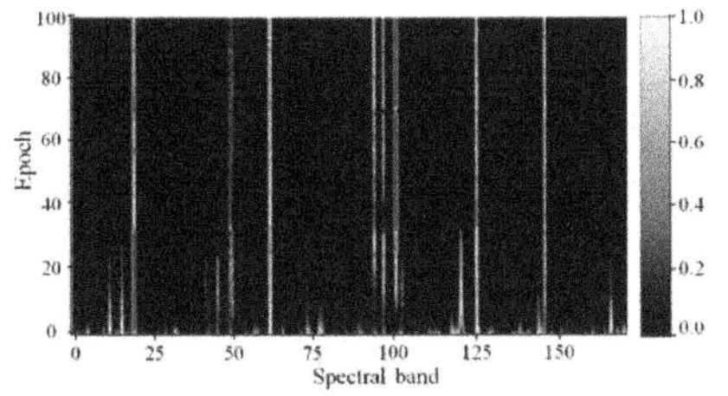

Figure 5. Visualization results of spectral–space attention for the GF-5 dataset.

From the spectral–space attention visualization results of the Pavia University dataset in Figure 2, it can be seen that for the 102 bands, the weights of individual bands are trained iteratively by the feature self-attention interpretable model to obtain a sparser distribution of band weights. In the initial stage of training, the lighter-colored bands have a small distribution in the whole range of bands, and after 100 rounds of iterative training, it can be seen from the color of the weights that the feature bands are significantly concentrated in the 26th, 43rd, etc., which indicates that the above bands contain the main spectral–space information. Although the weights of the other bands have higher weights at the initial stage, the colors tend to be closer to black and are given lower weights after several rounds of spectral–space information perception, which indicates that the other bands have the problems of redundancy of information and containing less information.

As can be seen from the spectral–space attention visualization results of the Pavia Centre dataset in Figure 3, for the 102 bands, the weights of each band obtain a sparser distribution of band weights after iterative training of the feature self-attention interpretable model. In the initial stage of training, the light grey-colored bands have a significant distribution reduction in the whole range of bands, and after 100 rounds of iterative training, the feature bands can be seen from the color of the weights to be significantly concentrated in the 81st, 39th, 78th, 57th, and 76th bands, which indicates that the above bands contain the main spectral–space information. Although the weights of the other bands have higher weights in the initial stage, the colors tend to be closer to black after several rounds of spectral–space information perception and are given lower weights, which indicates that the other bands have the problems of redundancy of information and containing less information.

From the visualization results of the spectral–space attention of the Washington DC dataset in Figure 4, it can be seen that for the 191 bands, the weights of individual bands have been trained iteratively by the feature self-attention interpretable model for the feature features, the weight distribution of the bands has a significant change, and the final white weighted bands with high weights are concentrated in the 6th, 34th, 132nd, 26th, 178th, 61st, 46th, and other bands, indicating that the above bands contain the main spectral–space information and have more important roles in the classification of features. The other bands tend to be dark grey and black, indicating that the other bands are given low weights.

From the visualization results of the spectral–space attention of the GF-5 dataset in Figure 5, it can be seen that for the 180 bands, the weights of the individual bands have significant changes in the distribution of the weights of the bands after the iterative training of the self-attention interpretable model of the feature features, and the bands with high weights of the white color eventually are concentrated in the 19th, 49th, 62nd, 94th, 96th, 105th, 106th, 125th, 146th, etc. bands, indicating that the above bands contain more important information for the classification of the features. and other bands, indicating that the above bands contain the main spectral–space information for classifying features and have a more important role in feature classification. The other bands tend to be dark grey and black as the training iteration progresses, indicating that the feature information in the other bands has been replaced by the main band representations, which is redundant information.

By analyzing the experimental results, the visual significance map results further illustrate the degree of contribution of the feature bands integrating the spectral–space information to the final feature classification, providing a clear and powerful explanation for the selection of spectral–space features for features in the "black box" process of deep learning.

4.2. Comparative Analysis of Classification Results

The classification results of different algorithms for the Pavia University dataset are shown in Figure 6.

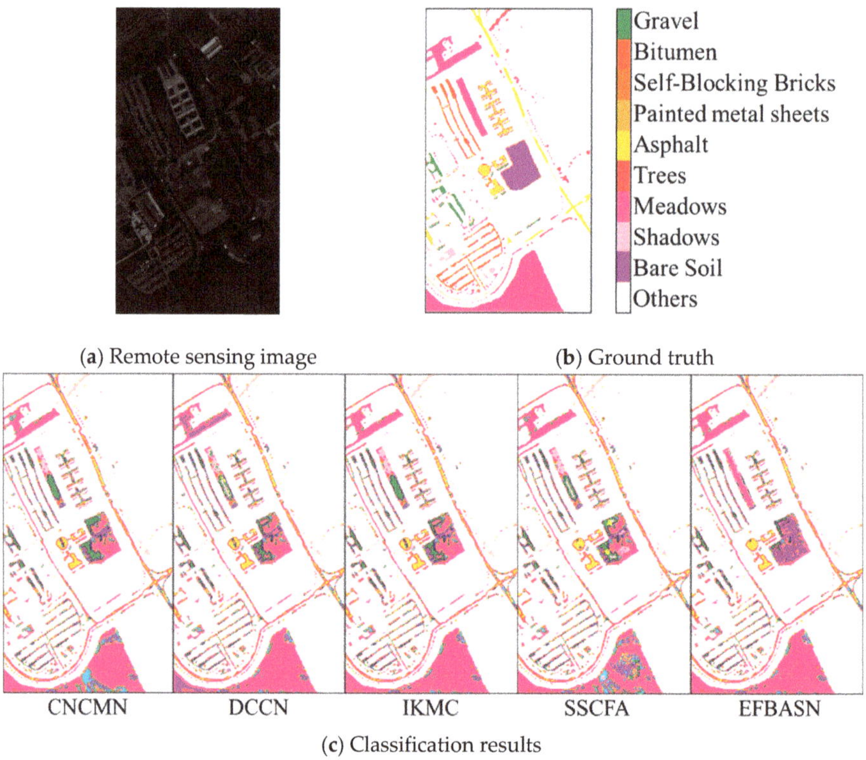

Figure 6. Classification Results of the Pavia University dataset.

From the classification results, it can be seen that the algorithm in this paper has a better distinction between natural and man-made environments. In the bare soil area and gravel area, the classification is more accurate, and the misrecognition rate is lower. The classification accuracy is shown in Table 2:

Table 2. Classification result of different algorithms for the Pavia University dataset.

Class Name	CNCMN	DCCN	IKMC	SSCFA	EFBASN
Gravel	93.84	95.77	92.66	95.96	99.46
Bitumen	98.76	98.22	96.88	99.08	99.38
Self-Blocking Bricks	95.97	99.14	97.81	99.08	99.45
Painted metal sheets	98.08	96.12	90.20	93.47	98.22
Asphalt	97.25	96.96	97.20	98.20	98.85
Trees	92.07	97.72	94.21	97.15	97.16
Meadows	91.94	95.03	96.34	90.02	95.92
Shadows	87.09	91.73	91.47	95.95	94.24
Bare Soil	84.04	87.06	80.88	81.65	95.88
OA	93.12	95.25	93.18	94.48	97.68
AA	93.22	95.30	93.07	94.50	97.61
$\mathcal{K}$	93.20	95.28	93.15	94.49	97.63

The classification results of different algorithms for the Pavia Centre dataset are shown in Figure 7.

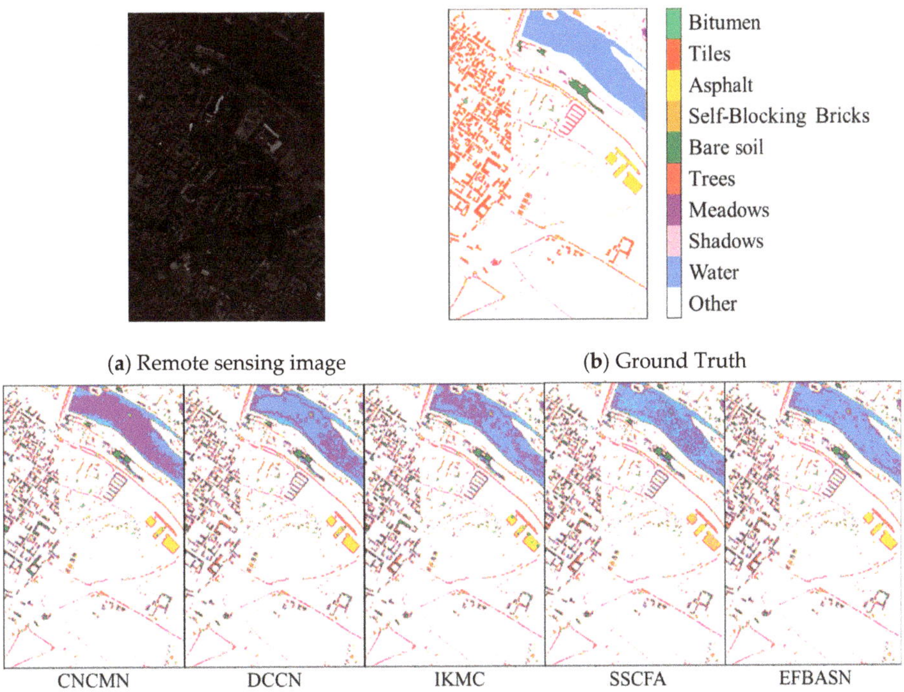

(**a**) Remote sensing image (**b**) Ground Truth

(**c**) Classification results

Figure 7. Classification results of the Pavia Centre dataset.

As can be seen from the classification results schematic diagram, due to the grass containing more water, the water body also contains more algae, so the two are easy to lead to more misclassification. This paper's algorithm through the feature band selection [40] for such easy-to-mix categories has very good classification results. The classification accuracy is shown in Table 3:

Table 3. Classification results of different algorithms for the Pavia Centre dataset.

Class Name	CNCMN	DCCN	IKMC	SSCFA	EFBASN
Bitumen	93.88	94.76	94.96	95.34	96.97
Tiles	88.76	84.31	85.76	88.68	89.32
Asphalt	95.97	98.14	94.44	89.45	99.45
Self-Blocking Bricks	80.44	80.93	81.33	82.54	83.03
Bare soil	90.56	91.55	90.08	91.38	92.78
Trees	90.98	91.39	91.55	91.03	91.86
Meadows	78.66	83.68	80.79	81.62	90.63
Shadows	80.78	80.66	80.48	80.43	80.98
Water	76.04	96.06	84.58	80.35	97.88
OA	86.32	88.93	87.07	86.78	91.41
AA	86.23	89.05	87.10	86.75	91.43
$\mathcal{K}$	86.20	89.00	87.09	86.77	91.42

The classification results of different algorithms for the Washington DC dataset are shown in Figure 8.

As can be seen from the classification results, the Washington DC dataset has a high difficulty in identifying paths and houses, and the algorithm in this paper has good classification results by selecting feature bands for paths and roofs, grass and trees, and other easy-to-confuse categories. The classification accuracy is shown in Table 4:

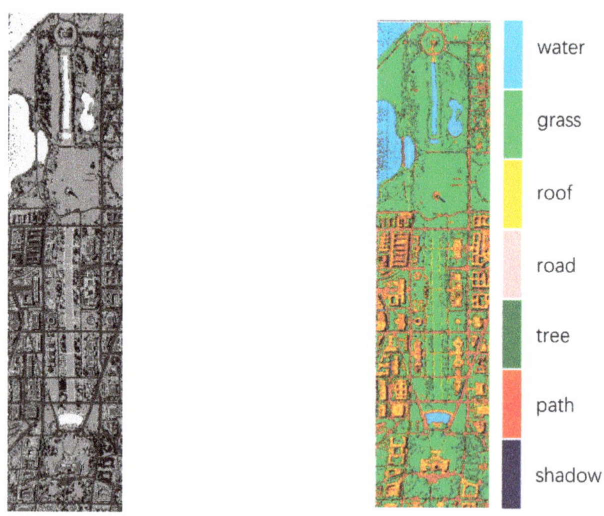

(**a**) Remote sensing image

(**b**) Ground Truth

Figure 8. *Cont.*

(c) Classification results

Figure 8. Classification Results of the Washington DC dataset.

Table 4. Classification results of different algorithms for the Washington DC dataset.

Class Name	CNCMN	DCCN	IKMC	SSCFA	EFBASN
Water	79.53	83.24	84.29	85.67	98.71
Grass	79.22	78.93	78.98	87.34	93.77
Roof	89.88	89.63	90.36	90.55	91.54
Road	92.79	93.04	93.53	92.91	94.22
Tree	83.25	81.67	81.81	82.53	87.79
Path	77.31	76.29	74.69	75.91	81.33
Shadow	81.22	82.03	81.97	80.93	86.61
OA	83.33	83.52	83.64	85.15	90.60
AA	83.31	83.54	83.66	85.12	90.56
$\mathcal{K}$	83.32	83.53	83.65	85.13	90.58

The classification results of different algorithms for the GF-5 dataset are shown in Figure 9.

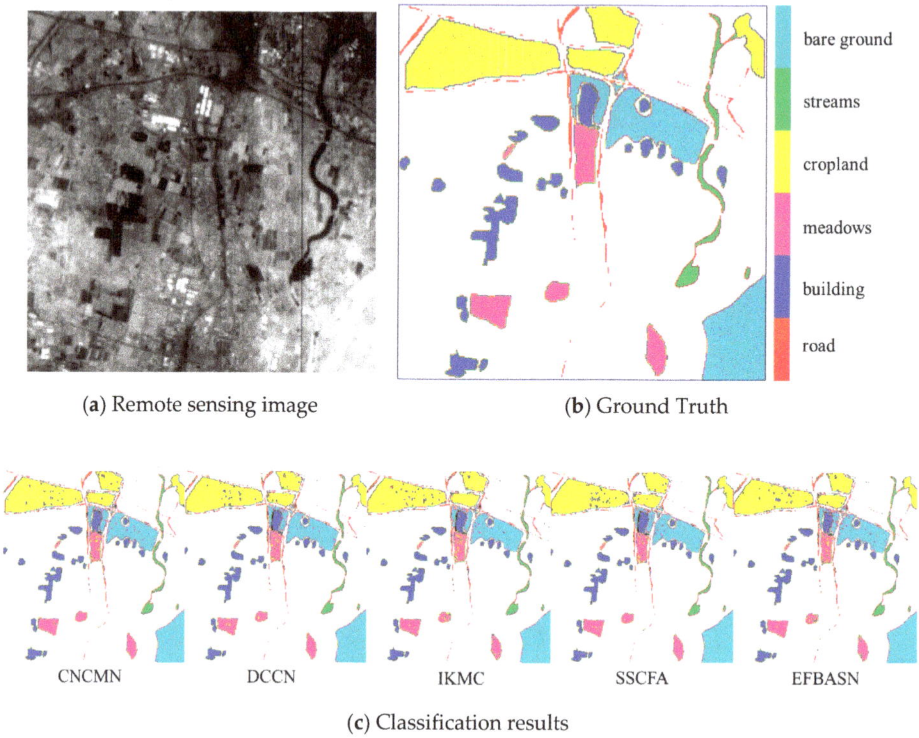

Figure 9. Classification results of the GF-5 dataset.

As can be seen from the classification results schematic diagram, the GF-5 data are concentrated, the distribution of features is relatively concentrated, and the distribution of artificial building targets and natural environments is intermingled. The algorithm in this paper can more clearly reproduce the true value edge contour and has better recognition and classification results on the easily mixed building image elements in the range of cultivated land. The classification accuracy is shown in Table 5:

Table 5. Classification results of different algorithms for the GF-5 dataset.

Class Name	CNCMN	DCCN	IKMC	SSCFA	EFBASN
bare ground	96.53	97.78	98.05	97.51	98.79
streams	99.54	99.35	99.89	99.91	99.93
cropland	98.90	97.69	98.40	97.93	98.88
meadows	96.73	97.21	97.16	97.51	97.29
building	97.29	98.62	97.78	97.89	97.51
road	92.79	93.26	93.58	93.93	94.98
OA	96.98	97.28	97.51	97.42	97.86
AA	96.96	97.31	97.47	97.44	97.89
$\mathcal{K}$	96.97	97.30	97.48	97.43	97.88

From Tables 2–5, it can be seen that the algorithm proposed in this paper has the highest classification accuracy compared with other algorithms, indicating that the algorithm proposed in this paper is not only suitable for hyperspectral data containing more categories of background and artificial feature targets but also for hyperspectral data with larger and smaller sizes, which can have good classification results, which fully explains the validity of the algorithm proposed in this paper in a quantitative form.

4.3. Ablation Experiments and Analysis of Results

In order to verify the validity of each module in EFBASN, we performed five ablation experiments on the GF-5 database using the optimal parameters obtained from previous experiments. The ablation results of EFBASN are shown in Table 6.

Table 6. Ablation experiments for the GF-5 dataset.

Case	Space Feature Self-Attention	Band Feature Self-Attention	Band Select	OA	AA
1	×	×	×	68.32	59.03
2	√	×	×	82.49	79.51
3	×	√	×	90.67	87.84
4	×	×	√	85.10	84.96
5	√	√	√	97.86	97.89

The ablation study quantitatively dissects the contributions of individual components through controlled experiments. When isolating the spatial attention module, the model achieves an overall accuracy improvement of 14.17% over the baseline CNN architecture. This enhancement is particularly evident in urban feature classes such as asphalt roads, which exhibit a 9.2% accuracy gain attributable to the module's emphasis on textural edge detection. In contrast, the spectral attention module demonstrates superior performance with a 22.35% OA increase, driven by its capacity to prioritize chlorophyll-sensitive wavelengths between 680 and 720 nm. For instance, band 43 at 732 nm—a known chlorophyll red-edge indicator—receives attention weights 3.1 times higher than less informative bands.

The integrated model's 97.86% OA not only surpasses standalone modules but also exceeds their theoretical additive performance when normalized. This nonlinear synergy emerges from complementary mechanisms: spatial attention resolves geometric ambiguities in urban–natural boundary regions, while spectral attention suppresses noise from overlapping reflectance profiles. A representative case occurs in mixed land-cover areas, where the combined model reduces misclassification between buildings and vegetation by 12.7% compared to individual modules. Such interactions validate the necessity of co-optimizing spatial and spectral interpretability throughout the network architecture.

5. Conclusions

This paper proposes a new method of feature self-awareness interpretable classification for hyperspectral remote sensing targets. The method is mainly divided into a target feature alignment model and a target feature self-awareness interpretable model. The proposed target feature alignment model uses deep neural networks to extract multidimensional features of hyperspectral remote sensing targets and maps the hyperspectral image features into the high-dimensional feature space for alignment metrics so that for different kinds of targets, the intra-class feature distances are closer, and the inter-class feature distances are further apart, which solves the problem of the large dimensional differences in the target features that are difficult to be aligned. The proposed target feature self-awareness interpretable model focuses on the joint salient target feature regions in hyperspectral images, visualizes the contribution of target features, reveals more intuitively the basis for selecting the interpretable target features of the aerial spectrum in the process of the "black box" of deep learning, parses the correlation decision-making process of the model, realizes the interpretability of the network, and selects target features with high contribution to the network. The target features with high contribution degrees are selected for classification experiments.

In the experiments, the proposed algorithm is validated using publicly available hyperspectral datasets, and the experimental results show that the algorithm in this paper is significantly better than the existing algorithms in terms of classification accuracy, and the model not only has a high accuracy but also has a good interpretability. In the subsequent work, we will combine the characteristics of remote sensing image data to design an interpretable algorithm adapted to multi-tasking and multi-remote sensing types and further improve the structure of the algorithm to improve the classification accuracy.

Author Contributions: Conceptualization, J.L. (Jirui Liu) and J.L. (Jinhui Lan); methodology, J.L. (Jirui Liu); software, Z.Z.; validation, J.L. (Jirui Liu), W.L. and Z.Z.; formal analysis, J.L. (Jirui Liu); investigation, Y.Z.; resources, J.L. (Jirui Liu); data curation, J.L. (Jirui Liu); writing—original draft preparation, J.L. (Jirui Liu); writing—review and editing, J.L. (Jinhui Lan); visualization, J.Z.; supervision, Y.Z.; project administration, J.Z.; funding acquisition, J.Z. All authors have read and agreed to the published version of the manuscript.

Funding: This research was funded by the 14th Five-Year Plan Funding of China (Grant No. 50916040401), the Fundamental Research Program (Grant No. 514010503-201), and the Natural Science Basic Research Program of Shaanxi Province (Grant No. 2023-JC-QN-0769).

Data Availability Statement: The datasets used in this study are publicly available as follows:Pavia University dataset and Pavia Centre dataset: Available at the IEEE GRSS Data and Algorithm Standard Evaluation (DASE) platform via http://www.ehu.eus/ccwintco/index.php/Hyperspectral_Remote_Sensing_Scenes (accessed on 10 January 2025). Washington DC dataset: Hosted by the U.S. Geological Survey (USGS) Earth Resources Observation and Science Center: https://earthexplorer.usgs.gov/ (accessed on 10 January 2025) (Search criteria: Sensor "AVIRIS", Location "Washington DC Mall"). Gaofen-5 (GF-5) dataset: Acquired from the China Centre for Resources Satellite Data and Application (CCRSDA). Due to data licensing restrictions, access requires registration and approval through the official platform: https://data.cresda.cn (accessed on 10 January 2025).

Conflicts of Interest: The authors declare no conflict of interest.

References

1. Wang, C.; Huang, J.; Lv, M.; Du, H.; Wu, Y.; Qin, R. A local enhanced mamba network for hyperspectral image classification. *Int. J. Appl. Earth Obs. Geoinf.* **2024**, *133*, 104092. [CrossRef]
2. Jiang, Y.; Zhang, Z.; Zhang, C.; Zhou, H.; Ma, Q.; Zhong, C. S2WaveNet: A novel spectral–spatial wave network for hyperspectral image classification. *Int. J. Appl. Earth Obs. Geoinf.* **2024**, *128*, 103754. [CrossRef]
3. Zhang, X.; Sun, Y.; Qi, W. Hyperspectral Image Classification Based on Extended Morphological Attribute Profiles and Abundance Information. In Proceedings of the 2018 9th Workshop on Hyperspectral Image and Signal Processing: Evolution in Remote Sensing, Amsterdam, The Netherlands, 23–26 September 2018; pp. 1–5.
4. He, Z.; Xia, K.; Ghamisi, P.; Hu, Y.; Fan, S.; Zu, B. HyperViTGAN: Semisupervised Generative Adversarial Network with Transformer for Hyperspectral Image Classification. *IEEE J. Sel. Top. Appl. Earth Obs. Remote Sens.* **2022**, *15*, 6053–6068. [CrossRef]
5. Paoletti, M.; Haut, J.; Plaza, J.; Plaza, A. Neighboring Region Dropout for Hyperspectral Image Classification. *IEEE Geosci. Remote Sens. Lett.* **2020**, *17*, 1032–1036. [CrossRef]
6. He, N.; Paoletti, M.; Haut, J.; Fang, L.; Li, S.; Plaza, A.; Plaza, J. Feature Extraction with Multiscale Covariance Maps for Hyperspectral Image Classification. *IEEE Trans. Geosci. Remote Sens.* **2019**, *57*, 755–769. [CrossRef]
7. Guo, Z.; Xin, J.; Wang, N.; Li, J.; Gao, X. External-Internal Attention for Hyperspectral Image Super-Resolution. *IEEE Trans. Geosci. Remote Sens.* **2022**, *60*, 1–14. [CrossRef]
8. Zhong, Z.; Li, J.; Luo, Z.; Chapman, M. Spectral–Spatial Residual Network for Hyperspectral Image Classification: A 3-D Deep Learning Framework. *IEEE Trans. Geosci. Remote Sens.* **2018**, *56*, 847–858. [CrossRef]
9. Hong, D.; Han, Z.; Yao, J.; Gao, L.; Zhang, B.; Plaza, A.; Chanussot, J. SpectralFormer: Rethinking hyperspectral image classification with transformers. *IEEE Trans. Geosci. Remote Sens.* **2022**, *60*, 5518615. [CrossRef]
10. Zhang, Q.; Zhu, S.C. Visual interpretability for deep learning: A survey. *Front. Inf. Technol. Electron. Eng.* **2018**, *19*, 27–39. [CrossRef]

11. Zhao, J.; Tian, S.; Geiß, C.; Wang, L.; Zhong, Y.; Taubenbock, H. Spectral-Spatial Classification Integrating Band Selection for Hyperspectral Imagery with Severe Noise Bands. *IEEE J. Sel. Top. Appl. Earth Obs. Remote Sens.* **2020**, *13*, 1597–1609. [CrossRef]
12. Samek, W.; Montavon, G.; Lapuschkin, S.; Anders, C.J.; Muller, K.-R. Explaining Deep Neural Networks and Beyond: A Review of Methods and Applications. *Proc. IEEE* **2021**, *109*, 247–278. [CrossRef]
13. Zeiler, M.; Fergus, R. *Visualizing and Understanding Convolutional Networks. European Conference on Computer Vision*; Springer: Cham, Switzerland, 2014; pp. 818–833.
14. Aksoy, S.; Sertel, E.; Roscher, R.; Tanik, A.; Hamzehpour, N. Assessment of soil salinity using explainable machine learning methods and Landsat 8 images. *Int. J. Appl. Earth Obs. Geoinf.* **2024**, *130*, 103879. [CrossRef]
15. Ishikawa, S.-N.; Todo, M.; Taki, M.; Uchiyama, Y.; Matsunaga, K.; Lin, P.; Ogihara, T.; Yasui, M. Example-based explainable AI and its application for remote sensing image classification. *Int. J. Appl. Earth Obs. Geoinf.* **2023**, *118*, 103215. [CrossRef]
16. Zaigrajew, V.; Baniecki, H.; Tulczyjew, L.; Wijata, A.M.; Nalepa, J.; Longépé, N.; Biecek, P. Red Teaming Models for Hyperspectral Image Analysis Using Explainable AI. *arXiv* **2024**, arXiv:2403.08017.
17. Zhong, S.; Chang, C.; Li, J.; Shang, X.; Chen, S.; Song, M.; Zhang, Y. Class Feature Weighted Hyperspectral Image Classification. *IEEE J. Sel. Top. Appl. Earth Obs. Remote Sens.* **2019**, *12*, 4728–4745. [CrossRef]
18. Sun, K.; Geng, X.; Ji, L. A New Sparsity-Based Band Selection Method for Target Detection of Hyperspectral Image. *IEEE Geosci. Remote Sens. Lett.* **2014**, *12*, 329–333.
19. Qing, Y.; Liu, W.; Feng, L.; Gao, W. Improved transformer net for hyperspectral image classification. *Remote Sens.* **2021**, *13*, 2216. [CrossRef]
20. Porta, C.; Bekit, A.; Lampe, B.; Chang, C. Hyperspectral Image Classification via Compressive Sensing. *IEEE Trans. Geosci. Remote Sens.* **2019**, *57*, 8290–8303. [CrossRef]
21. Shao, Y.; Lan, J.; Niu, B. Dual-Channel Networks with Optimal-Band Selection Strategy for Arbitrary Cropped Hyperspectral Images Classification. *IEEE Geosci. Remote Sens. Lett.* **2022**, *19*, 1–5. [CrossRef]
22. Guo, Y.; Xiao, Y.; Hao, F.; Zhang, X.; Chen, J.; de Beurs, K.; He, Y.; Fu, Y.H. Comparison of different machine learning algorithms for predicting maize grain yield using UAV-based hyperspectral images. *Int. J. Appl. Earth Obs. Geoinf.* **2023**, *124*, 103528. [CrossRef]
23. Liu, J.; Lan, J.; Zeng, Y. GL-Pooling: Global–Local Pooling for Hyperspectral Image Classification. *IEEE Geosci. Remote Sens. Lett.* **2023**, *20*, 5509305. [CrossRef]
24. Guo, X.; Hou, B.; Yang, C.; Ma, S.; Ren, B.; Wang, S.; Jiao, L. Visual explanations with detailed spatial information for remote sensing image classification via channel saliency. *Int. J. Appl. Earth Obs. Geoinf.* **2023**, *118*, 103244. [CrossRef]
25. Xin, Z.; Li, Z.; Xu, M.; Wang, L.; Ren, G.; Wang, J.; Hu, Y. Feature disentanglement based domain adaptation network for cross-scene coastal wetland hyperspectral image classification. *Int. J. Appl. Earth Obs. Geoinf.* **2024**, *129*, 103850. [CrossRef]
26. Haut, J.; Paoletti, M.; Plaza, J.; Plaza, A.; Li, J. Visual Attention-Driven Hyperspectral Image Classification. *IEEE Trans. Geosci. Remote Sens.* **2019**, *57*, 8065–8080. [CrossRef]
27. Chang, C.; Du, Q.; Sun, T.; Althouse, M.L.G. A Joint Band Prioritization and Band-Decorrelation Approach to Band Selection for Hyperspectral Image Classification. *IEEE Trans. Geosci. Remote Sens.* **1999**, *37*, 2631–2641. [CrossRef]
28. Wang, L.; Wang, Q. Fast spatial-spectral random forests for thick cloud removal of hyperspectral images. *Int. J. Appl. Earth Obs. Geoinf.* **2022**, *112*, 102916. [CrossRef]
29. He, X.; Chen, Y.; Lin, Z. Spatial–spectral transformer for hyperspectral image classification. *Remote Sens.* **2021**, *13*, 498. [CrossRef]
30. Sun, L.; Zhao, G.; Zheng, Y.; Wu, Z. Spectral–spatial feature tokenization transformer for hyperspectral image classification. *IEEE Trans. Geosci. Remote Sens.* **2022**, *60*, 5522214. [CrossRef]
31. Shu, Z.; Wang, Y.; Yu, Z. Dual attention transformer network for hyperspectral image classification. *Eng. Appl. Artif. Intell.* **2024**, *127*, 107351. [CrossRef]
32. Gong, M.; Zhang, M.; Yuan, Y. Unsupervised Band Selection Based on Evolutionary Multiobjective Optimization for Hyperspectral Images. *IEEE Trans. Geosci. Remote Sens.* **2015**, *54*, 544–557. [CrossRef]
33. Cai, Y.; Liu, X.; Cai, Z. BS-Nets: An End-to-End Framework for Band Selection of Hyperspectral Image. *IEEE Trans. Geosci. Remote Sens.* **2019**, *58*, 1969–1984. [CrossRef]
34. Ben Hamida, A.; Benoit, A.; Lambert, P.; Ben Amar, C. 3-D Deep Learning Approach for Remote Sensing Image Classification. *IEEE Trans. Geosci. Remote Sens.* **2018**, *56*, 4420–4434. [CrossRef]
35. Paoletti, M.E.; Haut, J.M.; Fernandez-Beltran, R.; Plaza, J.; Plaza, A.J.; Pla, F. Deep Pyramidal Residual Networks for Spectral–Spatial Hyperspectral Image Classification. *IEEE Trans. Geosci. Remote Sens.* **2019**, *57*, 740–754. [CrossRef]
36. Shi, K.; Liu, Q.; Zheng, Z.; Xiao, L. Efficient Implementation for Composite CNN-Based HSI Classification Algorithm with Huawei Ascend Framework. In Proceedings of the 2023 13th WHISPERS, Athens, Greece, 19 February 2023; pp. 1–5.
37. Yu, H.; Zhang, H.; Liu, Y.; Zheng, K.; Xu, Z.; Xiao, C. Dual-Channel Convolution Network with Image-Based Global Learning Framework for Hyperspectral Image Classification. *IEEE Geosci. Remote Sens. Lett.* **2022**, *19*, 6005705. [CrossRef]

38. Goud, O.S.C.; Sarma, T.H.; Bindu, C.S. Improved K-Means Clustering Algorithm for Band Selection in Hyperspectral Images. In Proceedings of the 2023 International Conference on Electrical, Electronics, Communication and Computers (ELEXCOM), Roorkee, India, 26–27 August 2023; pp. 1–6.
39. Li, W.; Yang, Y.; Zhang, M.; Mi, P.; Xiao, Z.; Xiang, J. Deep Spatial-Spectral Feature Extraction Network for Hyperspectral Image Classification. In Proceedings of the 2024 43rd Chinese Control Conference (CCC), Kunming, China, 28–31 July 2024; pp. 7955–7960.
40. Sun, W.; Du, Q. Hyperspectral Band Selection: A Review. *IEEE Geosci. Remote Sens. Mag.* **2019**, *7*, 118–139. [CrossRef]

remote sensing

MDPI

Article

Prediction of Future Land Use/Land Cover Changes Using a Coupled CA-ANN Model in the Upper Omo–Gibe River Basin, Ethiopia

Paulos Lukas [1], Assefa M. Melesse [2,*] and Tadesse Tujuba Kenea [1]

1 Faculty of Meteorology and Hydrology, Water Technology Institute, Arba Minch University, Arba Minch 4400, Ethiopia

2 Department of Earth and Environment, Florida International University, Miami, FL 33199, USA

* Correspondence: melessea@fiu.edu

Abstract: Land use/land cover change evaluation and prediction using spatiotemporal data are crucial for environmental monitoring and better planning and management of land use. The main objective of this study is to evaluate land use/land cover changes for the time period of 1991–2022 and predict future changes using the CA-ANN model in the Upper Omo–Gibe River basin. Landsat-5 TM for 1991, 1997, and 2004, Landsat-7 ETM+ for 2010, and Landsat-8 (OLI) for 2016 and 2022 were downloaded from the USGS Earth Explorer Data Center. A random forest machine learning algorithm was employed for LULC classification. The LULC classification result was evaluated using an accuracy assessment technique to assure the correctness of the classification method employing the kappa coefficient. Kappa coefficient values of the classification indicate that there was strong agreement between the classified and reference data. Using the MOLUSCE plugin of QGIS and the CA-ANN model, future LULC changes were predicted. Artificial neural network (ANN) and cellular automata (CA) machine learning methods were made available for LULC change modeling and prediction via the QGIS MOLUSCE plugin. Transition potential modeling was computed, and future LULC changes were predicted using the CA-ANN model. An overall accuracy of 86.53% and an overall kappa value of 0.82 were obtained by comparing the actual data of 2022 with the simulated LULC data from the same year. The study findings revealed that between 2022 and 2037, agricultural land (63.09%) and shrubland (5.74%) showed significant increases, and forest (−48.10%) and grassland (−0.31%) decreased. From 2037 to 2052, the built-up area (2.99%) showed a significant increase, and forest and agricultural land (−2.55%) showed a significant decrease. From 2052 to 2067, the projected LULC simulation result showed that agricultural land (3.15%) and built-up area (0.32%) increased, and forest (−1.59%) and shrubland (−0.56%) showed significant decreases. According to the study's findings, the main drivers of LULC changes are the expansion of built-up areas and agricultural land, which calls for a thorough investigation using additional data and models to give planners and policymakers clear information on LULC changes and their environmental effects.

Keywords: land use/land cover; machine learning; remote sensing; random forest; MOLUSCE plugin; artificial neural network; cellular automata

1. Introduction

Currently, the issues of land use/land cover changes have become part of a global agenda due to their association with human and environmental aspects. In this regard, it has attracted the attention of an enormous number of researchers and scholars throughout the world [1]. For example, [2] examined land use/land cover changes using different time data and revealed their impacts on the living environment and human life, thereby pinpointing essential solutions for planners and stakeholders. The alarmingly increasing population and attributed need for natural resources, such as agricultural land and land required for housing, are some of the factors responsible for perpetuating LULC changes in all corridors of the globe [3–5]. The dynamic nature of land use/land cover (LULC) changes [6,7] requires continuous assessment, analysis, and monitoring employing socioeconomic and geospatial data for integrated and more accurate results. LULC changes have multidimensional effects on the present and future ecosystem balance [8–10].

It is obvious that the dynamics and magnitude of the change become serious in developing countries such as Ethiopia because of the lack of awareness about the utilization of environmental resources and uncontrolled utilization of resources without taking accountability for future risk, which is sourced from the lack of a strong legal framework [11,12]. The dynamic nature of LULC changes has a significant effect on spatiotemporal environmental stability [13] due to its connection to local, regional, and international climate conditions, clean water supply, agricultural activities, stability of biodiversity, and food security. Therefore, it is imperative to understand the change exerted from land use/land cover dynamics and multidimensional effects either positively or negatively that should be treated in a collaborative manner by the community and governmental and nongovernmental organizations to achieve improved environmental stability [14]. To this end, LULC change analysis, monitoring, and evaluation need the integrated approach and mobilization of local communities for active participation in soil and water conservation practices [15].

The changes in LULC are derived from the interaction among socioeconomic, environmental, and institutional phenomena occurring on the land [16]. Factors such as limited livelihood options, limited technological approaches, and highly demanding social conditions aggravate the competition on land, and due to this, the changes from land, particularly the negative changes, for example deforestation, overgrazing, and informal settlements, adversely affect the community in particular and the country in general [17]. Various studies on LULC changes [18] have revealed that, in particular, highland areas are victims of deforestation and encroachment of cultivated land into marginal areas, which causes land degradation, soil loss through uncontrolled soil erosion, and environmental damage. For instance, as stated in [19,20], most LULC changes resulted from the areal growth of land for crop production and urbanization with the disbursement of forest cover in the present study area. In Ethiopia, deforestation and expansion of urbanization are continuous processes that cause loss of biodiversity, climate variability and change, desertification, and soil erosion [21]. The present and expected future changes from LULC dynamics, as indicated in [22], are analyzed using different machine learning and geospatial technologies [23]. The advancement in these technologies and their integration with others widen the horizon of their application by attracting many scholars and researchers globally to investigate the multidimensional effect of LULC on the environment and human life [24]. The identification of LULC class types requires local knowledge, i.e., knowledge about the research area due to its proximity and accessibility to the researcher, referring to topographic maps, Google Earth maps, and published documents [25,26] if the image data required for the classification are old images (satellite image acquired from 30 or 40 years ago). However, as stated in [27], the identification of LULC class types requires less effort because of the presence of high-resolution remote sensing satellite image data and globally classified LULC data provision by accredited international organizations for reference, such as MODIS global land cover classification from NASA (https://modis.gsfc.nasa.gov/data/dataprod/mod12.php (accessed on 15 September 2020)) and ESRI's Global LULC classified data [28].

Recently, technologies for the analysis and evaluation of LULC have become very advanced in terms of geospatial modeling, simulation, and prediction for future modifications [29]. The present and future determinants of LULC changes can be identified and examined using internationally accredited models [30]. Spatially distributed models such as CA-ANN [30], ANN-Markov Chain [31], and CLUE-S [32] have been recommended by scholars for the analysis and prediction of LULC changes. Models have their own specific performing capabilities and unique behavior in terms of approach and simulation complexities [26]. Neural network models are frequently applied models in LULC change analysis and prediction because they precisely represent complex geographically heterogeneous modifications of land use and land cover [21]. Artificial neural networks in combination with cellular automata models can effectively realize land use systems [33]. The coupled CA-ANN model utilizes "what-if" scenarios in LULC change simulation and modeling [34].

In this study, a QGIS plugin known as MOLUSCE (modules for land use change simulation and evaluation) [35] was employed to evaluate the future LULC condition and the spatial and temporal transition. The MOLUSCE plugin is a user-friendly plugin that is compatible only with QGIS versions 2.00 to 2.99 [36]. It was developed by the Asia Air Survey (AAS) in 2012 [37] for quick and convenient performance and analysis of land cover changes. Various algorithms are embedded in the MOLUSCE plugin, which is essential for a variety of techniques as necessitated in the respective objective of the study [38]. Future LULC predictions and spatiotemporal transition possibilities were simulated and modeled for 2037, 2052, and 2067 using the coupled CA-ANN model with suitable remote sensing data from 1991 to 2022 with seven-year intervals employing spatial variables such as altitude and slope, which were derived from the shuttle radar topographic mission (SRTM) 30 m DEM, and population density, distance from the urban center, distance from major roads, and distance from streams [39].

The present study has the following objectives: to analyze LULC changes for the period 1991–2022, identify spatial variables suitable for LULC change prediction, and predict future LULC changes from 2022 to 2067 using a coupled CA-ANN model in the Upper Omo–Gibe River basin.

2. Materials and Methods

2.1. Study Area

The Omo–Gibe River basin is one of the twelve river basins in Ethiopia. It covers 79,000 km^2 and is situated in three regional states: the Oromia regional state, Southern Nations, Nationalities and Peoples' Region, and Southwest Ethiopia regional states. The basin is conventionally divided into two large subbasins: the Upper Omo–Gibe River basin and the Lower Omo–Gibe River basin. The Upper Omo–Gibe River basin (hereafter called UOGRB) covers a total area of 34,150 km^2 [40]. Geographically, it is located between 6°15′00″ and 9°50′00″ north latitude and 35°30′00″ and 38°45′00″ east longitude (Figure 1). It constitutes three hydroelectric power dams from Gilgel Gibe I to Gilgel Gibe III. Topographically, the basin is characterized by rugged terrain, mountains and valleys. The climate in the area is subhumid, and the main rainy season is between June and September, with a mean annual rainfall of 1300 to 2000 mm and an average temperature of 20.4 °C [41].

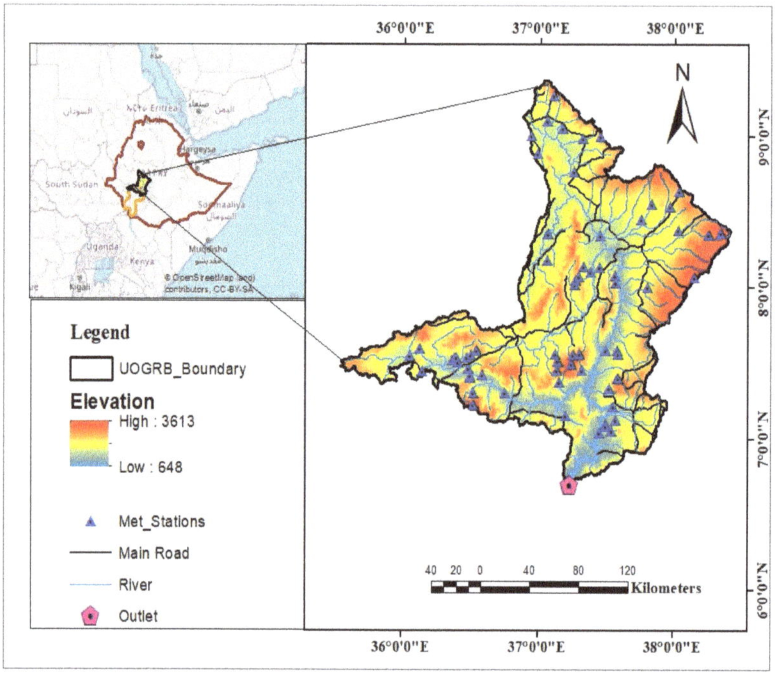

Figure 1. Location map of the Upper Omo–Gibe River basin.

2.2. Materials

2.2.1. Data Acquisition and Preprocessing

The UOGRB watershed boundary (Figure 1) was extracted from the SRTM 30 m DEM, which was downloaded from http://earthexplorer.usgs.org/ (accessed on 17 October 2021). Temporal satellite imagery data were obtained from Landsat satellite series for the years 1991, 1997, 2004, 2010, 2016, and 2022. Landsat-5 TM for 1991 and 2004, Landsat-7 ETM+ for 2010, and Landsat-8 operational land imager (OLI) for 2016 and 2022 were downloaded from the USGS Earth Explorer Data Center (https://earthexplorer.usgs.gov/ (accessed on 4 February 2022)). For compatibility issues, the data were projected to the Universal Transverse Mercator (UTM) with data from the World Geodetic System 84 (WGS84), the projection system zone 37 N (Table 1). The dry season (January to March) was chosen for the data selection to obtain the image data with the lowest cloud cover.

Table 1. Landsat satellite image acquisition and description.

S.N	Satellite	Sensor	Path	Row	Ground Resolution	Bands	Date of Acquisition
01	Landsat-5	TM *	169	054/055	30 m	7	10 January 1991
			170	054/055	30 m	7	18 February1991
02	Landsat-5	TM	169	054/055	30 m	7	14 January 2004
			170	054/055	30 m	7	15 February 2004
03	Landsat-7	ETM+ **	169	054/055	30 m	7	7 February 2010
			170	054/055	30 m	7	15 January 2010
04	Landsat-8	OLI ***	169	054/055	30 m	7	7 February 2016
			170	054/055	30 m	7	12 January 2016
04	Landsat-8	OLI	169	054/055	30 m	7	31 January 2022
			170	054/055	30 m	7	22 January 2022

* Thematic Mapper, ** Enhanced Thematic Mapper+, *** Operational Land Imager.

In this study, satellite images from different UTM zones (UTM Zone 36 N and UTM Zone 37 N) were used. To make image data compatible with each other, images were reprojected to the common geographic coordinate system using the world coordinate reference system (CRS: EPSG: 4326–WGS 84). Atmospheric, geometric, and radiometric corrections were performed in QGIS 3.22 and ERDAS IMAGINE 2014 software. To correct the scan line error (SLE) that occurred in the Landsat 7-ETM+ image due to the SLC-off [42] the Landsat toolbox was used.

2.2.2. Spatial Variable Identification and Preparation

Since they have a higher impact on the LULC change mechanisms, researchers pay particular attention to the physical and socioeconomic factors that contribute to LULC modifications [43]. It is obvious that geographical and climatic factors are thought to have the greatest influence on how people behave. The proximity to roads makes it easier to ascertain the motivations underlying the construction of the terrain. In this study, a variety of proximity and physical factors were considered.

The MOLUSCE plugin in QGIS offers several well-known techniques, including Pearson's correlation and Cramer's coefficient, for assessing the connection between LULC data and geographic factors [44]. Pearson's correlation was used in the present study to measure the relationship between variables because the spatial variables employed were not categorical. Pearson's correlation, r_p, is a parametric measure of the linear correlation between two variables, which was defined as the covariance of the two variables divided by the product of their standard deviations, as stated in [45]:

$$r_p\,(X,\,Y) = \frac{\frac{\sum_{i=1}^{n}(xi-\overline{x})\,(yi-\overline{y})}{n-1}}{\sqrt{\frac{\sum_{i=1}^{n}(xi-\overline{x})^2}{n-1}\frac{\sum_{i=1}^{n}(yi-\overline{y})^2}{n-1}}} = \frac{\sum_{i=1}^{n}(xi-\overline{x})(yi-\overline{y})}{\sqrt{\sum_{i=1}^{n}(xi-\overline{x})^2\sum_{i=1}^{n}(yi-\overline{y})^2}} \tag{1}$$

where the mean of each variable is subtracted from the raw scores in the numerator, and the cross-products of the centered variables are added up. The scales of the variables are changed by the denominator to have equal units. As a result, r is defined as the standardized and centered cross-product of two variables in Equation (1). Pearson's correlation coefficient, r_p ranges from -1 to 1. A value of 0 implies no linear association between the two variables, whereas values of 1 and -1 imply perfect positive and negative associations between the two variables, respectively. Spatial variables chosen for this study with reference to previous studies [46] in association with the study area existing and potential features include altitude, slope gradient, population density, distance from the main road, distance from the rivers, and distance from the urban center.

2.3. Methods

2.3.1. Land Use/Land Cover Classification

After the data were preprocessed and the quality enhancement techniques were conducted, LULC classification (Figure 2) was performed using seven land use/land cover classes. LULC classes were categorized as forest (riparian vegetation, deciduous woodlands, dense forest, woodland, mixed forest, open forest, evergreen broad leaved forests, evergreen needle leaved forests, deciduous forest, and afro-alpines), agricultural land (cropland, farmland, plantation, and fallow land), water bodies (rivers, ponds, swamps, wetlands, and reservoirs), built-up areas (infrastructures include asphalt roads, houses, buildings, and urban areas), shrubland (open and closed shrublands), grassland (range land, open areas with vegetation less than 2 m in height, and savannas), and bare land (bare soil with no vegetation or plantation, silt, gravel, degraded hillsides, rock outcrops, and open lands without natural forests or crops) (Table 2).

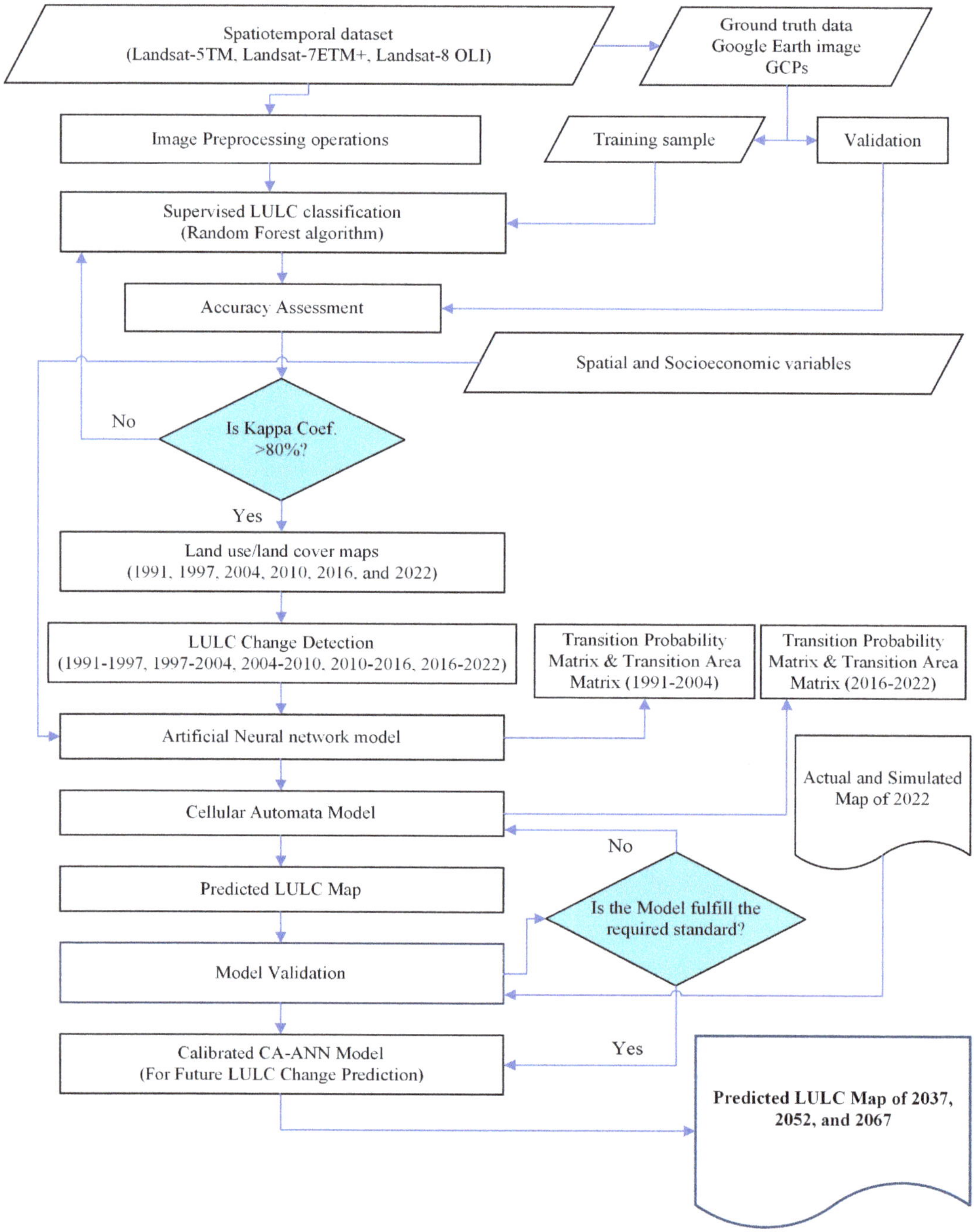

Figure 2. General flowchart of LULC change detection and prediction using the CA–ANN transition probability model (adopted from [47]).

Table 2. LULC classes and description.

Land Use/Land Cover Type	Description
Forest	Evergreen broad-leafed and evergreen needle-leafed forest, deciduous, woodland, open forest, dense forest, and afro-alpine.
Agricultural land	Lands occupied by crops, farmland, plantation, and fallow land.
Waterbody	Rivers, ponds, swamps, and reservoirs.
Built-up area	Infrastructures include houses, asphalt roads, buildings, and urban areas.
Shrubland	Include open and closed shrublands.
Grassland	Open grass lands, vegetation less than 2 m height, and range lands.
Bare land	Lands without vegetation, crops or grasses, and barren soils.

2.3.2. Change Evaluation and Modeling Transition Potential

In this study, the MOLUSCE plugin in QGIS was utilized to simulate the LULC transition between LULC classes and estimate spatiotemporal changes for the five time periods (1991–1997, 1997–2004, 2004–2010, 2010–2016, and 2016–2022), and five LULC maps were produced. Area change and transition probability matrixes were generated using 2010 and 2016 LULC. The artificial neural network (ANN) multilayer perception strategy was employed. Elevation, slope gradient, population density, distance from the main roads, distance from the urban center, and distance from streams were considered in the study as the determinant factors for future land use/land cover change prediction. In LULC change analysis and prediction, these variables are frequently used because they provide verifiable information on the impact of anthropogenic and natural factors on LULC dynamics [48].

3. Results

3.1. LULC Classification

Based on the classes identified prior to the classification process, LULC for the years 1991, 1997, 2004, 2010, 2016, and 2022 was performed using the random forest classification algorithm [49] and (Figure 3). The LULC classification method has advanced over time by employing various machine learning algorithms, such as random forest (RF) [50] support vector machine (SVM) [51] maximum likelihood classifier (MLC) [17], and kNN [52]. Among the machine learning approaches, random forest is preferable for various reasons: better capability of handling outliers and data noises, better performance with multidimensional datasets from different sources, relatively better accuracy than other commonly used classifiers, such as kNN, SVM, and MLC [53], and optimized processing speed due to selection of effective variables [50]. In the random forest classification method [49] the number of trees (Ntree) and individual split-based features (Mtry) are needed. Many studies have stated that the random forest classifier is effective in LULC classification [37], resulting in accurate classification output. According to [19] a large number of trees can provide a sound result for variables. Various studies [54,55] have stated that better results can be obtained with the default number of trees (Ntree = 500) as in the QGIS random forest classifier, which works in combination with Sentinel Application Platform (SNAP) software integrated into the classification algorithm.

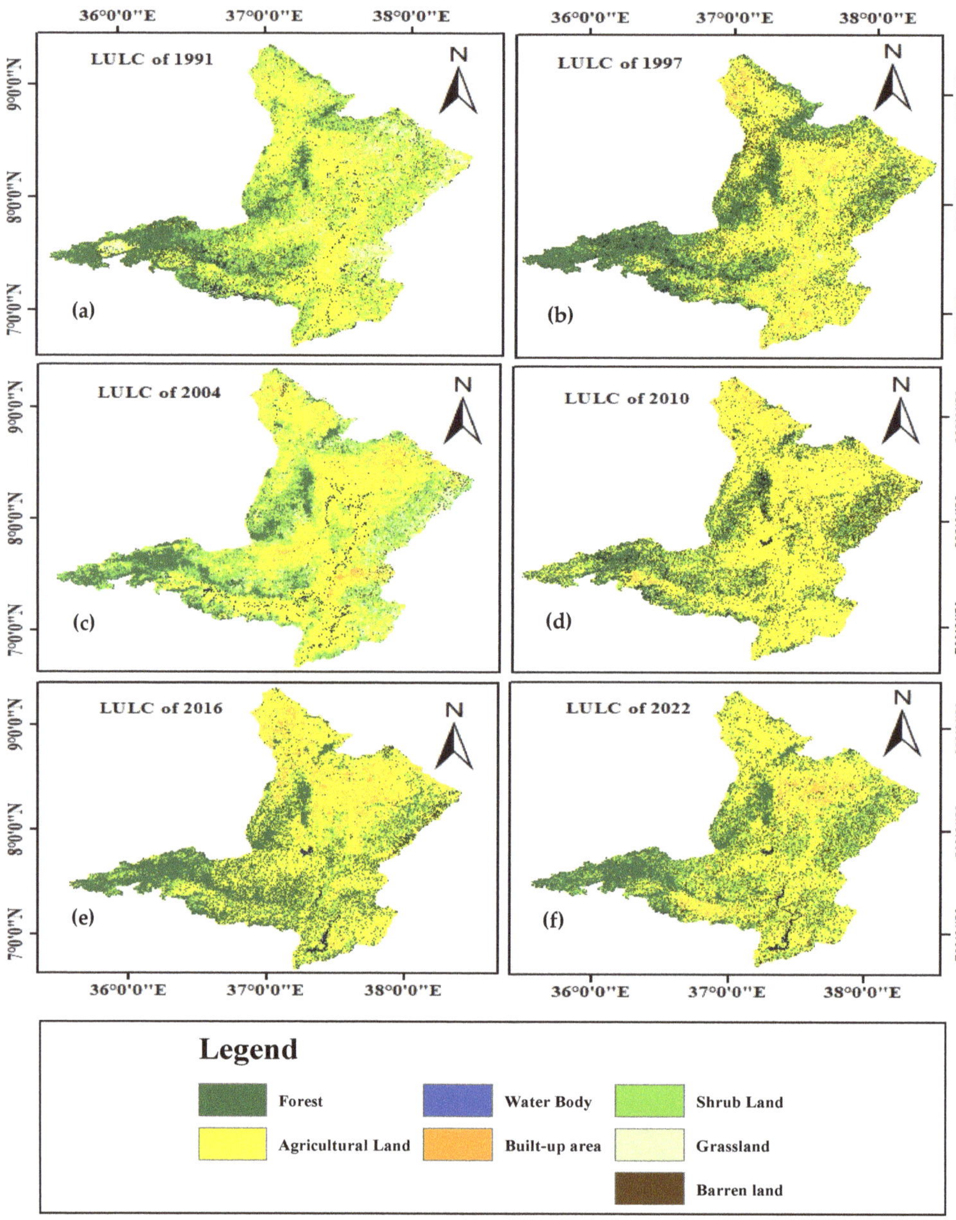

Figure 3. Land use/land cover (1991–2022): (**a**) LULC of 1991, (**b**) LULC of 1997, (**c**) LULC of 2004, (**d**) LULC of 2010, (**e**) LULC of 2016, and (**f**) LULC of 2022.

3.2. Classification Accuracy Assessment

After the images were classified using the random forest classification algorithm, the next step employed was image classification accuracy assessment [56]. Accuracy assessment is considered a component of the image classification method that enables the researcher to check whether the classification result has a strong, medium, or poor agreement with the ground truth [57]. It helps assess the classification methodology and is crucial for identifying any potential errors. In the form of a confusion matrix, the classifier's accuracy was evaluated (Table 3). The accuracy evaluation of each classification generated by the random forest-based classification was performed in the present study utilizing random points extracted using the classification and segmentation toolsets in the Spatial Analyst tools of ArcGIS and verified using historical Google Earth images.

Table 3. Confusion matrixes for LULC classification (i) 1991, (ii) 1997, (iii) 2004, (iv) 2010, (v) 2016, and (vi) 2022.

LULC 1991	(i)				Reference Data					
Classified Data	LULC Cat.	Forest	AgrL	WB	BuA	ShL	GrL	BrnL	RTotal	UA
	Forest	69	0	0	0	2	1	0	72	0.97
	AgrL	5	64	0	0	1	1	0	71	0.90
	WB	29	8	34	0	1	0	0	72	0.47
	BuA	5	0	0	65	0	1	1	72	0.92
	ShL	7	1	2	0	63	0	1	74	0.85
	GrL	1	2	0	1	0	68	0	72	0.94
	BrnL	4	4	0	0	4	2	57	71	0.80
	CTotal	120	79	36	66	71	73	59	504	
	PA	0.58	0.82	0.94	0.98	0.9	0.93	0.97		

LULC of 1997	(ii)				Reference Data					
Classified Data	LULC Cat.	Forest	AgrL	WB	BuA	ShL	GrL	BrnL	Total	UA
	Forest	57	14	0	0	2	0	0	73	0.79
	AgrL	0	69	1	0	1	1	0	72	0.96
	WB	22	3	46	0	0	0	0	71	0.65
	BuA	0	2	1	69	0	0	0	72	0.96
	ShL	0	6	0	0	66	0	0	72	0.92
	GrassL	0	7	0	0	0	65	0	72	0.90
	BarrenL	0	8	0	0	0	0	64	72	0.89
	Total	79	109	48	69	69	66	64	504	
	PA	0.72	0.63	0.96	1	0.97	0.98	1	0	0.87

LULC of 2004	(iii)				Reference Data					
Classified Data	LULC Cat.	Forest	AgrL	WB	BuA	ShL	GrL	BrnL	Total	UA
	Forest	72	0	0	0	0	0	0	72	1
	AgrL	3	67	0	0	1	0	0	71	0.94
	WB	6	13	53	0	0	0	0	72	0.74
	BuA	0	2	0	70	0	0	0	72	0.97
	ShL	0	0	0	1	71	1	0	73	0.99
	GrL	0	0	0	0	0	72	0	72	1
	BrnL	3	0	0	0	0	0	69	72	0.96
	Total	84	82	53	71	72	73	69	504	0
	PA	0.86	0.82	1	1	1	0.99	1	0	0.94

LULC 2010	(iv)					Reference Data				
Classified Data	LULC Cat.	Forest	AgrL	WB	BuA	ShL	GrL	BrnL	Total	UA
	Forest	66	4	1	0	2	0	0	73	0.90
	AgrL	3	67	0	0	1	0	0	71	0.94
	WB	17	0	54	0	1	0	0	72	0.76
	BuA	1	2	1	68	1	0	0	73	0.93
	ShL	1	1	0	1	71	0	0	74	0.96
	GrL	0	0	0	0	0	71	0	71	1
	BrnL	0	0	0	0	0	0	70	70	1
	Total	88	74	56	69	76	71	70	504	
	PA	0.77	0.91	0.96	1	0.93	1	1		

Table 3. *Cont.*

		(v)	Reference Data								
		LULC Cat.	Forest	AgrL	WB	BuA	ShL	GrL	BrnL	Total	UA
LULC 2016	Classified Data	Forest	66	6	0	0	0	0	0	72	0.92
		AgrL	0	69	0	1	1	1	0	72	0.96
		WB	3	3	66	0	0	0	0	72	0.92
		BuA	0	0	0	72	0	0	0	72	1
		ShL	0	4	0	0	68	0	0	72	0.94
		GrL	0	5	0	1	0	66	0	72	0.92
		BrnL	12	0	0	0	0	0	60	72	0.83
		Total	81	87	66	74	69	67	60	504	0
		PA	0.81	0.79	1	0.97	0.99	0.99	1	0	0.93
		(vi)	Reference Data								
			Forest	AgrL	WB	BuA	ShL	GrL	BrnL	Total	UA
LULC 2022	Classified Data	FRST	71	0	0	0	0	0	0	71	1
		AGRL	2	68	0	0	1	1	0	72	0.94
		WB	2	0	74	0	1	0	0	77	0.97
		BUA	0	10	0	58	2	1	0	71	0.82
		SHL	2	0	0	0	70	0	0	72	0.97
		GL	0	2	0	0	0	69	0	71	0.97
		BL	0	0	0	0	0	0	70	70	1
		Total	77	80	74	58	74	71	70	504	
		PA	0.93	0.85	1	1	0.96	0.97	1		

AgrL = agricultural land, WB = waterbody, BuA = built-up area, ShL = shrubland, GrL = grassland, BrnL = barn land, PA= producer's accuracy, UA = user's accuracy.

The accuracy for LULC classification was assessed using the statistical tool known as the Kappa coefficient, as shown in Table 4. According to [58], a classification has low agreement if the kappa coefficient value is less than 0.4, medium agreement if it is between 0.4 and 0.8, and good agreement if it is over 0.8. For the LULC classification in 1991, the kappa coefficient was 0.83 (83%) in 1997, 0.87 (87%) in 2004, and 0.94 (94%); in 2010, it was 0.93 (93%) in 2016 and 0.93 (93%); and in 2022, it was 0.95 (95%). According to the classification results, the 2022 LULC classification had the highest degree of agreement (95%), while the 1991 LULC classification had the lowest degree of agreement. The quality of the data was what caused the disparity. Comparatively, the 1991 Landsat-5 TM (Thematic Mapper) image has lower quality than the 2022 Landsat-8 (OLI) image, which influences the quality of LULC classification.

Table 4. Classification degree of agreement.

Year	Over All Accuracy (%)	Kappa Coefficient	Degree of Agreement (%)
1991	82	0.81698	Almost perfect agreement
1997	87	0.85296	Almost perfect agreement
2004	94	0.94348	Almost perfect agreement
2010	93	0.93179	Almost perfect agreement
2016	93	0.914352	Almost perfect agreement
2022	95	0.94825	Almost perfect agreement

3.3. LULC Change Analysis and Evaluation

After the accuracy assessment was conducted and the classification status was evaluated based on the kappa value for each year, change detection for 1991–1997, 1997–2004, 2004–2010, 2010–2016, and 2016–2022 was performed using QGIS 3.22 software. Change detection, as shown in Figure 4, was presented using maps and bar graphs. From 1991 to 1997, forest and agricultural land showed a high increase in area, and built-up area and shrubland showed less change in area coverage (Tables 5 and 6). Between 2016 and 2022, the forestland showed a dramatic change (decreased) in area, and the built-up area greatly increased (Table 6).

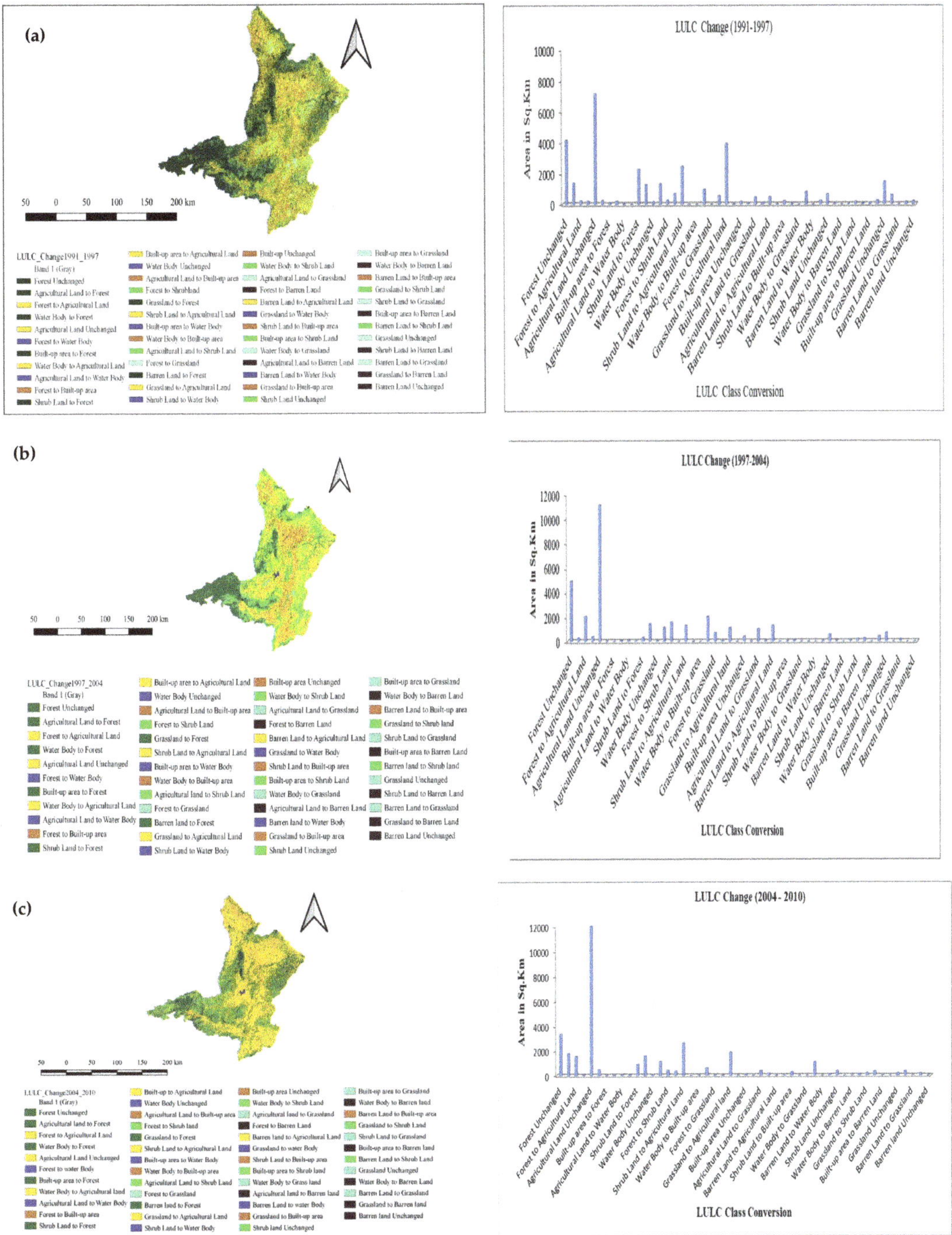

Figure 4. *Cont.*

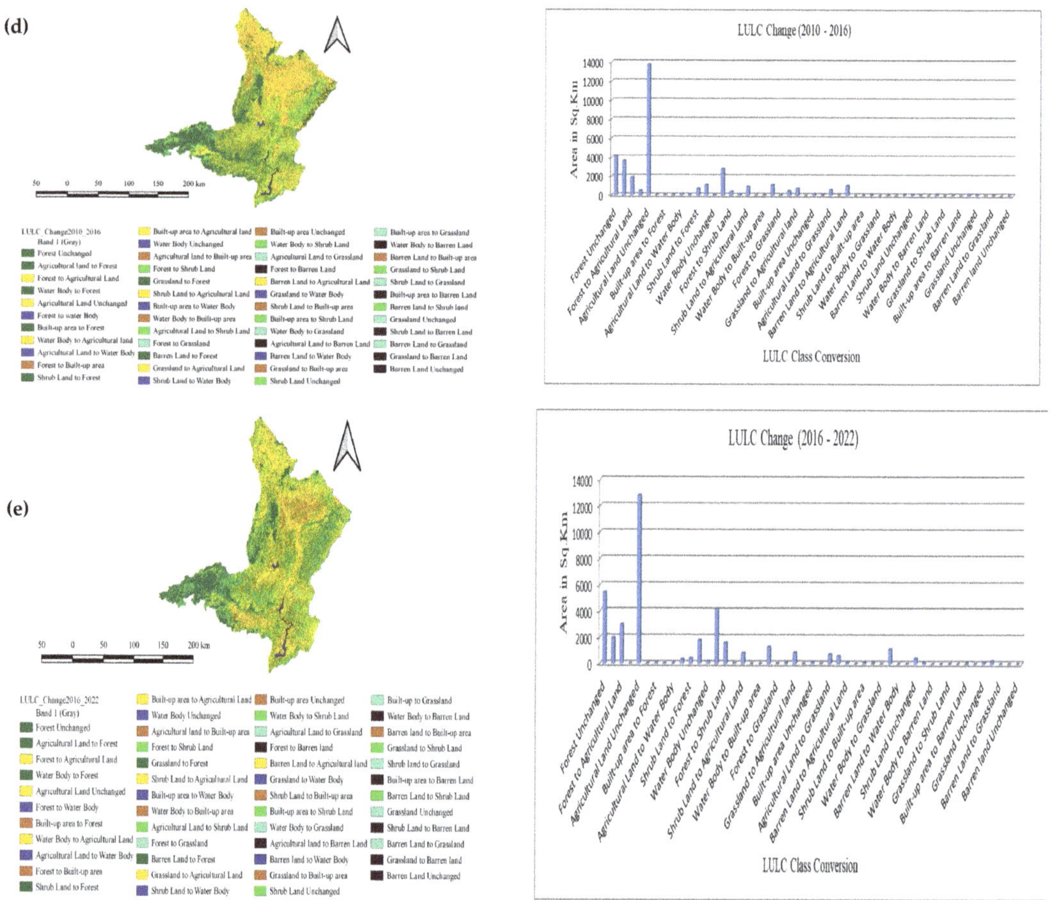

Figure 4. Change map from (1991–2022), (**a**) LULC changes (1991–1997), (**b**) LULC changes (1997–2004), (**c**) LULC changes (2004–2010), (**d**) 2010–2016, and (**e**) LULC changes (2016–2022).

Table 5. LULC area (km^2) from 1991 to 2022.

	1991		1997		2004		2010		2016		2022	
LULC Cat.	**Area (km^2)**	**Area (%)**	**Area (km^2)**	**Area (%)**	**Area (km^2)**	**Area (%)**	**Area (km^2)**	**Area (%)**	**Area (km^2)**	**Area (%)**	**Area (km^2)**	**Area (%)**
Forest	5026	15	9372	27	6028	18	6483	19	9768	28	5004	15
AgrL	12,032	36	15,736	46	18,275	53	21,219	62	19,536	57	22,147	65
WB	729	2	567	2	257	1	652	2	281	1	277	1
BuA	1776	5	2037	6	1812	5	1529	4	2025	6	2559	7
ShL	6255	18	2476	7	4841	14	1632	5	1791	5	2398	7
GrL	6629	19	2091	6	2903	8	981	3	772	2	859	3
BrnL	1515.87	4	2007	6	172	1	1791	5	113	0	1044	3

Table 6. LULC change analysis in 7-year intervals from 1991 to 2022.

LULC Cat.	1991–1997 Area (km²)	1991–1997 Change (%)	1997–2004 Area (km²)	1997–2004 Change (%)	2004–2010 Area (km²)	2004–2010 Change (%)	2010–2016 Area (km²)	2010–2016 Change (%)	2016–2022 Area (km²)	2016–2022 Change (%)
Forest	12,717	37	−3344	−10	456	1	3285	10	−4764	−14
AgL	13,197	38	2539	7	2945	9	−1684	−5	2611	8
WB	878	3	−311	−1	395	1	129	0	112	0
BuA	263	1	1775	5	−284	−1	497	1	1544	5
ShL	111	0	2365	7	−3209	−9	159	0	607	2
GrL	1279	4	812	2	−1922	−6	−209	−1	87	0
BrnL	3841	11	−1835	−5	1619	5	−1678	−5	931	3

3.4. Spatial Variables

The spatial variables selected for the current study are altitude, slope gradient, distance from the main roads, distance from the streams, population density, and the distance from the urban centers based on previous studies [59]. After relevant spatial variables (Figure 5) were selected and factor maps were created, geometry matching was conducted, which included the cell size of the raster data, NoData value, data extent, dimension, and coordinate reference system. The cell size, NoData value, and coordinate reference system are 30 m, 0, and WGS_1984_UTM_Zone_37 N, respectively, for all spatial variables.

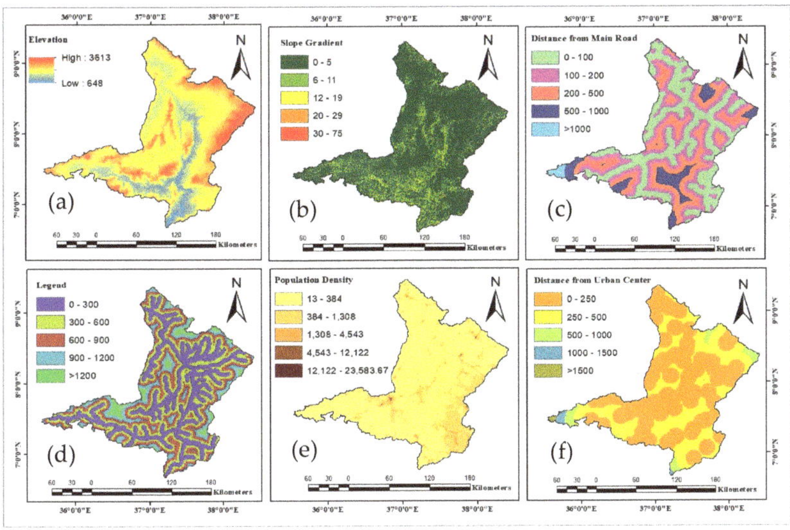

Figure 5. Spatial variables (explanatory variables): (**a**) elevation (**b**) slope gradient (**c**) distance from the main road, (**d**) distance from the streams, (**e**) population density, and (**f**) distance from the urban center.

In this study, Pearson's correlation was used to measure the correlation between variables. The MOLUSCE plugin in QGIS provided three options for correlation measurement: Pearson's correlation, Cramer's coefficient, and joint information uncertainty [60]. Pearson correlation was chosen because the variables used in this study are not categorical. Table 7 indicates that elevation, distance from the main road, and distance from the urban center are more strongly correlated than other variables, so these variables were selected for LULC change simulation and prediction.

Table 7. Evaluation of the correlation between variables using Pearson's correlation method.

Spatial Variables	Elev	Dmrd	SlpGt	DStr	Pdsty	DUC
Elevation (Elev)		0.038	0.005	0.525 *	0.037	0.109
DistMainRoad (Dmrd)			0.153	−0.025	0.045	0.609 *
Slope Gradient (SlpGt)				0.084	0.026	0.059
DistStreams (DStr)					0.031	0.016
Population Density (Pdsty)						0.072
DistUrbanCenter (DUC)						

DistMainRoad = distance from the main road, DistStreams = distance from the streams, DistUrbanCenter = distance from the urban center. * Are strongly correlated variables.

3.5. LULC Change Simulation and Prediction Using the CA-ANN Model

The QGIS MOLUSCE plugin provided artificial neural network (ANN) and cellular automata (CA) machine learning algorithms for LULC change simulation and prediction. ANN follows the nonparametric, nonlinear, and stochastic approach in LULC change simulation and prediction, which processes the potential to solve problems emanating from different variables [61]. Using its capacity to learn from a large number of datasets, ANN enabled the model to manage complex conditions during simulation. ANN basically uses the multilayer perception (MLP) approach [62] considering previously recorded LULC changes and spatial variables (explanatory variables) for LULC modification. In this study, the LULC of 1991, 1997, 2004, 2010, 2016, and 2022 (Table 8) and six spatial variables, including altitude, slope gradient, distance from the main road, distance from the streams, population density, and distance from the urban center, were considered as inputs for model simulation using the MOLUSCE plugin in QGIS. The initial and final rasters were the LULC maps of 2010 and 2016, respectively, to simulate 2022 LULC. The simulation model was constructed, and verification was performed for 2022. A change map was created using 2010 and 2016 LULC maps (Figure 6).

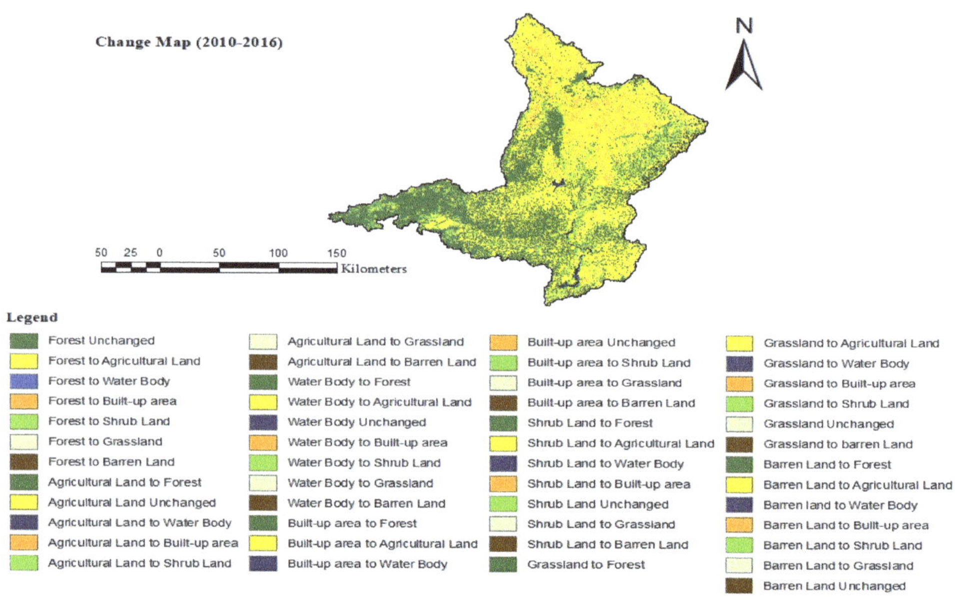

Figure 6. Change map (20102016).

Table 8. Transition matrix of 1991–2022.

		1997						
		Forest	AgrL	WB	BuA	ShL	GL	BL
1991	Forest	0.586	0.063	0.007	0.029	0.351	0.096	0.008
	AgrL	0.014	0.603	0.023	0.016	0.141	0.199	0.004
	WB	0.364	0.124	0.240	0.002	0.181	0.080	0.009
	BuA	0.005	0.641	0.001	0.027	0.061	0.252	0.012
	ShL	0.056	0.342	0.004	0.011	0.241	0.33	0.013
	GL	0.012	0.492	0.001	0.149	0.039	0.259	0.043
	BL	0.043	0.316	0.003	0.007	0.253	0.390	0.009
		2004						
		Forest	AgrL	WB	BuA	ShL	GL	BL
1997	Forest	0.472	0.352	0.221	0.003	0.034	0.004	0.023
	AgrL	0.049	0.809	0.003	0.070	0.019	0.018	0.030
	WB	0.435	0.462	0.052	0.018	0.015	0.003	0.014
	BuA	0.007	0.854	0.003	0.081	0.004	0.042	0.009
	ShL	0.373	0.389	0.016	0.010	0.113	0.022	0.076
	GL	0.123	0.591	0.004	0.036	0.073	0.058	0.107
	BL	0.078	0.714	0.007	0.045	0.056	0.241	0.073
		2010						
		Forest	AgrL	WB	BuA	ShL	GL	BL
2004	Forest	0.631	0.412	0.016	0.014	0.234	0.007	0.061
	AgrL	0.137	0.619	0.008	0.061	0.078	0.037	0.059
	WB	0.736	0.081	0.088	0.004	0.069	0.001	0.020
	BuA	0.070	0.760	0.001	0.060	0.038	0.044	0.037
	ShL	0.310	0.359	0.001	0.018	0.210	0.006	0.094
	GL	0.115	0.703	0.001	0.010	0.070	0.065	0.035
	BL	0.321	0.634	0.004	0.021	0.210	0.083	0.077
		2016						
		Forest	AgrL	WB	BuA	ShL	GL	BL
2010	Forest	0.633	0.271	0.007	0.011	0.067	0.007	0.003
	AgrL	0.174	0.647	0.008	0.087	0.052	0.029	0.002
	WB	0.757	0.115	0.083	0.004	0.034	0.003	0.005
	BuA	0.051	0.756	0.002	0.132	0.036	0.023	0.000
	ShL	0.448	0.445	0.002	0.012	0.074	0.014	0.004
	GL	0.199	0.640	0.003	0.024	0.026	0.098	0.010
	BL	0.277	0.578	0.006	0.031	0.076	0.019	0.013
		2022						
		Forest	AgrL	WB	BuA	ShL	GL	BL
2016	Forest	0.870	0.130	0	0	0	0	0
	AgrL	0.142	0.858	0	0	0	0	0
	WB	0.952	0.044	0.004	0	0	0	0
	BuA	0.006	0.987	0	0.007	0	0	0
	ShL	0.188	0.698	0	0	0.114	0	0
	GL	0.069	0.916	0	0	0	0.015	0
	BL	0.290	0.676	0	0	0	0	0.034
		2037						
		Forest	AgrL	WB	BuA	ShL	GL	BL
2022	Forest	0.999	0.001	0	0	0	0	0
	AgrL	0.001	0.999	0	0	0	0	0
	WB	0.002	0	0.998	0	0	0	0
	BuA	0.000	0.000	0	1.000	0	0	0
	ShL	0.026	0.004	0	0	0.970	0	0
	GL	0.001	0.002	0	0	0	0.998	0
	BL	0	0	0	0	0	0	1

AgrL = agricultural land, WB = waterbody, BuA = built-up Area, ShL = shrubland, GL = grassland, BL = bare land.

As indicated in Figure 7, agricultural land expanded, and forest cover, shrubland, and bare land decreased from 1991 to 1997. From 1997 to 2004, forest cover and agricultural land increased, but shrubland and grassland decreased. From 2016 to 2022, agricultural land and built-up areas showed a dramatic expansion, but forest cover, water bodies, shrubland, and bare land decreased in size (Figure 7).

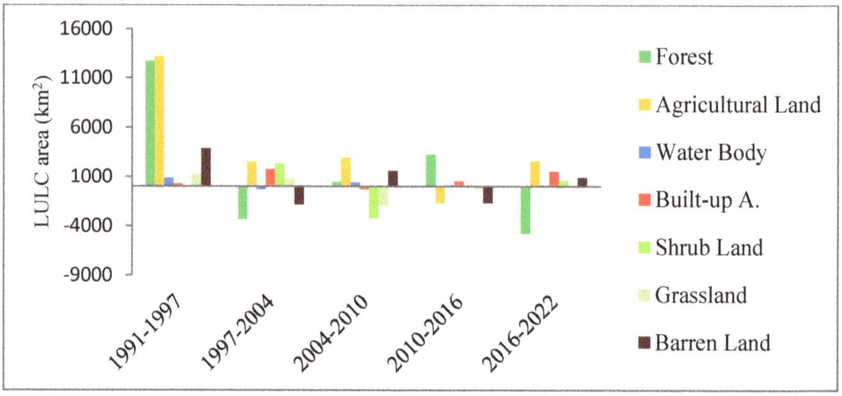

Figure 7. Gain and loss of LULC (1991–2022).

3.6. Artificial Neural Network (ANN)-Based LULC Change Transition Potential Modeling

Based on MOLUSCE's approach, transition potential modeling was conducted using an artificial neural network (ANN), which is ruled by the multilayer perception method employing neighborhood (1 px), learning rate (0.01) (Figure 8), maximum iteration (1000), hidden layer (12), momentum (0.05), fixed overall accuracy (−0.0018), minimum error for validation (0.0317), and validation kappa (0.9781). The model used 5000 randomly distributed samples to provide spatial representation for ANN. Based on the fixed value (0.01 learning rate), the neural network learned and simulated LULC changes for 2022.

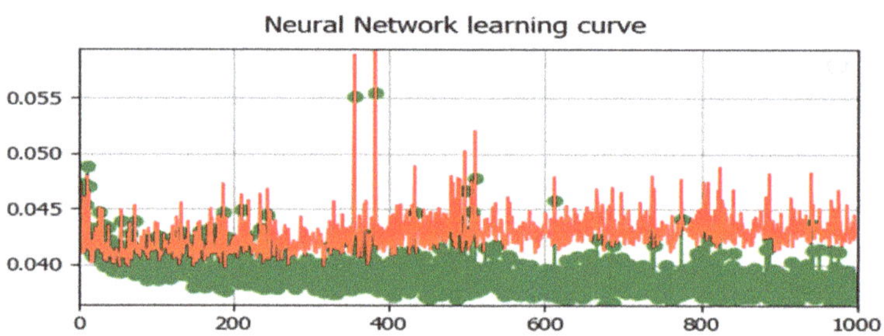

Figure 8. Neural network learning curve.

Transition potential modeling was computed, and future LULC changes were predicted using the CA-ANN model. The model utilized LULC data from 2010 to 2016 to predict the change in 2022, which resulted in a validation kappa value of 0.97. An overall accuracy of 87% and an overall kappa value of 0.82 were obtained by comparing the actual data of 2022 with the simulated LULC data from the same year. The actual and simulated LULC maps for 2022 are indicated in Figure 9 and their areal coverage was presented in Table 9.

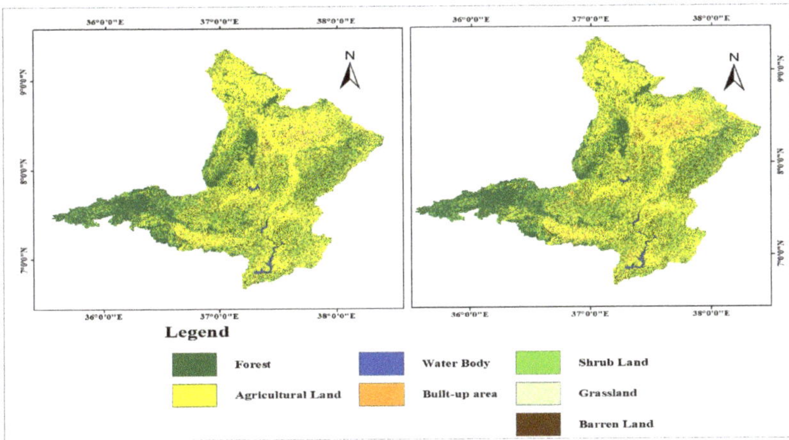

Figure 9. Actual and simulated LULC of 2022.

Table 9. Actual and simulated LULC of 2022.

LULC Category	Actual		Simulated	
	Area (km^2)	Area (%)	Area (km^2)	Area (%)
Forest	5004	13	4371	10
Agricultural land	22,147	58	28,314	63
Waterbody	893	2	938	2
Built-up area	5569	15	7658	17
Shrubland	2398	6	2149	5
Grassland	859	2	678	2
Barren land	1044	3	984	2

3.7. LULC Prediction

The LULC predictions for 2037, 2052, and 2067 (Figure 10 and Table 10) were performed after obtaining acceptable model validation results. The procedure of model validation combines the contents of two data sources and takes into consideration the features of actual and simulated LULC data. The LULC for 2037 was predicted using the LULC data between 2016 and 2022, the spatial variables selected for simulation, and the transition probability matrix, and a kappa value of 0.73 was attained. Consequently, LULC for 2052 was predicted using the LULCs of 2022 and 2037, employing the explanatory factors and transition matrix, and the kappa value computed was 0.68. Finally, using the projected data for 2052–2067 and the transition matrix, we forecasted the LULC for 2082 and arrived at a kappa value of 0.69 as a result.

Table 10. Area in km^2 for the LULC (2037, 2052, and 2067).

LULC Cat.	LULC2037		LULC2052		LULC 2067	
	Area (km^2)	Area (%)	Area (km^2)	Area (%)	Area (km^2)	Area (%)
Forest	5139	15	4872	14	4328	13
Agricultural land	22,146	64	21,273	62	22,353	65
Waterbody	273	1	272	1	293	1
Built-up area	1725	5	2749	8	2858	8
Shrubland	1859	5	1763	5	1572	5
Grassland	1787	5	1674	5	1561	5
Barren land	1895	5	1684	5	1473	4

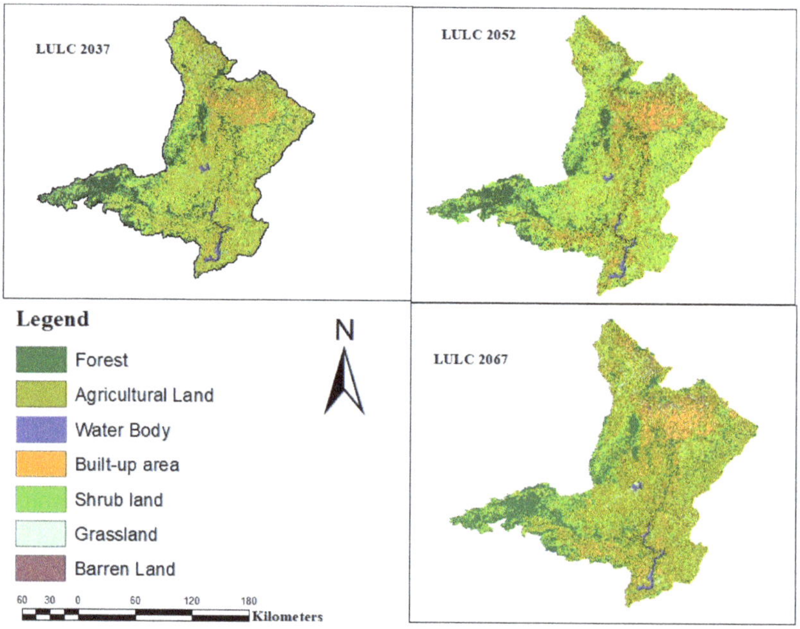

Figure 10. Prediction of LULC (2037–2067).

3.8. Change Prediction

The evaluation performed on LULC changes during the course of the investigation showed significant changes across land LULC classes. The results from 2022 to 2067 demonstrate a discernible increase in built-up areas and a decrease in forest cover and shrubland. The area coverage and percentage changes for each LULC category are presented in Figure 11 and Table 11. Inter-transitions between LULC classes and their contribution to other classes intensify changes in LULC at times. As indicated in Table 11, between 2022 and 2037, agricultural land (2163 km^2, 63%) and shrubland (1966 km^2, 5%) showed a significant increase, and forest (−1649, −48%) and grassland (−107 km^2, −0.31%) decreased. From 2037 to 2052, the built-up area of 1023 km^2 (3%) showed a significant increase, and forest −267 km^2 and agricultural land of −873 km^2 (−3%) showed a significant decrease. From 2052 to 2067, the projected LULC simulation result showed that agricultural land (1079 km^2, 3%) and built-up area (109 km^2, 0.32%) increased and forest −544 km^2 (−2%) and shrubland −191 km^2 (−1%) showed significant decreases.

Table 11. LULC changes from 2037 to 2067.

LULC Cat	2022–2037 km^2	2022–2037 %	2037–2052 km^2	2037–2052 %	2052–2067 km^2	2052–2067 %
Forest	−16,492	−48	−267	−1	−544	−2
Agricultural land	21,631	63	−873	−3	1079	3
Waterbody	515	2	−1	0	21	0
Built-up area	115	0	1024	3	109	0
Shrubland	1967	6	−96	−0	−191	−1
Grassland	−107	−0	−113	−0	−113	−0
Barren land	1895	6	−211	−1	−210	−1

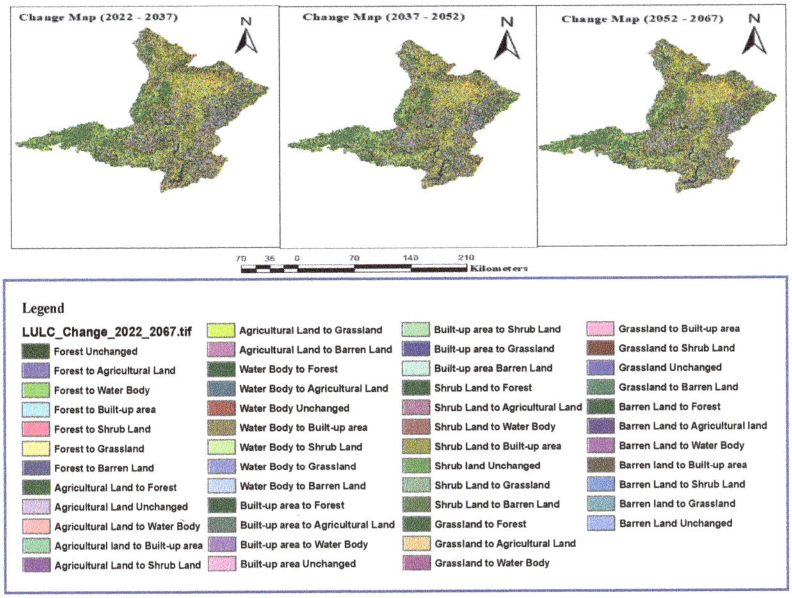

Figure 11. Change maps (2022–2067).

4. Discussion

Recently, most of the changes occurring in the physical and human environments have been directly related to the changes emanating from LULC dynamics. This condition has attracted the interest of scholars from academic and research institutes globally. In this regard, an ample number of studies have been conducted [63,64] dealing with the impact of LULC changes on various aspects of the environment. Water resource (surface and subsurface) availability, climate variability, soil fertility, and agricultural productivity are some of the phenomena directly or indirectly influenced by LULC changes. Climate variability may have a negative impact on land use and cover, as well as alter the environment's natural aesthetic. Lack of rainfall and temperatures that are abnormally high or low have detrimental impacts on the ecosystem [34].

The expansion of anthropogenic activities due to newly commenced national and international projects contributes greatly to environmental alteration, paving the way for the intensification of LULC changes. The study conducted in Linyi, China, considering land use/land cover change analysis, evaluation, and prediction using the QGIS MOLUSCE plugin and remote sensing big data revealed an increase in the population number and density; and the associated expansion of urban areas significantly caused a decrease in forest resource and grasslands [65].

In the Ethiopian context, population growth is the leading factor for LULC changes, which contributes to land fragmentation, forest loss, and loss of biodiversity. LULC dynamics can be primarily attributed to a variety of anthropogenic activities, including the encroachment of farmland into vegetated lands, the growth of farm plots at the expense of forestlands, the massive production of fuelwood and charcoal to support livelihood, overgrazing, and the expansion of farm plots into agricultural lands.

The study undertaken in Nashe Watershed, Upper Blue Nile Basin, Ethiopia, considered the prediction of land use/land cover changes using a land change modeler. This study used Landsat imageries for land use/land cover classification and predicted future LULC changes using the CA-Markov model. The study findings revealed that there will be a rapid conversion of forest land, range land, and grassland to agricultural land and built-up areas in the future and require a national policy implementation in order to maintain the goal of sustainable development [66]

The present study area, the Upper Omo–Gibe River basin, is known for its long mountain ranges, deep gorges, physically interconnected subbasins, lowland plains, woodlands, broad- and needle-leafed trees and grasslands, an enormous number of wild species, and mineral-rich soil [10]. Due to the existence of these environmental features in the study area, it was imperative to include spatial variables, such as elevation, slope gradient, distance from the river, and distance from the main road in the LULC change simulation and prediction.

According to the Gibe III hydroelectric project [67], the land cover of the study area comprises forest (open and dense forest, woodland, riverine forest, and plantation), agricultural land (cultivation, recession agriculture, fallow land, and croplands), water bodies, built-up areas, shrubland, grassland, and bare soil.

The present study considered seven land use/land cover classes, including forest, agricultural land, water bodies, built-up areas, shrubland, grassland, and barren land in reference to Landsat image time series data and some supportive sources, such as published and unpublished documents, topographic maps, and historical and current Google Earth images. Geospatial data for LULC classification were acquired from different sources. Digital elevation data with 30 m ground resolution were acquired from the shuttle radar topographic mission (SRTM). Landsat-5 TM for 1991 and 2004, Landsat-7 ETM+ for 2010, and Landsat-8 operational land imager (OLI) for 2016 and 2022 were downloaded from the USGS Earth Explorer Data Center. Spatial variables selected for the study were altitude, slope gradient, distance from the main roads, population density, distance from the streams, and distance from the urban center. Among the machine learning approaches, random forest is preferable for various reasons: better capability of handling outliers and data noises, better performance with multidimensional datasets from different sources, relatively better accuracy than other commonly used classifiers, such as kNN, SVM, and MLC, and optimized processing speed due to selection of effective variables.

The LULC classification result was evaluated using an accuracy assessment technique to assure the correctness of the classification method employing the kappa coefficient. Based on the classification evaluation outcome of the study, the kappa value for 1991 is 0.82 (82%), 1997 is 0.85 (85%), 2004 is 0.94 (94%), 2010 is 0.93 (93%), 2016 is 0.91 (91%), and 2022 is 0.95 (95%). Kappa coefficient values of the classification indicate that there was strong agreement between the classified and reference data.

The present study used the CA-ANN model for the evaluation and prediction of LULC changes in the future and found that the model is effective and reliable. However, it is suggested that further studies could employ more spatial and socioeconomic variables such as climate variability, economic development, technological advancement, political economy, and GDP (gross domestic product) to provide more sufficient and valuable information regarding the positive and negative impacts of LULC changes in the future.

5. Conclusions

LULC change detection was carried out for the time periods 1991–1997, 1997–2004, 2004–2010, 2010–2016, and 2016–2022, and the results were presented in maps and graphs. Landsat time series imageries were used to develop information classes based on the LULC categories. Raster data geometries were checked prior to simulation. The data extent, coordinate system, and NoData value were made to have the same content for simulation. Pearson's correlation was used to measure the correlation between variables. Future LULC changes were predicted employing spatial and socioeconomic variables, such as elevation, slope gradient, distance from the main road, distance from streams, and population density, respectively, using the MOLUSCE plugin of QGIS and the CA-ANN model. The following conclusions were made based on the study's findings.

Artificial neural network (ANN) and cellular automata (CA) machine learning methods were made available for LULC change modeling and prediction via the QGIS MO-

LUSCE plugin. ANNs have the potential to handle issues arising from many variables using nonparametric, nonlinear, and stochastic methods for LULC change modeling and prediction. The model was able to control complicated situations during simulation because ANN has the ability to learn from a large number of datasets. ANN primarily employs the multilayer perception (MLP) method while taking into account previously observed LULC changes and spatial variables (explanatory variables) for LULC modification. Transition potential modeling was computed, and future LULC changes were predicted using the CA-ANN model. The model utilized LULC data from 2010 to 2016 to predict the change in 2022, which resulted in a validation kappa value of 0.97. An overall accuracy of 86.53% and an overall kappa value of 0.82 were obtained by comparing the actual data of 2022 with the simulated LULC data from the same year and this result proved the reliability of the model.

The present study findings revealed that between 2022 and 2037, agricultural land increased by 2163 km^2 (or 63%) and shrubland by 1967 km^2 (or 6%) whereas grassland decreased by 107 km^2 (or 0.31%) and forest by 1650 km^2 (or 48%). From 2037 to 2052, the built-up area (1024 km^2) (3%) showed a considerable increase, while the forest (-267 km^2) and agricultural land (-873 km^2) (3%) showed a significant decline. The predicted LULC simulation result from 2052 to 2067 revealed that while built-up area 109.05 km^2 (0.32%) and agricultural land 1079.28 km^2 (3.15%) both increased, forest -544.11 km^2 (-2%) and shrubland -191 km^2 (-1%) significantly decreased.

The results of the study show that the future physical and human environment will be impacted by the quick and intense LULC transformation. Due to this condition, the study catchment (Gibe-III catchment) will experience climate variability, soil erosion, biodiversity loss, scarcity of water resources (both surface and subsurface), drought susceptibility, disruption of water balance components, and reservoir sedimentation in the future. In addition, increased human population density in the study area, particularly in urban areas and areas near main roads, has led to extraordinary levels of land pressure and increased demand for vast amounts of land for the building of residential homes both in urban and rural areas.

Therefore, it is crucial to call for the development of long-run local and national plans and their implementation regarding the utilization of natural resources and land use in order to safeguard the natural resources and maintain the water balance of the study area in particular and the entire basin (Omo–Gibe River basin) in general. Studies such as this one will significantly contribute to bringing about the best resource usage in the future by offering timely and relevant information in regard to LULC change evaluation and monitoring; and its impact on the physical and human environment.

Finally, the predicted conditions in this study particularly the negative impacts of LULC changes may be overturned through the implementation of integrated local and regional scale policies and strategies toward efficient resource utilization, land use, and physical and human environmental protection. In line with this, practicing effective soil and water conservation measures, climate-resilient agricultural activities, periodic afforestation, and reforestation programs will tackle the adverse impacts emanating from the LULC changes. It is recommended that future studies undertake multisource data, more variables (spatial and socioeconomic), and models to evaluate and predict LULC changes, and provide sound information in the entire basin (Omo–Gibe River basin) and other basins.

Author Contributions: Conceptualization, P.L. and A.M.M.; methodology, P.L.; software, P.L. and T.T.K.; model, P.L. and A.M.M.; formal analysis, P.L.; investigation, P.L.; writing—original draft preparation, P.L.; writing—review and editing, A.M.M. and T.T.K.; supervision, A.M.M. and T.T.K. All authors have read and agreed to the published version of the manuscript.

Funding: This study was funded by the Arba Minch University-Water Resource Research Center (GOV/AMU/SS/TH02/AWTI/WRRC/08/2015).

Data Availability Statement: The data presented in this study are available upon request from the corresponding author.

Conflicts of Interest: The authors declare no conflict of interest.

References

1. Ayele, G.T.; Demessie, S.S.; Mengistu, K.T.; Tilahun, S.A.; Melesse, A.M. *Multitemporal Land Uscape /Dynamics, Soils and Hydrological Processes in Varierd Change Detectlion for the Bmatena Watershed, Rift Valley Lakes Basin, Ethiopia Author Multitemporal Land Use/Land Cover Change Detection for the Batena Watershed, Rift Valley Lakes Basin, Ethiopia*; Springer International Publishing: Cham, Switzerland, 2016. [CrossRef]
2. Welde, K.; Gebremariam, B. Effect of land use land cover dynamics on hydrological response of watershed: Case study of Tekeze Dam watershed, northern Ethiopia. *Int. Soil Water Conserv. Res.* **2017**, *5*, 1–16. [CrossRef]
3. Halmy, M.W.A.; Gessler, P.E.; Hicke, J.A.; Salem, B.B. Land use/land cover change detection and prediction in the north-western coastal desert of Egypt using Markov-CA. *Appl. Geogr.* **2015**, *63*, 101–112. [CrossRef]
4. Hyandye, C.; Martz, L.W. A Markovian and cellular automata land-use change predictive model of the Usangu Catchment. *Int. J. Remote Sens.* **2017**, *38*, 64–81. [CrossRef]
5. Khan, Z.; Saeed, A.; Bazai, M.H. Land use / land cover change detection and prediction using the CA-Markov model: A case study of Quetta city, Pakistan. *J. Geogr. Soc. Sci.* **2020**, *2*, 164–182.
6. Da Cunha, E.R.; Santos, C.A.G.; da Silva, R.M.; Bacani, V.M.; Pott, A. Future scenarios based on a CA-Markov land use and land cover simulation model for a tropical humid basin in the Cerrado/Atlantic forest ecotone of Brazil. *Land Use Policy* **2021**, *101*, 105141. [CrossRef]
7. Yifru, B.A.; Chung, I.-M.; Kim, M.-G.; Chang, S.W. Assessing the Effect of Land/Use Land Cover and Climate Change on Water Yield and Groundwater Recharge in East African Rift Valley using Integrated Model. *J. Hydrol. Reg. Stud.* **2021**, *37*, 100926. [CrossRef]
8. Cherie, W.D. Assessing the Impacts of Land use / cover Change on the Hydrological Response of Temcha watershed, upper Blue Nile basin. *Ethiopia* **2018**, *2019*, 10–19.
9. Getachew, B.; Manjunatha, B.; Bhat, H.G. Modeling projected impacts of climate and land use/land cover changes on hydrological responses in the Lake Tana Basin, upper Blue Nile River Basin, Ethiopia. *J. Hydrol.* **2021**, *595*, 125974. [CrossRef]
10. Estifanos, T.H.; Gebremariam, B. Modeling-impact of Land Use/Cover Change on Sediment Yield (Case Study on Omo-gibe Basin, Gilgel Gibe III Watershed, Ethiopia). *Am. J. Mod. Energy* **2019**, *5*, 84. [CrossRef]
11. Belihu, M.; Tekleab, S.; Abate, B.; Bewket, W. Hydrologic response to land use land cover change in the Upper Gidabo Watershed, Rift Valley Lakes Basin, Ethiopia. *Hydroresearch* **2020**, *3*, 85–94. [CrossRef]
12. Näschen, K.; Diekkrüger, B.; Leemhuis, C.; Steinbach, S.; Seregina, L.S.; Thonfeld, F.; van der Linden, R. Hydrological Modeling in Data-Scarce Catchments: The Kilombero Floodplain in Tanzania. *Water* **2018**, *10*, 599. [CrossRef]
13. Abebe, G.; Getachew, D.; Ewunetu, A. Analysing land use/land cover changes and its dynamics using remote sensing and GIS in Gubalafito district, Northeastern Ethiopia. *SN Appl. Sci.* **2022**, *4*, 30. [CrossRef]
14. Anose, F.A.; Beketie, K.T.; Zeleke, T.T.; Ayal, D.Y.; Feyisa, G.L. Spatio-temporal hydro-climate variability in Omo-Gibe river Basin, Ethiopia. *Clim. Serv.* **2021**, 24. [CrossRef]
15. Aneseyee, A.B.; Elias, E.; Soromessa, T.; Feyisa, G.L. Land use/land cover change effect on soil erosion and sediment delivery in the Winike watershed, Omo Gibe Basin, Ethiopia. *Sci. Total. Environ.* **2020**, *728*, 138776. [CrossRef] [PubMed]
16. Ochoa, C.G.; Sierra, A.M.; Vives, L.; Zimmermann, E.; Bailey, R. Spatio-temporal patterns of the interaction between groundwater and surface water in plains. *Hydrol. Process.* **2019**, *34*, 1371–1392. [CrossRef]
17. Alam, A.; Bhat, M.S.; Maheen, M. Using Landsat satellite data for assessing the land use and land cover change in Kashmir valley. *Geojournal* **2020**, *85*, 1529–1543. [CrossRef]
18. Tadese, S.; Soromessa, T.; Bekele, T. Analysis of the Current and Future Prediction. *Sci. World J.* **2021**, *2021*, 6685045.
19. Phan, T.N.; Kuch, V.; Lehnert, L.W. Land Cover Classification using Google Earth Engine and Random Forest Classifier—The Role of Image Composition. *Remote Sens.* **2020**, *12*, 2411. [CrossRef]
20. Dagnachew, M.; Kebede, A.; Moges, A.; Abebe, A. Land Use Land Cover Changes and Its drivers in Gojeb River Catchment, Omo Gibe Basin, Ethiopia. *J. Agric. Environ. Int. Dev.* **2020**, *114*, 33–56. [CrossRef]
21. Ateka, M.; Agegn, L.; Belayneh, A. Evaluating the Effects of Land Use and Land Cover Change on Watershed Surface Runoff: Case of Abelti Watershed, Omo Basin, Ethiopia. *Int. J. Earth Sci. Knowl. Appl.* **2021**, *3*, 32–42.
22. Hasan, S.; Shi, W.; Zhu, X.; Abbas, S.; Khan, H.U. Future simulation of land use changes in rapidly urbanizing South China based on land change modeler and remote sensing data. *Sustainability.* **2020**, *12*, 4350. [CrossRef]
23. Maeda, E.E.; Formaggio, A.R.; Shimabukuro, Y.E. Impacts of Land Use and Land Cover Changes on Sediment Yield in a Brazilian Amazon Drainage Basin. *GIScience Remote. Sens.* **2008**, *45*, 443–453. [CrossRef]
24. Wang, S.W.; Munkhnasan, L.; Lee, W.-K. Land use and land cover change detection and prediction in Bhutan's high altitude city of Thimphu, using cellular automata and Markov chain. *Environ. Challenges* **2020**, *2*, 100017. [CrossRef]
25. Li, Y.; Chang, J.; Luo, L.; Wang, Y.; Guo, A.; Ma, F.; Fan, J. Spatiotemporal impacts of land use land cover changes on hydrology from the mechanism perspective using SWAT model with time-varying parameters. *Hydrol. Res.* **2019**, *50*, 244–261. [CrossRef]
26. Tessema, N.; Kebede, A.; Yadeta, D. Modeling land use dynamics in the Kesem sub-basin, Awash River basin, Ethiopia. *Cogent Environ. Sci.* **2020**, 6. [CrossRef]

27. Shih, H.-C.; Stow, D.A.; Chang, K.-C.; Roberts, D.A.; Goulias, K.G. From land cover to land use: Applying random forest classifier to Landsat imagery for urban land-use change mapping. *Geocarto Int.* **2021**, *37*, 5523–5546. [CrossRef]
28. Venter, Z.S.; Barton, D.N.; Chakraborty, T.; Simensen, T. Global 10 m Land Use Land Cover Datasets: A Comparison of Dynamic World, World Cover and Esri Land Cover. *Remote Sens.* **2022**, *14*, 4101. [CrossRef]
29. Engida, T.G.; Nigussie, T.A.; Aneseyee, A.B.; Barnabas, J. Land Use/Land Cover Change Impact on Hydrological Process in the Upper Baro Basin, Ethiopia. *Appl. Environ. Soil Sci.* **2021**, *2021*, 6617541. [CrossRef]
30. Singh, S.K.; Srivastava, P.K.; Gupta, M. Appraisal of land use/land cover of mangrove forest ecosystem using support vector machine. *Env. Earth Sci.* **2014**, *71*, 2245–2255. [CrossRef]
31. Patel, A.; Singh, V.; Kansara, B.; Kalubarme, M.; Panchal, B. Monitoring Land Use and Infrastructure Changes in Industrial Complex Using Geo-Informatics Technology in Gujarat State, India. *Int. J. Geosci.* **2016**, *7*, 1283–1298. [CrossRef]
32. Saputra, M.H.; Lee, H.S. Prediction of Land Use and Land Cover Changes for North Sumatra, Indonesia, Using an Artificial-Neural-Network-Based Cellular Automaton. *Sustainability* **2019**, *11*, 3024. [CrossRef]
33. Past, A.; Land, F.; Land, U.; Kumar, P. *Cellular Automata-Based Artificial Neural Network Model for Cover Dynamics*; Society for Computer Simulation International: San Diego, CA, USA, 2022; pp. 1–9.
34. Alqadhi, S.; Mallick, J.; Balha, A.; Bindajam, A.; Singh, C.K.; Hoa, P.V. Spatial and decadal prediction of land use/land cover using multi-layer perceptron-neural network (MLP-NN) algorithm for a semi-arid region of Asir, Saudi Arabia. *Earth Sci. Inform.* **2021**, *14*, 1547–1562. [CrossRef]
35. Kamaraj, M.; Rangarajan, S. Predicting the future land use and land cover changes for Bhavani basin, Tamil Nadu, India, using QGIS MOLUSCE plugin. *Environ. Sci. Pollut. Res.* **2022**, *29*, 86337–86348. [CrossRef] [PubMed]
36. Rahman, A.N.M.F. Predicting Changes in Land Use and Land Surface Temperature Using Cellular Automata Based Algorithm. Master's Thesis, Department of Civil Engineering, Dhaka, Bangladesh, 2020.
37. Kulithalai, P.; Sundar, S. Spatio-Temporal Classi cation and Prediction of Land Use and Land Cover Change for the Vembanad Lake System, Kerala—A Machine Learning Approach. *Environ. Sci. Pollut. Res.* **2022**, *29*, 86220–86236. [CrossRef] [PubMed]
38. Leta, M.K.; Demissie, T.A.; Tränckner, J. Modeling and Prediction of Land Use Land Cover Change Dynamics Based on Land Change Modeler (LCM) in Nashe Watershed, Upper Blue Nile Basin, Ethiopia. *Sustainability* **2021**, *13*, 3740. [CrossRef]
39. Kucsicsa, G.; Popovici, E.-A.; Bălteanu, D.; Grigorescu, I.; Dumitraşcu, M.; Mitrică, B. Future land use/cover changes in Romania: Regional simulations based on CLUE-S model and CORINE land cover database. *Landsc. Ecol. Eng.* **2019**, *15*, 75–90. [CrossRef]
40. Nesru, M.; Nagaraj, M.K.; Shetty, A. Assessment of consumption and availability of water in the upper Omo-Gibe basin, Ethiopia. *Arab. J. Geosci.* **2020**, *13*, 13. [CrossRef]
41. Chaemiso, S.E.; Abebe, A.; Pingale, S.M. Assessment of the impact of climate change on surface hydrological processes using SWAT: A case study of Omo-Gibe river basin, Ethiopia. *Model. Earth Syst. Environ.* **2016**, *2*, 1–15. [CrossRef]
42. Khaki, M.; Franssen, H.-J.H.; Han, S.C. Multi-mission satellite remote sensing data for improving land hydrological models via data assimilation. *Sci. Rep.* **2020**, *10*, 18791. [CrossRef]
43. Kuma, H.G.; Feyessa, F.F.; Demissie, T.A. Hydrologic responses to climate and land-use/land-cover changes in the Bilate catchment, Southern Ethiopia. *J. Water Clim. Chang.* **2021**, *12*, 3750–3769. [CrossRef]
44. Hakim, A.M.Y.; Baja, S.; Rampisela, D.A.; Arif, S. Modelling land use/land cover changes prediction using multi-layer perceptron neural network (MLPNN): A case study in Makassar City, Indonesia. *Int. J. Environ. Stud.* **2020**, *78*, 301–318. [CrossRef]
45. Rodgers, J.L.; Nicewander, W.A. Thirteen Ways to Look at the Correlation Coefficient. *Am. Stat.* **2008**, *42*, 59. [CrossRef]
46. El-tantawi, A.M.; Bao, A.; Chang, C.; Liu, Y. Monitoring and predicting land use / cover changes in the Aksu-Tarim River Basin, Xinjiang-China. *Environ. Monit. Assess.* **2019**, *191*, 1–18. [CrossRef] [PubMed]
47. Koko, A.; Yue, W.; Abubakar, G.; Hamed, R.; Noman, A.A. Monitoring and Predicting Spatio-Temporal Land Use/Land Cover Changes in Zaria City, Nigeria, through an Integrated Cellular Automata and Markov Chain Model (CA-Markov). *Sustainability* **2020**, *12*, 10452. [CrossRef]
48. Hamad, R.; Balzter, H.; Kolo, K. Predicting Land Use/Land Cover Changes Using a CA-Markov Model under Two Different Scenarios. *Sustainability* **2018**, *10*, 3421. [CrossRef]
49. Kulkarni, A.D.; Lowe, B.; Kulkarni, A.D.; Lowe, B. *Random Forest Algorithm for Land Cover Classification Random Forest Algorithm for Land Cover Classification*; University of Texas at Tyler: Tyler, TX, USA, 2016.
50. Ming, D.; Zhou, T.; Wang, M.; Tan, T. Land cover classification using random forest with genetic algorithm-based parameter optimization. *J. Appl. Remote. Sens.* **2022**, *10*, 035021. [CrossRef]
51. AlDahoul, N.; Ahmed, A.N.; Allawi, M.F.; Sherif, M.; Sefelnasr, A.; Chau, K.-W.; El-Shafie, A. A comparison of machine learning models for suspended sediment load classification. *Eng. Appl. Comput. Fluid Mech.* **2022**, *16*, 1211–1232. [CrossRef]
52. Reshmidevi, T.; Kumar, D.N.; Mehrotra, R.; Sharma, A. Estimation of the climate change impact on a catchment water balance using an ensemble of GCMs. *J. Hydrol.* **2018**, *556*, 1192–1204. [CrossRef]
53. Piao, Y.; Jeong, S.; Park, S.; Lee, D. Analysis of Land Use and Land Cover Change Using Time-Series Data and Random Forest in North Korea. *Remote. Sens.* **2021**, *13*, 3501. [CrossRef]
54. Wu, F.; Zhan, J.; Su, H.; Yan, H.; Ma, E. Scenario-Based Impact Assessment of Land Use / Cover and Climate Changes on Watershed Hydrology in Heihe River Basin of Northwest China. *Res. Artic.* **2015**, *2015*, 4101. [CrossRef]
55. Dibaba, W.T.; Demissie, T.A.; Miegel, K. Watershed Hydrological Response to Combined Land Use / Land Cover and Climate Change in Highland. *Water* **2020**, *12*, 1801. [CrossRef]

56. Feizizadeh, B. A Novel Approach of Fuzzy Dempster–Shafer Theory for Spatial Uncertainty Analysis and Accuracy Assessment of Object-Based Image Classification. *IEEE Geosci. Remote. Sens. Lett.* **2017**, *15*, 18–22. [CrossRef]
57. Liping, C.; YuJun, S.; Saeed, S. Monitoring and predicting land use and land cover changes using remote sensing and GIS techniques—A case study of a hilly area, Jiangle, China. *PLoS ONE* **2018**, *13*, e0200493. [CrossRef] [PubMed]
58. Pandey, S.; Kumari, N.; Al Nawajish, S. Land Use Land Cover (LULC) and Surface Water Quality Assessment in and around Selected Dams of Jharkhand using Water Quality Index (WQI) and Geographic Information System (GIS). *J. Geol. Soc. India* **2023**, *99*, 205–218. [CrossRef]
59. Satya, B.A. Future land use land cover scenario simulation using open source GIS for the city of Warangal, Telangana, India. *Appl. Geomat.* **2020**, *12*, 281–290. [CrossRef]
60. Akoglu, H. User's guide to correlation coefficients. *Turk. J. Emerg. Med.* **2018**, *18*, 91–93. [CrossRef]
61. Baig, M.F.; Mustafa, M.R.U.; Baig, I.; Takaijudin, H.B.; Zeshan, M.T. Assessment of Land Use Land Cover Changes and Future Predictions Using CA-ANN Simulation for Selangor, Malaysia. *Water* **2022**, *14*, 402. [CrossRef]
62. Shen, L.; Li, J.B.; Wheate, R.; Yin, J.; Paul, S.S. Multi-Layer Perceptron Neural Network and Markov Chain Based Geospatial Analysis of Land Use and Land Cover Change. *J. Environ. Inform. Lett.* **2020**, *3*, 29–39. [CrossRef]
63. Zaidi, S.M.; Akbari, A.; Abu Samah, A.; Kong, N.S.; Gisen, J.I.A. Landsat-5 Time Series Analysis for Land Use/Land Cover Change Detection Using NDVI and Semi-Supervised Classification Techniques. *Pol. J. Environ. Stud.* **2017**, *26*, 2833–2840. [CrossRef]
64. Kenea, U.; Adeba, D.; Regasa, M.S.; Nones, M. Hydrological Responses to Land Use Land Cover Changes in the Fincha'a Watershed, Ethiopia. *Land* **2021**, *10*, 916. [CrossRef]
65. Dos Santos, V.; Laurent, F.; Abe, C.; Messner, F. Hydrologic Response to Land Use Change in a Large Basin in Eastern Amazon. *Water* **2018**, *10*, 429. [CrossRef]
66. Xie, G.; Niculescu, S. Monitoring of Land Cover/Land Use (LCLU) Changes in the Crozon Peninsula (Brittany, France) from 2007 to 2018 by Machine Learning Algorithms (Support Vector Machine, Random Forest, and Convolutional Neural Network) and by Post-classification Comparison (PCC). *Remote Sens.* **2021**, *13*, 3899. [CrossRef]
67. Rahman, A.; Kumar, S.; Fazal, S. Assessment of Land use/land cover Change in the North-West District of Delhi Using Remote Sensing and GIS Techniques. *J. Indian Soc. Remote Sens.* **2012**, *40*, 689–697. [CrossRef]

remote sensing

MDPI

Article

Detecting, Analyzing, and Predicting Land Use/Land Cover (LULC) Changes in Arid Regions Using Landsat Images, CA-Markov Hybrid Model, and GIS Techniques

Salman A. H. Selmy [1,*], Dmitry E. Kucher [2], Gintautas Mozgeris [3], Ali R. A. Moursy [4], Raimundo Jimenez-Ballesta [5], Olga D. Kucher [6], Mohamed E. Fadl [7] and Abdel-rahman A. Mustafa [4]

1 Department of Soils and Water, Faculty of Agriculture, Assiut University, Assiut 71526, Egypt
2 Department of Environmental Management, Institute of Environmental Engineering, People's Friendship University of Russia Named after Patrice Lumumba (RUDN University), 6 Miklukho-Maklaya St., 117198 Moscow, Russia; kucher-de@rudn.ru
3 Department of Forest Sciences, Agriculture Academy, Vytautas Magnus University, Studentų Str. 11, LT-53361 Akademija, Kaunas Dstr., Lithuania; gintautas.mozgeris@vdu.lt
4 Department of Soil and Water, Faculty of Agriculture, Sohag University, Sohag 82524, Egypt; ali.refaat@agr.sohag.edu.eg (A.R.A.M.); a_mustafa32@agr.sohag.edu.eg (A.-r.A.M.)
5 Department of Geology and Geochemistry, Autónoma University of Madrid, 28049 Madrid, Spain; raimundo.jimenez@uam.es
6 Scientific Department, Rural Urban Framework Center, People's Friendship University of Russia Named after Patrice Lumumba (RUDN University), 6 Miklukho-Maklaya St., 117198 Moscow, Russia; kucher-od@rudn.ru
7 Division of Scientific Training and Continuous Studies, National Authority for Remote Sensing and Space Sciences (NARSS), Cairo 11769, Egypt; madham@narss.sci.eg
* Correspondence: salman.selmy@agr.aun.edu.eg

Abstract: Understanding the change dynamics of land use and land cover (LULC) is critical for efficient ecological management modification and sustainable land-use planning. This work aimed to identify, simulate, and predict historical and future LULC changes in the Sohag Governorate, Egypt, as an arid region. In the present study, the detection of historical LULC change dynamics for time series 1984–2002, 2002–2013, and 2013–2022 was performed, as well as CA-Markov hybrid model was employed to project the future LULC trends for 2030, 2040, and 2050. Four Landsat images acquired by different sensors were used as spatial–temporal data sources for the study region, including TM for 1984, ETM+ for 2002, and OLI for 2013 and 2022. Furthermore, a supervised classification technique was implemented in the image classification process. All remote sensing data was processed and modeled using IDRISI 7.02 software. Four main LULC categories were recognized in the study region: urban areas, cultivated lands, desert lands, and water bodies. The precision of LULC categorization analysis was high, with Kappa coefficients above 0.7 and overall accuracy above 87.5% for all classifications. The results obtained from estimating LULC change in the period from 1984 to 2022 indicated that built-up areas expanded to cover 12.5% of the study area in 2022 instead of 5.5% in 1984. This urban sprawl occurred at the cost of reducing old farmlands in old towns and villages and building new settlements on bare lands. Furthermore, cultivated lands increased from 45.5% of the total area in 1984 to 60.7% in 2022 due to ongoing soil reclamation projects in desert areas outside the Nile Valley. Moreover, between 1984 and 2022, desert lands lost around half of their area, while water bodies gained a very slight increase. According to the simulation and projection of the future LULC trends for 2030, 2040, and 2050, similar trends to historical LULC changes were detected. These trends are represented by decreasing desert lands and increasing urban and cultivated newly reclaimed areas. Concerning CA-Markov model validation, Kappa indices ranged across actual and simulated maps from 0.84 to 0.93, suggesting that this model was reasonably excellent at projecting future LULC trends. Therefore, using the CA-Markov hybrid model as a prediction and modeling approach for future LULC trends provides a good vision for monitoring and reducing the negative impacts of LULC changes, supporting land use policy-makers, and developing land management.

Keywords: remote sensing; LULC classification; LULC changes; CA-Markov model; LULC projection; land-use planning; GIS; IDRISI

1. Introduction

Land degradation is a global issue caused by demographic growth, inappropriate land management, deforestation, climate change, and others [1]. Recently, Egypt has experienced a very complicated dynamic built-up sprawl process due to the fast population increase, which is the key driver of urbanization and LULC changes in arid lands, converting productive agricultural lands to urbanized areas, prompting concern among researchers [2]. Furthermore, in Egypt, growing urbanization is linked with population increases, resulting in intense farming land use to fulfill food demand. Accordingly, some Nile Valley and Delta fertile soils have been vulnerable to degradation and have lost their fertility and productivity. Hence, the sustainable development of current agricultural lands with an expansion of reclaimed soils is necessary to achieve food security, which has become the primary concern [3]. Land reclamation programs for arable land outside the Nile Valley have accelerated rapidly, aiming to increase farmed areas while relieving the pressure on fertile agricultural soils [4]. Thus, for Egypt, as an arid region, the detection and projecting of LULC changes are essential for long-term management and land-use planning of natural resources.

Many factors influence LULC changes, including time, scale, politics, economics, and social and cultural factors [5]. At all spatial–temporal scales, LULC changes have been recognized as significant drivers of environmental changes [6]. Such changes, along with other negative changes (e.g., climate change, biodiversity loss, and water, soil, and air pollution), should be the top priorities for humans [7]. Therefore, studying and monitoring the changes in LULC over time is both urgent and necessary to obtain an accurate future vision for sustainable development. Monitoring and relieving the unfavorable consequences of LULC dynamics while sustaining essential resource generation has, thus, become a top emphasis for researchers and policy-makers worldwide. Land use and land cover (LULC) describe the physical land types such as forests, wetlands, impervious surfaces, agriculture, water types, and others, as well as how humans use these land types in a region. Furthermore, LULCs are spatially dispersed due to the dynamic interactions between human activities and natural factors of ecosystems [8–11]. Natural, social, and economic variables all influence the complex dynamics of LULC systems. Changes in LULC availability and distribution have a substantial influence on climate, environmental challenges, and natural ecosystem conditions [12–14]. Furthermore, changes in global LULC are a major cause of significant concern for future LULC trends [15–19]. Moreover, LULC changes are a critical aspect of sustainability and management of natural resources [20–23].

In many cases, human activities have unfavorable impacts on LULC in diverse territories of the world. Human-induced LULC changes are driven mainly by the requirements of communities for food and economic development due to the rapidly increasing population growth globally [24–28]. These LULC shifts, however, have severe consequences for climate change, environmental conservation, environmental pollution, soil erosion, agricultural land production, biodiversity, and water resources [29–37]. In the investigation region, as arid land, the main driver of LULC changes is human activities due to population increase. This is similar to the rest of Egypt, which requires supporting economic development to meet food and housing requirements.

Above all, understanding and identifying detailed historical and future LULC dynamics knowledge over a lengthy series of periods and analyzing their patterns and processes is critical for land development decision-makers and planners. LULC changes are rapidly dynamic processes worldwide. Thus, it is widely recognized that LULC trends documenting and modeling improve knowledge of historical and future LULC patterns and their repercussions and guide future land-use planning [27,38,39].

Geographic information systems (GIS) and remote sensing (RS) techniques provide helpful approaches for understanding, analyzing, and monitoring LULC over time in landscapes [40,41]. Many investigations on LULCs have been conducted using these tools [42–45]. Additionally, Landsat periodic imagery data for a specific area is a reliable data source that can be utilized to forecast LULC trends in that area. GIS and RS techniques are effective for displaying spatial modeling methods [46]. Thus, modeling, simulation, and predicting the LULC changes using temporal–spatial data are crucial for improving LULC planning and management. Although the complexity of developing models and simulations is great, they are necessary to detect the LULC changes and analyze the causes and consequences of this phenomenon [47]. Furthermore, various approaches for determining historical and prospective LULC have been established to assist stakeholders in economic improvement, land conservation, and land-use planning. In addition, several models for simulating future scenarios of LULC have been created. These models deliver appropriate approaches for detecting the spatial variability patterns in LULC. Moreover, model validation is required for an accurate assessment of LULC prediction in a particular area by comparing predicted and observed LULC changes [48].

The combined Cellular Automata (CA) and Markov Chain model (CA-Markov Model) is one widespread model used with high accuracy for analyzing LULU dynamics [27,36,49]. Furthermore, the CA-Markov model's robustness provides an opportune manner for modeling LULC change dynamics spatially and temporally in complex landscapes [50–52]. The flexibility of the CA-Markov hybrid model to combine spatial and remote sensing data, as well as biophysical and socio-economic data, promotes a more extensive, detailed, and accurate projection of LULC change transitions [53]. Several studies have employed this model successfully in LULC prediction, e.g., [36,54–58]. CA-Markov hybrid model, which was employed in the present investigation, is an effective and widely used model among several models by researchers to detect, predict, monitor, and simulate spatiotemporal change in LULC [59–62].

The most significant stage in the CA-Markov model is the rules of transition, which depend on the training data [63]. Furthermore, the applied model is affected by neighborhood class and cell size, which are also considered in obtaining optimum simulation or prediction outcomes [13]. CA-Markov hybrid model can effectively incorporate remotely sensed data and GIS. This model can convert the results into spatially explicit results necessary for LULC development [64]. For simulating and predicting land-use changes, several modeling approaches are used [65]. Some of them are applied based on the statistical matching approach of the spatio-temporal data using the main variables of prediction. The others utilize algorithms specifically designed to assess human–environment interactions [66]. However, the Markov model is the earliest statistical model that can be generated with minimal input data [67,68]. In the current study, assessing and analyzing historical LULC change patterns from 1984 to 2022 and simulating and predicting future LULC trends of 2030, 2040, and 2050 in the study area using the CA-Markov model is beneficial for implementing sustainable land management [69]. Furthermore, there is a scarcity of research on detecting LULC and forecasting future scenarios in LULC in arid lands such as the study region. Therefore, the main goal of this work was to detect, analyze, and evaluate the historical change dynamics in LULC from 1984 to 2022 and to simulate and predict LULC trends for 2030, 2040, and 2050 using CA-Markov spatial modeling.

2. Materials and Methods

2.1. Study Area

In the current study, Sohag Governorate represents the arid land study area. Sohag Governorate is located in Upper Egypt and encompasses a portion of the Nile Valley beside the eastern and western desert belts. It has an area of around 11,120 km^2 and a population of 5.45 million in 2021. The Nile River runs for about 125 km through the Sohag Governorate. The study region is situated at latitudes of 26°07′ and 26°57′N, as well as longitudes of 32°14′ and 31°20′E, and it is bounded on the south by Qena Governorate and on the north by Assiut Governorate (Figure 1). Climatically, the research region is located in North Africa's arid region, which is characterized by hot summers and mild winters with extremely low to nil rainfall. The monthly averages of temperatures, evaporation, and rainfall data for the study region are depicted in Figure 2. Temperature varies from 15.5 °C to 36.5 °C, and the relative humidity ranges between 35% and 61%. Following Soil Survey Staff [70], *Hyperthermic* is the soil temperature regime, and *Toric* is the moisture regime. The study area includes four LULC categories: urban areas; cultivated lands; deserts; and water bodies, as shown in Table 1.

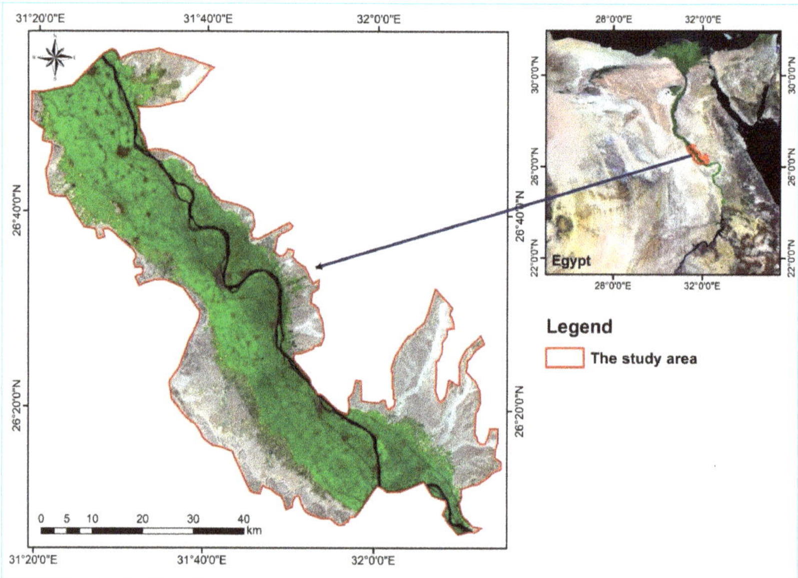

Figure 1. Map of the investigation region.

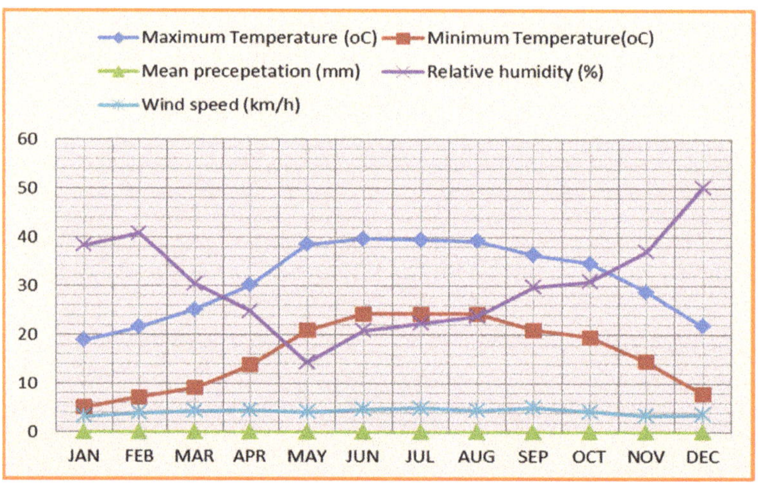

Figure 2. Climate data for the study region.

Table 1. Types of LULC in the study site.

LULC	Description
Water bodies	The Nile River, canals, drainage patterns, and wastewater treatment plants.
Desert lands	Bare lands, sand sheets, and rocky lands in the eastern and western parts outside the Nile Valley.
Cultivated lands	Old cultivated lands in the Nile Valley and newly reclaimed lands outside the Nile Valley.
Urban areas	Residential, commercial, industrial, and road constructions in cities and rural areas.

2.2. Data Acquisition

Remotely sensed and spatial data are trustworthy data for detecting and comprehending the dynamic causes of LULC in any terrain [71]. The primary remotely sensed and ancillary data utilized in this work were four multispectral images of Landsat. These images were acquired by the thematic mapper (TM), the enhanced thematic mapper (ETM+), and the operational land imager (OLI) sensors for the years 1984, 2002, 2013, and 2022, respectively. Each of the four images used was created by mosaicking three adjacent captured images to cover the investigation region. The four images' geo-coded cloud-free digital data were obtained from the Landsat archive of the United States Geological Survey (USGS) (https://earthexplorer.usgs.gov, accessed on 5 January 2022). The principal specifications of the images used are shown in Table 2. The study site boundary was subsetted using topographic maps at 1:50,000 scales obtained from the Egyptian Survey Authority (ESA) as a reference. For creating the digital elevation model (DEM), Advanced Spaceborne Thermal Emission and Reflection Radiometer (ASTER) data were used after radiometric and geometric corrections, as described in Table 3. Figure 3 displays the created DEM.

Table 2. The principal specifications of the used Landsat satellite images.

Satellite Image	Year of Acquisition	Path/Row	Resolution (m)	Image Type
Landsat 5 (TM)	1984	175/42, 176/42 and 176/41	15–30	Level-1
Landsat 7 (ETM+)	2002			
Landsat 8 (OLI)	2013			
Landsat 9 (OLI)	2022			

TM = thematic mapper. ETM+ = enhanced thematic mapper. OLI = operational land imager.

Table 3. The specifications of the ASTER utilized.

Sensor	Spectral Range	Bands	Resolution (m)	Swath Width (km)	Quantization Level (Bits)
ASTER	VNIR, SWIR, and TIR	14	15–90	60	8–12

VNIR = visible and near-infrared. SWIR = short-wave infrared. TIR = thermal infrared.

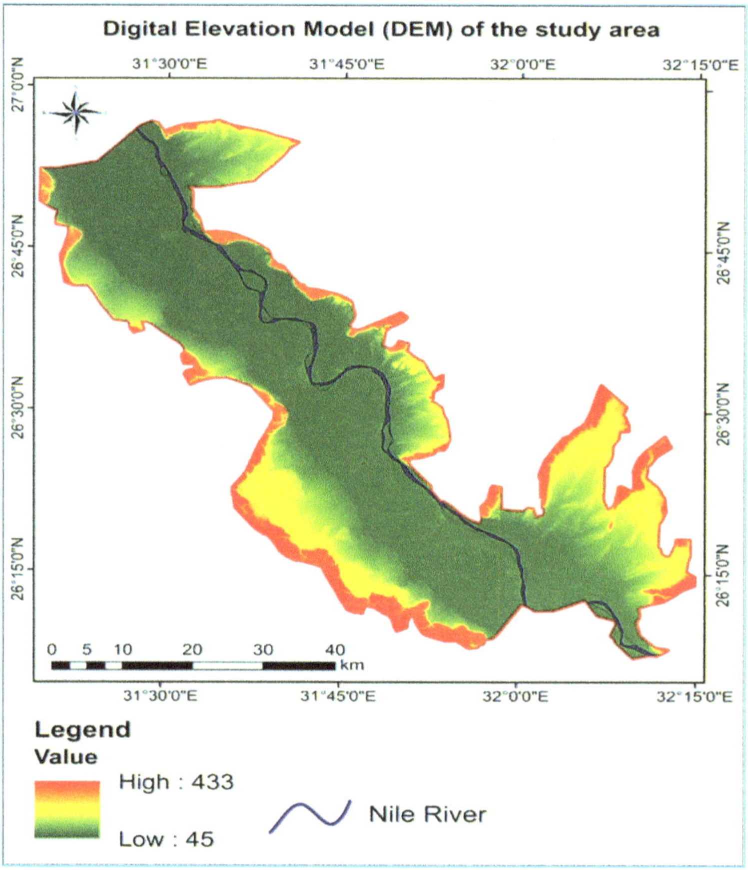

Figure 3. Digital elevation model (DEM) map generated from ASTER data.

2.3. Pre-Processing of Remote Sensing Data

Pre-processing remotely sensed data is required to reduce noise, prepare the data, and increase its suitability for further analysis. As previously stated, three images were employed to cover the study area, totaling twelve images covering the investigated time series from 1984 to 2022. Using the ENVI platform, all Landsat images underwent pre-processing, including layer stacking, geometric correction, mosaicking, and band extraction. First, the multi-temporal Landsat images were imported into the ENVI platform, and all spectral bands were layer-stacked. Concerning the geometric correction process, the entire datasets employed were resampled and projected to the World Geodetic System 1984 (WGS84) into the Universal Transverse Mercator (UTM), with a root mean square error (RMSE) of less than 0.5 pixels [43,72]. For all images of interest, a standard geographic coordinate system was used throughout the geometric rectification process. Furthermore, choosing ground control points (GCPs) is critical for geometric rectification. The digital data for 1984 were registered using 50 GCPs that were easily identifiable on both satellite images and on the ground (map-to-image registration). In addition, the first polynomial order and nearest neighborhood resampling approaches were used to set the pixel size at 30 m for image registration. The remaining images were rectified (image-to-image registration) using the corrected image of 1984 as a reference, following the same resampling method. Furthermore, the nearest neighborhood algorithm was applied over images for the resampling process without changing the original brightness magnitudes of pixels.

Before image classification, the temporal images must be radiometrically adjusted to normalize brightness variations induced by changing atmospheric conditions. Radiometric normalization was performed on the images using pseudo-invariant features (PIFs), which are spatially well-defined, spectrally stable, and radiometrically stable features. The image with the highest dynamic brightness value range, i.e., 2013, was chosen as a reference or base image for radiometric correction of other images from 1984, 2002, and 2022.

2.4. Image Classification and Accuracy Assessment

The main targeted LULC categories in the study site were built-up areas, cultivated lands (old farmlands and newly reclaimed farmlands), deserts, and water bodies. In terms of image classification, each image was classified individually before extracting primary LULC maps. Furthermore, image classification was performed employing supervised classification. For image classification, a decision tree classifier (DTC) approach was applied to integrate different remote sensing-derived indices. The DTC is defined as a machine-learning-based analysis technique comprising several classes of modeling algorithms using a tree structure, in which each node shows a test on attributes; the branch represents the test results, and the leaf node shows the target classes. In addition, the DTC is a multistage classifier that can be applied to a single image or a stack of images. It comprises a series of binary decisions to place pixels into classes. Each decision divides the pixels of the image into two categories based on the expression. Furthermore, the DTC is a hierarchical model composed of decision rules that recursively split independent variables into homogeneous zones. Therefore, we used it to determine the correct category for each pixel, with certain classes being separated during each step in the simplest manner possible (Figure 4). The DCT has been widely used for land cover classification, particularly with remote sensing technology approaches, and is often used to integrate multi-source data. In practice, the DTC concept is often translated using classification rules. Three stages are used to build the classification rules, including the following: (1) generating and perfecting knowledge and rules from experts; (2) extracting variables and rules using cognitive methods; and (3) automatically generating rules from observed data. As a result, the LULC in the study area is categorized into four LULC categories: urban areas; cultivated land; deserts; and water bodies, as defined in Table 1. The Normalized Difference Vegetation Index (NDVI), Modified Normalized Difference Water Index (MNDWI), Normalized Difference Built-Up Index (NDBI), and Dry Bare Soil Index (DBSI) were among the indices utilized. These indices were chosen according to the diverse land uses in the study area and were employed to improve the precision of the image classification process. The expressions of all indices utilized are shown in Table 4.

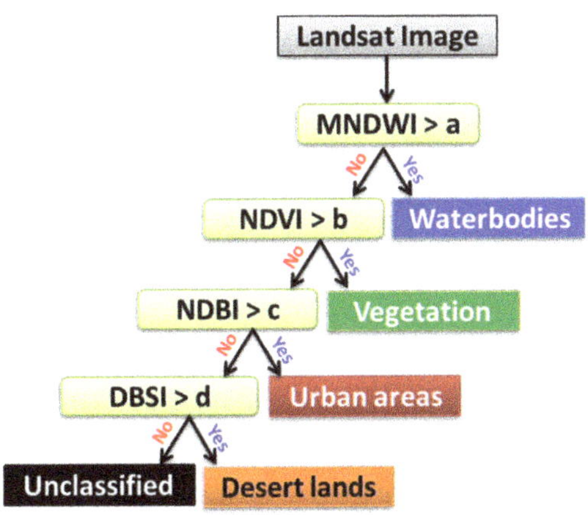

Figure 4. Decision tree for the LULC classification in the study area.

Table 4. The applied indices expressions.

The Index	Expression	Reference
NDVI	(NIR − R)/(NIR + R)	[73]
MNDWI	(G − SWIR)/(G + SWIR)	[74]
NDBI	(SWIR − NIR)/(SWIR + NIR)	[75]
DBSI	[(SWIR)/(SWIR + G)] − [(NIR − R)/(NIR + R)]	[76]

NIR = near-infrared. SWIR = short-wave infrared. R = red. G = green.

Accuracy evaluation processes decide the reliability of the spatial information generated from remote sensing images for accurate image classification [77,78]. Remotely sensed spatial information is both reliable and accurate when used in integration with ground control points that serve as a reference [79]. The preliminary geomorphologic units of the Sohag Governorate area were identified and verified through field observations after visual interpretation of satellite Landsat images. In this work, the accuracy of images from 1984, 2002, 2013, and 2022 was assessed using the visual interpretation of Landsat images [80]. A semi-detailed survey was completed in 2022 to obtain more detailed spatial information about the soil patterns, landforms, and landscape characteristics as ground truth data throughout the investigated area. Moreover, various statistical methods of the error matrix can be used for accuracy assessment [81]. Therefore, the accuracy of each categorized image was determined and evaluated by calculating the producer's accuracy, user's accuracy, overall accuracy, and Kappa coefficient values, as described in Equations (1)–(4) [79,82,83]. Furthermore, due to the nature of GIS and its integration with remote sensing (RS), the combination of GIS, RS data, and the Markov model was supported [53,84]. Therefore, thematic maps of different LULC for studied periods were created using ArcGIS 10.8 software.

$$A_p = p_{ii}/P_{+i} \tag{1}$$

$$A_u = p_{ii}/P_{i+} \tag{2}$$

$$A_O = \frac{\sum_{i=1}^{m} p_{ii}}{n} \times 100 \tag{3}$$

$$k = \frac{\mathrm{n}\sum_{i=1}^{m} \mathrm{P_{ii}} - \sum_{i=1}^{c} p_{i+}p_{+i}}{\mathrm{n}^2 - \sum_{i=1}^{m} p_{i+}p_{+i}} \tag{4}$$

where A_p is the accuracy of the producer; A_u is the accuracy user; A_o is the overall accuracy; K is the kappa coefficient; p_{ii} is the percentage of zones of categorized and reference categories, and P_{+i} = P_{i+} = $\sum_{i=1}^{m} p_{ii}$ is the percentage of zones of categorized categories and ground truth categories.

2.5. LULC Change Analysis Using CA-Markov Modeling

CA-Markov hybrid model was employed to simulate future LULC trends in this study. Using transition probability, the CA-Markov hybrid model, as a separate random process, forecasts the next scenario of LULC and all future scenarios depending on the current situation [85]. Furthermore, the CA-Markov analysis technique is a simple statistical tool that employs a transition probability matrix depending on the influences of the neighborhoods via a spatially influenced algorithm [86,87]. The CA-Markov hybrid model has been extensively and increasingly utilized in LULC prediction in recent years because of its ability to fit the complicated spatial nature [47,88–91]. However, the matrix of transition probability of each LULC class may be accurate, but the spatial distribution of the occurrences is unknown [46]. This is because the Markov chain model has a deficiency in projecting the spatial assignment and distribution of LULC categories [92]. Therefore, to fill this gap, the CA-Markov hybrid model was developed to enhance the precision of the change detection of different land-use categories by providing spatial dimension via cellular automata (CA) filter to the used model [93,94].

The cellular automaton (CA) model assumes that a grid cell's states are determined by the changes (dynamics) of the cell itself and its neighbors' grid cells [95,96]. Thus, a hybrid Markov–Cellular Automata model can be applied to detect the spatial and temporal changes in the Sohag Governorate's different land uses with a dynamic degree of estimation by setting a dataset of land uses at a specific time, then projecting the changing probabilities of these data for a future time [43,90]. Change detection in LULC changes, on the other hand, is accomplished by predicting changes in satellite image pixels from one land-use category at a time (t1) into the other at a time (t2). Equations (5)–(7) express the employed models [97].

$$P_{ij} = \begin{bmatrix} P_{11}\ P_{12} \ldots\ P_{1n} \\ P_{21}\ P_{22} \ldots\ P_{2n} \\ P_{n1}\ P_{n2} \ldots\ P_{nn} \end{bmatrix} \tag{5}$$

$$0 \le P_{ij} < 1\ and \sum_{j=1}^{n} P_{ij} = 1, i, j = 1, 2, \ldots, n \tag{6}$$

$$S_{t+1} = P_{ij} \times S_t \tag{7}$$

where P_{ij} is the transitional probability matrix of changing from a specific land use studied (n) to another, and its value ranges from 0 to 1; S_t is the current land-use status at time t, and S_{t+1} is the next land-use state at time t + 1. To obtain both the transition and probability matrix of LULC types, CA-Markov Chain analysis was performed using IDRISI 7.02 software for the 1984–2002, 2002–2013, and 2013–2022 time series.

2.6. Future LULC Prediction for 2030, 2040, and 2050

Markov Chain model, when combined with Cellular Automata, offers an excellent chance to forecast and simulates the spatial–temporal LULC changes. Furthermore, this integrated approach is also effective in modeling and forecasting the complex LULC categories [50,52,98]. Therefore, the CA-Markov hybrid model with two variables was employed in the process of LULC simulation and prediction. The two variables used are the discrete variable in time and space and the local variable, which they usually allocate to interactions [99]. Grid sizes, cell neighborhoods, cell spaces, time phases, and transition rules are the core components of the CA-Markov hybrid model [100,101]. Cells are objects in any dimensional space that are adjacent or close to one another. Each cell can only be in one of the states that define the system's attributes at a time. The state of any cell is determined by the states of other cells in its neighborhood, defined as the immediately adjacent set of cells that are "next" to the cell in question. Finally, transition rules drive changes in the state of each cell as a function of what exists or is happening in the cell's vicinity. Employing the inter-period change–transition probability matrix, the CA-Markov hybrid model integrates the idea of projecting the next time state depending on the previous period states [102,103]. Furthermore, this model can determine neighbors, and the greater the weight factor, the closer the distance between cells and their neighbors [43,90]. Therefore, the weight factor and transition probability can forecast the states of neighbor grid cells [44]. The land-use changes were calculated using Equation (8).

$$Si = \frac{LU_{(i,\ t2)} - LUA_i}{LU_{(i,\ t1)}} \times \frac{1}{t_2 - t_1} \times 100\% \tag{8}$$

where Si is land-use change; $LU_{(i,t1)}$ is land-use change at an earlier period; $LU_{(i,t2)}$ is land-use changed area at a later period, and LUA_i is the area with no change.

Future LULC trends for future dates studied were projected based on the historical patterns (from 1984 to 2022) in the study area. CA-Markov was employed in IDRISI 7.02 software to forecast the future LULC trends in 2030, 2040, and 2050. In the present study, LULC images of the Sohag Governorate area over two-time series in 1984 as earlier land-use (t1) and 2022 as later land-use (t2) were utilized, as described by Hyandye and Martz [50]. A standard contiguity filter (5 × 5 pixels) was applied to the images to identify the neighborhood cells of each land-use class. To achieve a good influence, each cell was surrounded by a matrix scope of 5 × 5 cells. This spatial filter occurs near an existing category and rules out changes in land use randomly [104]. Furthermore, the pixels close to the existent LULC class are more suitable than those far away [105]. In the current study, neighboring pixels were utilized to generate spatially neighboring weights to project LULC categories for 2030, 2040, and 2050. The contiguity filter used for LULC change analysis is shown below (Equation (9)).

$$\text{Contiguityfilter } 5 \times 5 = \begin{bmatrix} 0\,0\,1\,0\,0 \\ 0\,1\,1\,1\,0 \\ 1\,1\,1\,1\,1 \\ 0\,1\,1\,1\,0 \\ 0\,0\,1\,0\,0 \end{bmatrix} \tag{9}$$

The methodology of analyzing, simulating, and projecting LULCs employing the CA-Markov hybrid model is summarized into four main steps, which are as follows:

1. The Markov chain model was used to compute transition probabilities matrices for 1984, 2002, 2013, and 2022;
2. The computed transition matrices were used to generate a set of conditional probability data for the different land uses from 1984 to 2022;
3. The transition probabilities matrices of 1984–2002, 2002–2013, and 2013–2022 for each LULC category, as well as conditional probability data and LULC, classified maps of 2013 and 2020, were integrated using the CA-Markov spatial operator in IDRISI 7.02 software, which is based on Markov chain analysis and multi-criteria evaluation (MCE), to simulate the LULC maps of 2030, 2040, and 2050;
4. Predicted LULC maps for future dates were produced by overlapping the results obtained in the previous steps.

2.7. CA-Markov Model Validation

According to Eastman [106], the utility of any predictive change model is dependent on the outcome of the validation process. The CA-Markov model outputs had to be validated before modeling the future LULC trends for 2030, 2040, and 2050. Therefore, the CA-Markov model was validated using the validation module in IDRISI 7.02 software to compare the degree of agreement between the predicted and the classified maps. The validation process was performed by comparing the LULC predicted results (projected maps) for 2013 and 2022 to their corresponding observed datasets (classified maps), which were used as LULC reference data. Furthermore, Kappa coefficients were calculated and utilized to evaluate the performance precision of the applied model in the LULC projection maps. The current study applied the statistic Kappa indices, including the Kappa for no information (K_{no}), Kappa for location ($K_{location}$), Kappa for location strata ($K_{location\ strata}$), and Kappa standard ($K_{standard}$) [51,104]. Kappa statistical indicator can distinguish between quantity errors and location errors between two qualitative maps [52,107] and is computed employing Equation (10) [108].

$$Kappa = \frac{P_0 - P_C}{1 - P_C} \tag{10}$$

where P_0 is the portion of cells correctly classified, and P_C is the hypothetical probability of opportunity agreement between the actual LULC of 2013 and 2022 maps (Landsat images classification) and the projected LULC of 2013 and 2022 maps.

The employment of Kappa indices in the computation determines the overall accomplishment rate and provides insight into the true causes of the strength or weakness of the outcomes. The Kappa values were categorized according to Aliani et al. [93], as illustrated in Table 5. Furthermore, Figure 5 demonstrates the methodology for LULC analysis and LULC projection employing the CA-Markov hybrid model.

Table 5. Kappa coefficient values and agreement degrees.

Kappa Coefficient Value	Agreement Degree
<0.2	Weak
0.21–0.4	Acceptable
0.41–0.6	Moderate
0.61–0.8	Good
0.81–1.0	Very good

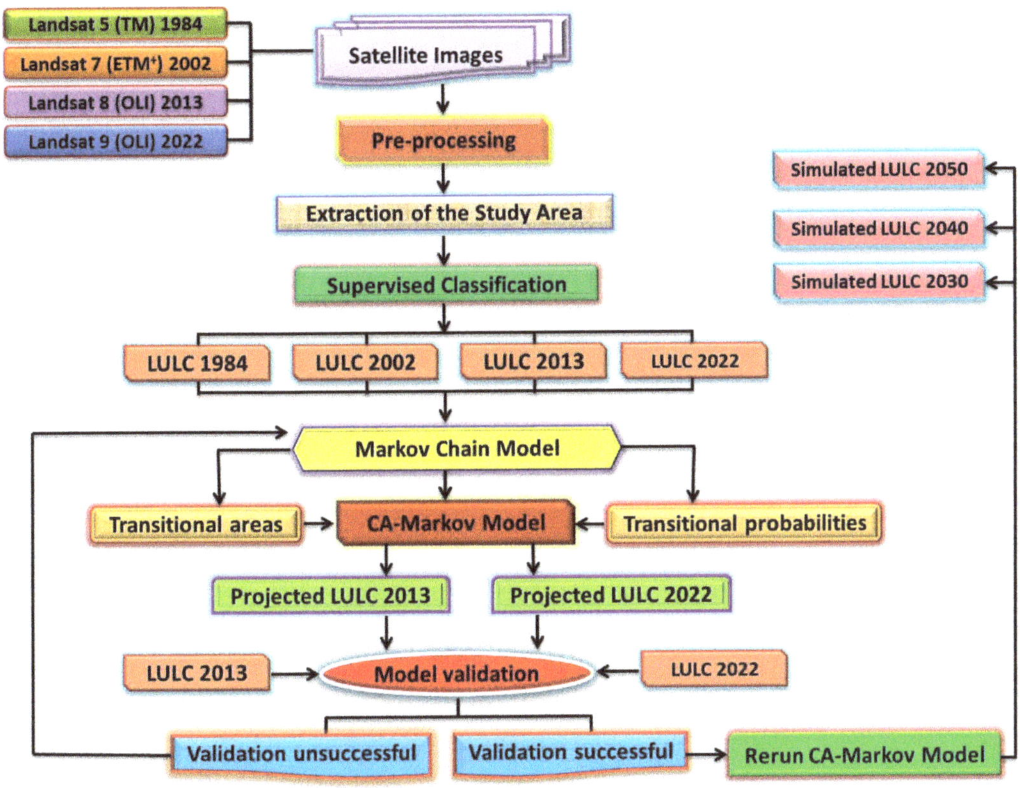

Figure 5. Simplified flowchart of LULC simulation methodology.

3. Results and Discussion

3.1. Accuracy Assessment of the LULC Classification

The obtained results showed that the overall accuracy and kappa statistics of all different land-use types in the Sohag Governorate area were acceptable for the different periods studied in 1984, 2002, 2013, and 2022 (Table 6). In the current study, the classification accuracy met the criterion that there must have been at least 80% accuracy for the sensor data [109]. The overall accuracy of the LULC classification ranged from 87.5% to 95.5% (Table 6). According to Anderson et al. [82], these percentages represent high accuracy; thus, the LULC maps can be utilized with reliability for LULC change dynamics analysis and forecasting. Furthermore, the kappa coefficient values indicated that the strength of agreement ranged from good to very good according to criteria adopted by Aliani et al. [93], as the Kappa coefficient values ranged from 0.71 to 0.95 (Table 6). Furthermore, Kappa values were described as a considerable agreement to near perfect agreement [98]. The assessment of overall accuracy and Kappa coefficient (Table 6) demonstrated the high capacity of the decision tree classifier (DTC) approach in integrating with different remote sensing-derived indices in producing multi-temporal LULC maps. Thus, this technique's findings are supposed to be compatible and reliable for modeling future LULC scenarios [27,110–112]. As a consequence, the categorized images are valid for examining and forecasting the changing dynamics of LULCs in the study area [50].

Table 6. Accuracy assessment of classified images for 1984, 2002, 2013, and 2022.

LULC Class	Landsat TM 1984		Landsat ETM^{+} 2002		Landsat ETM^{+} 2013		Landsat OLI 2022	
	PA	UA	PA	UA	PA	UA	PA	UA
	(%)							
Water bodies	97.1	100.0	99.4	98.9	98.7	96.3	82.2	97.4
Desert lands	99.7	91.1	97.8	99.2	96.4	92.5	88.6	45.9
Cultivated lands	72.2	91.2	91.7	86.4	86.1	95.5	94.6	92.3
Urban Areas	82.5	68.6	96.9	57.3	96.7	71.8	95.8	40.4
Overall accuracy	91.0		95.5		92.1		87.5	
Kappa coefficient	0.71		0.94		0.84		0.79	

PA = producer accuracy. UA = user accuracy.

3.2. The LULC Classification

According to the results of the maximum likelihood algorithm of Landsat images supervised classification, there were four LULC categories recognized in the study region: urban areas (residential, commercial, industrial, and road constructions); cultivated lands (old cultivated lands, and newly reclaimed lands); deserts (bare lands, sand sheets, and rocky land); and water bodies (the Nile River, canals, drainage patterns, and wastewater treatment plants). These findings are in line with earlier studies [113–121], which found the same LULC categories in other areas of Egypt, such as the Northwestern Coast, Nile Delta, and others. Table 7 displays the measurable data of the four LULC categories over the different time nodes. Furthermore, Figure 6 illustrates the spatial distribution of LULC types (classified maps) derived from satellite images of the Sohag Governorate area for the examined dates of 1984, 2002, 2013, and 2022. On the created maps, the different LULC categories were tagged with different colors.

The LULC class percentages of the total study area for 1984, 2002, 2013, and 2022 showed that deserts and cultivated lands were consistently the most extensive, followed by built-up and water bodies, which were far less extensive (Table 7). Similarly, previous investigations have reported that the LULC classes in the investigated areas of Egypt are dominated by deserts and cultivated regions [113,116,117,119]. Thus, in arid land regions, these LULC categories are dominant. Furthermore, throughout the 38 years from 1984 to 2022, the area of each LULC category in the Sohag Governorate changed significantly. According to the current study, cultivated lands and desert areas have been predominant from 1984 until 2022. According to Table 7, deserts and cultivated land constituted the majority of LULCs in the study region, accounting for 1398.9 km^2 (46.8%) and 1361.0 km^2 (45.5%) of the total area in 1984, respectively, and 703.8 km^2 (23.6%) and 1814.5 km^2 (60.7%) in 2022 (Figure 6). Moreover, urban and water bodies represented the lowest proportion of LULCs, as shown in Table 7. Also, urban areas more than doubled from 165 km^2 (5.5%) in 1984 to 369 km^2 (12.3%) in 2022 (Table 7).

Table 7. LULC classification for 1984, 2002, 2013, and 2022.

Land Use	1984		2002		2013		2022	
	Area							
	km^2	%	km^2	%	km^2	%	km^2	%
Water body	64.2	2.2	60.6	2.0	68.6	2.3	101.8	3.4
Desert lands	1398.9	46.8	1181.8	39.5	861.3	28.8	703.8	23.6
Cultivated lands	1361.0	45.5	1514.0	50.7	1754.7	58.7	1814.5	60.7
Urban	165.0	5.5	232.8	7.8	304.5	10.2	369.0	12.3
Total	2989.1	100.0	2989.1	100.0	2989.1	100.0	2989.1	100.0

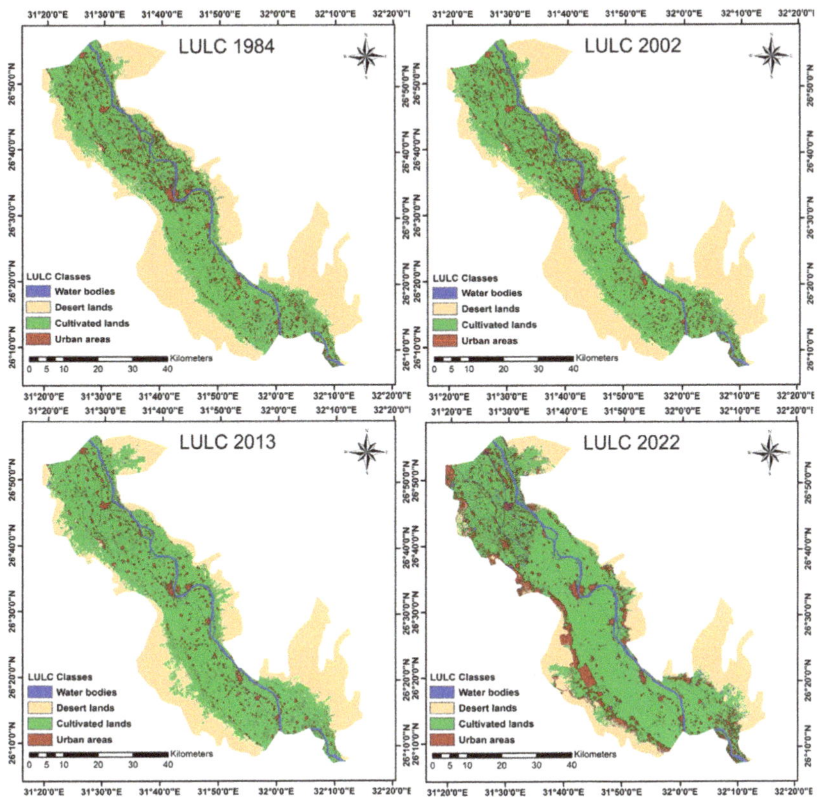

Figure 6. Classified maps of LULCs for 1984, 2002, 2013, and 2022.

3.3. LULC Change Dynamics

The results indicated that the built-up areas had a net gain of 67.74 km^2 between 1984 and 2002 due to the construction of new housing as the population increased (Figure 7). Similarly, cultivated lands have gained 152.99 km^2. Furthermore, during this period (1984–2002), many desert land areas were subjected to reclamation processes. As a result, cultivated lands increased while desert lands decreased. This is obvious in Figure 7, which showed a decrease in desert lands with a loss of 217.13 km^2. Concerning water bodies, some islands in the study area had vanished completely, while new islands had appeared in other locations. Furthermore, depositional processes cause the formation of new floodplains on the river's convex sides, as well as new islands and sand bars. As a result, between 1984 and 2002, the Nile River lost 3.6 km^2.

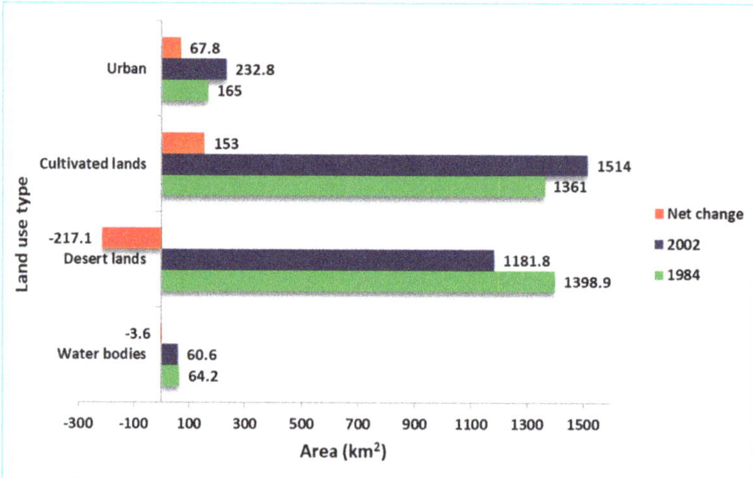

Figure 7. The changes in LULC categories between 1984 and 2002 (km^2).

During the time series from 2002 to 2013, water bodies in the study area increased by 8.04 km^2 (Figure 8). This could be attributed to the construction of numerous wastewater treatment plants. These treatment units are located on the outskirts of desert areas outside the Nile Valley. Furthermore, this treated wastewater is destined to irrigate crops, forest trees, decorative plants, and landscapes of greenbelts and is not intended for food. Moreover, soil reclamation activities increased dramatically during this period, consequently increasing cultivated lands by 240.73 km^2. Therefore, the desert lands had a net loss of 320.5 km^2 in the same period (2002–2013). Additionally, during this period, a substantial increase in built-up settlements occurred at the cost of the most productive land in the investigation area (Figure 8). According to Figure 8, there was a high sprawl rate of built-up areas on cultivated lands, which is one of the main problems that threaten the limited fertile lands in the Sohag Governorate. During this period (2002–2013), urban settlements expanded by 71.73 km^2. Figure 8 depicts the changes in LULC categories in the study area from 2002 to 2013.

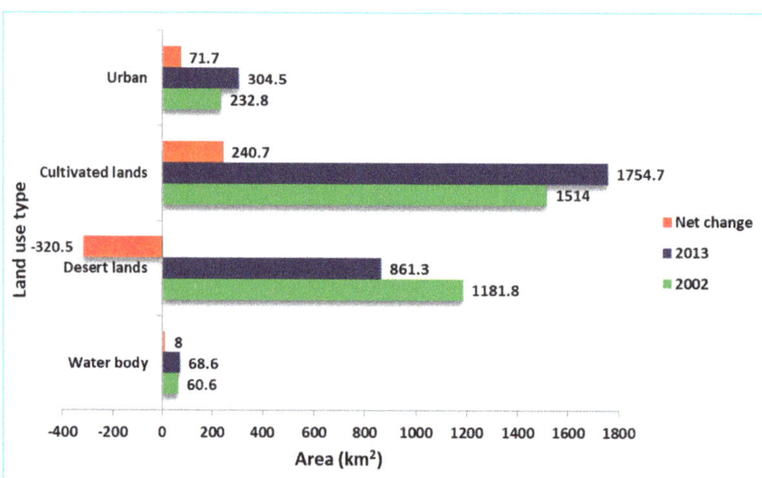

Figure 8. The changes in LULC categories between 2002 and 2013 (km^2).

During the third investigated period (2013–2022), water bodies in the study area increased by 33.21 km^2, as shown in Figure 9. At this period, the population of the Sohag Governorate was estimated to be around 5,193,052, accounting for 5.2 percent of the total population of Egypt. This could result in increased water consumption and, as a result, increased wastewater pumped to treatment plants. Both soil reclamation activities and urbanization increased during this period but to a lesser extent than in previous periods. The cultivated lands and urban areas recorded an increase of 59.73 km^2 and 64.52 km^2, respectively. In addition, the desert lands experienced a net loss of 157.47 km^2 (Figure 9).

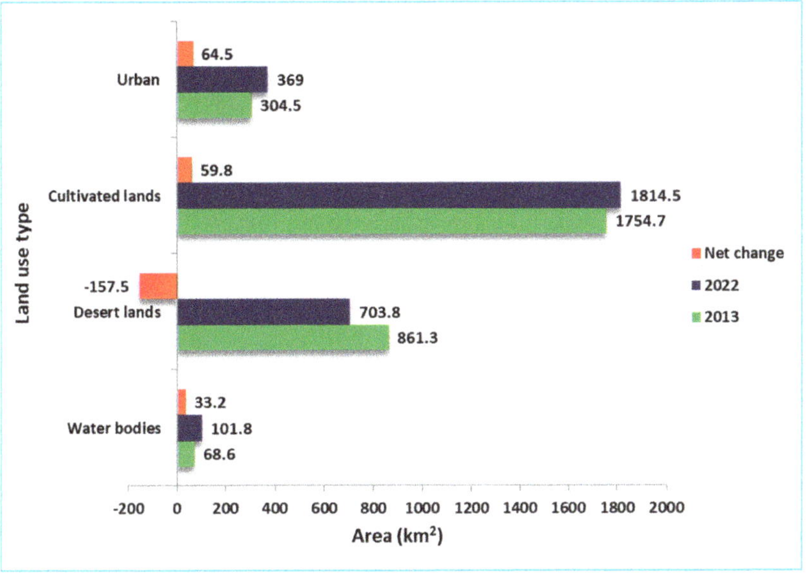

Figure 9. The changes in LULC categories between 2013 and 2022 (km^2).

Overall, the findings of this study revealed a continuous increase in urban sprawl and expansion of newly reclaimed cultivated lands, with a reduction in desert areas over the time series analyzed. These findings are in agreement with previous studies, which reported similar LULC change trends [113–121].

3.4. Markov Chain Model Analysis

The data displaying the decreases and increases in transition probabilities over time series from 1984 to 2022 are shown in Tables 8–10. The gains were acquired by subtracting the persistence from the entire column for each group, while the losses were received by deducting the persistence from the whole row [122]. For instance, the probability of remaining built-up areas as built-up areas from 1984 to 2002 is 69.75%, while the future change possibility of cultivated lands to urban areas is 20.36%, and so on for the rest of the LULC types. Desert lands and water bodies had high probabilities of 76.33 and 76.49%, respectively, as they did in 2002.

Table 8. Transition probability matrix derived from LULC-classified categories in Sohag Governorate from 1984 to 2002.

	Probability of Changing from 1984 to 2002				Subtotals	
	Water Bodies	**Desert Lands**	**Cultivated Lands**	**Urban Areas**	**Total**	**Loss**
Water bodies	0.7649	0.0005	0.1626	0.072	1	0.2351
Desert lands	0.0009	0.7633	0.2277	0.0081	1	0.2367
Cultivated lands	0.0043	0.0079	0.7842	0.2036	1	0.2158
Urban Areas	0.0356	0.2528	0.0141	0.6975	1	0.3025
Total	0.8057	1.0245	1.1886	0.9812	4	
Gain	0.0408	0.2612	0.4044	0.2837		

Table 9. Transition probability matrix derived from LULC-classified categories in Sohag Governorate from 2002 to 2013.

	Probability of Changing from 2002 to 2013				Subtotals	
	Water Bodies	**Desert Lands**	**Cultivated Lands**	**Urban Areas**	**Total**	**Loss**
Water bodies	0.8135	0.0016	0.172	0.0129	1	0.1865
Desert lands	0.001	0.7423	0.2467	0.01	1	0.2577
Cultivated lands	0.0146	0.0523	0.8071	0.126	1	0.1929
Urban Areas	0.0145	0.0125	0.5095	0.4635	1	0.5365
Total	0.8436	0.8087	1.7353	0.6124	4	
Gain	0.0301	0.0664	0.9282	0.1489		

Table 10. Transition probability matrix derived from LULC-classified categories in Sohag Governorate from 2013 to 2022.

	Probability of Changing from 2013 to 2022				Subtotals	
	Water Bodies	**Desert Lands**	**Cultivated Lands**	**Urban Areas**	**Total**	**Loss**
Water bodies	0.8237	0.0030	0.1642	0.0092	1	0.1763
Desert lands	0.0052	0.5909	0.2003	0.2036	1	0.4091
Cultivated lands	0.0362	0.0181	0.7820	0.1638	1	0.218
Urban Areas	0.0306	0.0289	0.4610	0.4796	1	0.5205
Total	0.8957	0.6409	1.6075	0.8559	4	
Gain	0.072	0.05	0.8255	0.3764		

For further understanding of the dynamic changes in LULC types in the Sohag Governorate, the annual change in these types was calculated for the three studied periods (1984–2002, 2002–2013, and 2013–2022), as shown in Table 11. According to the data in Table 11, cultivated lands increased annually by 8.5 km^2, 21.88 km^2, and 6.64 km^2 for the three studied periods, respectively. The greatest annual increase in cultivated land was recorded in the second period (2002–2013), during which increasing desert land reclamation activities were noticed. Furthermore, over the three time periods, the desert lands shrank by 12.06 km^2, 29.14 km^2, and 17.50 km^2 per year, respectively (Table 11). Moreover, the urban zones expanded at an annual rate of 3.76 km^2, 6.52 km^2, and 7.17 km^2 for the three studied periods, respectively. Water bodies decreased by 0.2 km^2 per year from 1984 to 2002, then increased by 0.73 km^2 in the second period, and the highest increase (3.69 km^2) occurred between 2013 and 2022 due to the construction and startup of wastewater treatment units in the study area (Table 11).

Table 11. Annual changes in LULC classes of Sohag Governorate.

Land Use	Annual Change km²		
	1984–2002	2002–2013	2013–2022
Water bodies	−0.20	0.73	3.69
Desert lands	−12.06	−29.14	−17.50
Cultivated lands	8.50	21.88	6.64
Urban Areas	3.76	6.52	7.17

The results of the LULC class changes for different LULCs in the Sohag Governorate over the studied periods are displayed in Table 12. Furthermore, Figure 10 demonstrates the temporal distribution of LULC class changes in the Sohag Governorate in square kilometers over the years studied. The findings indicate that urban areas increased during the studied periods, and there was also a noticeable reduction in desert lands. The increase in urbanized areas can be attributed to population growth. The growth in urban areas (urbanization), which is associated with population increases, infrastructure development, and growing domestic product, is one of the primary drivers of LULC changes [24,123–127]. Moreover, cultivated lands, water bodies, and urban areas increased during the studied periods (Figure 10).

Table 12. Temporal distribution in km² of LULC distribution.

Land Use	1984		2002		Change from 1984 to 2002	2013		Change from 2002 to 2013	2022		Change from 2013 to 2022
	Areas										
	km²	%	km²	%	km²	km²	%	km²	km²	%	km²
WB	64.2	2.2	60.6	2.0	−3.6	68.6	2.3	8.04	101.8	3.4	33.21
DL	1398.9	46.8	1181.8	39.5	−217.13	861.3	28.8	−320.5	703.8	23.6	−157.47
CL	1361.0	45.5	1514.0	50.7	152.99	1754.7	58.7	240.73	1814.5	60.7	59.73
Urban	165.0	5.5	232.8	7.8	67.74	304.5	10.2	71.73	369.0	12.3	64.52
Total	2989.1	100.0	2989.1	100.0		2989.1	100.0		2989.1	100.0	

WB = water bodies. DL = desert lands. CL = cultivated lands.

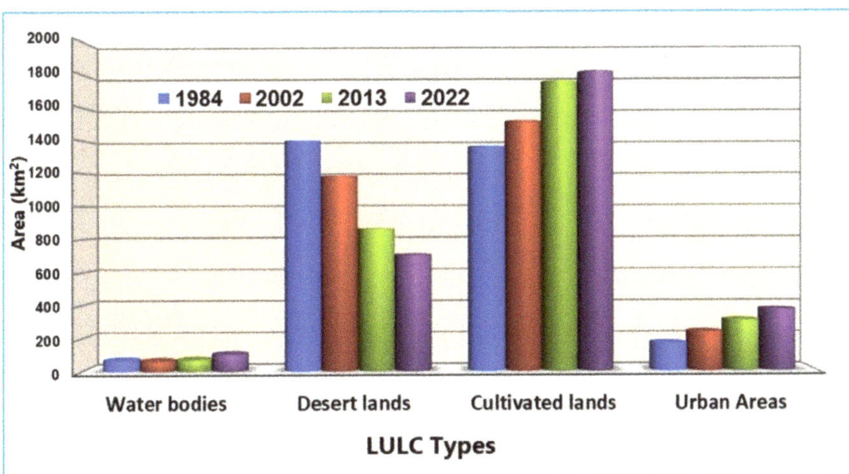

Figure 10. Temporal distribution of LULC change distribution in 1984, 2002, 2013, and 2022 (km²).

3.5. CA-Markov Model Validation for Predicting Future LULC Scenarios

The Kappa coefficients for the quantity and location of correct cells were derived based on a comparison of the projected and classified LULC maps at a resolution of 30 m × 30 m for model validation employing the IDRISI program. According to k-indicator statistics for the LULCs in 2013, the $K_{standard}$, K_{no}, $K_{location}$, and $K_{location\text{-}strata}$ were 0.8703, 0.9142, 0.9336, and 0.9336, respectively, as shown in Table 13. Similarly, for 2022, k-indicator statistics revealed that the $K_{standard}$, K_{no}, $K_{location}$, and $K_{location\text{-}strata}$ were 0.8402, 0.8942, 0.9012, and 0.9012, respectively (Table 13). These findings demonstrated high agreement between the predicted and observed LULC maps for 2013 and 2022. This revealed that there were minor quantification and location errors between the projected and actual maps [104,128–130]. Therefore, employing the CA-Markov model can precisely specify the quantity and location of the LULC changes [100,118,128,131]. Furthermore, based on the satisfactory agreement between the predicted and observed LULC maps, the data in Table 13 and Figure 11 established that the accuracy appraisal of categorized data for 2013 and 2022 is satisfactory and reasonable for simulation and prediction applications [132]. As a result, the validated CA-Markov model is nearly perfect for simulating and projecting future LULC trends in arid land regions. These findings are in agreement with the results obtained from prior investigations [27,110,113,114,133,134].

Table 13. Kappa indices for 2013 and 2022.

K-Index	2013	2022
$K_{standard}$	0.8703	0.8402
K_{no}	0.9142	0.8942
$K_{location}$	0.9336	0.9012
$K_{location\ strata}$	0.9336	0.9012

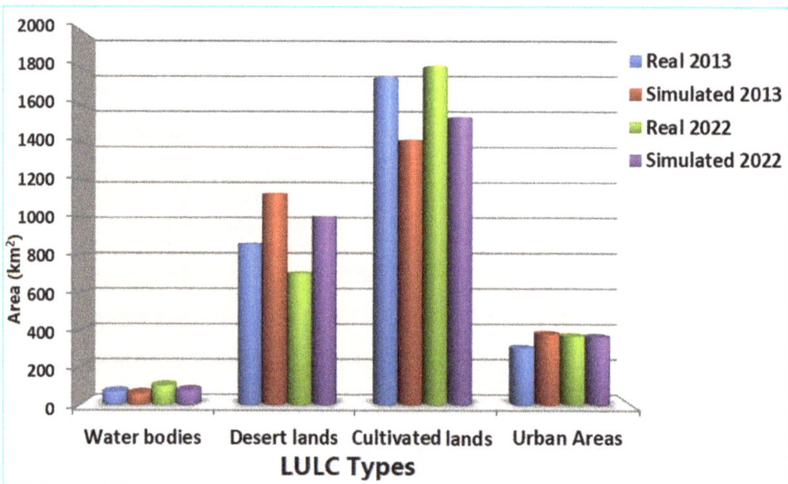

Figure 11. The comparison between simulated and actual LULCs in km^2 in 2013 and 2022.

3.6. Prediction of the Future LULC Scenarios

CA-Markov hybrid model was used to simulate and project future LULC trends for 2030, 2040, and 2050. The future scenarios of LULC classes for 2030, 2040, and 2050 are illustrated in Figure 12 as thematic maps of the predicted LULC spatial distributions. Furthermore, Table 14 displays the percentages and the areas in km^2 of the predicted LULC categories for 2030, 2040, and 2050. During all of the predicted periods, there would be increases in water bodies, cultivated lands, and urban areas. Desert lands, on the other hand, would decline in the study area during the same periods.

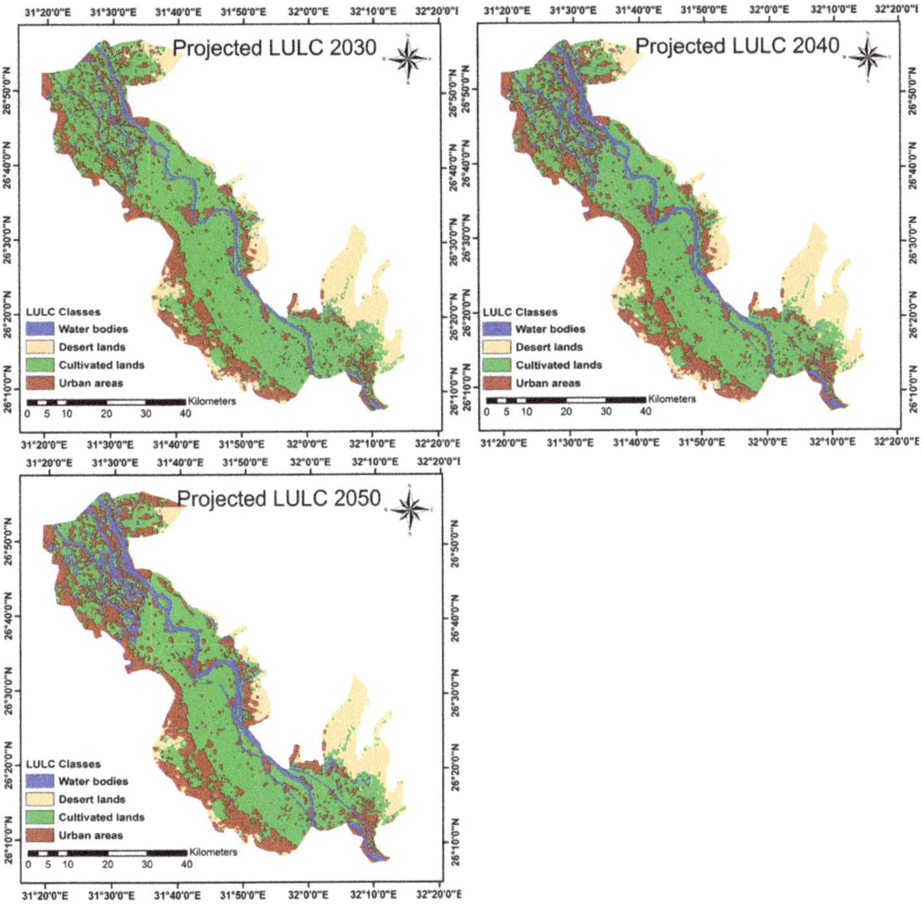

Figure 12. Spatial distribution maps of predicted LULCs for 2030, 2040, and 2050.

Table 14. The future predicted changes in areas of LULC classes in km^2.

Land Use	Projected 2030		Change from 2022 to 2030	Projected 2040		Change from 2030 to 2040	Projected 2050		Change from 2040 to 2050
	Areas								
	km^2	%	km^2	km^2	%	km^2	km^2	%	km^2
WB	164.44	5.50	62.64	194.1	6.49	29.66	200.8	6.72	6.7
DL	459.71	15.38	−244.09	312.4	10.45	−147.31	245.4	8.21	−67
CL	1746.63	58.43	−67.87	1808.3	60.50	61.67	1851.5	61.94	43.2
Urban	618.35	20.69	249.35	674.3	22.56	55.95	691.4	23.13	17.1
Total	2989.13	100	-	2989.1	100	-	2989.1	100	-

WB = water bodies. DL = desert lands. CL = cultivated lands.

The modeling results showed that built-up areas would increase significantly compared to their zones in the past and present, expanding from 165 km^2 (5.5%) in 1984 to 369 km^2 (12.3%) in 2022 to 618 km^2 (20.7%) in 2030 and 691 km^2 (23.1%) in 2050, as shown in Table 14. This predicted increase in urban areas might be related to the existing expansion in the built-up areas due to population growth [27,49,124,125,135]. Furthermore, cultivated lands would decrease slightly in 2030 by 67.9 km^2 and then increase slightly to reach 1851 km^2 (61.9%) in 2050 (Table 14). The predicted expansion in cultivated lands may be associated with the expansion of newly reclaimed soils due to ongoing land reclamation projects on arable lands. Moreover, desert land would continue to decline considerably from 703.8 km^2 (23.6%) in 2022 to 459.7 km^2 (15.4%), 312.4 km^2 (10.5%), and 245.4 km^2 (8.2%) in 2030, 2040, and 2050, respectively, as displayed in Table 14. Desert lands would be reduced due to urban sprawl and cultivated land expansion outside the Nile Valley through the reclamation of arable land in desert areas. Also, according to modeling results, the water bodies would increase slightly and gradually to reach 6.7% of the entire area of the study region in 2050.

Results of the simulation indicated that future scenarios represented by urban expansion growing, cultivated lands and water bodies increasing, and desert lands declining followed the past and current trends of LULC in the study region. This may be because the CA-Markov approach does not contain the variables that could influence or prevent the event of LULC changes. Thus, the resultant LULC trend updates in the coming years are expected to be linear. Projected maps can be identical to LULCs in 2030, 2040, and 2050 if the rate of the current changes is the same. However, because the current study was limited in its use of LULC change drivers and future LULC trends are unsure, additional factors should be considered to enrich our knowledge of LULC change trends [136]. Therefore, future research studies are required to assess the accuracy of these predicted scenarios and identify the variables that cause the LULC changes. Furthermore, this study's findings indicate that the Egyptian government should focus on LULC changes, specifically urban sprawl on old fertile farmlands, to ensure sustainable development in the study area.

Sustained farming should use resources in such a manner that they can renew their productivity potential while minimizing detrimental effects on ecosystems [137]. In this regard, FAO [4] has indicated that rapid population growth is causing increased land degradation to meet food demands. Overall, the current study provided a future vision of LULC trends that would be beneficial in land use management and offered a comprehensive understanding of ecosystem functions for establishing land use sustainability [138].

4. Conclusions

The current study was carried out to assess the performance of the CA-Markov hybrid model in predicting and modeling future LULC trends in arid regions using conventional remote sensing data. Therefore, this study focused on detecting and analyzing LULC change in the Sohag Governorate region over the different periods examined (1984, 2002, 2013, and 2022), as well as projecting and modeling future LULC trends for 2030, 2040, and 2050. The findings of the current study provided evidence that the CA-Markov hybrid model is an effective model for predicting and modeling the future LULC trends for 2030, 2040, and 2050 in the Sohag Governorate as an arid region using conventional remotely sensed data. Furthermore, results proved that the cellular automata filter was critical for improving the spatial distribution of various LULC classes in the study area. Concerning the accuracy assessment of the used model, an accuracy of greater than 70% was obtained in all stages. According to this study, CA-Markov modeling has delivered promisingly precise and reliable outcomes.

Moreover, the increased population and industrial and commercial activities in the Sohag Governorate are causing considerable urban sprawl. Although the newly reclaimed agricultural lands with moderate agricultural production potential are increasing, their productivity is insufficient to maintain sustainable food security, as the old farming lands with high production potential are dwindling. This investigation has shown that changing land use and land cover is a pervasive, rapid, and significant trend in Sohag Governorate. The results revealed that cultivated lands, built-up areas, and water bodies expanded from 1984 to 2022 at the expense of decreasing deserts. Furthermore, the most considerable increase in the area of cultivated lands and built-up occurred between 2002 and 2013, while the most significant decline in desert lands happened over the same period. The predicted 2030, 2040, and 2050 LULC results also presented that the trend from historical to future LULC change will be extended to be ongoing in the future.

Furthermore, this work exhibited the flexibility of integrating remote sensing, GIS, and CA-Markov modeling that can be utilized as an effective technique for mapping and tracking the LULC changes. Moreover, the change detection and prediction modeling of future LULC trends achieved in this work has some limitations. These limitations stem from the adoption of the CA-Markov approach, which does not contain the variables or factors that could influence or hinder the event of the LULC change in the study area. Therefore, future research studies are required to assess the accuracy of these expected scenarios and identify the variables that cause the LULC changes. Furthermore, future research is needed to evaluate the performance of the CA-Markov hybrid model with diverse land use classes in various arid regions. The Egyptian government should work hard to prevent the loss of productive old lands due to urban sprawl and to reclaim more unused arable land.

Author Contributions: Conceptualization, S.A.H.S., D.E.K., G.M., A.R.A.M., R.J.-B., O.D.K., M.E.F. and A.-r.A.M.; methodology, S.A.H.S., A.R.A.M., O.D.K., M.E.F. and A.-r.A.M.; software, S.A.H.S., D.E.K., A.R.A.M., O.D.K., M.E.F. and A.-r.A.M.; validation, S.A.H.S., D.E.K., G.M., A.R.A.M., R.J.-B., O.D.K., M.E.F. and A.-r.A.M.; formal analysis, S.A.H.S., A.R.A.M., M.E.F. and A.-r.A.M.; investigation, S.A.H.S., D.E.K., G.M, A.R.A.M., O.D.K., M.E.F. and A.-r.A.M.; resources, S.A.H.S., D.E.K., G.M., A.R.A.M., R.J.-B., O.D.K., M.E.F. and A.-r.A.M.; data curation, S.A.H.S., D.E.K., G.M., A.R.A.M., O.D.K., M.E.F. and A.-r.A.M.; writing—original draft preparation, S.A.H.S., A.R.A.M., O.D.K., M.E.F. and A.-r.A.M.; writing—review and editing, S.A.H.S., D.E.K., G.M., R.J.-B. and A.-r.A.M.; visualization, S.A.H.S., D.E.K., G.M., A.R.A.M., R.J.-B., O.D.K., M.E.F. and A.-r.A.M.; supervision, S.A.H.S., D.E.K. and A.-r.A.M.; project administration, S.A.H.S. and A.-r.A.M.; funding acquisition, S.A.H.S., D.E.K., A.R.A.M., G.M., R.J.-B., O.D.K., M.E.F. and A.-r.A.M. All authors have read and agreed to the published version of this manuscript.

Funding: This research received no external funding.

Data Availability Statement: Data are contained within the article.

Conflicts of Interest: The authors declare no conflict of interest.

References

1. Biro, K.; Pradhan, B.; Buchroithner, M.; Makeschin, F. Land use/land cover change analysis and its impact on soil properties in the northern part of Gadarif region, Sudan. *Land Degrad. Dev.* **2013**, *24*, 90–102. [CrossRef]
2. Ayele, A.; Tarekegn, K. The impact of urbanization expansion on agricultural land in Ethiopia: A review. *Environ. Socio-Econ. Stud.* **2020**, *8*, 73–80. [CrossRef]
3. Abdulaziz, A.M.; Hurtado, J.M.; Al-Douri, R. Application of multitemporal Landsat data to monitor land cover changes in the Eastern Nile Delta region, Egypt. *Int. J. Remote Sens.* **2009**, *30*, 2977–2996. [CrossRef]
4. FAO. *The Future of Food and Agriculture: Trends and Challenges*; FAO: Rome, Italy, 2017; ISBN 978-9-25109-551-5.
5. Calicioglu, O.; Flammini, A.; Bracco, S.; Bellù, L.; Sims, R. The Future Challenges of Food and Agriculture: An Integrated Analysis of Trends and Solutions. *Sustainability* **2019**, *11*, 222. [CrossRef]
6. Adepoju, M.O.; Millington, A.C.; Tansey, K.T. Land Use/Land Cover Change Detection in Metropolitan Lagos (Nigeria): 1984–2000. In Proceedings of the AASPRS 2006 Annual Conference, Reno, NV, USA, 1–5 May 2006.
7. Zabihi, M.; Moradi, H.; Gholamalifard, M.; Khaledi Darvishan, A.; Fürst, C. Landscape Management through Change Processes Monitoring in Iran. *Sustainability* **2020**, *12*, 1753. [CrossRef]
8. Dadashpoor, H.; Azizi, P.; Moghadasi, M. Analyzing spatial patterns, driving forces and predicting future growth scenarios for supporting sustainable urban growth: Evidence from Tabriz metropolitan area, Iran. *Sustain. Cities Soc.* **2019**, *47*, 101502. [CrossRef]
9. Dewan, A.M.; Yamaguchi, Y.; Ziaur Rahman, M. Dynamics of land use/cover changes and the analysis of landscape fragmentation in Dhaka Metropolitan, Bangladesh. *GeoJournal* **2012**, *77*, 315–330. [CrossRef]
10. Liu, J.; Zhang, Z.; Xu, X.; Kuang, W.; Zhou, W.; Zhang, S.; Li, R.; Yan, C.; Yu, D.; Wu, S.; et al. Spatial patterns and driving forces of land use change in China during the early 21st century. *J. Geogr. Sci.* **2010**, *20*, 483–494. [CrossRef]
11. Shi, G.; Jiang, N.; Yao, L. Land Use and Cover Change during the Rapid Economic Growth Period from 1990 to 2010: A Case Study of Shanghai. *Sustainability* **2018**, *10*, 426. [CrossRef]
12. Cihlar, J. Land cover mapping of large areas from satellites: Status and research priorities. *Int. J. Remote Sens.* **2000**, *21*, 1093–1114. [CrossRef]
13. Wang, S.Q.; Zheng, X.; Zang, X. Accuracy assessments of land use change simulation based on Markov-cellular automata model. *Procedia Environ. Sci.* **2012**, *13*, 1238–1245. [CrossRef]

14. Yan, W.Y.; Shaker, A.; El-Ashmawy, N. Urban land cover classification using airborne LiDAR data: A review. *Remote Sens. Environ.* **2015**, *158*, 295–310. [CrossRef]
15. Karki, S.; Thandar, A.M.; Uddin, K.; Tun, S.; Aye, W.M.; Aryal, K.; Kandel, P.; Chettri, N. Impact of land use land cover change on ecosystem services: A comparative analysis on observed data and people's perception in Inle Lake, Myanmar. *Environ. Syst. Res.* **2018**, *7*, 25. [CrossRef]
16. Long, H.; Tang, G.; Li, X.; Heilig, G.K. Socio–economic driving forces of land–use change in Kunshan, the Yangtze River Delta economic area of China. *J. Environ. Manag.* **2007**, *83*, 351–364. [CrossRef] [PubMed]
17. Ning, J.; Liu, J.; Kuang, W.; Xu, X.; Zhang, S.; Yan, C.; Li, R.; Wu, S.; Hu, Y.; Du, G.; et al. Spatiotemporal patterns and characteristics of land-use change in China during 2010–2015. *J. Geogr. Sci.* **2018**, *28*, 547–562. [CrossRef]
18. Prestele, R.; Alexander, P.; Rounsevell, M.D.A.; Arneth, A.; Calvin, K.; Doelman, J.; Eitelberg, D.A.; Engstrom, K.; Fujimori, S.; Hasegawa, T.; et al. Hotspots of uncertainty in land-use and land-cover change projections: A global-scale model comparison. *Glob. Change Biol.* **2016**, *22*, 3967–3983. [CrossRef]
19. Srivastava, P.K.; Han, D.; Rico-Ramirez, M.A.; Bray, M.; Islam, T. Selection of classification techniques for land use/land cover change investigation. *Adv. Space Res.* **2012**, *50*, 1250–1265. [CrossRef]
20. Degife, A.W.; Zabel, F.; Mauser, W. Assessing land use and land cover changes and agricultural farmland expansions in Gambella Region, Ethiopia, using Landsat 5 and Sentinel 2a multispectral data. *Heliyon* **2018**, *4*, e00919. [CrossRef]
21. Hou, H.; Wang, R.; Murayama, Y. Scenario-based modelling for urban sustainability focusing on changes in cropland under rapid urbanization: A case study of Hangzhou from 1990 to 2035. *Sci. Total Environ.* **2019**, *661*, 422–431. [CrossRef]
22. Motlagh, Z.K.; Lotfi, A.; Pourmanafi, S.; Ahmadizadeh, S.; Soffianian, A. Spatial modeling of land-use change in a rapidly urbanizing landscape in central Iran: Integration of remote sensing, CA-Markov, and landscape metrics. *Environ. Monit. Assess.* **2020**, *192*, 695. [CrossRef]
23. Ren, Y.; Lü, Y.; Comber, A.; Fu, B.; Harris, P.; Wu, L. Spatially explicit simulation of land use/land cover changes: Current coverage and future prospects. *Earth-Sci. Rev.* **2019**, *190*, 398–415. [CrossRef]
24. Berihun, M.L.; Tsunekawa, A.; Haregeweyn, N.; Meshesha, D.T.; Adgo, E.; Tsubo, M.; Masunaga, T.; Fenta, A.A.; Sultan, D.; Yibeltal, M. Exploring land use/land cover changes, drivers and their implications in contrasting agro-ecological environments of Ethiopia. *Land Use Policy* **2019**, *87*, 104052. [CrossRef]
25. Deng, X.; Huang, J.; Rozelle, S.; Uchida, E. Economic Growth and the Expansion of Urban Land in China. *Urban Stud.* **2010**, *47*, 813–843. [CrossRef]
26. Gan, B.R.; Liu, X.N.; Yang, X.G.; Wang, X.K.; Zhoua, J.W. The impact of human activities on the occurrence of mountain flood hazards: Lessons from the 17 August 2015 flash flood/debris flow event in Xuyong County, South-Western China. *Geomat. Nat. Hazards Risk* **2018**, *9*, 816–840. [CrossRef]
27. Munthali, M.G.; Mustak, S.; Adeola, A.; Botai, J.; Singh, S.K.; Davis, N. Modelling Land Use and Land Cover Dynamics of Dedza District of Malawi Using Hybrid Cellular Automata and Markov Model. *Remote Sens. Appl. Soc. Environ.* **2020**, *17*, 100276. [CrossRef]
28. You, H. Agricultural landscape dynamics in response to economic transition: Comparisons between different spatial planning zones in Ningbo region, China. *Land Use Policy* **2017**, *61*, 316–328. [CrossRef]
29. Apollonio, C.; Balacco, G.; Novelli, A.; Tarantino, E.; Piccinni, A.F. Land use change impact on flooding areas: The case study of Cervaro Basin (Italy). *Sustainability* **2016**, *8*, 996. [CrossRef]
30. Carvalho-Santos, C.; Nunes, J.P.; Monteiro, A.T.; Hein, L.; Honrado, J.P. Assessing the effects of land cover and future climate conditions on the provision of hydrological services in a medium-sized watershed of Portugal. *Hydrol. Process.* **2016**, *30*, 720–738. [CrossRef]
31. Hossain, F.; Moniruzzaman, M. Environmental Change Detection through Remote Sensing Technique: A Study of Rohingya Refugee Camp Area (Ukhia and Teknaf Sub-district), Cox's Bazar, Bangladesh. *Environ. Chall.* **2021**, *2*, 100024. [CrossRef]
32. Lambin, E.F.; Turner, B.L.; Geist, H.J.; Agbola, S.B.; Angelsen, A.; Bruce, J.W.; Coomes, O.T.; Dirzo, R.; Fischer, G.; Folke, C.; et al. The causes of land-use and land-cover change: Moving beyond the myths. *Glob. Environ. Change* **2001**, *11*, 261–269. [CrossRef]
33. Lambin, E.F.; Geist, H.J.; Lepers, E. Dynamics of Land-Use and Land-Cover Change in Tropical Regions. *Annu. Rev. Environ. Resour.* **2003**, *28*, 205–241. [CrossRef]
34. Mezger, G.; De Stefano, L.; Gonz, M. Analysis of the Evolution of Climatic and Hydrological Variables in the Tagus River Basin, Spain. *Water* **2022**, *14*, 818. [CrossRef]
35. Shang, X.; Jiang, X.; Jia, R.; Wei, C. Land use and climate change effects on surface runoff variations in the upper Heihe River basin. *Water* **2019**, *11*, 344. [CrossRef]
36. Sibanda, S.; Ahmed, F. Modelling historic and future land use/land cover changes and their impact on wetland area in Shashe sub-catchment, Zimbabwe. *Model. Earth Syst. Environ.* **2021**, *7*, 57–70. [CrossRef]
37. Yang, D.; Kanae, S.; Oki, T.; Koike, T.; Musiake, K. Global potential soil erosion with reference to land use and climate changes. *Hydrol. Process.* **2003**, *17*, 2913–2928. [CrossRef]
38. Behera, M.D.; Borate, S.N.; Panda, S.N.; Behera, P.R.; Roy, P.S. Modelling and analyzing the watershed dynamics using Cellular Automata (CA)–Markov model—A geo-information based approach. *J. Earth Syst. Sci.* **2012**, *121*, 1011–1024. [CrossRef]
39. Rahman, M.T.U.; Tabassum, F.; Rasheduzzaman, M.; Saba, H.; Sarkar, L.; Ferdous, J.; Uddin, S.Z.; Islam, A.Z. Temporal dynamics of land use/land cover change and its prediction using CA-ANN model for southwestern coastal Bangladesh. *Environ. Monit. Assess.* **2017**, *189*, 565. [CrossRef]

40. Armin, M.; Majidian, M.; Kheybari, V.G. Land Use/Land Cover Change Detection and Prediction in the Yasouj City Suburbs in Kohgiluyeh Va Boyerahmad Province in Iran. *Arid. Ecosyst.* **2020**, *10*, 203–210. [CrossRef]
41. Kotaridis, I.; Lazaridou, M. Environmental Change Detection Study in the Wider Area of Lignite Mines. *Civil Eng. Archit.* **2018**, *6*, 108–114. [CrossRef]
42. Hua, A.K. Land use land cover changes in detection of water quality: A study based on remote sensing and multivariate statistics. *J. Environ. Public Health* **2017**, *2017*, 7515130. [CrossRef]
43. Liping, C.; Yujun, S.; Saeed, S. Monitoring and predicting land use and land cover changes using remote sensing and GIS techniques—A case study of a hilly area, Jiangle, China. *PLoS ONE* **2018**, *13*, e0200493. [CrossRef] [PubMed]
44. Mannan, A.; Liu, J.; Zhongke, F.; Khan, T.U.; Saeed, S.; Mukete, B.; ChaoYong, S.; Yongxiang, F.; Ahmad, A.; Amir, M.; et al. Application of land-use/land cover changes in monitoring and projecting forest biomass carbon loss in Pakistan. *Glob. Ecol. Conserv.* **2019**, *17*, e00535. [CrossRef]
45. Rawat, J.S.; Kumar, M. Monitoring land use/cover change using remote sensing and GIS techniques: A case study of Hawalbagh block, district Almora, Uttarakhand, India. *Egypt. J. Remote Sens. Space Sci.* **2015**, *18*, 77–84. [CrossRef]
46. Nouri, J.; Gharagozlou, A.; Arjmandi, R.; Faryadi, S.; Adl, M. Predicting urban land use changes using a CA–Markov model. *Arab. J. Sci. Eng.* **2014**, *39*, 5565–5573. [CrossRef]
47. Yang, X.; Zheng, X.-Q.; Lv, L.-N. A spatiotemporal model of land use change based on ant colony optimization, Markov chain and cellular automata. *Ecol. Model.* **2012**, *233*, 11–19. [CrossRef]
48. Nath, B.; Wang, Z.; Ge, Y.; Islam, K.; Singh, R.P.; Niu, Z. Land Use and Land Cover Change Modeling and Future Potential Landscape Risk Assessment Using Markov-CA Model and Analytical Hierarchy Process. *ISPRS Int. J. Geo-Inf.* **2020**, *9*, 134. [CrossRef]
49. Maviza, A.; Ahmed, F. Analysis of past and future multi-temporal land use and land cover changes in the semi-arid UpperMzingwane sub-catchment in the Matabeleland south province of Zimbabwe. *Int. J. Remote Sens.* **2020**, *41*, 5206–5227. [CrossRef]
50. Hyandye, C.; Martz, L.W. A Markovian and cellular automata land-use change predictive model of the Usangu Catchment. *Int. J. Remote Sens.* **2017**, *38*, 64–81. [CrossRef]
51. Memarian, H.; Balasundram, S.K.; Talib, J.B.; Sung, C.T.B.; Sood, A.M.; Abbaspour, K. Validation of CA-Markov for simulation of land use and cover change in the Langat Basin, Malaysia. *Sci. Res.* **2012**, *4*, 542–554. [CrossRef]
52. Singh, S.K.; Mustak, S.; Srivastava, P.K.; Szabó, S.; Islam, T. Predicting spatial and decadal LULC changes through cellular automata Markov chain models using earth observation datasets and geo-information. *Environ. Process.* **2015**, *2*, 61–78. [CrossRef]
53. Weng, Q. Land use change analysis in the Zhujiang Delta of China using satellite remote sensing, GIS and stochastic modelling. *J. Environ. Manag.* **2002**, *64*, 273–284. [CrossRef] [PubMed]
54. Bashir, O.; Bangroo, S.A.; Guo, W.; Meraj, G.; Ayele, G.T.; Naikoo, N.B.; Shafai, S.; Singh, P.; Muslim, M.; Taddese, H.; et al. Simulating Spatiotemporal Changes in Land Use and Land Cover of the North-Western Himalayan Region Using Markov Chain Analysis. *Land* **2022**, *11*, 2276. [CrossRef]
55. Jia, W.; Gu, X.; Mi, X.; Yang, J.; Zang, W.; Liu, P.; Yan, J.; Zhu, H.; Zhang, X.; Zhang, Z. Multi-Scale Spatiotemporal Pattern Analysis and Simulation (MSPAS) Model with Driving Factors for Land Cover Change and Sustainable Development Goals: A Case Study of Nepal. *Remote Sens.* **2022**, *14*, 6295. [CrossRef]
56. Naboureh, A.; Moghaddam, M.H.R.; Feizizadeh, B.; Blaschke, T. An integrated object-based image analysis and CA-Markov model approach for modeling land use/land cover trends in the Sarab plain. *Arab. J. Geosci.* **2017**, *10*, 259. [CrossRef]
57. Pereira, C.O.; Escanilla-Minchel, R.; González, A.C.; Alcayaga, H.; Aguayo, M.; Arias, M.A.; Flores, A.N. Assessment of Future Land Use/Land Cover Scenarios on the Hydrology of a Coastal Basin in South-Central Chile. *Sustainability* **2022**, *14*, 16363. [CrossRef]
58. Wu, X.; Zhu, C.; Yu, J.; Zhai, L.; Zhang, H.; Yang, K.; Hou, X. Ecological Vulnerability in the Red Soil Erosion Area of Changting under Continuous Ecological Restoration: Spatiotemporal Dynamic Evolution and Prediction. *Forests* **2022**, *13*, 2136. [CrossRef]
59. Dey, N.N.; Al Rakib, A.; Kafy, A.A.; Raikwar, V. Geospatial Modelling of Changes in Land Use/Land Cover Dynamics Using Multi-Layer Perception Markov Chain Model in Rajshahi City, Bangladesh. *Environ. Chall.* **2021**, *4*, 100148. [CrossRef]
60. Fan, F.; Wang, Y.; Wang, Z. Temporal and spatial change detecting (1998–2003) and predicting of land use and land cover in Core corridor of Pearl River Delta (China) by using TM and ETM+ images. *Environ. Monit. Assess.* **2007**, *137*, 127–147. [CrossRef]
61. Hamad, R.; Balzter, H.; Kolo, K. Predicting land use/land cover changes using a CA-Markov model under two different scenarios. *Sustainability* **2018**, *10*, 3421. [CrossRef]
62. Yirsaw, E.; Wu, W.; Shi, X.; Temesgen, H.; Bekele, B. Land Use/Land Cover Change Modeling and the Prediction of Subsequent Changes in Ecosystem Service Values in a Coastal Area of China, the Su-Xi-Chang Region. *Sustainability* **2017**, *9*, 1204. [CrossRef]
63. Al-shalabi, M.; Billa, L.; Pradhan, B.; Mansor, S.; Al-Sharif, A.A.A. Modelling urban growth evolution and land-use changes using GIS based cellular automata and SLEUTH models: The case of Sana'a metropolitan city, Yemen. *Environ. Earth Sci.* **2013**, *70*, 425–437. [CrossRef]
64. Guan, D.; Li, H.; Inohae, T.; Su, W.; Nagaie, T.; Hokao, K. Modeling urban land use change by the integration of cellular automaton and markov model. *Ecol. Model.* **2011**, *222*, 3761–3772. [CrossRef]
65. Cheong, S.M.; Brown, D.; Kok, K.; López-Carr, D. Mixed Methods in Land Change Research: Towards Integration. *Trans. Inst. Br. Geogr.* **2012**, *37*, 8–12. [CrossRef]
66. Brown, D.G.; Walker, R.; Manson, S.; Seto, K.C. Modeling land use and land cover change. In *Land Change Science: Observing, Monitoring, and Understanding Trajectories of Change on the Earth's Surface*; Gutman, G., Janetos, A., Justice, C., Moran, E., Mustard, J., Rindfuss, R., Skole, D., Turner, B.L., II, Cochrane, M.A., Eds.; Kluwer Academic Publishers: Dordrecht, The Netherlands, 2004; pp. 395–409.

67. Bracchetti, L.; Carotenuto, L.; Catorci, A. Land-cover changes in a remote area of central Apennines (Italy) and management directions. *Landsc. Urban Plan.* **2012**, *104*, 157–170. [CrossRef]
68. Tattoni, C.; Ciolli, M.; Ferretti, F. The fate of priority areas for conservation in protected areas: A fine-scale markov chain approach. *Environ. Manag.* **2011**, *47*, 263–278. [CrossRef] [PubMed]
69. Belay, T.; Mengistu, D.A. Impacts of land use/land cover and climate changes on soil erosion in Muga watershed, Upper Blue Nile basin (Abay), Ethiopia. *Ecol. Process.* **2021**, *10*, 68. [CrossRef]
70. Soil Survey Staff. *Keys to Soil Taxonomy*, 12th ed.; USDA-Natural Resources Conservation Service: Washington, DC, USA, 2014.
71. Hansen, M.C.; Defries, R.S.; Townshend, J.R.G.; Sohlberg, R. Global land cover classification at 1 km spatial resolution using a classification tree approach. *Int. J. Remote Sens.* **2000**, *21*, 1331–1364. [CrossRef]
72. Jensen, J.R. *Introductory Digital Image Processing*, 3rd ed.; Prentice Hall: Upper Saddle River, NJ, USA, 2005.
73. Weier, J.; Herring, D. *Measuring Vegetation: NDVI and EVI*; NASA Earth Observatory: Washington, DC, USA, 2000.
74. Xu, H.Q. Modification of normalised difference water index (NDWI) to enhance open water features in remotely sensed imagery. *Int. J. Remote Sens.* **2006**, *27*, 3025–3033. [CrossRef]
75. Zha, Y.; Gao, J.; Ni, S. Use of normalized difference built-up index in automatically mapping urban areas from TM imagery. *Int. J. Remote Sens.* **2003**, *24*, 583–594. [CrossRef]
76. Zhao, H.; Chen, X. Use of normalized difference bareness index in quickly mapping bare areas from TM/ETM+. In Proceedings of the 2005 IEEE International Geoscience and Remote Sensing Symposium, Seoul, Republic of Korea, 25–29 July 2005; Volume 3, pp. 1666–1668. [CrossRef]
77. Ibharim, N.; Mustapha, M.A.; Lihan, T.; Mazlan, A. Mapping mangrove changes in the Matang Mangrove Forest using multi temporal satellite imageries. *Ocean Coast. Manag.* **2015**, *114*, 64–76. [CrossRef]
78. Lin, C.; Wu, C.-C.; Tsogt, K.; Ouyang, Y.-C.; Chang, C.-I. Effects of atmospheric correction and pansharpening on LULC classification accuracy using WorldView-2 imagery. *Inf. Processing Agric.* **2015**, *2*, 25–36. [CrossRef]
79. Congalton, R.G.; Green, K. *Assessing the Accuracy of Remotely Sensed Data*; CRC Press: Boca Raton, FL, USA, 2009; ISBN 0-87-37198-67.
80. Sulieman, H.M. Mapping and modelling of vegetation changes in the Southern Gadarif region, Sudan, using remote sensing. Ph.D. Thesis, Dresden University of Technology, Dresden, Germany, 2008.
81. Zhang, Z.; Liu, S.; Wei, J.; Xu, J.; Guo, W.; Bao, W.; Jiang, Z. Mass change of glaciers in Muztag Ata–Kongur Tagh, Eastern Pamir, China from 1971/76 to 2013/14 as derived from remote sensing data. *PLoS ONE* **2016**, *11*, e0147327. [CrossRef] [PubMed]
82. Anderson, J.R.; Hardy, E.E.; Roach, J.T.; Witmer, R.E. *A Land Use and Land Cover Classification System for Use with Remote Sensor Data*; Government Printing Office: Washington, DC, USA, 1976.
83. Stehman, S. Estimating the kappa coefficient and its variance under stratified random sampling. *Photogramm. Eng. Remote Sens.* **1996**, *62*, 401–407.
84. Kumar, S.; Radhakrishnan, N.; Mathew, S. Land use change modelling using a Markov model and remote sensing. *Geomat. Nat. Hazards Risk* **2014**, *5*, 145–156. [CrossRef]
85. Guan, D.; Gao, W.; Watari, K.; Fukahori, H. Land use change of Kitakyushu based on landscape ecology and Markov model. *J. Geogr. Sci.* **2008**, *18*, 455–468. [CrossRef]
86. Chen, Y.; Li, X.; Liu, X.; Ai, B. Analyzing land-cover change and corresponding impacts on carbon budget in a fast developing sub-tropical region by integrating MODIS and Landsat TM/ETM+ images. *Appl. Geogr.* **2013**, *45*, 10–21. [CrossRef]
87. Myint, S.W.; Wang, L. Multicriteria decision approach for land use land cover change using Markov chain analysis and a cellular automata approach. *Can. J. Remote Sens.* **2006**, *32*, 390–404. [CrossRef]
88. Biswas, M.; Banerji, S.; Mitra, D. Land-use–land-cover change detection and application of Markov model: A case study of Eastern part of Kolkata. *Environ. Dev. Sustain.* **2019**, *22*, 4341–4360. [CrossRef]
89. Hu, Y.; Zheng, Y.; Zheng, X. Simulation of land-use scenarios for Beijing using CLUE-S and Markov composite models. *Chin. Geogr. Sci.* **2013**, *23*, 92–100. [CrossRef]
90. Sang, L.; Zhang, C.; Yang, J.; Zhu, D.; Yun, W. Simulation of land use spatial pattern of towns and villages based on CA-Markov model. *Math. Comput. Model.* **2011**, *54*, 938–943. [CrossRef]
91. Wu, N.; Silva, E. Artificial intelligence solutions for urban land dynamics: A review. *J. Plan. Lit.* **2010**, *24*, 246–265. [CrossRef]
92. Eastman, J.R. *IDRISI Selva Tutorial*; Clark University: Worcester, MA, USA, 2012; pp. 267–275.
93. Aliani, H.; Malmir, M.; Sourodi, M.; Kafaky, S.B. Change detection and prediction of urban land use changes by CA–Markov model (case study: Talesh County). *Environ. Earth Sci.* **2019**, *78*, 546. [CrossRef]
94. Nadoushan, M.A.; Soffianian, A.; Alebrahim, A. Modeling land use/cover changes by the combination of markov chain and cellular automata markov (CA-Markov) models. *J. Earth Environ. Health Sci.* **2015**, *1*, 16–21. [CrossRef]
95. Clarke, K.C.; Hoppen, S.; Gaydos, L. A self-modifying cellular automaton model of historical urbanization in the San Francisco Bay area. Environ. *Plan. B Plan. Des.* **1997**, *24*, 247–261. [CrossRef]
96. Yang, Q.S.; Li, X.; Shi, X. Cellular automata for simulating land use changes based on support vector machines. *Comput. Geosci.* **2008**, *34*, 592–602. [CrossRef]
97. Ma, C.; Zhang, G.; Zhang, X.; Zhao, Y.; Li, H. Application of Markov model in wetland change dynamics in Tianjin Coastal Area, China. *Procedia Environ. Sci.* **2012**, *13*, 252–262. [CrossRef]
98. Irwin, E.G.; Jayaprakash, C.; Munroe, D.K. Towards a comprehensive framework for modeling urban spatial dynamics. *Landsc. Ecol.* **2009**, *24*, 1223–1236. [CrossRef]

99. He, D.; Zhou, J.; Gao, W.; Guo, H.Y.U.S.; Liu, Y. An integrated CA–markov model for dynamic simulation of land use change in Lake Dianchi watershed. *Acta Sci. Nat. Univ. Pekin.* **2014**, *50*, 1095–1105. [CrossRef]
100. Keshtkar, H.; Voigt, W. A spatiotemporal analysis of landscape change using an integrated Markov chain and cellular automata models. *Model. Earth Syst. Environ.* **2016**, *2*, 1–13. [CrossRef]
101. Liang, X.; Liu, X.; Li, D.; Zhao, H.; Chen, G. Urban growth simulation by incorporating planning policies into a CA–based future land–use simulation model. *Int. J. Geogr. Inf. Sci.* **2018**, *32*, 2294–2316. [CrossRef]
102. Aksoy, H.; Kaptan, S. Simulation of future forest and land use/cover changes (2019–2039) using the cellular automata-Markov model. *Geocarto Int.* **2020**, *37*, 1183–1202. [CrossRef]
103. Wang, R.; Murayama, Y. Change of Land Use/Cover in Tianjin City Based on the Markov and Cellular Automata Models. *ISPRS Int. J. Geo-Inf.* **2017**, *6*, 150. [CrossRef]
104. Singh, S.K.; Laari, P.B.; Mustak, S.; Srivastava, P.K.; Szabó, S. Modelling of land use land cover change using earth observation data-sets of Tons River Basin, Madhya Pradesh, India. *Geocarto Int.* **2018**, *33*, 1202–1222. [CrossRef]
105. Tali, J.A.; Divya, S.; Murthy, K. Influence of urbanization on the land use change: A case study of Srinagar City. *Am. J. Res. Comm.* **2013**, *1*, 271–283.
106. Eastman, J.R. *IDRISI Taiga Guide to GIS and Image Processing*; Clark Labs Clark University: Worcester, MA, USA, 2009.
107. Subedi, P.; Subedi, K.; Thapa, B. Application of a Hybrid Cellular Automaton—Markov (CA-Markov) Model in Land-Use Change Prediction: A Case Study of Saddle Creek Drainage Basin, Florida. *Appl. Ecol. Environ. Sci.* **2013**, *1*, 126–132. [CrossRef]
108. Pan, S.; Liu, D.; Wang, Z.; Zhao, Q.; Zou, H.; Hou, Y.; Liu, P.; Xiong, L. Runoff Responses to Climate and Land Use/Cover Changes under Future Scenarios. *Water* **2017**, *9*, 475. [CrossRef]
109. Tilahun, A.; Teferie, B. Accuracy assessment of land use land cover classification using Google Earth. *Am. J. Environ. Prot.* **2015**, *4*, 193–198. [CrossRef]
110. Da Cunha, E.R.; Santos, C.A.G.; da Silva, R.M.; Bacani, V.M.; Teodoro, P.E.; Panachuki, E.; de Souza Oliveira, N. Mapping LULC types in the Cerrado-Atlantic Forest ecotone region using a Landsat time series and object-based image approach: A case study of the Prata River Basin, Mato Grosso do Sul, Brazil. *Environ. Monit. Assess.* **2020**, *192*, 136. [CrossRef]
111. Li, X.; Liu, X.; Yu, L. A systematic sensitivity analysis of constrained cellular automata model for urban growth simulation based on different transition rules. *Int. J. Geogr. Inf. Syst.* **2014**, *28*, 1317–1335. [CrossRef]
112. Rimal, B.; Zhang, L.; Keshtkar, H.; Haack, B.N.; Rijal, S.; Zhang, P. Land Use/Land Cover Dynamics and Modeling of Urban Land Expansion by the Integration of Cellular Automata and Markov Chain. *ISPRS Int. J. Geo-Inf.* **2018**, *7*, 154. [CrossRef]
113. Abd El-Hamid, H.T.; El-Alfy, M.A.; Elnaggar, A. Prediction of future situation of land use/cover change and modeling sensitivity to pollution in Edku Lake. Egypt based on geospatial analyses. *GeoJournal* **2020**, *86*, 1895–1913. [CrossRef]
114. Abd El-Hamid, H.T.A.; Nour-Eldin, H.; Rebouh, N.Y.; El-Zeiny, A.M. Past and Future Changes of Land Use/Land Cover and the Potential Impact on Ecosystem Services Value of Damietta Governorate, Egypt. *Land* **2022**, *11*, 2169. [CrossRef]
115. Abd El-Kawy, O.R.; Rød, J.K.; Ismail, H.A.; Suliman, A.S. Land use and land cover change detection in the western Nile delta of Egypt using remote sensing data. *Appl. Geogr.* **2011**, *31*, 483–494. [CrossRef]
116. Elagouz, M.; Abou-Shleel, S.; Belal, A.; El-Mohandes, M. Detection of land use/cover change in Egyptian Nile Delta using remote sensing. *Egypt. J. Remote Sens. Space Sci.* **2019**, *23*, 57–62. [CrossRef]
117. El-Zeiny, A.M.; Effat, H.A. Environmental monitoring of spatiotemporal change in land use/land cover and its impact on land surface temperature in El-Fayoum governorate, Egypt. *Remote Sens. Appl. Soc. Environ.* **2017**, *8*, 266–277. [CrossRef]
118. Halmy, M.W.A.; Gessler, P.E.; Hicke, J.A.; Salem, B.B. Land use/land cover change detection and prediction in the north-western coastal desert of Egypt using Markov-CA. *Appl. Geogr.* **2015**, *63*, 101–112. [CrossRef]
119. Kamel, M. Monitoring of Land Use and Land Cover Change Detection Using Multi-temporal Remote Sensing and Time Series Analysis of Qena-Luxor Governorates (QLGs), Egypt. *J. Indian Soc. Remote Sens.* **2020**, *48*, 1767–1785. [CrossRef]
120. Mohamed, E.S.; Baroudy, A.A.E.; El-beshbeshy, T.; Emam, M.; Belal, A.A.; Elfadaly, A.; Aldosari, A.A.; Ali, A.M.; Lasaponara, R. Vis-NIR Spectroscopy and Satellite Landsat-8 OLI Data to Map Soil Nutrients in Arid Conditions: A Case Study of the Northwest Coast of Egypt. *Remote Sens.* **2020**, *12*, 3716. [CrossRef]
121. Said, M.E.S.; Ali, A.M.; Borin, M.; Abd-Elmabod, S.K.; Aldosari, A.A.; Khalil, M.M.N.; Abdel-Fattah, M.K. On the Use of Multivariate Analysis and Land Evaluation for Potential Agricultural Development of the Northwestern Coast of Egypt. *Agronomy* **2020**, *10*, 1318. [CrossRef]
122. Biondini, M.; Kandus, P. Transition matrix analysis of land-cover change in the accretion area of the Lower Delta of the Paraná River (Argentina) reveals two succession pathways. *Wetlands* **2006**, *26*, 981–991. [CrossRef]
123. Bai, X.; Chen, J.; Shi, P. Landscape Urbanization and Economic Growth in China: Positive Feedbacks and Sustainability Dilemmas. *Environ. Sci. Technol.* **2012**, *46*, 132–139. [CrossRef]
124. Huang, G.; Jiang, Y. Urbanization and Socioeconomic Development in Inner Mongolia in 2000 and 2010: A GIS Analysis. *Sustainability* **2017**, *9*, 235. [CrossRef]
125. Deng, X.; Huang, J.; Rozelle, S.; Zhang, J.; Li, Z. Impact of urbanization on cultivated land changes in China. *Land Use Policy* **2015**, *45*, 1–7. [CrossRef]
126. Marques, A.; Martins, I.S.; Kastner, T.; Plutzar, C.; Theurl, M.C.; Eisenmenger, N.; Huijbregts, M.A.J.; Wood, R.; Stadler, K.; Bruckner, M.; et al. Increasing impacts of land use on biodiversity and carbon sequestration driven by population and economic growth. *Nat. Ecol. Evol.* **2019**, *3*, 628–637. [CrossRef] [PubMed]

127. Pandey, B.K.; Khare, D. Analyzing and modeling of a large river basin dynamics applying integrated cellular automata and Markov model. *Environ. Earth Sci.* **2017**, *76*, 779. [CrossRef]
128. Gashaw, T.; Tulu, T.; Argaw, M.; Worqlul, A.W. Evaluation and prediction of land use/land cover changes in the Andassa watershed, Blue Nile Basin, Ethiopia. *Environ. Syst. Res.* **2017**, *6*, 17. [CrossRef]
129. Pontius, G.J.; Schneider, L.C. Land-cover change model validation by an ROC method for the Ipswich watershed, Massachusetts, USA. *Agric. Ecosyst. Environ.* **2001**, *85*, 239–248. [CrossRef]
130. Pontius, R.; Millones, M. Death to Kappa: Birth of quantity disagreement and allocation disagreement for accuracy assessment. *Int. J. Remote Sens.* **2011**, *32*, 4407–4429. [CrossRef]
131. Luo, G.; Amuti, T.; Zhu, L.; Mambetov, B.T.; Maisupova, B.; Zhang, C. Dynamics of landscape patterns in an inland river delta of Central Asia based on a cellular automata-Markov model. *Reg. Environ. Chang.* **2015**, *15*, 277–289. [CrossRef]
132. Mondal, M.S.; Sharma, N.; Kappas, M.; Garg, P.K. CA Markov modeling of land use land cover change predictions and effect of numerical iterations, image interval (time steps) on prediction results. Int. Arch. Photogramm. *Remote Sens. Spatial Inf. Sci.* **2020**, *43*, 713–720. [CrossRef]
133. Huang, Y.C.; Yang, B.G.; Wang, M.; Liu, B.W.; Yang, X.D. Analysis of the future land cover change in Beijing using CA-Markov chain model. *Environ. Earth Sci.* **2020**, *79*, 60. [CrossRef]
134. Khwarahm, N.R.; Najmaddin, P.M.; Ararat, K.; Qader, S. Past and future prediction of land cover land use change based on earth observation data by the CA–Markov model: A case study from Duhok governorate, Iraq. *Arab. J. Geosci.* **2021**, *14*, 1–14. [CrossRef]
135. Gidey, E.; Dikinya, O.; Sebego, R.; Segosebe, E.; Zenebe, A. Cellular automata and Markov Chain (CA_Markov) model-based predictions of future land use and land cover scenarios (2015–2033) in Raya, northern Ethiopia. *Modeling Earth Syst. Environ.* **2017**, *3*, 1245–1262. [CrossRef]
136. Matlhodi, B.; Kenabatho, P.K.; Parida, B.P.; Maphanyane, J.G. Analysis of the Future Land Use Land Cover Changes in the Gaborone Dam Catchment Using CA-Markov Model: Implications on Water Resources. *Remote Sens.* **2021**, *13*, 2427. [CrossRef]
137. Brodt, S.; Six, J.; Feenstra, G.; Ingels, C.; Campbell, D. Sustainable Agriculture. *Nat. Educ. Knowl.* **2011**, *3*, 1.
138. Dewan, A.; Yamaguchi, Y. Land use and land cover change in Greater Dhaka, Bangladesh: Using remote sensing to promote sustainable urbanization. *Appl. Geogr.* **2009**, *29*, 390–401. [CrossRef]

remote sensing

MDPI

Communication

Mapping Bedrock Outcrops in the Sierra Nevada Mountains (California, USA) Using Machine Learning

Apoorva Shastry [1], Corina Cerovski-Darriau [2,3,*], Brian Coltin [4,5] and Jonathan D. Stock [2,4]

1 Universities Space Research Association, Contractor to the U.S. Geological Survey, Moffett Field, CA 94035, USA
2 U.S. Geological Survey, Moffett Field, CA 94035, USA
3 U.S. Geological Survey, Golden, CO 80401, USA
4 NASA Ames Research Center, Moffett Field, CA 94035, USA
5 KBR Inc., Moffett Field, CA 94035, USA
* Correspondence: ccerovski-darriau@usgs.gov

Abstract: Accurate, high-resolution maps of bedrock outcrops can be valuable for applications such as models of land–atmosphere interactions, mineral assessments, ecosystem mapping, and hazard mapping. The increasing availability of high-resolution imagery can be coupled with machine learning techniques to improve regional bedrock outcrop maps. In the United States, the existing 30 m U.S. Geological Survey (USGS) National Land Cover Database (NLCD) tends to misestimate extents of barren land, which includes bedrock outcrops. This impacts many calculations beyond bedrock mapping, including soil carbon storage, hydrologic modeling, and erosion susceptibility. Here, we tested if a machine learning (ML) model could more accurately map exposed bedrock than NLCD across the entire Sierra Nevada Mountains (California, USA). The ML model was trained to identify pixels that are likely bedrock from 0.6 m imagery from the National Agriculture Imagery Program (NAIP). First, we labeled exposed bedrock at twenty sites covering more than 83 km^2 (0.13%) of the Sierra Nevada region. These labels were then used to train and test the model, which gave 83% precision and 78% recall, with a 90% overall accuracy of correctly predicting bedrock. We used the trained model to map bedrock outcrops across the entire Sierra Nevada region and compared the ML map with the NLCD map. At the twenty labeled sites, we found the NLCD barren land class, even though it includes more than just bedrock outcrops, accounted for only 41% and 40% of mapped bedrock from our labels and ML predictions, respectively. This substantial difference illustrates that ML bedrock models can have a role in improving land-cover maps, like NLCD, for a range of science applications.

Keywords: machine learning; neural network; quaternary geology; bedrock mapping; aerial photos; land cover

1. Introduction

Advances in imaging Earth's surface are making our ability to map surficial geology an attractive, and increasingly possible, reality. This is already being performed for land cover, but only poorly for features of geologic interest, like bedrock and soil [1–3]. Since Landsat satellite imagery became widely available, a consortium of federal agencies has been maintaining the National Land Cover Database (NLCD), which classifies the 30 m resolution data into surface cover categories, like forests, wetlands, barren land, etc., (e.g., [4]). These thematic categories are broad and focused on vegetation, so the most relevant category to geologists—barren land—includes soil cover and rock without distinction. Publicly available, sub-meter imagery for the United States presents an opportunity to improve the resolution and extent of surface cover maps—particularly for bedrock outcrops, as we tested here.

One could improve existing land cover maps by manually mapping bedrock outcrops using sub-meter imagery. While still interpretive, that map would arguably be the most accurate and useful but could introduce variability due to individual mapper interpretations and be prohibitively time-consuming over a large area (e.g., [5–8]). A less expensive and efficient way to overcome these two challenges is to use machine learning techniques to capture the visual characteristics that a human expert sees and automate them at scales that would not be mappable by a single human [8]. Automated methods are increasingly common for classifying and making land cover maps (e.g., [9–14]), but their use for bedrock mapping is a newer application with limited examples [2,15–19]. The major challenge is successfully training an algorithm because of potentially large regional variations in soils, geology, and ecology. Coupling high-resolution imagery and expert-guided mapping with machine learning improves our ability to interpret large areas of the Earth's surface, which could expand Quaternary mapping capacity as an entry point for mineral assessments, ecosystem mapping, and hazard mapping [19]. Better differentiation between bedrock outcrops and barren soil will conversely improve estimates of soil cover, which could improve carbon cycling, hydrologic, landslide susceptibility, and other models reliant on accurate soil maps [1,19]. Therefore, the goal of this study is to create a regional bedrock map using high-resolution imagery and machine learning.

In this paper we train DELTA (Deep Earth Learning, Tools, and Analysis), an open-source machine learning tool developed by NASA [20,21], to identify surface exposures of bedrock without including mobile, fragmental material (e.g., talus, boulders, etc.), or bare soil. We compare the DELTA accuracy to the existing U.S. Geological Survey (USGS) National Land Cover Database (NLCD). We use the Sierra Nevada Mountains, California (USA) as our study area. We test if we can produce an improved map of bedrock outcrops using expert mapping at local scales for the training input that captures natural geologic variability at a regional scale.

2. Data

2.1. Study Area

We mapped the exposed bedrock in the ~63,000 km^2 Sierra Nevada region (Figure 1, red outline). The Sierra Nevada is a mountain range in the Western United States between the latitudes of 35.10°N and 39.80°N and the longitudes of 117.87°W and 120.85°W. The region experiences a Mediterranean climate with dry summers and annual precipitation between 500 mm and 2000 mm. The mountains are predominantly high-albedo, glacially carved granitic rocks, with localized outcrops of volcanic, sedimentary, and metamorphic rocks. The vegetated areas include grasslands, chaparral, woodlands, subalpine forests, and alpine meadows [4].

2.2. Remote Sensing Images

We used publicly available 2016 National Agriculture Imagery Program (NAIP) imagery (Figure 1). NAIP regularly acquires aerial imagery during the agricultural growing seasons across the United States. The 1411 image tiles used have a spatial resolution of 0.6 m, with an 8-bit radiometric resolution in four spectral bands: blue, green, red, and near-infrared (NIR).

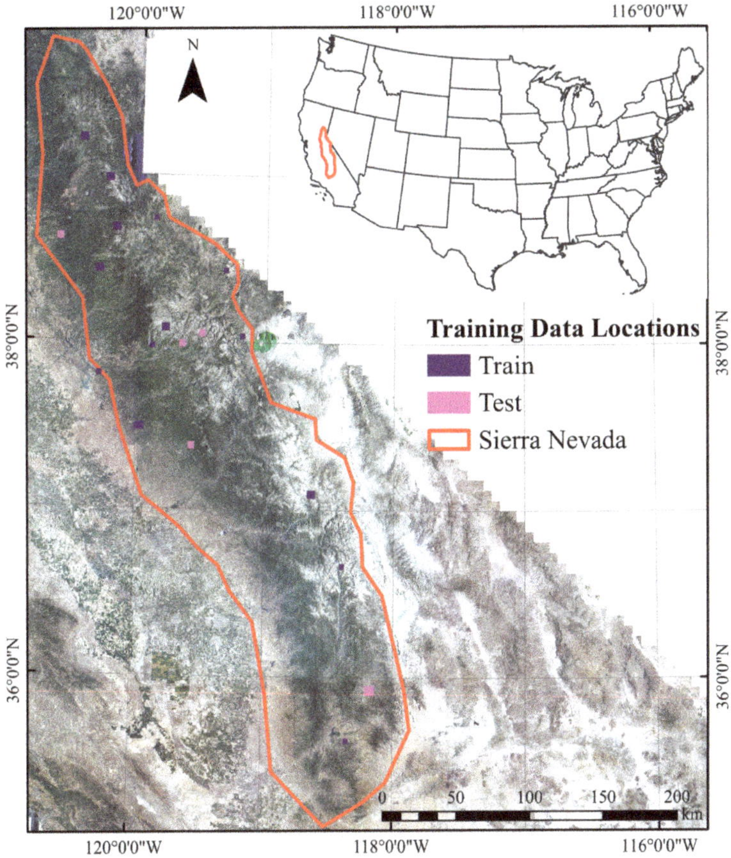

Figure 1. Extent of the Sierra Nevada Mountains (USA) study area (red outline) on the 2016 National Agriculture Imagery Program (NAIP) imagery in North American Datum of 1983 (NAD 83) and Universal Transverse Mercator (UTM) Zones 10N and 11N. Inset map shows the study area location in California (USA). Squares show twenty locations of mapped data used to train (dark purple) and test (pink) the model.

Training Data

We selected twenty sites to train and test the machine learning model that visually captures the regional variability of rock types, outcrop morphology, and vegetation cover (Figure 1). Eight existing labeled sites were available from an earlier effort [1]. We further revised those labels for improved accuracy and created an additional twelve reference sites where Petliak et al. [1] performed poorly, following the same protocol of using the landscape classification tool [22] and manual corrections to map "rock" and "not rock" from the 2016 NAIP imagery. All data are available in the associated data release [23]. In total, the twenty mapped areas cover 83.2 km^2 (Figure 1). Within these areas, ~16.3 km^2 is labeled rock, or bedrock outcrops, and the other 66.9 km^2 was not bedrock.

2.3. National Land Cover Database

We compared our results to the NLCD [24,25] barren land class. NLCD provides land cover data at 30 m spatial resolution for the whole of the United States based on Landsat satellite data [4]. NLCD land cover data were classified from various temporal, spectral, spatial, and terrain data [26]. We used 2016 NLCD maps to directly compare with the mapping from the 2016 NAIP imagery. NLCD contains twenty distinct land cover classes, but the bedrock is not an exclusive class. Instead, outcrops are included in the barren class, which includes bedrock, desert pavement, scarps, talus, landslides, volcanic material, glacial debris, sand dunes, strip mines, gravel pits, and other accumulations of earthen material. NLCD barren land class accounts for ~2035 km^2 of the Sierra Nevada, or about 3% of the land cover in Figure 1.

3. Methods

3.1. Machine Learning Training and Implementation

The two key components for successfully running a machine learning model are: (1) creating accurate and representative reference input, and (2) splitting the input data into representative train and test subsets [1]. We used reference labels at twenty sites for bedrock outcrops across diverse rock types and rock appearances as our representative sample. We used a larger, fifteen-site subset, covering 65.9 km^2, to sufficiently train the model. We used the remaining five sites, covering 17.3 km^2, as the testing subset to calculate model accuracy. We ensured a balanced bedrock percentage in the training data (~19%), testing (~22%), and the entire dataset (~20%). In Figure 1, we show the sites used for training the machine learning model (dark purple) and the sites used for testing the machine learning model (pink).

We used the open-source toolkit, DELTA [21], to train a neural network model to segment, or map, bedrock. A neural network is a machine learning technique that trains a computer to recognize patterns in data by analyzing training examples, in this case, 4-band NAIP imagery guided by our input reference labels. Neural networks are shown to provide improved accuracy for image and pixel classification over traditional supervised methods [1,14]. We followed the protocol developed by Shastry et al. [20] to segment flood water using DELTA and adapted the neural network to identify bedrock outcrops using our fifteen training sites. The DELTA neural network is based on an existing network for spinal cord gray matter [27] that has previously worked well for the segmentation of binary features, such as the presence of water versus no water [20]. Further details on the model structure and parameters are included in Shastry et al. [20].

Once the model was trained, and before running it for the entire region, we validated the accuracy by comparing the model predictions to our five reserved testing sites. The model output gives the likelihood (from 0 to 255) that each pixel is bedrock, not a direct binary classification. We used the standard binary classification threshold of 50%, or pixel value 127, where pixel values higher than 127 are classified as rock and pixel values less than 127 as not rock. This 50% threshold tends to strike a balance between false positives and false negatives but can be adjusted to shift the balance as needed. DELTA-predicted bedrock refers to pixel values 127 or higher. Figure 2 shows a visual comparison of the testing labels (orange), DELTA-predicted bedrock (purple), areas of overlap (dark pink), and the statistics derived to evaluate the performance. Overlapping labels and DELTA predictions (dark pink) indicate agreement, or true positives (*TP*). Purple indicates areas where DELTA predicts bedrock, but training labels do not or give false positives (*FP*). Orange indicates areas where DELTA missed bedrock that is present in reference labels, or false negatives (*FN*). The white areas are not bedrock according to both our labels and DELTA predictions, or true negatives (*TN*). These four combinations are generally visualized as a confusion matrix (Figure 2B). The performance of the trained model is evaluated by commonly used statistics shown below the confusion matrix.

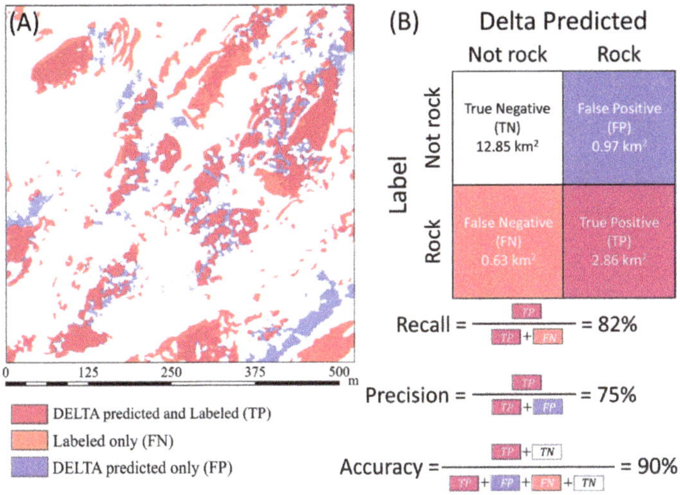

Figure 2. (**A**) Example showing comparison between reference labels and Deep Earth Learning, Tools, and Analysis (DELTA) predictions of bedrock. Purple is overprediction (false positives (FP)), orange is underprediction (false negatives (FN)), and dark pink is both true and predicted (true positives (TP)). These results are used to calculate (**B**) the confusion matrix for the testing set and the corresponding statistics derived from comparing reference labels with DELTA predictions of bedrock.

Precision is the fraction of correctly predicted bedrock among all the predicted bedrock, or:

$$Precision = \frac{TP}{TP + FP} \tag{1}$$

where *TP* and *FP* are true and false positives. Precision accounts for overprediction by the neural network. Recall (or hit rate) is the fraction of predicted bedrock among "true" bedrock from reference labels. It is defined as:

$$Recall = \frac{TP}{TP + FN} \tag{2}$$

where *TP* is true positives and *FN* is false negatives. This accounts for underprediction by the network. The overall accuracy is defined as:

$$Accuracy = \frac{TP + TN}{TN + FP + FN + TP} \tag{3}$$

where *TP* and *TN* are true positives and negatives, and *FP* and *FN* are false positives and negatives.

We implemented the trained machine learning model for the entire study area (Figure 1), iterating through the remaining 1411 NAIP image tiles to produce a probability map of likely bedrock for the entire region. The bedrock map, for applications like mineral resource identification, can be improved by masking out urban areas. Therefore, we decided to mask out urban areas by removing areas classified as any of NLCD's four "developed" classes to limit false positives in the final map.

3.2. Comparison with NLCD

3.2.1. Sites with Reference Data

To estimate the accuracy of the various methods, we compared the NLCD barren class to our reference labels at the twenty sites used for training and validating the machine learning model, as well as the DELTA-predicted bedrock at those twenty sites. We calculated the precision, recall, and accuracy of both DELTA predictions and the NLCD barren class against our NAIP-derived reference labels, or "truth". In addition, we also calculated the miss rate:

$$Miss\ rate = \frac{FN}{TP + FN} = 1 - Recall \tag{4}$$

where TP are true positives, and FN are false negatives. We did this in order to quantify the amount of bedrock missed by both DELTA predictions and by the NLCD barren class.

3.2.2. Entire Sierra Nevada

Lastly, using the DELTA-predicted bedrock results for the entire Sierra Nevada, we compared the presence of bedrock with the NLCD barren class to estimate how much NLCD underestimates the amount of bedrock. We used the urban-masked DELTA prediction layer for this analysis. Since the NLCD barren class includes more than just bedrock outcrops, we only considered underprediction by NLCD and ignored any overprediction by NLCD.

4. Results and Discussion

4.1. Machine Learning Training

We evaluated the DELTA-trained machine learning model on the five test sites, shown in pink (Figure 1). Figure 2B shows the confusion matrix for validation. Compared to the testing subset, the model performed well, with an overall accuracy of 90%, a recall of 82%, and a precision of 75%. The comparatively low precision means that the model has some problems with the overprediction of bedrock (Figure 3A). Qualitatively, we observed that these areas mostly belonged to two groups: urban infrastructure and loose, fragmental rock. Therefore, removing urban areas from the DELTA-predicted bedrock results improves the accuracy, and the urban-masked map is used for the comparisons below.

Table 1. Resulting metrics for the three reference sites shown in Figure 3A.

	DELTA			NLCD		
	Recall	**Precision**	**Accuracy**	**Recall**	**Precision**	**Accuracy**
Row 1 (SLT) *	98%	88%	90%	9%	84%	44%
Row 2 (FRS) *	81%	57%	76%	35%	23%	45%
Row 3 (TUO2) *	40%	36%	99%	0%	0%	99%

* Row number refers to image position in Figure 3. The name in parenthesis refers to the naming convention used in the data repository [23].

4.2. Comparison with NLCD

4.2.1. Sites with Reference Data

Figure 3 shows the NAIP imagery and corresponding reference labels, DELTA-predicted bedrock, and the NLCD barren class at three sample locations from the twenty sites used for training and validating the machine learning model. At these three sites, the DELTA predictions very closely follow our reference labels. However, the NLCD barren class seems to heavily under- or over-predict bedrock in these three examples (Table 1).

We compared the performance of urban-masked DELTA predictions and the NLCD barren class against the reference labels, which were considered "truth". Figure 3B shows the confusion matrix for the comparison between DELTA predictions and reference labels, and Figure 3C shows the same for the comparison between the NLCD barren class and reference labels. DELTA predictions compared to reference labels have an overall accuracy of 91%, recall of 84%, and precision of 72%. The miss rate for DELTA predictions is 16%. In contrast, when the NLCD barren labels are compared with reference labels, recall is

18%, and the miss rate is 82%. NLCD incorrectly predicts most bedrock as shrub/scrub (68%), evergreen forests (7%), and herbaceous (6%) at our twenty labeled sites. We tested changing the rock pixel threshold for the training sites. We found it did not significantly change the success metrics, so continued to use the traditional 50% threshold for simplicity. We do not calculate accuracy or precision since the barren class contains more areas than just bedrock and cannot accurately estimate overprediction or false positives.

Figure 3. (**A**) 2016 National Agriculture Imagery Program (NAIP) images showing three examples from the twenty reference sites (**top**: high-albedo, glacially smoothed granitic rocks, middle: abundant talus downslope of bedrock ridges, **bottom**: exposed, weathered granitic bedrock below a reservoir), corresponding reference labels (light blue), Deep Earth Learning, Tools, and Analysis (DELTA) predictions (dark blue), and National Land Cover Database (NLCD) Barren class (red), with resulting metrics in Table 1, (**B**) Confusion matrix of DELTA predicted bedrock with respect to reference labels, and (**C**) Confusion matrix of NLCD Barren class with respect to reference labels.

4.2.2. Entire Sierra Nevada

Using the trained DELTA model, we mapped bedrock across the study area to produce a map of likely bedrock exposure that does not include talus or other fragmental rock cover (Figure 4; [23]). We used the urban-masked DELTA predictions of bedrock for comparison with the NLCD barren class and considered this masked layer as "truth" in the confusion matrix. When compared to the NLCD barren class, NLCD had a miss rate of 85%. Each of the NAIP image tiles took less than 30 s to process, compared to the estimated 4–12 h for a human expert to map each of the twenty reference sites.

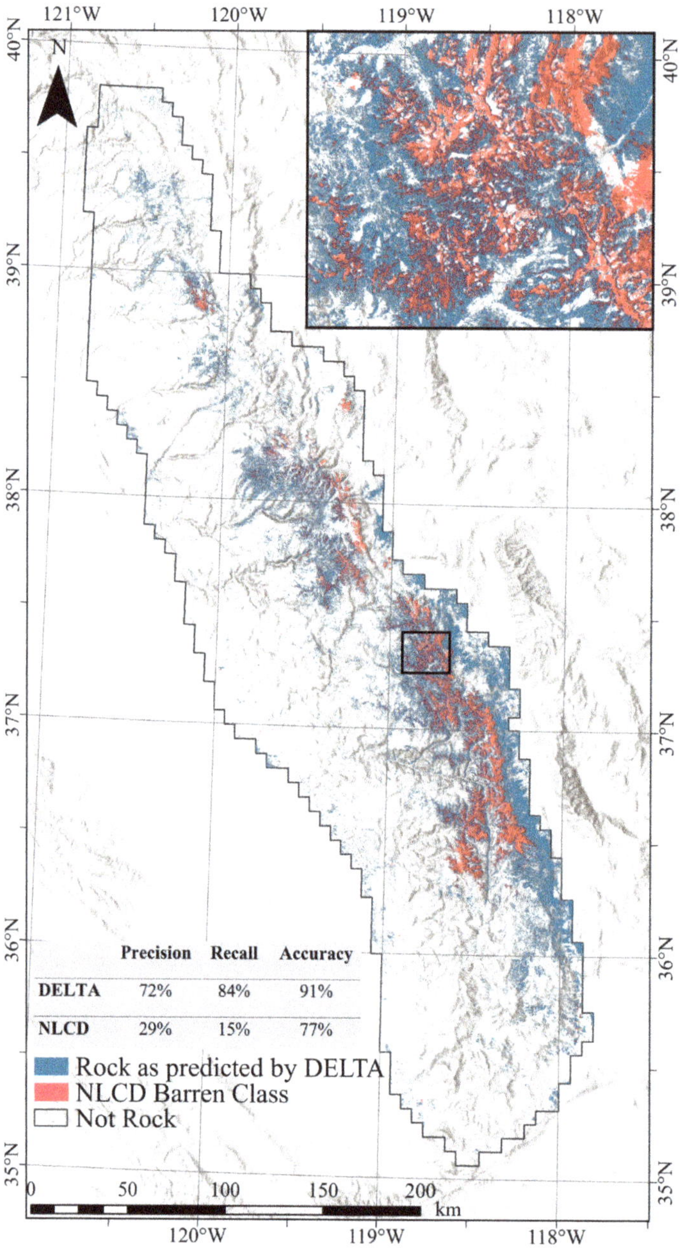

Figure 4. Deep Earth Learning, Tools, and Analysis (DELTA) predicted bedrock map (blue) compared to the National Land Cover Database (NLCD) Barren class (red) in the study area (black outline). The shaded refiled basemap is derived from the USGS 30 m DEM. Inset shows enlarged DELTA and NLCD predictions from the central study area (black box). The overlay and resulting metrics show that NLCD vastly underpredicts exposed bedrock visible in high-resolution imagery (NAD83, UTM Zones 10–11N).

4.3. Implications and Limitations

DELTA accurately differentiated in-place bedrock from bare soil or mobile rock and far outperformed the NLCD miss rate of 82%. The high accuracy was likely due to the careful selection of the training areas and expert input into the mapping [8]. The reference sites were rigorously selected and mapped to represent a range of bedrock (lithology type, color, and texture) and vegetation covers present across the study area, and bolster training where the previous model underperformed (e.g., dry grasslands and loose rock-dominated areas) [1]. While DELTA outperformed NLCD, loose rock-dominated areas as shown in the middle of Figure 3A continued to have a lower accuracy (Table 1). Future work could test the portability of this DELTA model using validation sites in other regions to verify whether this DELTA model could achieve similar accuracy without additional training. More testing would be needed to know how much training data are needed to achieve similar, or better, accuracy more broadly. We suspect that additional training would be necessary if the landscape includes visually different land cover or bedrock [28]. However, our initial study area did cover a wide range of ecosystems and visually different bedrock, and other deep-learning studies have shown good transferability (e.g., [14]). Therefore, it may be transferrable to other mountainous regions with no to minimal additional training.

The model accuracy would possibly be improved with high-resolution Digital Elevation Model (DEM) data (slope, elevation, curvature) [2,7] or with additional imagery bands or spectral analysis (e.g., Normalized Digital Vegetation Index, NDVI) [29–31] to separate the non-rock class into more precise surface covers (e.g., water and vegetation types). However, DELTA performed remarkably well using only a single, readily available data input (NAIP) and twenty well-chosen training sites. We did find DELTA overpredicted bedrock in urban areas, so masking with land-use maps as performed in this study, is likely necessary for successful performance in regions with more buildings or infrastructure.

Many previous studies have shown that neural networks like DELTA are better at image classification compared to traditional supervised models like Random Forest (e.g., [1,14]). However, deep-learning models like DELTA do require high-performance computing resources as they are more computationally expensive than traditional methods. So, the trade-off between computational resources and accuracy could be further tested for large-scale applications. Improved accuracy in locating surficial exposure of bedrock, or inversely soil, has implications for mineral resource to hazard mapping [1,19]. In natural landscapes, soil (or mobile regolith) exists where bedrock is not exposed. Therefore, accurately distinguishing between bedrock outcrops and barren soil greatly improves estimates of soil-covered landscapes, which is a key input for models such as carbon storage, infiltration, and shallow-landslide susceptibility [1]. Land cover, particularly the presence of bedrock or soil, has implications for models used by scientists to decision-makers for land-use planning, conservation or resource allocation, and climate adaptation.

5. Conclusions

We explored the use of machine learning tools to map exposed bedrock for the Sierra Nevada Mountains. We trained the DELTA model using 2016 NAIP images, labeled as rock and not rock at 0.6 m resolution using data from fifteen out of twenty sites. Reference labels from the remaining five sites were used for validation and reporting model performance. We found DELTA performed well with 90% accuracy, 82% recall, and 75% precision. The DELTA map of bedrock outcrops outperformed the existing NLCD barren class, with a miss rate of only 16% compared to 82% for NLCD. When the NLCD barren class was compared with the DELTA bedrock map for the entire Sierra Nevada (Figure 1), the NLCD barren class missed 85% of bedrock, despite barren being a broader class that includes bedrock as well as fragmental rock and bare soil. Our DELTA-trained model predicts bedrock outcrops much more accurately over a wide study area (63,000 km^2), even with minimal training data (83.2 km^2), across a range of rock types and morphology. We conclude that bedrock maps can be substantially improved using a combination of high-resolution imagery and machine learning, and that these improvements can impact many land-cover-dependent models, and allow for Sierra-wide analyses to test uplift or landscape models (e.g., [32]). However, more model training with additional labeled images could be needed to accurately expand the model to other regions.

Author Contributions: Conceptualization, A.S., C.C.-D. and J.D.S.; methodology, A.S.; software, B.C.; validation, A.S.; formal analysis, A.S.; investigation, A.S. and C.C.-D.; resources, J.D.S.; data curation, A.S. and C.C.-D.; writing—original draft preparation, A.S. and C.C.-D.; writing—review and editing, A.S., C.C.-D., B.C. and J.D.S.; visualization, A.S.; supervision, C.C.-D. and J.D.S.; project administration, C.C.-D.; funding acquisition, J.D.S. All authors have read and agreed to the published version of the manuscript.

Funding: This work was funded by the U.S. Geological Survey (USGS) Mineral Resources Program, National Cooperative Geologic Mapping Program, Landslide Hazards Program, and National Innovation Center and run on the USGS Tallgrass high-performance computer.

Data Availability Statement: Associated data are available at the USGS ScienceBase Data Repository https://doi.org/10.5066/P9UQDIDE. The DELTA code is available to download from https://github.com/nasa/delta (accessed on 21 September 2021). The National Land Cover Database (NLCD) is available online https://www.mrlc.gov (accessed on 27 January 2025). NAIP imagery is available to download from various online portals including https://earthexplorer.usgs.gov/ (accessed on 27 January 2025).

Acknowledgments: We thank the USGS reviewer for their thoughtful comments during the internal review, and the four anonymous journal reviewers for their helpful feedback that greatly improved this manuscript. Additionally, we thank the USGS Tallgrass high-performance computing staff. Any use of trade, firm, or product names is for descriptive purposes only and does not imply endorsement by the U.S. Government.

Conflicts of Interest: Author Brian Coltin was employed by the company KBR Inc. The remaining authors declare that the research was conducted in the absence of any commercial or financial relationships that could be construed as a potential conflict of interest.

References

1. Petliak, H.; Cerovski-Darriau, C.; Zaliva, V.; Stock, J. Where's the Rock: Using Convolutional Neural Networks to Improve Land Cover Classification. *Remote Sens.* **2019**, *11*, 2211. [CrossRef]
2. Ganerød, A.J.; Bakkestuen, V.; Calovi, M.; Fredin, O.; Rød, J.K. Where Are the Outcrops? Automatic Delineation of Bedrock from Sediments Using Deep-Learning Techniques. *Appl. Comput. Geosci.* **2023**, *18*, 100119. [CrossRef]
3. Lary, D.J.; Alavi, A.H.; Gandomi, A.H.; Walker, A.L. Machine Learning in Geosciences and Remote Sensing. *Geosci. Front.* **2016**, *7*, 3–10. [CrossRef]
4. Wickham, J.; Stehman, S.V.; Sorenson, D.G.; Gass, L.; Dewitz, J.A. Thematic Accuracy Assessment of the NLCD 2016 Land Cover for the Conterminous United States. *Remote Sens. Environ.* **2021**, *257*, 112357. [CrossRef] [PubMed]
5. Hillier, J.K.; Smith, M.J.; Armugam, R.; Barr, I.; Boston, C.M.; Clark, C.D.; Ely, J.; Fankl, A.; Greenwood, S.L.; Gosselin, L.; et al. Manual Mapping of Drumlins in Synthetic Landscapes to Assess Operator Effectiveness. *J. Maps* **2015**, *11*, 719–729. [CrossRef]
6. Guzzetti, F.; Mondini, A.; Cardinali, M.; Fiorucci, F.; Santangelo, M.; Chang, K.-T. Landslide Inventory Maps: New Tools for an Old Problem. *Earth-Sci. Rev.* **2012**, *112*, 42–66. [CrossRef]
7. Odom, W.; Doctor, D. Rapid Estimation of Minimum Depth-to-Bedrock from Lidar Leveraging Deep-Learning-Derived Surficial Material Maps. *Appl. Comput. Geosci.* **2023**, *18*, 100116. [CrossRef]
8. van der Meij, W.M.; Meijles, E.W.; Marcos, D.; Harkema, T.T.L.; Candel, J.H.J.; Maas, G.J. Comparing Geomorphological Maps Made Manually and by Deep Learning. *Earth Surf. Process. Landf.* **2022**, *47*, 1089–1107. [CrossRef]
9. Zhang, C.; Sargent, I.; Pan, X.; Li, H.; Gardiner, A.; Hare, J.; Atkinson, P.M. Joint Deep Learning for Land Cover and Land Use Classification. *Remote Sens. Environ.* **2019**, *221*, 173–187. [CrossRef]

10. Zhang, P.; Ke, Y.; Zhang, Z.; Wang, M.; Li, P.; Zhang, S. Urban Land Use and Land Cover Classification Using Novel Deep Learning Models Based on High Spatial Resolution Satellite Imagery. *Sensors* **2018**, *18*, 3717. [CrossRef] [PubMed]
11. Al-Najjar, H.A.H.; Kalantar, B.; Pradhan, B.; Saeidi, V.; Halin, A.A.; Ueda, N.; Mansor, S. Land Cover Classification from Fused DSM and UAV Images Using Convolutional Neural Networks. *Remote Sens.* **2019**, *11*, 1461. [CrossRef]
12. Maxwell, A.E.; Warner, T.A.; Fang, F. Implementation of Machine-Learning Classification in Remote Sensing: An Applied Review. *Int. J. Remote Sens.* **2018**, *39*, 2784–2817. [CrossRef]
13. Wang, J.; Bretz, M.; Dewan, M.A.A.; Delavar, M.A. Machine Learning in Modelling Land-Use and Land Cover-Change (LULCC): Current Status, Challenges and Prospects. *Sci. Total Environ.* **2022**, *822*, 153559. [CrossRef] [PubMed]
14. Heydari, S.S.; Mountrakis, G. Meta-Analysis of Deep Neural Networks in Remote Sensing: A Comparative Study of Mono-Temporal Classification to Support Vector Machines. *ISPRS J. Photogramm. Remote Sens.* **2019**, *152*, 192–210. [CrossRef]
15. Leverington, D.W.; Moon, W.M. Landsat-TM-Based Discrimination of Lithological Units Associated with the Purtuniq Ophiolite, Quebec, Canada. *Remote Sens.* **2012**, *4*, 1208–1231. [CrossRef]
16. Kahle, A.B.; Gillespie, A.R. Thermal Inertia Imaging: A New Geologic Mapping Tool. *Geophys. Res. Lett.* **1976**, *3*, 26–28. [CrossRef]
17. Asano, Y.; Yamaguchi, Y.; Kodama, S. Geological Mapping by Thermal Inertia Derived from Long-Term Maximum and Minimum Temperatures in ASTER Data. *Q. J. Eng. Geol. Hydrogeol.* **2022**, *56*, 1–9. [CrossRef]
18. Cracknell, M.J.; Reading, A.M. Geological Mapping Using Remote Sensing Data: A Comparison of Five Machine Learning Algorithms, Their Response to Variations in the Spatial Distribution of Training Data and the Use of Explicit Spatial Information. *Comput. Geosci.* **2014**, *63*, 22–33. [CrossRef]
19. Scarpone, C.; Schmidt, M.G.; Bulmer, C.E.; Knudby, A. Semi-Automated Classification of Exposed Bedrock Cover in British Columbia's Southern Mountains Using a Random Forest Approach. *Geomorphology* **2017**, *285*, 214–224. [CrossRef]
20. Shastry, A.; Carter, E.; Coltin, B.; Sleeter, R.; McMichael, S.; Eggleston, J. Mapping Floods from Remote Sensing Data and Quantifying the Effects of Surface Obstruction by Clouds and Vegetation. *Remote Sens. Environ.* **2023**, *291*, 113556. [CrossRef]
21. NASA. *DELTA (Deep Earth Learning, Tools, and Analysis)*; NASA Ames Intelligent Robotics Group: Moffett Field, CA, USA, 2021.
22. Buscombe, D.; Ritchie, A.C. Landscape Classification with Deep Neural Networks. *Geosciences* **2018**, *8*, 244. [CrossRef]
23. Shastry, A.; Cerovski-Darriau, C. *Data from "Mapping Bedrock Outcrops in the Sierra Nevada Mountains (California, USA) Using Machine Learning": U.S. Geological Survey Data Release*; U.S. Geological Survey: Reston, VA, USA, 2023. [CrossRef]
24. Jin, S.; Homer, C.; Yang, L.; Danielson, P.; Dewitz, J.; Li, C.; Zhu, Z.; Xian, G.; Howard, D. Overall Methodology Design for the United States National Land Cover Database 2016 Products. *Remote Sens.* **2019**, *11*, 2971. [CrossRef]
25. Yang, L.; Jin, S.; Danielson, P.; Homer, C.; Gass, L.; Bender, S.M.; Case, A.; Costello, C.; Dewitz, J.; Fry, J.; et al. A New Generation of the United States National Land Cover Database: Requirements, Research Priorities, Design, and Implementation Strategies. *ISPRS J. Photogramm. Remote Sens.* **2018**, *146*, 108–123. [CrossRef]
26. Homer, C.; Dewitz, J.; Jin, S.; Xian, G.; Costello, C.; Danielson, P.; Gass, L.; Funk, M.; Wickham, J.; Stehman, S.; et al. Conterminous United States Land Cover Change Patterns 2001–2016 from the 2016 National Land Cover Database. *ISPRS J. Photogramm. Remote Sens.* **2020**, *162*, 184–199. [CrossRef] [PubMed]
27. Perone, C.S.; Calabrese, E.; Cohen-Adad, J. Spinal Cord Gray Matter Segmentation Using Deep Dilated Convolutions. *Sci. Rep.* **2018**, *8*, 5966. [CrossRef] [PubMed]
28. Rafique, M.U.; Zhu, J.; Jacobs, N. Automatic Segmentation of Sinkholes Using a Convolutional Neural Network. *Earth Sp. Sci.* **2022**, *9*, e2021EA002195. [CrossRef]
29. Pei, J.; Wang, L.; Huang, N.; Geng, J.; Cao, J.; Niu, Z. Analysis of Landsat-8 OLI Imagery for Estimating Exposed Bedrock Fractions in Typical Karst Regions of Southwest China Using a Karst Bare-Rock Index. *Remote Sens.* **2018**, *10*, 1321. [CrossRef]
30. Ruan, O.; Liu, S.; Zhou, X.; Luo, J.; Hu, H.; Yin, X.; Yuan, N. LANDSAT Multispectral Image Analysis of Bedrock Exposure Rate in Highly Heterogeneous Karst Areas through Mixed Pixel Decomposition Considering Spectral Variability. *Land Degrad. Dev.* **2023**, *34*, 2880–2895. [CrossRef]
31. Chen, Y.; Wang, Y.; Zhang, F.; Dong, Y.; Song, Z.; Liu, G. Remote Sensing for Lithology Mapping in Vegetation-Covered Regions: Methods, Challenges, and Opportunities. *Minerals* **2023**, *13*, 1153. [CrossRef]
32. Hahm, W.J.; Riebe, C.S.; Lukens, C.E.; Araki, S. Bedrock Composition Regulates Mountain Ecosystems and Landscape Evolution. *Proc. Natl. Acad. Sci. USA* **2014**, *111*, 3338–3343. [CrossRef]

remote sensing

MDPI

Article

Reconstruction of a Monthly 1 km NDVI Time Series Product in China Using Random Forest Methodology

Mengmeng Sun [1,2], Adu Gong [1,2,*], Xiang Zhao [1,2], Naijing Liu [1,2], Longping Si [1,2] and Siqing Zhao [1,2]

1 State Key Laboratory of Remote Sensing Science, Jointly Sponsored by Beijing Normal University and Aerospace Information Research Institute of Chinese Academy of Sciences, Faculty of Geographical Science, Beijing Normal University, Beijing 100875, China; smm2021@mail.bnu.edu.cn (M.S.); zhaoxiang@bnu.edu.cn (X.Z.); liunj@mail.bnu.edu.cn (N.L.); silongp@bnu.edu.cn (L.S.); zhaosiqing@mail.bnu.edu.cn (S.Z.)

2 Beijing Engineering Research Center for Global Land Remote Sensing Products, Institute of Remote Sensing Science and Engineering, Faculty of Geographical Science, Beijing Normal University, Beijing 100875, China

* Correspondence: gad@bnu.edu.cn; Tel.: +86-010-58806178

Abstract: The normalized difference vegetation index (NDVI) is one of the most common metrics used to describe vegetation dynamics. Unfortunately, low-quality pixels resulting from contamination (by features including clouds, snow, aerosols, and mixed factors) have impeded NDVI products' widespread application. Researchers have thought of several ways to improve NDVI quality when contamination occurs. However, most of these algorithms are based on the noise-negative deviation principle, which aligns low-value NDVI products to an upper line but ignores cases where absolute surface values are low. Consequently, to fill in these research gaps, in this article, we use the random forest model to produce a set of high-quality NDVI products to represent actual surface characteristics more accurately and naturally. Climate and geographical products are used as model inputs to describe environmental factors. They represent the random forest (RF) model that establishes relationships between MODIS NDVI products and meteorological products in high-quality areas. In addition, auxiliary data and empirical knowledge are employed to meet filling requirements. Notably, the random forest (RF) algorithm exhibits a mean absolute error (MAE) of 0.024 and a root mean squared error (RMSE) of 0.034, in addition to a coefficient of determination (R^2) value of 0.974. Furthermore, the MAE and RMSE of the RF-based method decreased by 0.014 and 0.019, respectively, when compared to those of the STSG (spatial–temporal Savitzky–Golay) plan and by 0.013 and 0.015, respectively, when compared to the LSTM (long short-term memory) method. R^2 increased by 0.039 and 0.027, respectively, compared to the STSG and LSTM methods. We introduced a novel series of NDVI products that demonstrated consistent spatial and temporal connectivity. The novel product exhibits enhanced adaptability to intricate environmental conditions and promises the potential for utilization in investigating vegetation dynamics within the Chinese region.

Keywords: MODIS NDVI time series product; contaminated areas; random forest; reconstruction

1. Introduction

The use of NDVI is widespread due to its simplicity and stability and its close correlation with and high sensitivity to vegetation greenness. More specifically, NDVI reduces certain types of band-correlated noise (noise related positively) and the effects of changes in direct and diffuse irradiance, clouds and cloud shadows, sun and view angles, topography, atmospheric correction, and other factors. Rationalization can also minimize calibration and instrument-related errors to a certain extent. The time at which rationing can reduce noise depends on the noise correlation between red and NIR responses. In addition, it depends on responses as to the degree to which the surface exhibits Lambertian (angle-independent) behavior [1,2]. MODIS NDVI time series products have become widespread in monitoring vegetation growth in the face of global change. Through the quantitative analysis of the

NDVI, changes in vegetation growth patterns can be accurately discerned and observed over time. Such monitoring provides valuable information about global change processes and their impact on vegetation growth [3–5]. Nevertheless, specific pixels contaminated by clouds, snow, aerosols, and other compounding factors are found in these products, thereby precluding their practical application. As a result, the presence of the factors above can compromise the efficacy of these products. These factors can ultimately diminish their overall suitability for usage. Although preprocessing steps such as using a maximum value composite can remove some noise, a considerable amount remains in MODIS NDVI time series products [6]. Therefore, developing an efficient reconstruction method for postprocessing and obtaining high-quality NDVI time series products is necessary to remove the noise that the preprocessing steps cannot clearly achieve [7–9].

Previous studies have proposed reconstruction methods based on three principles for NDVI noise reduction: filter-based, temporal, and hybrid. First, many methods have used filters to fill the gaps caused by contaminated areas, such as the Savitzky–Golay filter and the mean value iterative filter [10–12]. Second, the temporal method is based on a mathematical function that aims to fit the daytime growth curve, such as the asymmetric Gaussian or double logistic part [13,14]. Temporal methods usually entirely use the relationship between vegetation phenology and long time series products to reduce the noise deviation on the time series curve. These curves are of the same category and standard libraries [15–17]. Third, the hybrid method considers temporal continuity and spatial correlation to correct long-term data gaps, such as the material–spatial iteration method and the spatiotemporal tensor method [18,19]. In addition, Cao et al. introduced a technique for reducing noise that incorporates spatiotemporal information with the SG filter, called spatial–temporal Savitzky–Golay (STSG) [20]. The results showed that STSG performed significantly better than the four previously widely used methods (i.e., the asymmetric Gaussian, double logistic, Fourier-based, and Savitzky–Golay filter). One obvious advantage was that STSG addressed the problem of continuous NDVI gaps. STSG effectively increased local low NDVI values and avoided overcorrecting common NDVI values during crop harvest. The long short-term memory network (LSTM) is a specific type of recurrent neural network (RNN) that has been modified to enhance its performance in time series prediction [21]. The unique neuron structure of LSTM selectively remembers information and converges on the optimal solution more efficiently than traditional RNNs. The LSTM achieves this through several controllable gates (namely the forgetting gate, input gate, and output gate) that facilitate the passage of information between different units within the hidden layer. Although these methods follow other principles, they all obey one rule. This mechanism enables the extraction of temporal features from time series data, creating a prediction model capable of accurately predicting long-term trends. A reconstructed NDVI product can correct for noise by iteratively updating deviations from the time series curve of the noisy point. The angle from the standard library is compared to the one in the same category [21,22].

While these existing methods have been widely used to reconstruct low-quality NDVI areas, two challenges still need to be addressed. The first challenge is filling in the remaining gaps (e.g., temporally determined and continuous spatial–temporal gaps) in the MODIS NDVI time series due to cloud and snow conditions. To solve this problem, the interpolation method is usually used [23]. However, how to fill in continuous gaps or sudden changes in the time series is yet to be discovered. Therefore, finding a robust method to estimate missing values more accurately is critical [12]. The second challenge is identifying the noise level, which may indicate over-alignment [24]. Many machine learning (ML) methods have been used to produce high-quality NDVI products [25–27]; however, these methods need to pay more attention to the unique periodic characteristics of multiyear accumulations. Several attempts have been made to address their gaps, e.g., applying hundreds of time series images to reconstruct the long-term NDVI products [17,19,28] to obtain a smooth curve that ignores short-term changes based on vegetation growth.

Despite the efforts regarding the method mentioned earlier, the NDVI products were primarily reconstructed without considering the impact of temperature, rainfall, and other information on the vegetation growth process in the MODIS NDVI time series product [18,29–31]. Therefore, finding a more stable gap-filling method that can effectively determine how to process low NDVI values, for example, whether to maintain or directly correct common NDVI values before processing them, is imperative. Considering this gap, we describe a novel technique to reconstruct NDVI products in contaminated areas in the present article. This technique was developed by making spatial and temporal features, continuous multiyear time series data, and multiple products [32]. This method can be used to reconstruct NDVI products using multisource quantitative remote sensing products, like FAPAR and LAI, as well as factors related to vegetation growth, like longitude, latitude, and elevation. This is carried out to avoid the effect of NDVI products alone when several low-quality products are in a row. This study reconstructed NDVI data from low-quality regions using the random forest (RF) method, environmental factors, and historical data from remote sensing products. This led to spatiotemporal continuous monthly average NDVI products. The product can monitor vegetation growth, ecological change, and species evolution.

The rest of this article is structured as follows: Section 2 covers the data sources. Section 3 describes the algorithm and process in detail. Section 4 presents the model's performance results and a comparison with those of other products. Sections 5 and 6 discuss and conclude.

2. Study Area and Datasets

2.1. Study Area

China is a vast country with 34 administrative regions covering 9.6 million square kilometers. The research region considered in this study spans from longitude 73°40′ to 135°2′E and latitude 3°52′ to 53°33′N and encompasses extensive grasslands, agricultural land, and woodland. The eastern region of China has more cultivated and forested land than the western region, mainly green and desert. Due to its geographical location, China experiences many climate conditions. These include moderate continental climates in the north, northeast, and south and highland mountain climates in the Tibetan Plateau. The northwest region is arid and semi-arid, while the southeast region is semi-humid and humid. Moreover, China's elevation and precipitation levels divide the country into eight subregions, each with unique climatic conditions.

2.2. Datasets

This study used several products as input data, and post reconstruction NDVI products were exported as output products. Table 1 lists the input product data sources. Nine products in the Global Land Surface Satellite (GLASS) product series were u in NDVI reconstruction, including albedo products [33], such as Albedo BSA VIS, Albedo WSA NIR, Albedo WSA SWIR, and Albedo WSA VIS, evapotranspiration (ET) [34], photosynthetically active radiation absorbed fraction (FAPAR) [35], fractional vegetation cover (FVC) [36], gross primary production (GPP) [37], and leaf area index (LAI) [38]. The meteorological products used in this study were the product of temperature and precipitation, which were derived from a dataset with a monthly temporal resolution and 0.5° × 0.625° spatial resolution called the Modern-Era Retrospective Analysis for Research and Applications version 2 (MERRA-2) reanalysis [39]. In addition, the product was available from 1982 to 2015. Obtaining precise surface elevation information for remote sensing information retrieval was imperative. Therefore, we also considered using NASA DEM in this study for reconstructing NDVI, which was developed for long-range postprocessing for the Topographic Radar Mission [40].

Table 1. The introduction of the products used in this study.

Data	Spatial Resolution	Time Resolution	Data Source
NDVI	1 km	Monthly	MOD13A3
Albedo (BSA VIS/WSA NIR/WSA SWIR/WSA VIS)	1 km	Eight days	GLASS
ET	1 km	Eight days	GLASS
FAPAR	1 km	Eight days	GLASS
FVC	1 km	Eight days	GLASS
GPP	1 km	Eight days	GLASS
LAI	1 km	Eight days	GLASS
Precipitation	0.5°	Monthly	MERRA2
Temperature	0.5°	Monthly	MERRA2
DEM	30 m		NASADEM
Longitude	1 km		According to the longitude range of China
Latitude	1 km		According to the latitude range of China

3. Methodology

Reconstructing high-quality MODIS NDVI products is quite challenging due to continuous missing areas. Due to the influence of clouds, snow, and aerosols on the NDVI product, there are some temporal and spatial series gaps. More sophisticated processing methods are required to improve MODIS NDVI time series products. Based on the combination of NDVI acquisition methods and their correlation with other products, we considered using multiple products to extract NDVI to complete the repair and filling tasks.

The proposed method was to generate continuous, long-span products with a monthly temporal resolution and a 1 km spatial resolution from 2001 to 2015. In addition, the high-quality areas in MOD13A3 were used as a reference to evaluate the performance of the proposed method. The quality flag layer illustrates the reliability of MODIS NDVI time series products and determines whether a value is good or contaminated by clouds, snow, ice, etc. It indicates that RI = 0 represents good NDVI quality, RI = 1 means some uncertainty in the NDVI, and multiple RI values indicate contaminated area pixels. The machine learning method divides datasets into training, validation, and testing. The training set is used to train the machine learning model. This involves the machine learning algorithm learning from the dataset and adjusting its parameters to arrive at the most effective predictive model. Through training the model on the training set, it can discern patterns and connections within the dataset. This ultimately facilitates its ability to accurately predict outcomes from unfamiliar data.

The model's parameters are adjusted using the validation set to optimize its performance during training. The selection of hyperparameters, such as the number of layers or decision tree depth, is required when training a neural network. The model's performance is evaluated on a validation set with varying parameter settings to identify the most effective ones that prevent overfitting. Overfitting occurs when the model performs exceptionally well on training data but needs to improve on new, unobserved data. The final step in evaluating model performance is using a distinct testing set. This comprises data that the model still needs to observe. The model's performance on unseen data can be assessed by making predictions on this testing set and comparing the predicted and actual labels. Such an evaluation plays a crucial role in determining the model's generalization ability, demonstrating its proficiency in processing novel data.

To summarize, the process of constructing a machine learning model involves the utilization of a training set for model training. In addition, there is a validation set for parameter adjustment to prevent overfitting and a testing set for model performance evaluation. We can develop high-performance machine learning models and robust generalization capabilities by adequately partitioning the datasets.

The flowchart of this paper is shown in Figure 1. Initially, gathering pertinent information required for the investigation, including ecological commodities, meteorological data, and geographical data, was imperative. Subsequently, the data needed to undergo various procedures such as format conversion, scaling, resampling, clipping, and synthesis of monthly average data (averaging data every eight days a month). Incorporating MOD13A3 long and short bit rules, the second step involved the selection of high-quality samples that met the specified requirements for a given month. Additionally, combining the pieces from every 12 months yielded high-quality samples for the entire year. To ensure representativeness, a specific number of high-quality models were randomly and uniformly selected for each category across various provinces in China, covering the period from 2001 to 2015. In the third step of the process, model parameters were established for training based on the samples created in step 2. Machine learning models, LSTM and RF, were trained. In the fourth step, LSTM, STSG, and RF algorithms were evaluated using verification data. The R^2 and RMSE of high-quality MOD13A3 were used for comparison purposes. Afterward, the trained model was used with the data to forecast the low-quality region. Finally, the model's projection results for the low-quality areas were employed to fill in the original low-quality-region data, resulting in a time series dataset of superior quality.

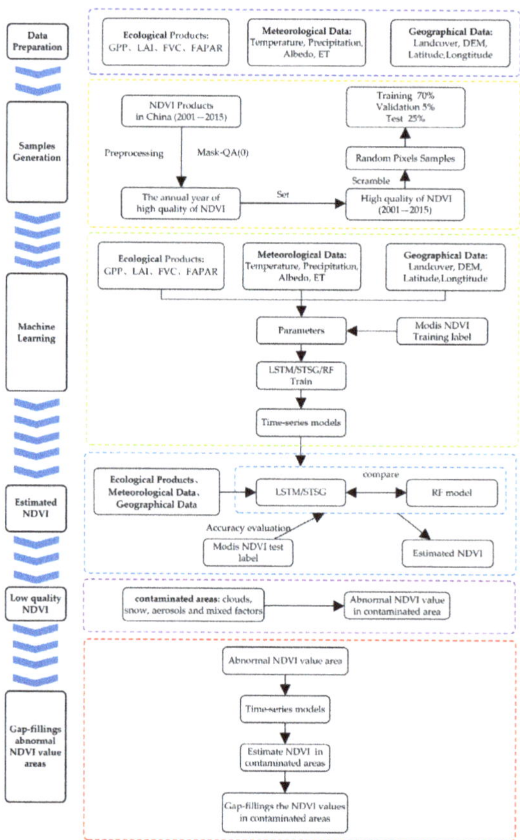

Figure 1. The flowchart of the proposed estimation method for the MODIS NDVI product in the contaminated areas of China. RF, LSTM, and STSG abbreviations represent random forest, long short-term memory, and the spatiotemporal Savitzky–Golay filter.

We thoroughly analyzed and examined the problems from two perspectives. In addition, we used spatial distribution images and histograms to illustrate MODIS NDVI quality. Figure 2a shows the spatial distribution of low- and high-quality products caused by clouds, snow, aerosols, and mixed factors. We found that high-quality samples were mainly concentrated in the central and northwestern regions, while low-quality areas due to aerosols, clouds, and snow were primarily located in the south and northwest. This distribution pattern was aligned with the area's industrial development and essential meteorological conditions site.

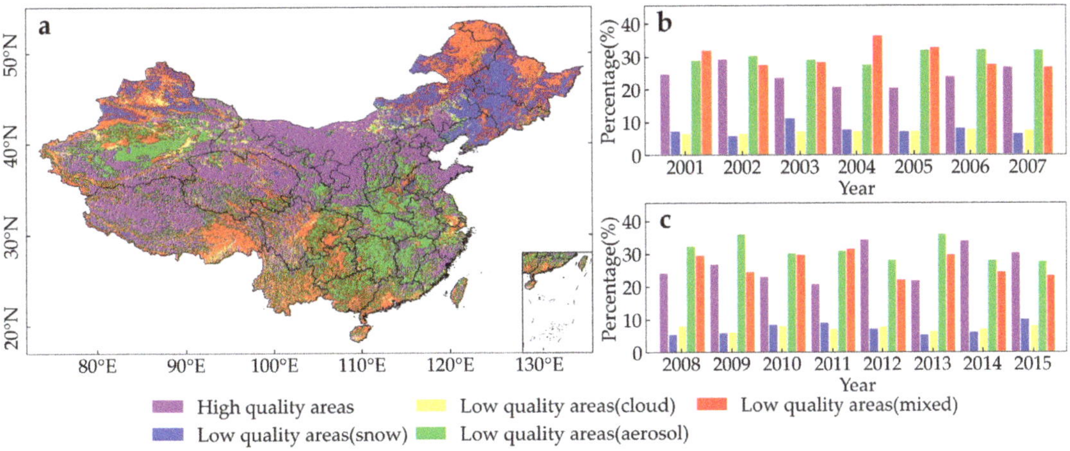

Figure 2. The spatial distribution maps and bar graphs of low-quality areas in China from 2001 to 2015 are based on the quality assurance (QA) of MODIS NDVI products. (**a**) illustrates the spatial distribution of pollutant factors, namely snow, cloud, aerosol, and mixed elements. (**b**) presents the annual percentage of incidents in various regions from 2001 to 2007. (**c**) displays the rates for different areas from 2008 to 2015.

To systematically analyze the distribution of NDVI products, we presented the results as simple statistical proportions. Our approach included using a histogram to objectively describe and understand the spatial distribution of NDVI products (Figure 2b,c). Clouds, snow, aerosols, and various factors caused low-quality areas. Multiple factors, such as clouds, snow, aerosols, and mixed factors, caused various factors in low-quality areas. Snow contributed to a relatively small proportion of low-quality NDVI. In contrast, clouds, aerosols, and a combination of factors contributed to a large proportion. MODIS NDVI time series products had many contaminated areas, resulting in gaps and limited applications. Adverse weather conditions and cloud cover caused decreased lower NDVI values in contaminated sites. Reconstructing high-quality MODIS NDVI time series data and developing research methods to repair gaps in contaminated regions remain challenging. Investigating suitable methods for reconstructing NDVI products and generating continuous spatial and temporal NDVI product series is essential.

3.1. Data Preprocessing

Initially, creating a time series was necessary to consolidate the conversion rate, resampling, clipping, and monthly average data synthesis. "Upscaling" refers to increasing an image's size while decreasing its spatial resolution. We utilised three rounds of convolution to transform NASA's Digital Elevation Model (NASA DEM) from a 30 m scale to a 1 km scale. Standard interpolation methods include a nearest neighbour, bilinear, and cubic convolution techniques to scale images. Among these, cubic convolution interpolation represents a more advanced approach. This algorithm performs linear interpolation using 16 neighbouring grayscale values around the target sample point. This method produces superior results by considering both the direct grayscale influence of the four adjacent points and the rate at which grayscale values change among neighbouring points. Although bicubic interpolation imposes a significant computational burden, it represents the best approach to image scaling. Therefore, we used an ascending scale with three convolutions to adjust the resolution from high to low. This was to convert a 30 m NASA DEM to a 1 km DEM with three interpolations.

In the second phase, high-quality samples were created by synthesising data and selecting areas based on NDVI product distribution. This resulted in a sample set that satisfied the model requirements. From 2001 to 2015, sample insertion and concatenation were performed, and the NDVI product was prioritised as label data.

3.2. Training Samples from High-Quality Areas of MODIS NDVI in China

Owing to the constraints imposed by the prevailing climate conditions and geographical location, the dataset employed for the model training fell short of good diversity. Diversity would have meant encompassing all of the climatic zones across the globe. Moreover, some areas needed more samples to be improved. Additionally, some land samples exhibited low numbers and their distribution focused on specific regions. For instance, forests dominated coastal regions. At the same time, other areas were scarce while the other parts were limited, leading to a need for more meaningful characteristic learning. As a result, it is imperative to delve deeper into developing algorithms that facilitate sample balancing and land classification to enhance estimates' precision.

Faced with the above problems, this paper adopted some strategies. For the first time, we relied on quality control documentation when initially evaluating high-quality samples. Then, based on this documentation, the pieces needed to meet a QA value of 0 or 1 for the short rule and a QA of 6–7 (aerosol), which should be 00, 01, 10, and 14–15 (clouds, snow). Secondly, to address the problem of unbalanced samples, which may have led to the overfitting of models with a large proportion, we selected samples based on the categorised products. We set a fixed percentage for selecting other categories in the respective states. This enabled us to address the challenges posed by overfitting and ensure that our samples were representative and diverse.

As illustrated in Figure 3, each province exhibited unique distribution and category characteristics. Fortunately, the samples met the criteria for balanced treatment conditions in China, covering a broad range of categories, quantities, and geographical features. Our selected samples were evenly distributed across their respective provinces, representing the overall population. Furthermore, we recommend employing a more sophisticated algorithm to learn the linear relationship between MODIS NDVI and estimated NDVI. This enhanced the accuracy and reliability of our findings, leading to more robust results.

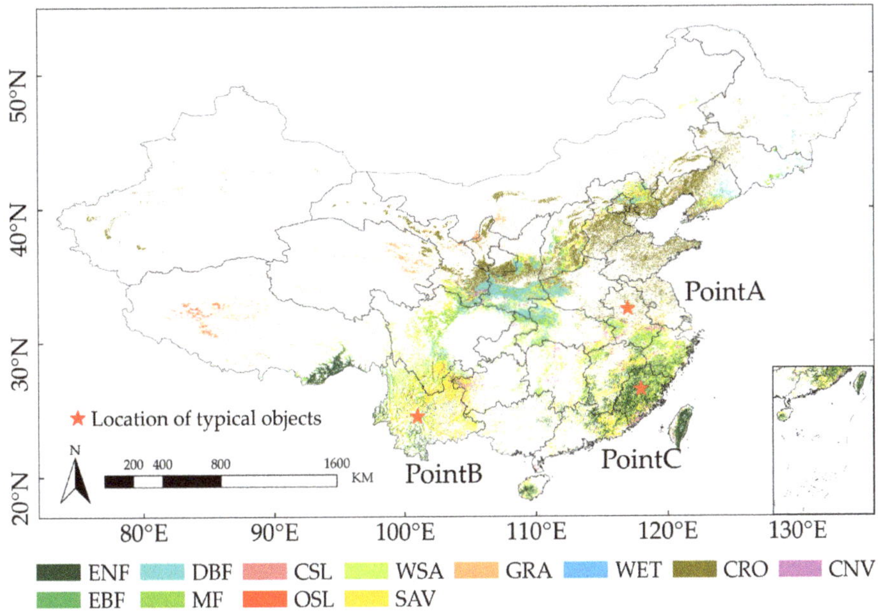

Figure 3. High-quality sample distribution map for 2001–2015. ENF presents evergreen needleleaf forest; EBF: evergreen broadleaf forest; DBF: deciduous broadleaf forest; MF: mixed forest; CSL: closed shrubland; OSL: open shrubland; WSA: woody savanna; SAV: savanna; GRA: grassland; WET: permanent wetland; CRO: cropland; CNV: cropland/natural vegetation mosaic.

3.3. Reconstruction Models

3.3.1. RF Model and Optional Parameters

RF is a continuous value prediction algorithm based on multiple decision trees, which can solve regression problems [41,42] and be applied to supervised classification [43–45]. It requires using bootstrap samples to generate multiple sub-training datasets, then using decision trees to fit each dataset, and finally taking the average of the prediction results of numerous decision trees [43]. In addition, the number of decision trees was based on the distribution of the samples via voting or averaging, which determined the method's performance [46]. Specifically, the rule reliability was also based on the splitting number, and the splitting numbers were based on the Gini index positively associated with rule reliability, which is relatively complex. Specifically, there are 12 products in this paper, taking place over 15 years. The year requires a significant workload, requiring a multiprocessing approach to handle it. Based on the above reasons, we used a default set of decision trees and (300, 4), which is an empirical location used by many researchers [42,43,46].

In the study, we constructed the relationship between the predicted value and the response variables to estimate the NDVI value. We used two main steps, the learning model acquisition phase and gap-filling in the contaminated areas of the MODIS NDVI time series product (Figure 1), to deal with these problems. The acquisition of training models primarily involved generating training samples, determining the generation of training samples, and choosing the optimal model. The trained model can be developed with high-quality sample illustrations and be a reference for forecasting contaminated areas. A high-quality place was obtained based on quality control documents during the sample acquisition stage. The proportion of samples in each category was fixed according to a principle that ensured that the samples were dispersed across provinces. The optimal model acquisition stage obtained a stable model based on the training samples and then selected the optimal training model based on the validation and test samples. The filling phase of the contaminated area mainly included obtaining low-quality data areas and estimating and filling the NDVI values of the regions. The NDVI of the infected areas was inputted into the optimal model, and the raw NDVI values in the contaminated areas were replaced with the estimated NDVI values. Moreover, we selected the estimated NDVI based on quantitative and qualitative evaluation results.

3.3.2. LSTM Model and Optional Parameters

Because the long short-term memory network (LSTM) can control the forgetting and updating gate, it can minimize the escalation, disappearance, or expansion during training. Therefore, it has many applications and excellent performance in natural language processing, speech recognition, and time sequence prediction [47–49]. Moreover, it has robust long-term dependency capabilities and can process sequential products [21,50].

In this study, we developed a five-layer deep neural network model for estimating NDVI, including an input layer, two LSTM layers, a dense layer, and an output layer. Specifically, the number of samples was N, and the number of time steps was 12. Our method's batch size was m, which increased the calculation speed. Moreover, the most effective performance was determined to be obtained with a batch size of 64. A dropout mechanism was introduced to the dense layer to prevent overfitting, and the dropout rates were empirically fixed at 0.5. As a result, the performance of the LSTM model was evaluated by utilizing two LSTM layers with 50 and 30 hidden units, respectively. Moreover, some parameters existed, such as Adam, a gradient-descending-based optimizer, and a learning rate of 0.001.

3.3.3. STSG Model and Optional Parameters

The STSG algorithm principle is a time-sequence-based approach used to fill the missing value in the MODIS NDVI time series product [20]. Currently, it exploits time sequence similarities between the same pixel class and spatial correlations between different types for long-term correction in the interval data gaps. The NDVI time sequence for a specific pixel was estimated based on the NDVI values observed in the neighbor pixels. Thus, NDVI calculations may give essential previous knowledge about the NDVI growth trajectory of the target pixel. In addition, we updated the NDVI time series by concatenating the original NDVI estimates with the pixel's raw NDVI time series. Subsequently, we calculated the weights of each NDVI in this updated time series and used a weighted SG filter to smooth the NDVI time series.

To validate the proposed RF method, we conducted a simulation experiment comparing LSTM and STSG and evaluated their performance through statistical histogram and spatial pattern comparisons. Furthermore, we tested the ability of the RF-based method to capture local phenological features for harvesting in the Taiwan cultivated area (TCA) in the North China Plain. This area consists of double-cropped croplands and broad-leaved deciduous forests, resulting in complex NDVI profiles throughout the year, making it a suitable case for comparing different reconstruction models. We verified whether the RF-based method could outperform the LSTM and STSG methods by retaining low NDVI actual values, while successfully simulating the harvest characteristics. Two methods were used to conduct the comparison of different NDVI reconstruction products. The first method compared the reconstructed NDVI with high-quality data across China's entire high-quality NDVI area. With the second method, the long-term robustness of the products was assessed under different climate conditions from 2001 to 2015. We used RMSE, R^2, MAE, and bias to evaluate the performance of the models. We also examined the high-quality-area MAE, the spatial distribution chart of RMSE, and the year-on-year change trend to determine whether the product had real phenology.

Finally, we found that NDVI reconstruction was necessary for cloudy areas to minimize missing data and reduce the negative impact of poor atmospheric conditions on NDVI values, which was crucial for addressing the effects of climate change. To enhance the reliability of our study, we used 15 years of products for the visual assessment of various vegetation types to determine general vegetation characteristics. We preserved high-quality NDVI values and substituted contaminated values with reconstructed NDVI values. We compared the performance of the RF, LSTM, and STSG models in correcting the contaminated pixels. To investigate phenological characteristics, we focused on mixed forests in the Yunnan, Fujian, and Chongqing provinces.

4. Results

4.1. Product Intercomparison

A comparison of the original NDVI products and reconstructed products generated using the STSG, LSTM, and RF models is presented in Figure 4. The RF and LSTM predictions exhibited high R^2 values, and most data points were clustered tightly around y = x. The NDVI output from the RF-based model displayed the highest accuracy among the models. The RF-based model yielded the highest R^2 value (0.965) and the lowest RMSE (0.0428) and MAE (0.0291) values for high-quality areas, indicating its superior performance in comparison to the LSTM and STSG methods. Consequently, the RF model was considered to be the most effective technique for reconstructing NDVI products.

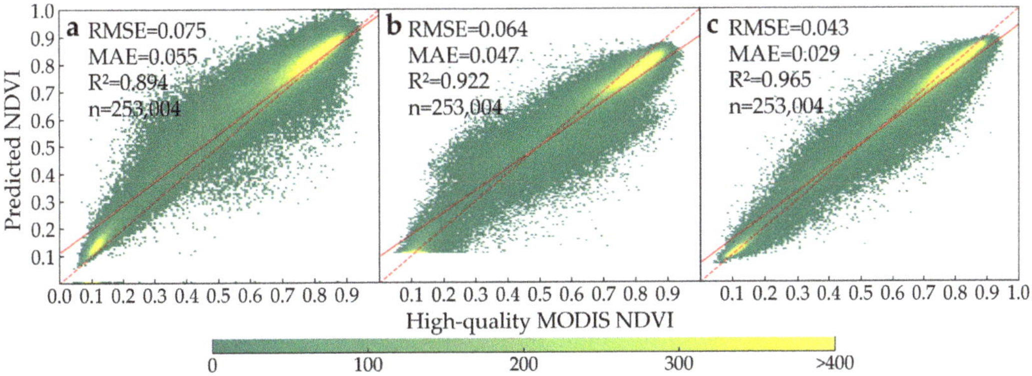

Figure 4. The comparative analysis of NDVI estimates using various methods in June, based on pixels in high-quality areas. The three methods used in the study were short-term satellite imagery gap-filling (STSG), long short-term memory (LSTM), and random forest (RF), as shown in panels (**a**), (**b**), and (**c**), respectively. Note: The red solid line depicts the forecast line, while the red dashed line represents the 1:1 line.

The results of the reconstruction of the long-time series curves in high-quality regions in China from 2001 to 2015 are presented in Figure 5. Statistical comparison outcomes such as relative RMSE, R^2, MAE, and relative bias are highlighted in the figure. The study findings indicated that the RF-based method generated more accurate and precise results for these high-quality areas. This discovery is significant, as it implies that the RF-based approach outperformed the STSG and LSTM methods regarding its ruggedness and adaptability in consistently reconstituting NDVI.

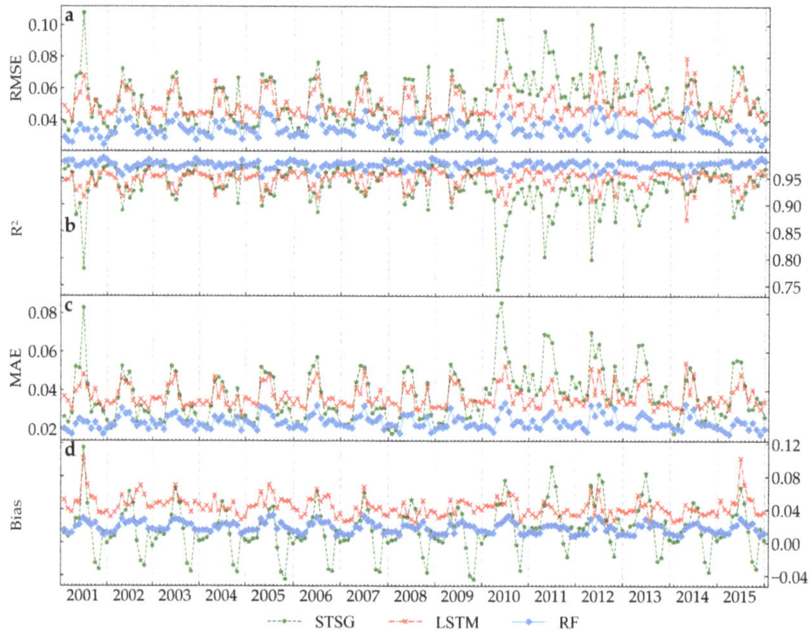

Figure 5. The comparisons of different methods based on the seasonal characteristics of NDVI in high-quality areas from 2001 to 2015. Comparison of the other methods based on seasonal factors of NDVI in high-quality regions during 2001–2015. The four indexes used in the analysis were RMSE, R^2, MAE, and bias comparisons of monthly NDVI estimation, as shown in panels (**a**–**d**), respectively.

A comparison of methods for estimating MODIS NDVI using quantitative and qualitative approaches is presented in Figure 6. Lower values of RMSE and MAE suggest greater accuracy in reconstruction performance. The findings indicated that the RF model was the most effective, with mean RMSE and MAE values of 0.034 and 0.026, respectively, on a national scale. Regarding spatial distribution, areas with an RMSE and MAE below 0.032 accounted for over 50% and 70% of the total area, respectively. The RF reconstruction result exhibited exceptional performance in locations such as the Inner Mongolia Plateau, Sichuan Basin, Loess Plateau, North China Plain, Southeast Coast, eastern Yunnan–Guizhou Plateau, and northern Qinghai–Tibet. However, some local regions, such as the Northern Plains and coastal areas, had RMSEs and MAEs greater than 0.064 due to various crop varieties and insufficient samples of corresponding vegetation types. Last, limitations intrinsic to the study include inadequate characterization and forest concentration in coastal areas.

The observations depicted in Figure 7 delineate the changes in China's high-quality areas in the past 15 years. Linear regression adjustment lines were utilized in this study to detail the trends in vegetation change. Notably, the post reconstruction NDVI, derived from different approaches, exhibited consistent trends. Since 2007, the NDVI values have demonstrated a general increase. This indicates an increase in vegetation and an overall improvement in the ecological environment in China.

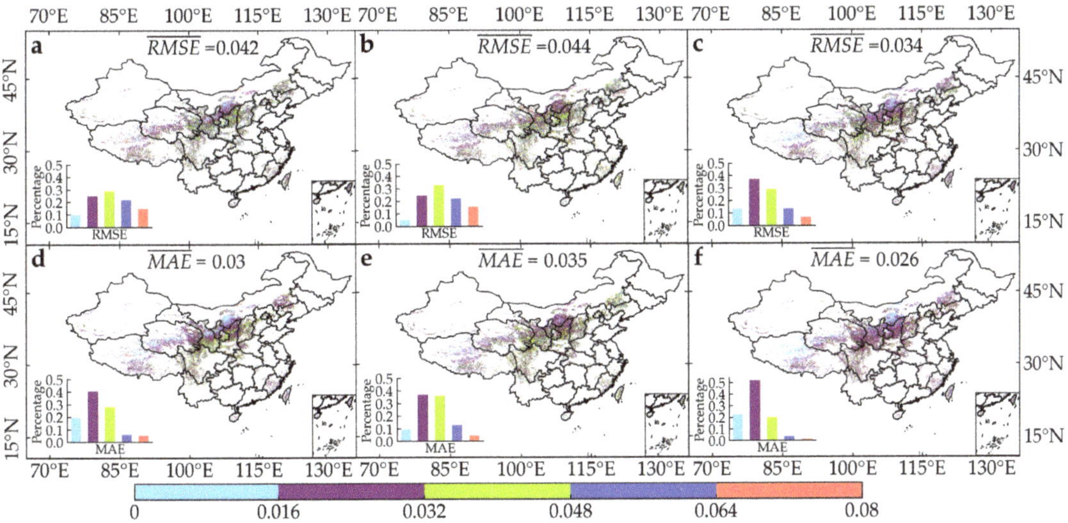

Figure 6. The performance of the reconstruction methods, namely STSG, LSTM, and RF, in high-quality areas. The evaluation was conducted by calculating the average root mean square error (RMSE) and mean absolute error (MAE) between the reconstructed normalized difference vegetation index (NDVI) and the reference NDVI for all pixels. The inset in each panel provides a bar graph depicting the percentage distributions of RMSE and MAE. Note: Quantitative assessments were based on the average of RMSE (shown at (**a**) STSG; (**b**) LSTM; (**c**) RF); and MAE between the reconstructed NDVI and the reference NDVI for high-quality pixels (shown at (**d**) STSG; (**e**) LSTM; (**f**) RF).

4.2. The Comparison of Biome Categories

The reconstructive process using RF alone is depicted in Figure 8. The figure shows that the raw product displayed noticeable spatially continuous gaps due to numerous inherent factors, such as rain, clouds, aerosols, and other impediments (left). The corrected figure represents using the estimated NDVI to fill in the low-quality values in the original NDVI. The ensuing reconstructed NDVI significantly improved the product's overall quality by mitigating the limitations caused by these factors. The reference curve embodies the average normalized difference vegetation index (NDVI) curve pertinent to a specific category within a selected province, for instance, Fujian Province. This reference curve is a product of meticulously chosen high-quality samples, whereby the selection of which is contingent on the quality assurance (QA) file quality identification bits. This QA file serves as a critical tool in ensuring the reliability and validity of the data used, thereby enhancing the accuracy of the resulting NDVI curve. It is paramount to highlight that these high-quality samples hold across all twelve months of the year. This implies that the reference curve, and hence the average NDVI curve, maintain their relevance and applicability throughout the year, providing a consistent and reliable measure of vegetation for the specified category within the chosen province.

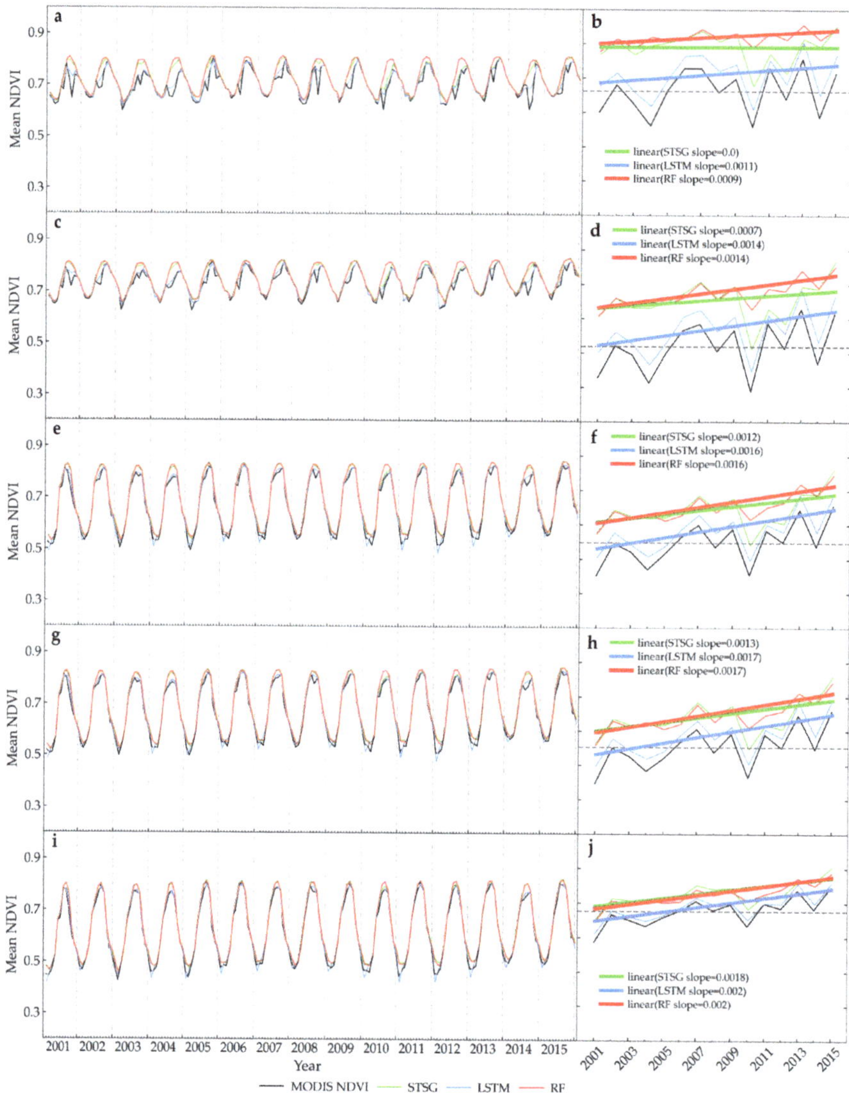

Figure 7. The NDVI trends in high-quality regions of China from 2001 to 2015. The time series of typical characteristics of the points affected by clouds, snow, aerosols, and mixed factors are represented by (**a**,**c**,**e**,**g**,**i**); the comparison of the time series between the original MODIS NDVI product and the estimated NDVI product in 2015 for evergreen needleleaf forests, evergreen broadleaf forests, and mixed forests are represented by (**b**,**d**,**f**,**h**,**j**).

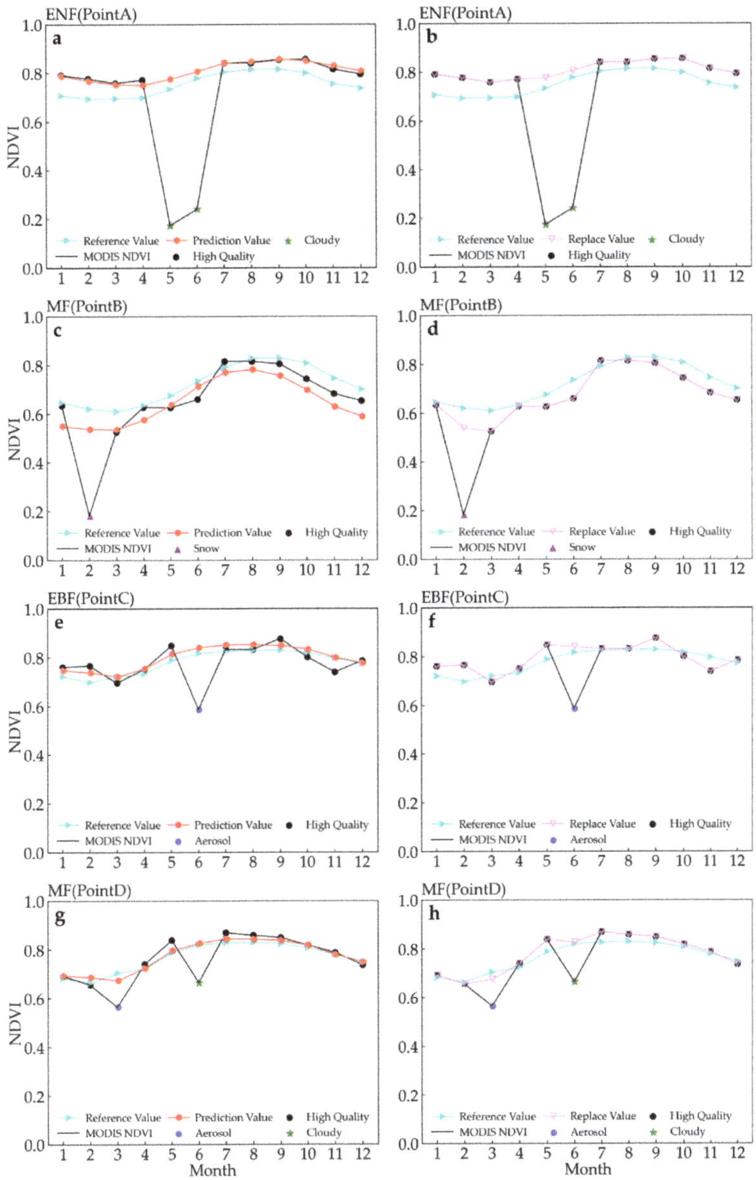

Figure 8. The comparisons between the original MODIS NDVI product and the rectified points of the estimated NDVI product affected by cloud, snow, aerosol, and mixed factors on the distinct biomes. The points affected by cloud in the evergreen needleleaf forests during 2015 are represented by (**a**,**b**); the points affected by snow in the mixed forests during the same year are represented by (**c**,**d**); the points affected by aerosol in the evergreen broadleaf forests are represented by (**e**,**f**); and the points affected by mixed factors in the mixed forests are represented by (**g**,**h**).

Figure 9 displays a comparison of different methodologies. Upon reviewing the temporal trends between 2001 and 2015, it can be inferred that the RF model and STSG are well suited for crop characteristics across two seasons and seed-sowing variations across diverse regions. Despite the ability of RF and STSG to tackle challenging topographical and climatic conditions, the LSTM model failed to account for low-lying areas and inclement weather. Consequently, the RF model exhibited greater resilience across multiple biomes.

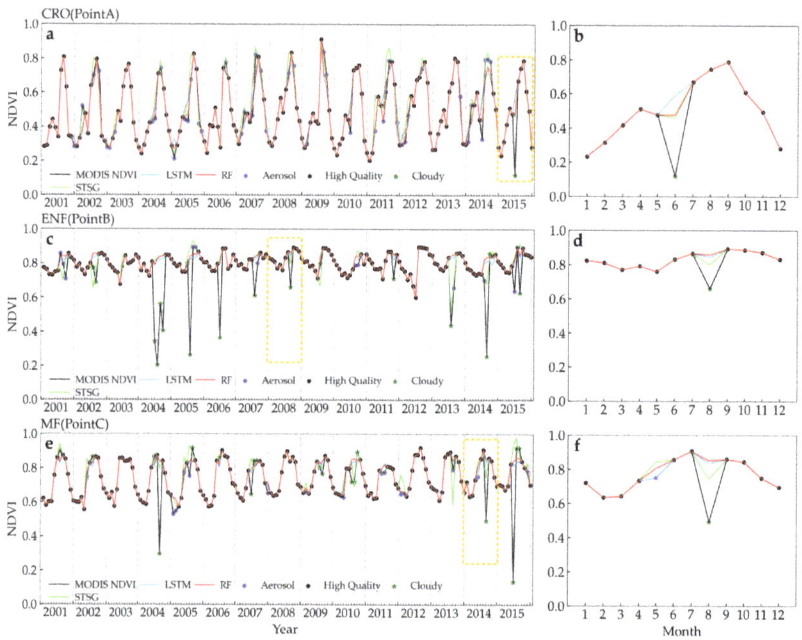

Figure 9. Three-point time series curves were estimated using various methods from 2001 to 2005. (**a**,**b**) CRO, (**c**,**d**) ENF, and (**e**,**f**) MF. The locations of the three points are depicted as a pentagon star (shown in Figure 3), and the dashed rectangle indicates the selected year (shown at (**b**,**d**,**f**)).

Figure 10 displays the MAE and RMSE of the estimated NDVI and MODIS NDVI for various land cover types on the horizontal axis, including CSL, SAV, and others. The left panels a, c, e, and g represent MAE, while the right panels b, d, f, and h represent RMSE. Blue, red, and green correspond to the STSG, LSTM, and RF methods. MODIS NDVI was chosen as the validation value of these methods based on high-quality NDVI selected via QA screening. Our findings stemmed from several factors. The random forest algorithm performs well on the MAE and RMSE of most distinct biome categories, but it could not be more effective in a few cases (Figure 10). Second, the estimated products of Figure 11 exhibited similar characteristics to those of specific biomes such as crops and forests, demonstrating greater accuracy in reflecting natural surface phenological characteristics. Thus, it is feasible for RF to estimate different terrains in different parts of China with remarkable precision. Figure 11 depicts the localized spatial distribution of the RF reconstruction product, along with the time series curves that represent the distinguishing features of a specific biome. All three methods can correct NDVI in low-quality areas to a certain extent, such as March in Anhui Province, June in the Fujian and Zhejiang provinces, and November in Chongqing. To study the characteristics of different biomes, various methodologies were evaluated based on spatial distribution data obtained from the Anhui, Fujian, Zhejiang, and Chongqing regions and their corresponding biome types, such as evergreen broadleaf forest (EBF), evergreen needleleaf forest (ENF), deciduous broadleaf forest (DBF), mixed forest (MF), and crop (CRO) (Figure 12). Using the RF method, the horizontal axis represents the boxplot of the estimated NDVI and MODIS NDVI for other months, such as January and February. The blue denotes a specific region's original observed NDVI values, including Anhui Province and Fujian Province. In contrast, the red represents the replaced values of the original low-quality NDVI using the RF estimated values of the same region. These figures provide valuable insights into various methods' performance and suitability for different categories and areas (Figure 12). A comprehensive analysis revealed that the RF model outperformed the STSG and LSTM models, as the former produced more realistic results with finer texture information. The efficacy of different approaches was further verified using growth curves, highlighting the features of different biomes.

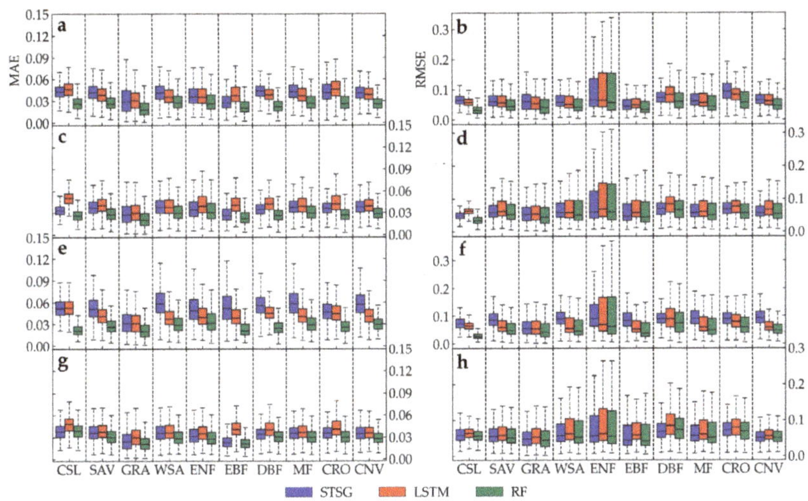

Figure 10. The comparison of different methods in different biome categories. The MAE of the estimated results for the selected years 2001, 2005, 2010, and 2015 is represented by (**a–d**), respectively. The RMSE of the estimated results for the same years is presented by (**e–h**), respectively. Note: the abbreviations of CSL, SAV, GRA, WSA, ENF, EBF, DBF, MF, CRO, and CNV stand for closed shrubland, savanna, grassland, woody savanna, evergreen needleleaf forest, evergreen broadleaf forest, deciduous broadleaf forest, mixed forest, cropland, and cropland/natural vegetation mosaic, respectively.

4.3. Model Validation

The simulated area in China was used to evaluate the accuracy of various methods for estimating NDVI against the standard NDVI in a reliable testing ground. This specific area was selected based on its low presence of low-quality pixels, which can be affected by factors such as clouds, snow, aerosols, and other possible sources of interference. The quality assessment process confirmed the determination above as being more significant. Furthermore, as approved by the QA evaluation, the simulation area also exhibited a scarcity of low-quality pixels impacted by clouds, snow, aerosols, etc. Therefore, the MODIS NDVI data obtained from this simulation area were considered to be appropriate for validation in this study.

As shown in Figure 13, the spatial pattern of the simulated NDVI was underestimated in the regions where the original high-quality NDVI was less than 0.4 and overestimated in those regions where the original NDVI was more significant than 0.4, using the STSG method. Compared with the STSG method, the estimated NDVI derived from the LSTM method differed less from the original high-quality NDVI regarding the spatial pattern, as shown in Figure 13a,b. In addition, this was also true for the histogram distribution, as seen in Figure 13f,g; however, the LSTM algorithm still yielded a problematic estimation result. As shown in Figure 13g, for the regions where the NDVI was in the range of

0.2 < NDVI < 0.4, LSTM underestimated the impact. When NDVI > 0.4, LSTM overestimated the impact. Unlike the other two algorithms, we found that the RF reconstruction method had a better overlap with the original high-quality NDVI in the statistical histogram (Figure 13h) and spatial pattern (Figure 13d). Through the separability analysis comparison, we also found that the RF reconstruction method proposed in this study outperformed the other two mainstream algorithms (shown in Table 2). Therefore, the MODIS NDVI data obtained from this simulation area were considered to be appropriate for validation in this study.

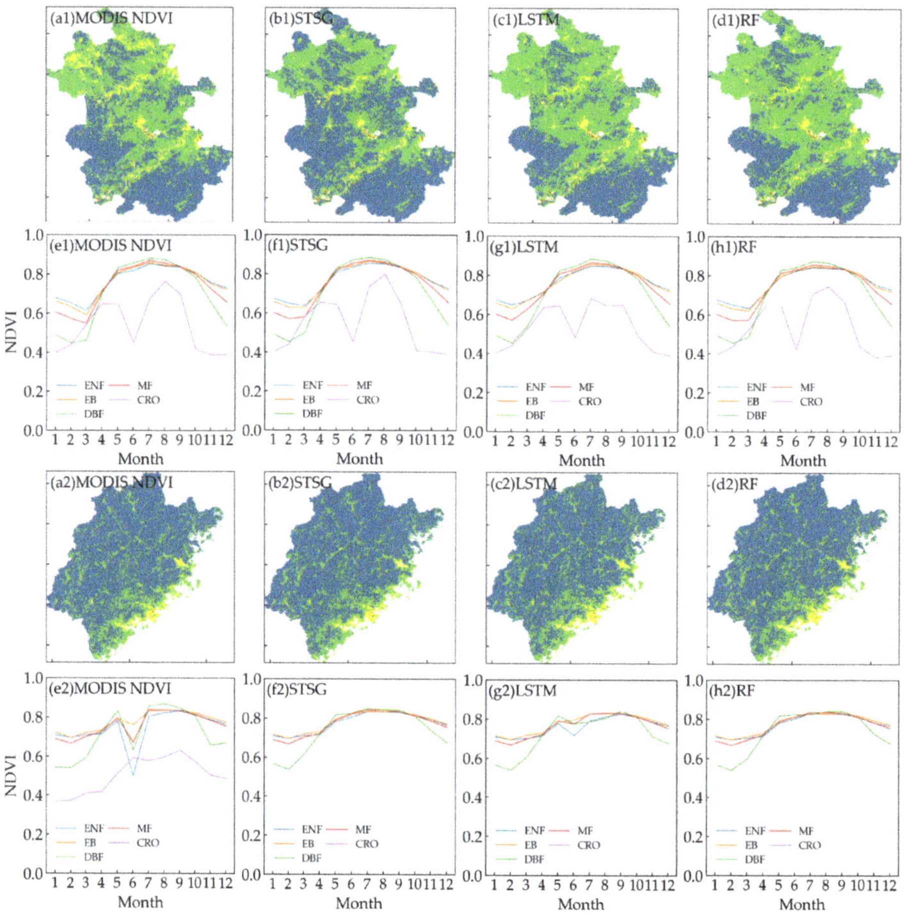

Figure 11. *Cont.*

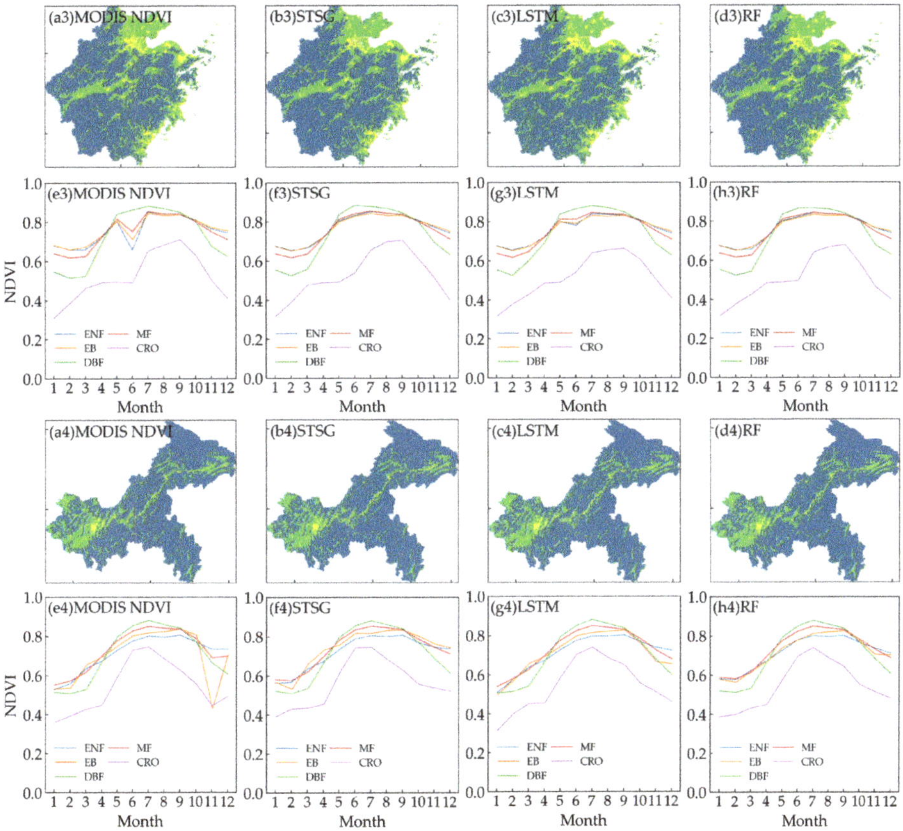

Figure 11. The reconstruction images and local spatial and phenological characteristics of specific biomes in Anhui Province, Fujian Province, Zhejiang Province, and Chongqing Province using STSG, LSTM, and RF. Note: (**a1–a4**,**e1–e4**): MODIS NDVI; (**b1–b4**,**f1–f4**): STSG; (**c1–c4**,**g1–g4**): LSTM; (**d1–d4**,**h1–h4**): RF.

As shown in Figure 14, the feasibility of the RF-based estimation result was compared with that of other models by displaying the spatial distribution and the time sequence curve of the TCA points. We found that all three methods could successfully capture the harvest condition of TCA, that is, the winter wheat is harvested in June in the TCA, and summer corn or soybeans are planted. This cultivation scheme forms a 'peak–valley–peak' pattern in the NDVI time series that reflects this condition (Figure 14e). All three methods could delineate this pattern. All three methods (i.e., LSTM, STSG, and RF) improved the contaminated NDVI from either spatial results or time sequence results (see Figure 14(b1,c1,d1,e)); however, the results of the RF-based estimates (the solid red line of Figure 14e) were closer to both the MODIS NDVI and Landsat data, with a higher consistent degree of good fitness with the data of the original high-quality areas. Compared with the LSTM and STSG methods, the RF-based process had the highest spatial congruence with the MODIS NDVI images (Figure 14(a1,d1)). We verified that the RF product accurately captured the ground's texture and actual natural land characteristics, providing higher accuracy. In other words, the RF-based model was more effective and stable in estimating low-quality NDVI areas while retaining the capacity to learn phenological information successfully.

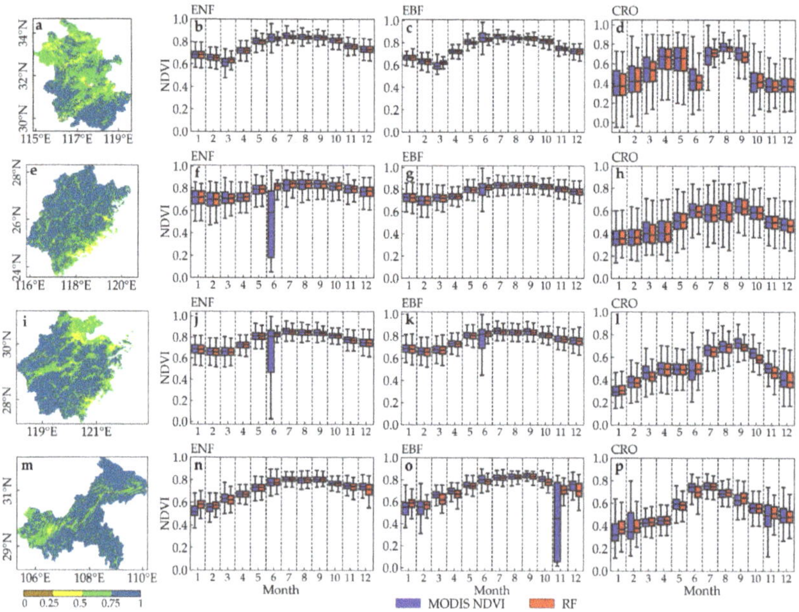

Figure 12. The spatial characteristics and phenological patterns of evergreen needleleaf forests, evergreen broadleaf forests, and croplands in Anhui Province, Fujian Province, Zhejiang Province, and Chongqing Province from 2001 to 2015 presented by (**a**–**p**).

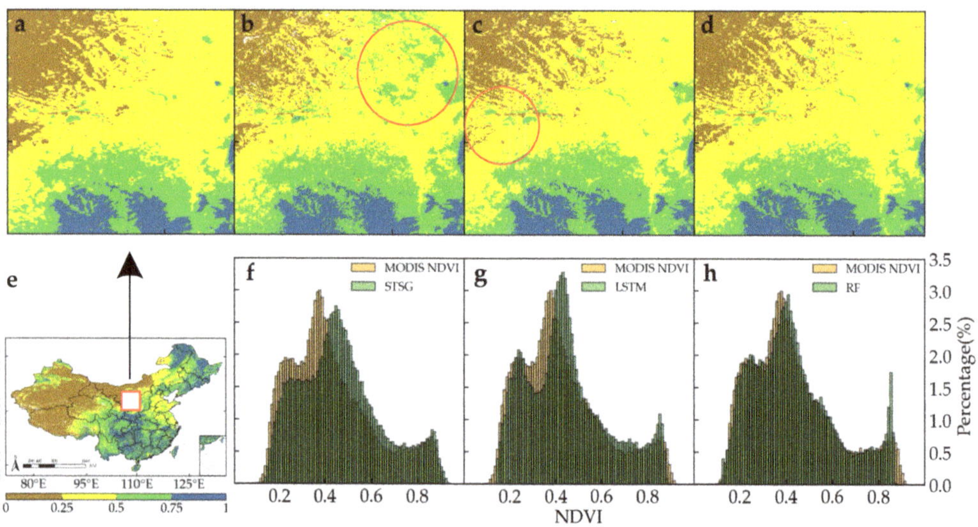

Figure 13. The simulations, the simulation performance of different methods, and the two regions marked by the red circle are other high-quality areas. (**a**) MODIS NDVI; (**b**) STSG; (**c**) LSTM; (**d**) RF; (**e**) the simulated areas of MODIS NDVI in China; (**f**) the different values of MODIS NDVI and STSG; (**g**) the different values of MODIS NDVI and LSTM; and (**h**) the different values of MODIS NDVI and RF. Note that the red circles in (**b**,**c**) denote the problematic estimated result (mainly overestimated).

Table 2. Comparisons of separability of the used methods.

	STSG	LSTM	RF
separability	0.051	0.013	0.010

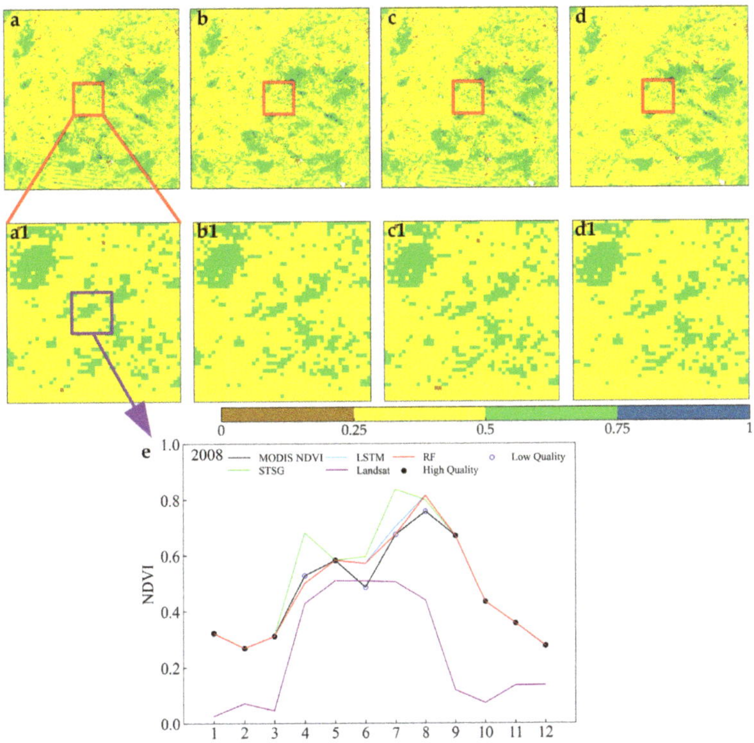

Figure 14. The comparison of different methods at the TCA site. (**a**) MODIS NDVI image (acquired June 20, 2008, for TCA), (**b–d**) reconstructed NDVI images using different methods: (**b**) STSG; (**c**) LSTM; and (**d**) RF (**a1–d1**) are enlarged views of the MODIS NDVI image, STSG simulation result, LSTM simulation result, and RF simulation result to show the spatial details, and (**e**) the time series curves of reconstructed NDVI for a small area.

This paper has proposed a novel method for reconstructing the low-quality areas of MODIS NDVI using the random forest model, specifically in the research area of China. The technique uses the relationship between MODIS NDVI products and auxiliary data, combined with long-term meteorological and geographical data, to fill in cloud-covered regions and produce a seamless spatiotemporal representation of the actual surface. The machine learning approach was tested on the complex terrain of the North China Plain using newly generated MODIS NDVI and Landsat 8 data from different sources. The product of the random forest model is closer to the actual surface representation with finer texture information than the other methods, as shown through spatial and temporal analyses. The method is also more effective at learning the phenological characteristics of typical land cover types. The results demonstrate the algorithm's universality and robustness under different climatic conditions. Overall, the random forest method outperforms the other methods in terms of accuracy and reliability in reconstructing NDVI data.

5. Discussion

It reflected the ability of the newly generated products to learn the phenological characteristics of typical ground biomes, such as evergreen needleleaf forests, evergreen broadleaf forests, deciduous needleleaf forests, deciduous broadleaf forests, mixed forests, and croplands (shown in Figures 6 and 9). Based on the results, the product is consistent with the phenological characteristics of the original high-quality area [51,52]. According to the analysis, different methods estimate the effects of varying land categories differently (shown in Figures 5 and 7). The random forest approach provided the most accurate estimation of crops, providing a basis for further research on crop yield [42,53]. The sample distribution is uniform and involves more climatic zones, which can support the model to learn enough features and produce sound learning effects for these objects.

5.1. Differences between the RF Method and Previous Methods

Compared with traditional inversion methods, such as STSG [20] and LSTM [21], RF [54] has made significant breakthroughs. Specifically, the nonlinear interactions between different factors still need to be explained to simulate them even with a state-of-the-art model, which usually results in underestimated values in these areas compared to when using a dynamic model [6]. In addition, it is challenging to identify nonlinear relationships in climate systems using linear statistical models. Moreover, the forecast targets of previous studies were not directly focused on extreme events but on the time series of rainfall anomalies. Furthermore, it took time to determine whether the extreme event had been successfully predicted, even if the expected mechanism had the same sign as the observed anomaly. Owning to the above reasons, a model that can integrate various nonlinear physical mechanisms into the model and detect weak signals that traditional linear methods can easily miss is needed. It is concluded that the RF could satisfy these requirements, and it is a high-accuracy, high-robustness, and high-interpretability model for estimating the NDVI in low-quality areas, based on the results shown in Figures 3, 13 and 14. In addition, it can thoroughly combine multiple relationships between various remote sensing products and NDVI products to achieve the project's goal in particular (shown in Figure 1), and it was established that a stable model could be developed for reconstruction. Moreover, based on the RF model, a continuous, long-term, and high-quality NDVI product could be designed after estimating NDVI values in contaminated areas. In addition, the input products had already processed the atmospheric or radiometric correction before applying the random forest algorithm, effectively avoiding uncertainties caused by error accumulation.

Based on the analysis presented in Figures 9 and 10, it can be inferred that the RF model is better suited for estimating crop characteristics across diverse terrains and regions than the LSTM model. The RF algorithm also exhibited greater resilience across multiple biomes and is effective in most cases except for a few scenarios. The TCA results depicted in Figure 14 also support the accuracy of the RF model in capturing ground texture and actual land characteristics. Moreover, the estimated products in Figure 11 demonstrate the ability of the RF model to accurately reflect the surface phenological characteristics of specific biomes such as crops and forests. Therefore, the RF model is feasible for estimating different terrains in different parts of China with remarkable precision. Finally, the reconstructed NDVI shows similar trends and patterns to the previous related works using the same data sources. Specifically, the overall upward movements of the random forest algorithm were observed, and the annual fluctuations were more noticeable, especially between 2009 and 2015 (shown in Figure 7). These fluctuations could be attributed to years of deforestation, forest protection measures, southern water irrigation, and large-scale farming soil management, where all of which have contributed to enhancing vegetation coverage. These findings of the random forest algorithm agree with those of several other research studies [55,56].

There is a significant difference in performance between these two methods in the case of the same environment. STSG requires pre-classification of MODIS NDVI time series products, which may introduce classification errors and boundary effects [23]. It requires selecting appropriate parameters, such as window size, threshold, etc., which may affect the algorithm's performance and stability. It cannot fill in missing data for multiple continuous years or seasons. LSTM is a neural network that processes data sequences and remembers values over long time intervals. It has a cell and three gates (input, output, and forget) that regulate the flow of information in and out of the cell. Moreover, it is helpful for handwriting recognition, speech recognition, machine translation, and robot control tasks.

The RF approach is better at gap restoration since it uses spatial–temporal information from all years. This applies to isolated cracks, continuous holes, and gaps in peaks and valleys. STSG can fix vegetation growth period cracks (Figure 9d) and narrow gaps (Figure 9c). Nevertheless, these do not address the gaps present during the peak growing season, as seen in Figure 9c,d and Figure 15e,k,m, or the gaps present consistently in Figure 15a,c,e,m. STSG outperforms LSTM in almost every circumstance because it uses spatial information. This supports Cao et al. [20]. However, the STSG technique is unstable at various points in the time series depicted in Figure 15b,d,f,j,l,n. The RF technique handles adverse marginal value distributions better than the others. It only eliminates somewhat unusual versions. Figure 15f,l shows that the initial time series includes marginal noise and frequent temporal fluctuations. Both figures exhibit these traits. The techniques can reduce anomalous volatility and smooth the curve if marginal data have a high negative deviation. The procedures produce this. LSTM rebuilds it in a slightly worse way. LSTM also overestimates average declines almost annually. The STSG and RF techniques best handle marginal data with low values in this context. The SG filter values may explain the STSG approach's modest oscillations, especially during the growing season. RF modelling is stable in peak and valley periods. The LSTM technique performs poorly during these periods due to excessive smoothing (Figure 15d,f,l,n). In Figure 15f, the STSG approaches reach the upper envelope of all values, while the RF approach usually only reaches the upper envelope of the best offers. This slight variation may prevent peak overestimation.

Comparing the spatial distribution and phenological characteristics of specific biomes in Yunnan Province, this section compares the results of reconstructions using the RF method and the other two methods. Figure 16 depicts the spatial distribution of the proposed RF method in July 2015, during the summer season, with extensive vegetation cover and severe cloud contamination, as evidenced by the comprehensive coverage of vegetation and cloud contamination. This is for the province of Yunnan. The RF method raises low NDVI values to high levels and achieves spatially continuous reconstruction. Figure 16a depicts the original image with high NDVI values before it was degraded by dense cloud cover. Even though LSTM can remedy low values, the results still exhibit negatively biased noise and spatial discontinuities (illustrated in Figure 16(b,b1)). The STSG procedure performs better, but the results still have some noise (Figure 16(c,c1)). Figure 16(d,d1) demonstrates that the illustrated RF method offers the best noise reduction and spatial continuity performance. Figure 16b depicts a region dominated by BNF as the date approaches the apex of BNF growth. The mean values of the original NDVI image and the three resulting images indicate that all methods increased the low NDVI values that had been corrupted. Only the proposed RF and STSG methods restored the common NDVI values to their optimum values, whereas LSTM underestimated the NDVI values (illustrated in Figure 16(b,b1)). Overall, the results demonstrate the optimal spatial performance of the RF method for maintaining critical points.

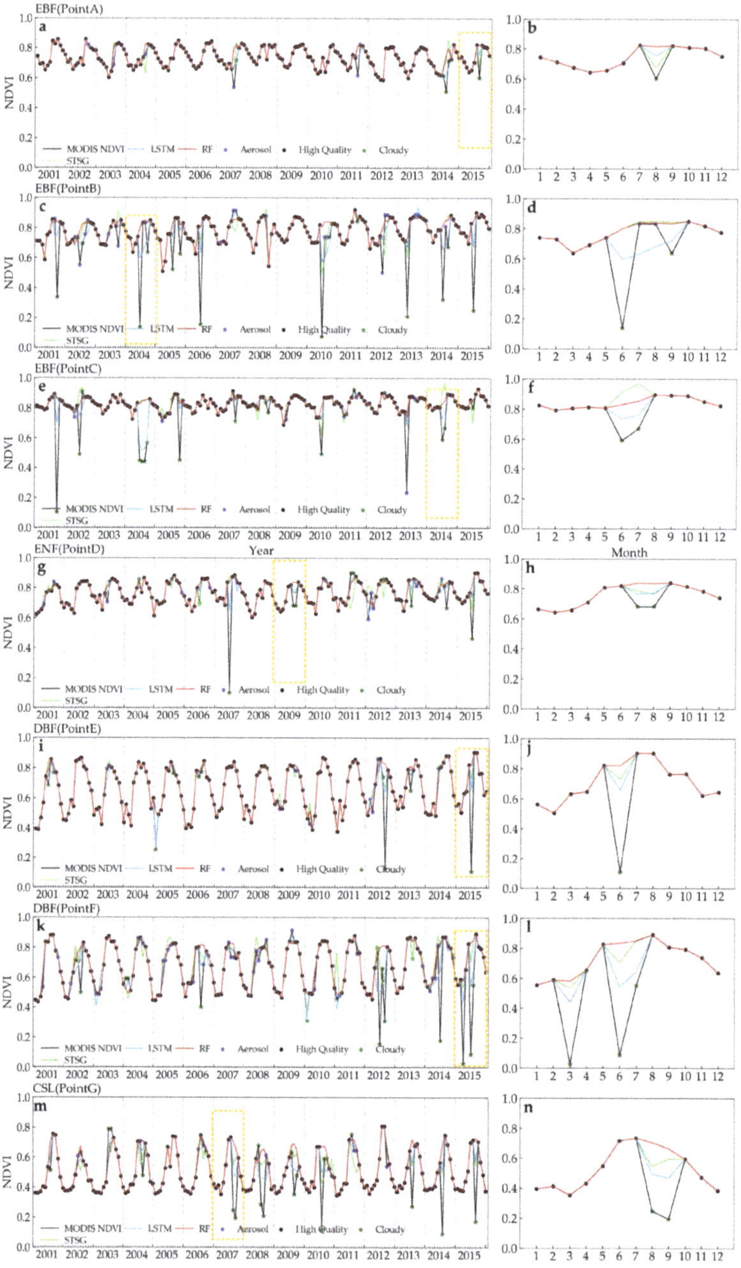

Figure 15. The performance of different methods over different vegetation types for long time series. A dashed rectangle indicates the chosen year. Note: ENF presents evergreen needleleaf forest; EBF: evergreen broadleaf forest; DBF: deciduous broadleaf forest; MF: mixed forest; CSL: closed shrubland. (**a–f**) EBF, (**g**,**h**) ENF, (**i–l**) DBF, and (**m**,**n**) CSL. The locations of the three points are depicted as a pentagon star (shown in Figure 3), and the dashed rectangle indicates the selected year (shown at (**b**,**d**,**f**,**h**,**j**,**l**,**n**)).

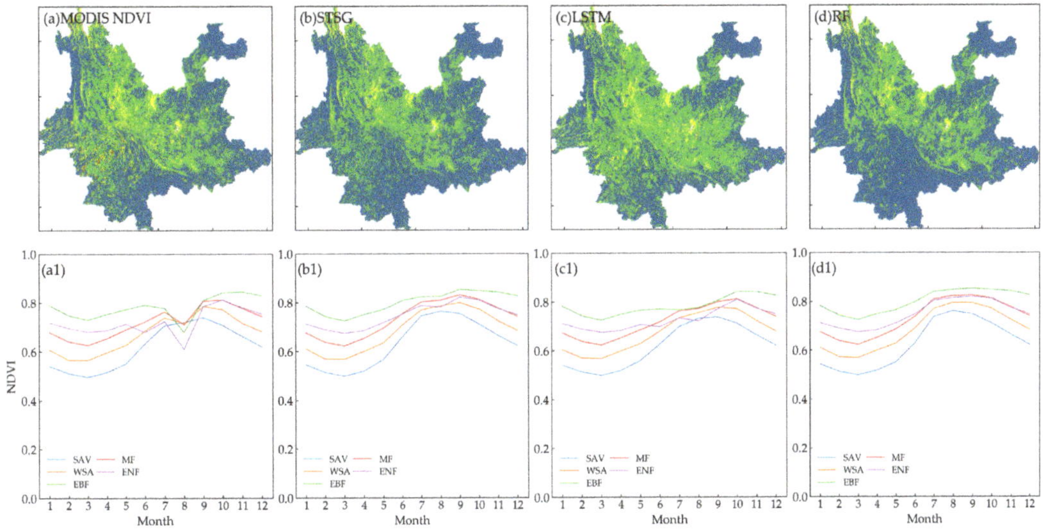

Figure 16. The reconstruction images, local spatial characteristics, and phenological characteristics of specific biomes in Yunnan Province using STSG, LSTM, and RF in July 2015. Note: (**a**,**a1**): MODIS NDVI; (**b**,**b1**): STSG; (**c**,**c1**): LSTM; and (**d**,**d1**): RF. Note: SAV, GRA, WSA, EBF, and MF are savanna, grassland, woody savanna, evergreen broadleaf, and mixed forest, respectively.

Long-time series NDVI data is widespread across various practical applications, such as vegetation monitoring, drought monitoring, land use classification, ecological assessment, climate change studies, and agricultural management. These data aid in monitoring vegetation growth, drought occurrence, and moisture status, as well as in assessing health and ecological changes. Additionally, using them facilitates the classification of land use/cover types, evaluation of ecological quality, and examination of the impacts of climate change on vegetation growth and distribution. Furthermore, long-term NDVI data enable farmers to make informed decisions regarding fertilisation and irrigation practices. The importance of NDVI data in ensuring the health and sustainability of diverse ecosystems cannot be overstated. Consequently, correcting long-time series NDVI data is of great interest, and the random forest algorithm is a valuable tool in this endeavour.

5.2. The Limitation of the RF Method

There are still some limitations to this study. First, the method may be better for MODIS sources, but the effects of other data sources require further experimentation and investigation efforts. At the same time, it only applies to China, involving smaller regions, and it may have the potential to process global products. In addition, we used default parameter values because of the large amount of experimental data and the limitations of the experimental conditions and time. Specifically, there were 125 products in this study, and it took place over 15 years. The workload was significant, requiring a multiprocessing approach to handle it. A more cost-effective algorithm is recommended in future work, and multiprocessing may be implemented to complete the process.

Moreover, the representative sample is another problem. The number and distribution of pieces are key factors affecting the learning of land features. Due to climate conditions and geographical location limitations, the model training sample needed to be sufficiently diverse in categories. It did not involve all climate zones worldwide; some regions lacked pieces to different extents. Furthermore, some land samples had fewer numbers, and some land distribution was concentrated in certain areas, such as forests, mainly in coastal areas. A few other sites may also be distributed, making characteristic learning insufficient. Thus, further exploring algorithms for sample balancing and land class identification, which will improve estimation results, is necessary.

When this method is used for estimation on a global scale, there are two problems: one is the increase in sample size; the other is that the more complex nonlinear relationship between more complex remote sensing products and NDVI requires more decisions, which will lead to an increase in the model training cycle. Therefore, in later research, multiple processes can be considered. At the same time, the data processing and model training can be run on the GPU, which has more computing cores than the CPU, which will significantly improve the computing efficiency.

6. Conclusions

Due to the nature of clouds, snow, and aerosols, they are discontinuous in space and time, severely limiting the broad application of the current NDVI products. To solve this problem, we used the random forest model to construct multisource quantitative remote sensing products such as FAPAR and LAI, as well as to find the relationship between the longitude, latitude, elevation, and other factors related to the vegetation growth process and NDVI, to avoid the influence of NDVI products alone when there continuously are multiple low-quality products. We found that when compared with the spatial–temporal Savitzky–Golay (STSG) method, our approach based on random forest (RF) methodology decreased the MAE and RMSE by 0.014 and 0.019, respectively. Compared with the long short-term memory (LSTM) method, our RF-based approach decreased the MAE and RMSE by 0.013 and 0.015, respectively. The results showed that the RF model had the best estimation performance compared to traditional linear methods such as STSG and LSTM.

Multiple verification approaches were conducted:

1. The performance of the models correcting the contaminated pixels was tested using the site-based method, which included the NDVI simulation in TCA and the actual ground observations (points A, B, and C). The results showed that the RF model had a lower MAE and RMSE than LSTM and STSG, and the correction was very close to the high-quality areas regardless of the biome type.
2. The performance of the models was tested using the time sequence test, and we found that the reconstructed NDVI had the lowest MAE and RMSE in the time series from 2001 to 2015.
3. By comparing the spatial pattern of different models, we found that the RF method could produce a product with more continuous spatial textures.

Using the RF method proposed in this study, we reconstructed a monthly NDVI product with a spatial resolution of 1 km in China from 2001 to 2015. The reconstruction product could be helpful for the yield estimation and dynamic monitoring of vegetation growth and could shed light on the processing of future cost-effective algorithms.

Author Contributions: Conceptualization, M.S., A.G. and X.Z.; formal analysis, M.S., A.G. and X.Z.; funding acquisition, X.Z.; methodology, X.Z.; writing—original draft, M.S.; writing—review and editing, M.S., A.G., X.Z., N.L., L.S. and S.Z. All authors have read and agreed to the published version of the manuscript.

Funding: This research was funded by the National Natural Science Foundation of China under grant (42090012).

Data Availability Statement: Not applicable.

Conflicts of Interest: The authors declare no conflict of interest.

References

1. Matsushita, B.; Yang, W.; Chen, J.; Onda, Y.C.; Qiu, G.Y. Sensitivity of the enhanced vegetation index (EVI) and normalised difference vegetation index (NDVI) to topographic effects: A case study in high-density cypress forest. *Sensors* **2007**, *7*, 2636–2651. [CrossRef] [PubMed]
2. Kumari, N.; Saco, P.M.; Rodriguez, J.F.; Johnstone, S.A.; Srivastava, A.; Chun, K.P.; Yetemen, O. The grass is not always greener on the other side: Seasonal reversal of vegetation greenness in aspect-driven semiarid ecosystems. *Geophys. Res. Lett.* **2020**, *47*, e2020GL088918. [CrossRef]
3. Chu, H.S.; Venevsky, S.; Wu, C.; Wang, M.H. NDVI-based vegetation dynamics and its response to climate changes at Amur-Heilongjiang River Basin from 1982 to 2015. *Sci. Total Environ.* **2019**, *650*, 2051–2062. [CrossRef] [PubMed]
4. Zhao, J.; Huang, S.Z.; Huang, Q.; Wang, H.; Leng, G.Y.; Fang, W. Time-lagged response of vegetation dynamics to climatic and teleconnection factors. *Catena* **2020**, *189*, 104474. [CrossRef]
5. Hu, T.; Myers Toman, E.; Chen, G.; Shao, G.; Zhou, Y.; Li, Y.; Zhao, K.; Feng, Y. Mapping fine-scale human disturbances in a working landscape with Landsat time series on Google Earth Engine. *ISPRS J. Photogramm. Remote Sens.* **2021**, *176*, 250–261. [CrossRef]
6. Mao, Y.J.; Van Niel, T.G.; McVicar, T.R. Reconstructing cloud-contaminated NDVI images with SAR-Optical fusion using spatio-temporal partitioning and multiple linear regression. *ISPRS J. Photogramm. Remote Sens.* **2023**, *198*, 115–139. [CrossRef]
7. Michishita, R.; Jin, Z.Y.; Chen, J.; Xu, B. Empirical comparison of noise reduction techniques for NDVI time-series based on a new measure. *ISPRS J. Photogramm. Remote Sens.* **2014**, *91*, 17–28. [CrossRef]
8. Shao, Y.; Lunetta, R.S.; Wheeler, B.; Iiames, J.S.; Campbell, J.B. An evaluation of time-series smoothing algorithms for land-cover classifications using MODIS-NDVI multi-temporal data. *Remote Sens. Environ.* **2016**, *174*, 258–265. [CrossRef]
9. Hird, J.N.; McDermid, G.J. Noise reduction of NDVI time series: An empirical comparison of selected techniques. *Remote Sens. Environ.* **2009**, *113*, 248–258. [CrossRef]
10. Chen, J.; Jönsson, P.; Tamura, M.; Gu, Z.; Matsushita, B.; Eklundh, L. A simple method for reconstructing a high-quality NDVI time-series data set based on the Savitzky–Golay filter. *Remote Sens. Environ.* **2004**, *91*, 332–344. [CrossRef]
11. Ma, M.; Veroustraete, F. Reconstructing pathfinder AVHRR land NDVI time-series data for the Northwest of China. *Adv. Space Res.* **2006**, *37*, 835–840. [CrossRef]
12. Tang, L.; Zhao, Z.M.; Tang, P.; Yang, H.J. SURE-based optimum- length S-G filter to reconstruct NDVI time series iteratively with outliers removal. *Int. J. Wavel. Multiresolut. Inf. Process.* **2020**, *18*, 2050001. [CrossRef]
13. Jonsson, P.; Eklundh, L. Seasonality extraction by function fitting to time-series of satellite sensor data. *IEEE Trans. Geosci. Remote Sens.* **2002**, *40*, 1824–1832. [CrossRef]
14. Beck, P.S.; Atzberger, C.; Høgda, K.A.; Johansen, B.; Skidmore, A.K. Improved monitoring of vegetation dynamics at very high latitudes: A new method using MODIS NDVI. *Remote Sens. Environ.* **2006**, *100*, 321–334. [CrossRef]
15. Liu, R.G.; Shang, R.; Liu, Y.; Lu, X.L. Global evaluation of gap-filling approaches for seasonal NDVI with considering vegeta-tion growth trajectory, protection of key point, noise resistance and curve stability. *Remote Sens. Environ.* **2017**, *189*, 164–179. [CrossRef]
16. Yang, Y.P.; Luo, J.C.; Huang, Q.T.; Wu, W.; Sun, Y.W. Weighted double-logistic function fitting method for reconstructing the high-quality Sentinel-2 NDVI time series data set. *Remote Sens.* **2019**, *11*, 18. [CrossRef]
17. Li, X.H.; Shen, R.P.; Chen, R.X. Improving time series reconstruction by fixing invalid values and its fidelity evaluation. *IEEE Access* **2020**, *8*, 7558–7572. [CrossRef]
18. Xu, L.L.; Li, B.L.; Yuan, Y.C.; Gao, X.Z.; Zhang, T. A temporal-spatial iteration method to reconstruct NDVI time series datasets. *Remote Sens.* **2015**, *7*, 8906–8924. [CrossRef]
19. Chu, D.; Shen, H.; Guan, X.; Chen, J.M.; Li, X.; Li, J.; Zhang, L. Long time-series NDVI reconstruction in cloud-prone regions via spatio-temporal tensor completion. *Remote Sens. Environ.* **2021**, *264*, 112632. [CrossRef]
20. Cao, R.; Chen, Y.; Shen, M.; Chen, J.; Zhou, J.; Wang, C.; Yang, W. A simple method to improve the quality of NDVI time-series data by integrating spatiotemporal information with the Savitzky-Golay filter. *Remote Sens. Environ.* **2018**, *217*, 244–257. [CrossRef]
21. Graves, A.; Graves, A. Long short-term memory. In *Supervised Sequence Labelling with Recurrent Neural Networks*; Springer: Berlin/Heidelberg, Germany, 2012; pp. 37–45.
22. Li, S.; Xu, L.; Jing, Y.H.; Yin, H.; Li, X.H.; Guan, X.B. High-quality vegetation index product generation: A review of NDVI time series reconstruction techniques. *Int. J. Appl. Earth Obs. Geoinf.* **2021**, *105*, 102640. [CrossRef]
23. Do Thi, N.; Pham, V.D.; Bui, Q.T.; Pham, V.M. Study Model for Information Reconstruction on Cloud Contaminated Area for Single Multispectral Remote Sensing Sentinel-2 Imagery using Generative Adversarial Network. *VNU J. Sci. Earth Environ. Sci.* **2022**, *38*, 32–44.
24. Liu, H.; Huang, B.; Cai, J.J. Thick Cloud Removal under Land Cover Changes Using Multisource Satellite Imagery and a Spatiotemporal Attention Network. *IEEE Trans. Geosci. Remote Sens.* **2023**, *61*, 1–18. [CrossRef]
25. Zhang, H.; Ma, J.Y.; Chen, C.; Tian, X. NDVI-Net: A fusion network for generating high-resolution normalized difference vegetation index in remote sensing. *ISPRS J. Photogramm. Remote Sens.* **2020**, *168*, 182–196. [CrossRef]
26. Li, X.Q.; Yuan, W.P.; Dong, W.J. A machine learning method for predicting vegetation indices in China. *Remote Sens.* **2021**, *13*, 1147. [CrossRef]
27. Roy, B. Optimum machine learning algorithm selection for forecasting vegetation indices: MODIS NDVI & EVI. *Remote Sens. Appl. Soc. Environ.* **2021**, *23*, 100582.

28. Padhee, S.K.; Dutta, S. Spatiotemporal reconstruction of MODIS land surface temperature with the help of GLDAS product using kernel-based nonparametric data assimilation. *J. Appl. Remote Sens.* **2020**, *14*, 18. [CrossRef]
29. Zheng, B.J.; Myint, S.W.; Thenkabail, P.S.; Aggarwal, R.M. A support vector machine to identify irrigated crop types using time-series Landsat NDVI data. *Int. J. Appl. Earth Obs. Geoinf.* **2015**, *34*, 103–112. [CrossRef]
30. Ng, M.K.; Yuan, Q.Q.; Yan, L.; Sun, J. An Adaptive Weighted Tensor Completion Method for the Recovery of Remote Sensing Images with Missing Data. *IEEE Trans. Geosci. Remote Sens.* **2017**, *55*, 3367–3381. [CrossRef]
31. Zhu, X.L.; Helmer, E.H. An automatic method for screening clouds and cloud shadows in optical satellite image time series in cloudy regions. *Remote Sens. Environ.* **2018**, *214*, 135–153. [CrossRef]
32. Malamiri, H.R.G.; Zare, H.; Rousta, I.; Olafsson, H.; Verdiguier, E.I.; Zhang, H.; Mushore, T.D. Comparison of harmonic analysis of time series (HANTS) and multi-singular spectrum analysis (M-SSA) in reconstruction of long-gap missing data in NDVI time series. *Remote Sens.* **2020**, *12*, 22.
33. Feng, Y.B.; Liu, Q.; Qu, Y.; Liang, S.L. Estimation of the Ocean Water Albedo from Remote Sensing and Meteorological Reanalysis Data. *IEEE Trans. Geosci. Remote Sens.* **2016**, *54*, 850–868. [CrossRef]
34. Yao, Y.J.; Liang, S.L.; Li, X.L.; Hong, Y.; Fisher, J.B.; Zhang, N.N.; Chen, J.Q.; Cheng, J.; Zhao, S.H.; Zhang, X.T.; et al. Bayesian multimodel estimation of global terrestrial latent heat flux from eddy covariance, meteorological, and satellite observations. *J. Geophys. Res. Atmos.* **2014**, *119*, 4521–4545. [CrossRef]
35. Xiao, Z.Q.; Liang, S.L.; Sun, R.; Wang, J.D.; Jiang, B. Estimating the fraction of absorbed photosynthetically active radiation from the MODIS data based GLASS leaf area index product. *Remote Sens. Environ.* **2015**, *171*, 105–117. [CrossRef]
36. Jia, K.; Liang, S.L.; Wei, X.Q.; Yao, Y.J.; Yang, L.Q.; Zhang, X.T.; Liu, D.Y. Validation of Global LAnd Surface Satellite (GLASS) fractional vegetation cover product from MODIS data in an agricultural region. *Remote Sens. Lett.* **2018**, *9*, 847–856. [CrossRef]
37. Yuan, W.P.; Cai, W.W.; Xia, J.Z.; Chen, J.Q.; Liu, S.G.; Dong, W.J.; Merbold, L.; Law, B.; Arain, A.; Beringer, J. Global comparison of light use efficiency models for simulating terrestrial vegetation gross primary production based on the LaThuile database. *Agric. For. Meteorol.* **2014**, *192*, 108–120. [CrossRef]
38. Xiao, Z.Q.; Liang, S.L.; Jiang, B. Evaluation of four long time-series global leaf area index products. *Agric. For. Meteorol.* **2017**, *246*, 218–230. [CrossRef]
39. Gelaro, R.; Mccarty, W.; Suárez, M.J. The modern-era retrospective analysis for research and applications, version 2 (MERRA-2). *J. Clim.* **2017**, *30*, 5419–5454. [CrossRef]
40. Simard, M.; Neumann, M.; Buckley, S. Validation of the New SRTM Digital Elevation Model (NASADEM) with ICESAT/GLAS over the United States. In Proceedings of the IEEE International Geoscience and Remote Sensing Symposium (IGARSS) 2016, Beijing, China, 10–15 July 2016; pp. 3227–3229.
41. Zamani Joharestani, M.; Cao, C.; Ni, X.; Bashir, B.; Talebiesfandarani, S. $PM_{2.5}$ prediction based on random forest, XGBoost, and deep learning using multisource remote sensing data. *Atmosphere* **2019**, *10*, 373. [CrossRef]
42. Jui, S.J.J.; Ahmed, A.M.; Bose, A.; Raj, N.; Sharma, E.; Soar, J.; Chowdhury, M.W. Spatiotemporal hybrid random forest model for tea yield prediction using satellite-derived variables. *Remote Sens.* **2022**, *14*, 805. [CrossRef]
43. Belgiu, M.; Drăguţ, L. Random forest in remote sensing: A review of applications and future directions. *ISPRS J. Photogramm. Remote Sens.* **2016**, *114*, 24–31. [CrossRef]
44. Gibson, R.; Danaher, T.; Hehir, W.; Collins, L. A remote sensing approach to mapping fire severity in south-eastern Australia using sentinel 2 and random forest. *Remote Sens. Environ.* **2020**, *240*, 111702. [CrossRef]
45. Adugna, T.; Xu, W.B.; Fan, J.L. Comparison of random forest and support vector machine classifiers for regional land cover mapping using coarse resolution FY-3C images. *Remote Sens.* **2022**, *14*, 574. [CrossRef]
46. Breiman, L. Random forests. *Mach. Learn.* **2001**, *45*, 5–32. [CrossRef]
47. Malhotra, P.; Vig, L.; Shroff, G.; Agarwal, P. Long Short Term Memory Networks for Anomaly Detection in Time Series. In Proceedings of the 23rd European Symposium on Artificial Neural Networks, Computational Intelligence and Machine Learning, ESANN 2015, Bruges, Belgium, 22–24 April 2015; p. 89.
48. Reddy, D.S.; Prasad, P.R. Prediction of vegetation dynamics using NDVI time series data and LSTM. *Model. Earth Syst. Environ.* **2018**, *4*, 409–419. [CrossRef]
49. Muruganantham, P.; Wibowo, S.; Grandhi, S.; Samrat, N.H.; Islam, N. A systematic literature review on crop yield prediction with deep learning and remote sensing. *Remote Sens.* **2022**, *14*, 1990. [CrossRef]
50. Hochreiter, S.; Schmidhuber, J. Long short-term memory. *Neural Comput.* **1997**, *9*, 1735–1780. [CrossRef]
51. Pan, S.; Zhao, X.W.; Yue, Y.J. Spatiotemporal changes of NDVI and correlation with meteorological factors in northern china from 1985–2015. In *E3S Web of Conferences 2019*; EDP Sciences: Les Ulis, France, 2019; Volume 131, p. 1040.
52. Zhao, Q.Q.; Zhang, J.P.; Zhao, T.B.; Li, J.H. Vegetation changes and its response to climate change in China Since 2000. *Plateau Meteorol.* **2021**, *40*, 292–301.
53. Li, F.; Song, G.; Liujun, Z.; Yanan, Z.; Di, L. Urban vegetation phenology analysis using high spatio-temporal NDVI time series. *Urban For. Urban Green.* **2017**, *25*, 43–57. [CrossRef]
54. Cutler, A.; Cutler, D.R.; Stevens, J.R. Random forests. In *Ensemble Machine Learning: Methods and Applications*; Springer: Berlin/Heidelberg, Germany, 2012; pp. 157–175.

55. Bryan, B.A.; Gao, L.; Ye, Y.Q.; Sun, X.F.; Connor, J.D.; Crossman, N.D.; Stafford-Smith, M.; Wu, J.G.; He, C.Y.; Yu, D.Y. China's response to a national land-system sustainability emergency. *Nature* **2018**, *559*, 193–204. [CrossRef]
56. Zastrow, M. China's tree-planting could falter in a warming world. *Nature* **2019**, *573*, 474–475. [CrossRef] [PubMed]

remote sensing

MDPI

Article

Integrating Otsu Thresholding and Random Forest for Land Use/Land Cover (LULC) Classification and Seasonal Analysis of Water and Snow/Ice

Xuexia Sun [1,2], Xiaoyao Li [1,2,*], Bingxiang Tan [1,2], Jian Gao [3], Lei Wang [3] and Shimei Xiong [1,2]

[1] Institute of Forest Resource Information Techniques, Chinese Academy of Forestry, Beijing 100091, China; sunxuexia20@mails.ucas.ac.cn (X.S.); tan@caf.ac.cn (B.T.); xiongshimei21@mails.ucas.ac.cn (S.X.)

[2] Key Laboratory of Forestry Remote Sensing and Information System, National Forestry and Grassland Administration, Beijing 100091, China

[3] Institute of Modern Forestry, Xinjiang Academy of Forestry, Urumqi 830063, China; gaojian21@mails.ucas.ac.cn (J.G.); wanglei21@mails.ucas.ac.cn (L.W.)

* Correspondence: xiaoyaoli@ifrit.ac.cn

Abstract: Accurate land use/land cover (LULC) classification and the detection of seasonal dynamics are crucial for effective environmental monitoring and resource management. To improve the precision and temporal resolution of regional LULC classification products, this study combined the Otsu threshold method and Random Forest algorithm to generate a 10 m-resolution land cover classification map for Wensu County based on Sentinel-2 imagery, with a particular focus on orchard categories, and investigated the seasonal dynamics of LULC between winter and summer. The results show that the overall accuracy (OA) of the water and snow/ice models was 85.50%, with a Kappa coefficient of 0.8088; for the vegetation model, the OA was 93.77%, with a Kappa coefficient of 0.8755. Feature importance analysis indicated that terrain features were key factors in improving classification performance. Seasonal dynamics analysis showed that the snow/ice coverage area in winter increased by 6379.18 square kilometers compared to that in summer, with 5252.85 km^2 of bare land and 910.66 km^2 of grassland being covered by snow/ice. Meteorological data analysis revealed that land cover changes caused by winter snowfall were primarily concentrated in areas where temperatures exceeded −8 °C, while land cover changes were smaller in areas with either low or high precipitation. These findings provide valuable data support for regional resource management and agricultural development.

Keywords: LULC; random forest; Otsu thresholding; combined model; seasonal dynamics

1. Introduction

Global climate change constitutes a primary challenge for contemporary society, exerting extensive effects on both environmental systems and human activities [1]. In particular, rising temperatures, altered precipitation patterns, and the increasing frequency of extreme weather events have profound impacts on ecosystems, agriculture, and water resource management [2]. These impacts are especially pronounced in arid and semi-arid regions, where complex topography and significant climatic heterogeneity exacerbate the vulnerability of ecosystems and human livelihoods to climate variability. For example, in regions with high dependence on seasonal snow accumulation and spring snowmelt, such as those characterized by significant elevation gradients, the availability of water resources is closely tied to winter snow cover and its subsequent melting. These dynamics directly influence agricultural practices, grassland dynamics, and the transformation between bare land and vegetated areas, making the study of land use/land cover (LULC) changes in these regions particularly critical [3]. Simultaneously, annual LULC data are increasingly inadequate for contemporary research and practical needs, as they do not effectively capture rapid changes or seasonal variations in vegetation growth, water body freezing, and snow cover [4,5]. To address these limitations, it is essential to improve the temporal resolution of LULC data and enhance classification accuracy [6–8]. Investigating seasonal variations in LULC is therefore becoming increasingly important [9,10]. In the context of climate change, assessing the seasonal responses of water bodies and snow cover is essential for effective agricultural management and water resource planning [11–13].

With the rapid advancement of artificial intelligence, the application of machine learning in land cover classification has gained increasing prominence, particularly in managing high-dimensional and multi-source data. Among various methodologies, deep learning methods—especially Convolutional Neural Networks (CNNs)—have demonstrated remarkable efficacy in pixel-level analysis of high-resolution imagery [14]. CNNs are highly effective in automatic feature extraction, particularly for high-resolution imagery such as that in urban environment monitoring and intricate terrain analysis [15,16]. In contrast, the Random Forest (RF) algorithm has also found widespread application in land cover classification, thanks to its unique advantages as an ensemble learning method. By constructing a multitude of decision trees, RF exhibits substantial adaptability and robustness, particularly in scenarios involving high-dimensional remote sensing data and relatively small, labeled datasets [17]. Moreover, RF offers clear feature importance metrics, facilitating a comprehensive understanding of each feature's contribution to classification outcomes while maintaining a relatively high level of computational efficiency [18,19]. Nevertheless, reliance solely on RF may pose challenges, including suboptimal initial classifications and diminished computational efficiency. In the context of high-dimensional data, the construction and evaluation of decision trees can incur significant computational costs, with initial classification results often necessitating further refinement to enhance overall accuracy. To address these challenges, combining thresholding methods with Random Forest (RF) presents an effective strategy. As an unsupervised classification technique, thresholding enables rapid preliminary classification by applying designated thresholds, effectively distinguishing between different land cover types [20]. The Otsu thresholding method automatically determines the optimal threshold by maximizing the difference between classes [21]. Compared to other traditional thresholding methods, it reduces the subjectivity of manually setting thresholds, making the image segmentation process more objective and consistent [22–24]. Therefore, integrating the results of Otsu thresholding with the RF model allows for further refinement of the classification, significantly improving accuracy [25]. This synergistic approach not only enhances classification efficiency but also strengthens the adaptability of the RF model to complex surface environments. Additionally, by developing multiple RF models tailored to specific feature types, complex classification tasks can be decomposed into more manageable sub-tasks. Each model focuses on particular features, thereby reducing bias and errors while enhancing the overall robustness of the classification process.

Effective feature selection is crucial for achieving accurate and robust results in land cover classification [26]. Research indicates that spectral, texture, and topographical features significantly influence classification outcomes across diverse environmental contexts [27,28]. Spectral features, especially those from the near-infrared and red light bands, are particularly effective in distinguishing vegetation types [29]. High-resolution data from sources like Sentinel-2 enhance classification in complex terrains [30]. Spectral indices such as the Normalized Difference Vegetation Index (NDVI) and the Normalized Difference Water Index (NDWI) are invaluable tools for enhancing specific land cover characteristics. The NDVI is widely used to monitor vegetation health, while the NDWI aids in identifying water bodies [31,32]. These indices improve the separability of land cover types, particularly in regions with intricate LULC intermingling. However, relying solely on spectral features can be insufficient in complex environments where spectral overlap occurs. Texture features provide essential information about spatial relationships among pixels, aiding in differentiating land cover types with similar appearances but different spectral characteristics [33]. Metrics derived from the Gray-level Co-occurrence Matrix (GLCM) demonstrate high discriminative power in land cover types with insignificant spectral differences [34]. Topographical features, such as elevation, slope, and aspect, are also vital in regions with complex terrain, enhancing model adaptability [35]. Therefore, a multi-scale and multi-dimensional approach to feature selection is essential for constructing a scientific dataset. This approach significantly impacts the accuracy of classification results by leveraging the diverse and complementary strengths of different feature types.

Our study area features important agricultural zones with complex landscapes. Compared to other cultivated regions in northern China, this area features extensive orchards and relies heavily on irrigation. Therefore, accurately identifying key areas such as arable land, orchards, and water bodies is essential for optimizing agricultural development, managing watershed water resources effectively, and improving irrigation infrastructure. Existing 10 m-resolution LULC products often struggle to differentiate between orchards and arable land, leading to inconsistent accuracy across different regions. Additionally, these products generally offer annual data, which do not capture the seasonal variations in LULC, particularly in areas with substantial glaciers and snow cover. The objectives of this paper are as follows: (1) develop an integrated method using Sentinel-2 data for accurate LULC classification in areas with complex landscapes; (2) enhance the differentiation between arable land and orchards, as well as the monitoring of seasonal changes in glaciers, snow cover, and water bodies; and (3) produce high-resolution LULC products that provide valuable data for ecosystem assessments, industry evaluations, and agricultural planning. The proposed approach is expected to be applicable in other regions, particularly those with similar landscape complexities.

2. Materials and Methods

2.1. Study Area

Wensu County, located between the latitudes 40°52′ and 42°15′N and longitudes 79°28′ and 81°30′E in the Aksu Prefecture of the Xinjiang Uygur Autonomous Region, China, spans an area of 14,569.3 km^2 (Figure 1). The region's land cover is primarily dominated by extensive stretches of bare land and scattered oases. The oases are vital ecological and agricultural zones, situated mainly along river systems originating from the glacier-covered mountains in the northern part of the county. These fertile areas support vegetation and agricultural activities, standing in stark contrast to the surrounding arid landscapes. The area experiences a typical continental climate, with average annual precipitation below 200 mm, largely confined to the summer months. Temperature fluctuations are significant, with spring (March to May) marked by rising temperatures and occasional precipitation, while summer (June to August) is hot and dry. In autumn (September to November), temperatures gradually decrease, and winters (December to February) are cold and dry. This climatic variability, combined with complex topography, leads to substantial heterogeneity in land cover across the county.

The study area features a distinct elevation gradient, transitioning from flat arid plains to glacier-covered mountains, leading to significant spatiotemporal heterogeneity in land cover. This topographic variability complicates classification, particularly due to the interwoven distribution of orchards and croplands, diverse tree species, and their overlapping spectral characteristics. These factors create indistinct class boundaries, reducing classification accuracy and stability. To address these challenges, it is essential to develop classification methods that account for complex terrain and high environmental heterogeneity, improving both precision and applicability.

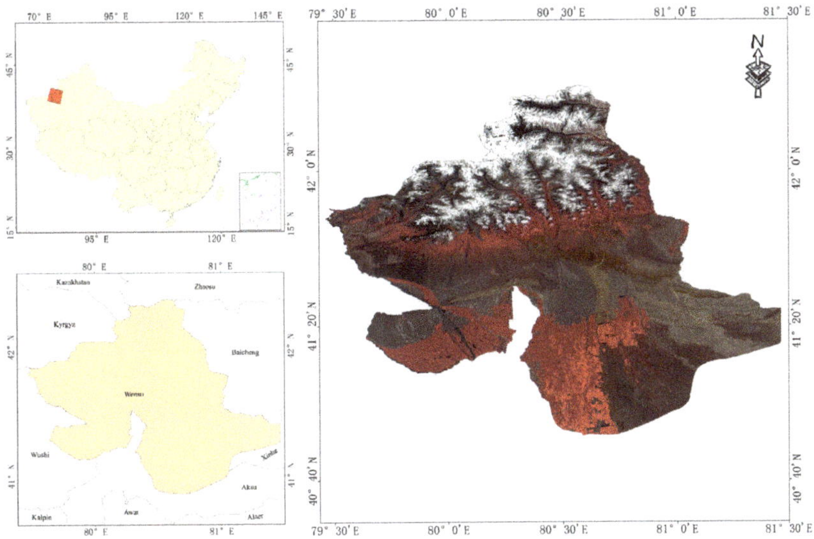

Figure 1. The study area in Sentinel-2 (Standard False Color Composite, B8/B4/B3).

2.2. LULC Classification Scheme

LULC products, whether evaluated on a global or national scale, generally exhibit high overall accuracy. However, their generalized classification schemes often lack the local precision required to address the specific needs of various study areas [36,37]. This limitation is particularly evident in annual-scale products, which may not adequately capture the temporal variability of LULC changes within specific regions [38,39]. In regions dominated by irrigated agriculture, precise LULC classification and the analysis of seasonal variations are crucial for effective agricultural planning and yield estimation. The accurate delineation of LULC types and understanding their seasonal dynamics are essential for optimizing local agricultural practices and forecasting crop yields. In this study, a detailed classification system was developed, identifying key LULC categories within the study area, including water, snow/ice, forests, grasslands, croplands, orchards, impervious surfaces, and bare land. In constructing and validating this classification, an initial sample set was created through the visual interpretation of summer Sentinel-2 imagery.

This sample set underwent rigorous cross-validation and was reviewed by multiple experts to ensure both accuracy and representativeness. To capture the spatial variability of the study area, a stratified random sampling approach was employed. The region was divided into distinct strata based on land cover types and topographic features, ensuring the comprehensive coverage of all land cover types and topographic variations. The sample size was carefully selected, taking into account the spatial heterogeneity of the area, with particular attention to key land cover types such as croplands, orchards, grasslands, and snow/ice areas. Furthermore, the representativeness of the sample points was validated through cross-validation, confirming that the selected samples effectively reflected the overall variability of the study area. After thorough validation, a total of 1213 sample points were confirmed, representing diverse LULC types, as outlined in Table 1.

Table 1. Classification system and samples.

LULC Type	Sentinel-2	Description	Samples
Water	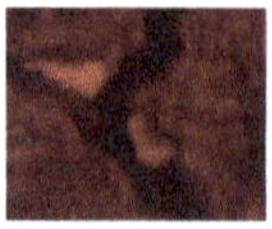	This class covers all types of water bodies, including rivers, lakes, and reservoirs.	160
Snow/Ice		This LULC type includes ice bodies and snow.	164
Forest		Forests are areas covered by dense trees.	176
Grassland		Grasslands are covered primarily by herbaceous plants, including natural prairies and artificially planted pastures.	170
Cropland	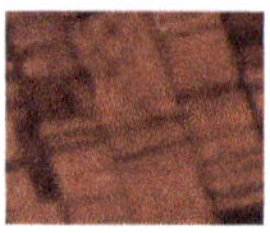	Cropland refers to areas used for the cultivation of food, fiber, and other crops, including both grain crops and commercial crops.	180
Orchard		Orchards include areas specifically planted with fruits, nuts, vegetables, and other economic crops.	183
Bare		Bare land refers to areas without vegetation cover, such as deserts, rocky outcrops, and wastelands.	180
Impervious		Impervious surfaces include all surfaces that do not allow water infiltration, such as urban roads, buildings, parking lots, and paved areas.	-

2.3. *Data*

This research utilized a multi-source dataset, which primarily included the following categories: (1) Sentinel-2 surface reflectance data, employed for extracting spectral and textural features of the study area; (2) Digital Elevation Model (DEM) data, used for deriving topographic features; (3) Global 10m Land Use/Land Cover (LULC) product data, utilized for extracting impervious surface information within the study area; (4) meteorological data (including temperature and precipitation data), applied for analyzing the seasonal variations in land cover and their relationships with climatic factors; and (5) GF-2 data used as supplementary data, mainly used for subsequent classification validation and not involved in the classification process itself.

2.3.1. Sentinel-2 Surface Reflectance Data

The Sentinel-2 mission comprises two satellites, Sentinel-2A and Sentinel-2B, which together provide a revisit period of 5 days [40]. Due to its high spatial and temporal resolution, advanced multispectral imaging capabilities, and open access data, Sentinel-2 is widely utilized in various fields [41]. This study utilized Sentinel-2 imagery from 2021, accessed via the Google Earth Engine platform. Cloud masking was conducted using the QA60 band, and seasonal images were processed into median composites. The data spanned from 1 March 2021 to 28 February 2022. For the classification process, seasonal composite images were created from the Sentinel-2 data. The use of median composites was selected because they effectively reduced the influence of anomalous data that could have distorted surface feature representation. By applying medians, we obtained a more accurate representation of the typical surface state, as it was less affected by short-term fluctuations and provided greater stability.

2.3.2. Digital Elevation Model Data

The DEM data and their derived variables were highly correlated with LULC distribution, highlighting their importance as supplementary data in LULC classification [42]. This study utilized Shuttle Radar Topography Mission (SRTM) data, which offered a spatial resolution of 30 m. From this dataset, elevation, slope, aspect, and hillshade variables were extracted and used as features for classification.

2.3.3. Global 10m LULC Product

The ESA WorldCover 10m v200 dataset provides global land cover classifications at a 10 m resolution, based on Sentinel-1 and Sentinel-2 imagery, ensuring high spatial accuracy [43]. Similarly, the ESRI 10m Annual Land Cover dataset applies artificial intelligence techniques to Sentinel-2 data to generate reliable and detailed land cover information [44]. Both datasets, generated in 2021, were utilized to extract impervious surface areas within the study region.

2.3.4. Meteorological Data

Meteorological data provides a crucial background for analyzing seasonal land cover changes, particularly variations in temperature and precipitation. The data we used were sourced from researcher Peng Shouzhang, shared on the National Tibetan Plateau Data Center website [45]. In this study, the meteorological data were derived from the CRU global 0.5° climate dataset and the WorldClim high-resolution climate dataset. It was generated through the Delta spatial downscaling scheme and validated using 496 independent meteorological observation points, ensuring reliable results. The summer seasonal average data were derived from observations between June and August 2021, while the winter seasonal average data were based on observations from December 2021 to February 2022.

2.3.5. High-Resolution Auxiliary Data

The GF-2 satellite is an important satellite in China's high-resolution remote sensing satellite series, equipped with both panchromatic and multispectral sensors. The spatial resolution of the panchromatic band is 2 m, while the spatial resolution of the multispectral bands is 8 m, covering blue, green, red, near-infrared, and shortwave infrared bands. To enhance the spatial resolution of the imagery, this study fused the panchromatic and multispectral bands, improving the resolution to 2 m, thus enhancing the ability to identify fine details of ground features. In this study, we used GF-2 imagery data from June to August 2021 as a reference for selecting Sentinel-2 image sample points. Additionally, we selected 30 independent sample points for each land cover type on the GF-2 imagery. These sample points were solely used for the validation of subsequent classification results and were not involved in the training of the classification model.

2.4. Preparation of Image Features

This study selected a combination of spectral bands, remote sensing indices, texture features, and terrain characteristics as input features for land cover classification. Specifically, the features included spectral bands from Sentinel-2, remote sensing indices, texture features extracted using the Gray-level Co-occurrence Matrix (GLCM), and terrain features derived from SRTM data. All the features selected—spectral, index-based, texture, and terrain ones—are summarized in Table 2.

Table 2. Initial image features serving as potential predictors.

Types	Features	Description	Number
Bands	Blue, Green, Red, Red Edge1, Red Edge2, Red Edge3, NIR, SWIR 1, SWIR 2	Spectral bands in remote sensing imagery exhibit significant sensitivity to various land cover types, encompassing key spectral characteristics related to vegetation, soil, water bodies, and atmospheric components.	9
Indices	ARVI, NDVI, BSI, CARI, EVI, GNDVI, MNDWI, MSAVI, NDBI, NDREI, NDWI, MSI	Atmospheric Resistant Vegetation Index (ARVI); Normalized Difference Vegetation Index (NDVI); Bare Soil Index (BSI); Chlorophyll Absorption Ratio Index (CARI); Enhanced Vegetation Index (EVI); Green Normalized Difference Vegetation Index (GNDVI); Modified Normalized Difference Water Index (MNDWI), Modified Soil-Adjusted Vegetation Index (MSA-VI); Normalized Difference Built-Up Index (NDBI); Normalized Difference Red-Edge Index (NDREI); Normalized Difference Water Index (NDWI); Moisture Stress Index (MSI).	12
Texture	pc1_asm, pc1_contrast, pc1_corr, pc1_ent, pc1_idm, pc1_var, pc2_asm, pc2_contrast, pc2_corr, pc2_ent, pc2_idm, pc2_var, pc3_asm, pc3_contrast, pc3_corr, pc3_ent, pc3_idm, pc3_var	Texture features derived from first three principal components include measures like Angular Second Moment (asm), contrast, correlation (corr), entropy (ent), Inverse Difference Moment (idm), and variance (var) for each component (pc1, pc2, pc3).	18
Terrain	Elevation, Elevation stdDev, Slope, Aspect, Hillshade, Ruggedness, Northness, Eastness	Topographic attributes derived from SRTM data provide crucial context for geomorphological analysis and understanding of land cover distribution.	8
Total number of features considered for classification			47

From Sentinel-2, nine spectral bands were selected, including those from the visible, near-infrared, and shortwave infrared regions (B2, B3, B4, B5, B6, B7, B8, B11, B12). These bands reflect various land cover characteristics, such as vegetation health, soil moisture, and water reflectance, and are particularly useful for distinguishing land cover types such as vegetation, water bodies, and bare soil. In addition, this study presents the spectral curves of the selected bands for different land cover types (Figure 2). These spectral curves illustrate the reflectance variations in each land cover type across the selected bands, providing a visual representation of the spectral characteristics. The analysis of these curves offered valuable insights for the subsequent selection of index features.

Additionally, multiple remote sensing indices were calculated to enhance land cover differentiation. These indices included the Atmospheric Resistance Vegetation Index (ARVI), Normalized Difference Vegetation Index (NDVI), Bare Soil Index (BSI), Chlorophyll Absorption Ratio Index (CARI), Enhanced Vegetation Index (EVI), Green NDVI (GNDVI), Modified Normalized Difference Water Index (MNDWI), Modified Soil-Adjusted Vegetation Index (MSAVI), Normalized Difference Built-Up Index (NDBI), Normalized Difference Red-Edge Index (NDREI), Normalized Difference Water Index (NDWI), and Moisture Stress Index (MSI). These indices facilitated the differentiation of vegetation, water, bare soil, and urban areas.

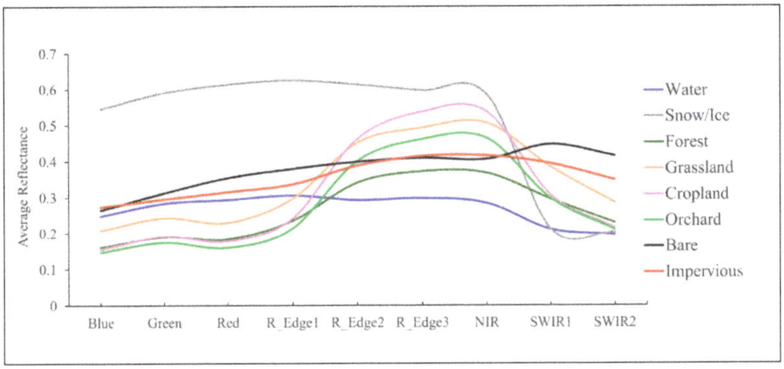

Figure 2. Average spectral reflectance curves of LULC based on sample points.

Principal Component Analysis (PCA) was applied to reduce the dimensionality of the key spectral bands from Sentinel-2. The first three principal components explained 89.14%, 6.82%, and 3.58% of the total variance, respectively, accounting for over 98% in total. This reduction preserved the most representative spectral features while minimizing redundancy. For texture feature extraction, the GLCM method was used to derive multiple classic texture features from the PCA-reduced image. These features included contrast, correlation, variance, idm and ent, which reflected the spatial structure and geometry of land cover types. To optimize texture extraction, different window sizes (3 × 3; 5 × 5; 7 × 7; and 9 × 9) were compared, and the 5 × 5 window was found to capture more texture information, which was selected for final texture feature extraction.

Eight terrain features were extracted from the SRTM DEM, including elevation, Elevation_stdDev, slope, aspect, hillshade, ruggedness, northness, and eastness. These terrain features provided valuable information on the surface morphology of the study area and assisted in distinguishing different land cover types, particularly in mountainous or sloping areas where terrain characteristics play a significant role.

2.5. Methods

The LULC classification in this study was based on the extracted spectral, textural, and topographic features. Following feature extraction, a feature subset selection process was conducted using the Recursive Feature Elimination (RFE) method to identify the most informative features, thereby optimizing computational efficiency while maintaining classification performance. The classification process integrated the Otsu thresholding method and the Random Forest algorithm to produce the final LULC classification results, ensuring the robustness and reliability of land cover mapping. Accuracy assessment was performed using five-fold cross-validation to evaluate the model's classification performance.

2.5.1. Integrating Otsu Thresholding and Random Forest

The Otsu thresholding method is a widely used automatic segmentation technique in remote sensing image classification, which determines the optimal threshold by maximizing the variance between classes, making it effective for distinguishing land cover types with significant spectral differences, particularly for categories such as water bodies and snow/ice with well-defined boundaries [46]. In contrast, the Random Forest (RF) algorithm, an ensemble learning method based on decision trees, is capable of handling high-dimensional and multi-class data, and its inherent randomness enhances the model's generalization ability and stability [47,48]. However, in environments with complex land cover distributions, traditional RF classification methods may suffer from the effects of data noise and class imbalance, which can significantly limit classification accuracy.

In the study area, the overall land cover types exhibit clear boundaries, but at a local scale, fine-grained land cover classes are highly interspersed. The northern region is predominantly covered by snow and glaciers, characterized by well-defined boundaries, while the southern oasis area is mainly agricultural, with orchards and croplands intricately interwoven. The diversity in crop types, fruit tree varieties, and age structures results in highly mixed spectral features, increasing classification uncertainty. Furthermore, bare land is extensively distributed, with significant variability in surface features across different regions, leading to considerable reflectance changes and challenging spectral-based classification. In such complex conditions, traditional RF classification methods often struggle to accurately distinguish fine-grained land cover types, particularly in high-dimensional, multi-class data, where noise and data imbalance exacerbate the classification challenges.

The classification process is shown in Figure 3. The Otsu method was first employed to automatically determine the optimal segmentation threshold for water bodies and snow/ice regions, followed by the differentiation between bare land and vegetation in non-water and non-snow areas. This segmentation strategy effectively masked land cover classes with clear boundaries, reducing their interference in the classification process and allowing the RF model to focus on distinguishing spectrally similar and difficult-to-identify land cover types. The classification process consisted of three stages: (1) initial feature extraction, (2) segmentation threshold determination, and (3) feature selection and model construction.

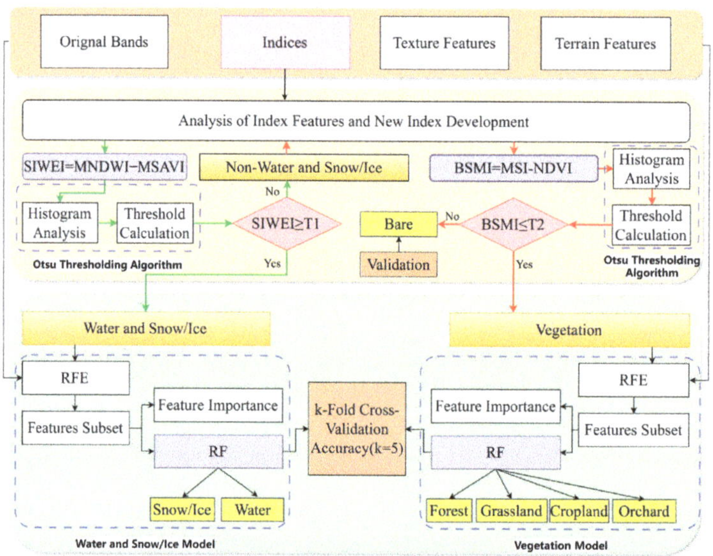

Figure 3. LULC classification workflow based on threshold methods and RF, including threshold selection, RF classification model building, and feature optimization through RFE. Green path: initial implementation. Red path: subsequent steps after green path.

In this study, to determine the appropriate index thresholds, we performed a comparative analysis of multiple indices based on boxplots of sample points (Figure 4). The results indicated the following: (1) the MNDWI and NDWI effectively distinguished between water/snow and ice and non-water/snow and -ice areas; (2) the NDVI, GNDVI, MSAVI, and NDREI all exhibited strong differentiation ability in distinguishing between vegetation and non-vegetation areas; (3) the MSI effectively differentiated between bare soil and other land cover types.

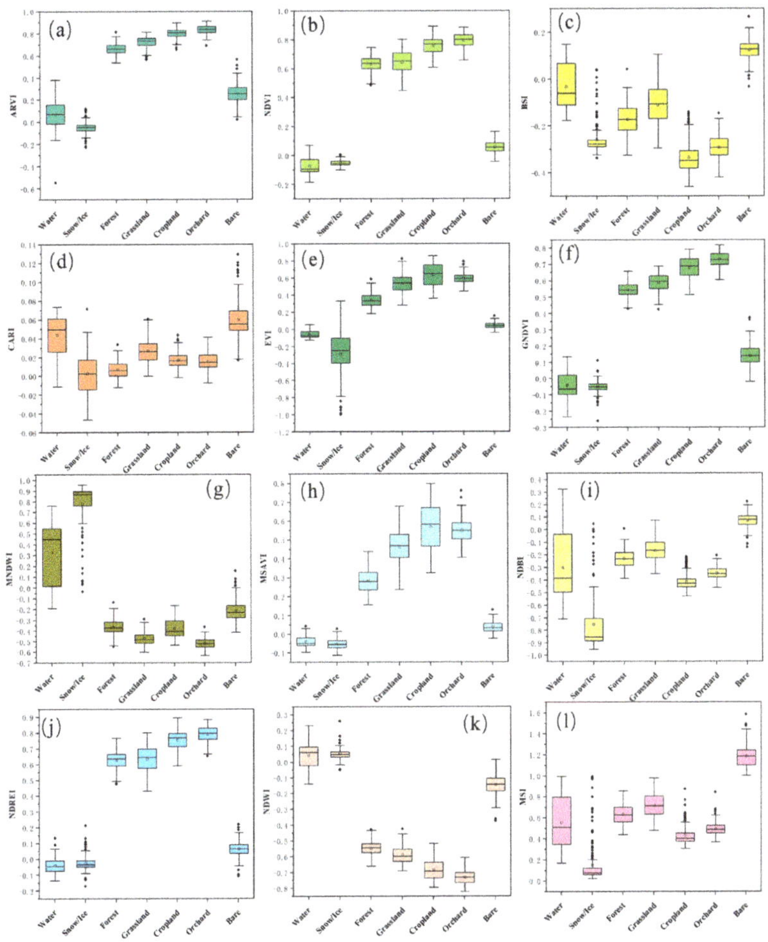

Figure 4. Variability analysis of spectral indices for land cover sample points and threshold determination (**a–l**: various indices).

To improve classification accuracy and enhance the differentiation of land cover types, we constructed new indices based on the performance of the aforementioned indices. These new indices were designed to more precisely capture the characteristics of specific land cover types and optimize segmentation results. For distinguishing between water/ice and non-water/ice areas, despite the potential of other index combinations, the Snow/Ice/Water Enhancement Index (SIWEI) performed the best. It effectively suppressed background interference, particularly from bare soil and other non-water classes, and accurately differentiated between water/ice and non-water/ice areas. Similarly, for non-water and -ice areas, the Bare Soil Moisture Index (BSMI) was effective in distinguishing between bare soil and vegetation. Specifically, the SIWEI was constructed using the combination of the MNDWI and MSAVI, expressed as

$$SIWEI = MNDWI - MSAVI \quad (1)$$

and the BSMI was formed using the MSI and NDVI, expressed as

$$BSMI = MSI - NDVI \quad (2)$$

Through the histogram analysis of both indices (Figure 5), we applied Otsu's method to determine the optimal threshold for the SIWEI as 0.211 and for the BSMI as 0.532.

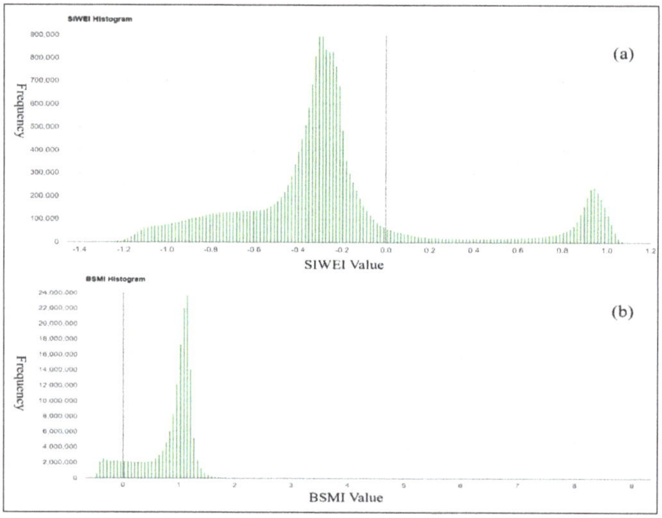

Figure 5. Histograms of SIWEI and BSMI for threshold determination: (**a**) SIWEI and (**b**) BSMI.

2.5.2. Recursive Feature Elimination Method

In this study, we first standardized all features and employed Recursive Feature Elimination (RFE) to select the most influential features, aiming to reduce overfitting and enhance model accuracy [49]. Multidimensional datasets often contain highly correlated redundant features, which provide no additional information and may introduce noise, thus increasing model uncertainty. By applying the RFE method, we effectively eliminated these redundant features, enabling the model to focus on inputs that were truly relevant to the target variable. This approach reduced error and improved classification performance. Furthermore, the reduction in feature dimensions lowered model complexity, enhancing its robustness to noise and minimizing the risk of overfitting. Ultimately, this feature selection process allowed the model to concentrate on the most critical variables, thereby improving its stability and precision.

2.5.3. Model Validation

To assess the classification accuracy of the RF model, k-fold cross-validation, a widely recognized technique for model validation, was applied [50]. In this study, k was set to 5. The dataset was partitioned into 5 subsets, and in each iteration, 4 subsets were used for model training, while the remaining subset served as the test set. This process was repeated 5 times, with each subset serving as the test set once [51]. The average performance metrics across all iterations were then calculated. This approach provided a robust evaluation of the model's stability and performance across various training and testing splits, thereby minimizing potential biases introduced by random data partitioning.

Additionally, several key metrics derived from the confusion matrix, such as the producer's accuracy (PA), user's accuracy (UA), overall accuracy (OA), and Kappa coefficient, were employed in this study to rigorously assess the model's classification performance. These metrics provided a detailed evaluation of both the model's accuracy and its ability to differentiate between different land cover types.

$$PA = \frac{TP}{TP + FN} \tag{3}$$

$$UA = \frac{TP}{TP + FP} \tag{4}$$

$$OA = \frac{TP + TN}{TP + TN + FP + FN} \tag{5}$$

$$Kappa = \frac{p_o - p_e}{1 - p_e} \tag{6}$$

Here, TP represents True Positives, FN stands for False Negatives, FP denotes False Positives, and TN refers to True Negatives. The term p_o indicates the observed agreement, which is the proportion of correctly classified samples along the diagonal of the confusion matrix. In contrast, p_e represents the expected agreement, which is the probability of random agreement based on the marginal probabilities of each class.

3. Results

3.1. Feature Importance for Enhanced LULC Classification

After performing RFE separately for the vegetation and non-vegetation regions, the optimal number of features was determined, as illustrated in Figure 6. These feature subsets, when combined, yielded the highest classification accuracy. The results demonstrate that, for the WSI_RF model evaluation, only 20 features were identified as critical to the classification process, with the optimal point occurring at (20, 0.8550). In the VGE-RF model evaluation, 23 features were selected, with the optimal point at (23, 0.9377). Beyond the optimal points, the accuracy exhibited fluctuations as the number of features increased, indicating that the presence of irrelevant or redundant features did not contribute to the improvement in classification performance, thus necessitating their removal.

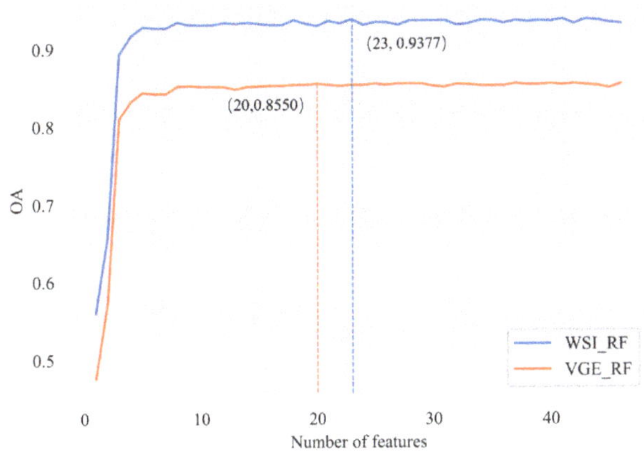

Figure 6. Plot of overall accuracy vs. number of features.

The features used for classification included nine original image bands, twelve spectral indices, eighteen texture features, and eight terrain features. Figure 7 illustrates the relative importance scores of these final input features in differentiating land cover categories. Figure 7a presents the importance ranking of the input features for the VGE model. Among the 23 features, the dataset included nine spectral bands, 11 indices, and three terrain features but no texture features. Elevation and elevation_stdDev were identified as the most critical features, followed by Bands 5 and 11, with Band 3 also showing notable importance. Among the indices, the MNDWI and NDBI were particularly valuable for distinguishing vegetation types. Figure 7b illustrates the importance ranking of input features for the WSI model. Of the 20 features, the dataset comprised seven spectral bands, five indices, four texture features, and four terrain features. Band 3 was the most influential, followed by slope and elevation. Other significant bands included Bands 2, 7, 5, and 4. The most critical index feature was the BSI, while the texture features, all derived from the first principal component, included metrics such as corr, var, asm, and ent.

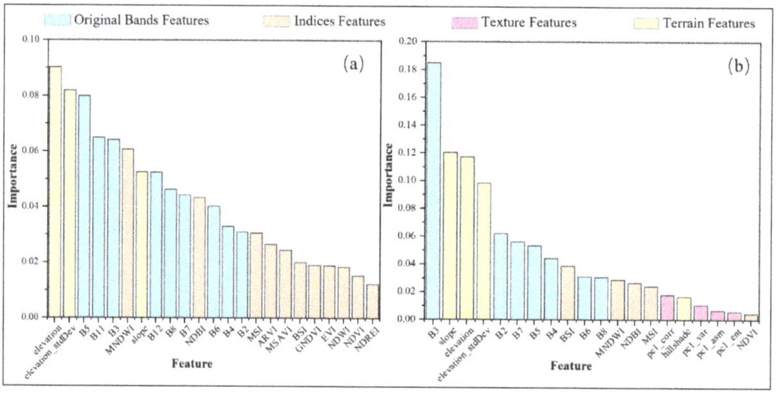

Figure 7. Feature importance scores of selected features following RFE. (**a**) Feature importance for VGE_RF model. (**b**) Feature importance for WSI_RF model.

3.2. Classification Accuracy

In the model training process, we utilized the Python 3.12.2 programming language and the Scikit-learn library to implement the Random Forest (RF) model, accompanied by hyperparameter optimization. Specifically, we employed the GridSearchCV function for grid search, allowing for a systematic exploration of the hyperparameter space and evaluation of various parameter combinations to identify the optimal configuration. In both the WSI_RF and VGE_RF models, key parameters such as the number of trees (n_estimators) and the maximum depth of the trees (max_depth) were carefully tuned. After optimization, the number of trees for the WSI_RF model was set to 80, with a maximum depth of 20, while for the VGE_RF model, the number of trees was set to 100, with a maximum depth of 50.

We applied five-fold cross-validation, generating a confusion matrix for each fold. The final classification confusion matrix, presented in Table 3, was derived by aggregating the matrices from each fold. Based on these matrices, the producer's accuracy (PA) and user's accuracy (UA) for the water body class were 94.87% and 92.50%, respectively, demonstrating a high level of classification accuracy. For the snow/ice class, the PA and UA were 92.86% and 95.12%, respectively. The PA and UA for the forest class were 84.81% and 90.91%, reflecting relatively low misclassification within this category. The grassland class exhibited PA and UA values of 83.93% and 85.29%, indicating a robust classification performance and the model's ability to effectively distinguish grassland from other land cover types. The PA and UA for the cropland class were 84.95% and 88.89%, respectively, indicating satisfactory classification accuracy. However, the orchard category displayed a PA of 88.74% but a relatively lower UA of 77.16%, indicating suboptimal user accuracy for this class. This discrepancy may be attributed to the spectral similarity between orchards and other classes, such as cropland and grassland, leading to higher misclassification rates, particularly where orchards were misclassified as cropland or grassland.

Table 3. Confusion matrix of water and snow/ice.

	Classification Result							
	Water	**Snow/Ice**	**Forest**	**Grassland**	**Cropland**	**Orchard**	**PA (%)**	**UA (%)**
Water	148		12				94.87	92.50
Snow/Ice	8		156				92.86	95.12
Forest			160	8	5	3	84.81	90.91
Grassland			10	145	8	7	83.93	85.29
Cropland			7	5	160	8	84.95	88.89
Orchard			12	15	15	141	88.74	77.16

To further assess the performance of the classification models, we calculated the OA and Kappa coefficient (Table 4). Both models demonstrated high classification accuracy, with the WSI_RF model achieving an OA of 85.50% and the VGE_RF model reaching an OA of 93.77%. Additionally, the Kappa coefficient for the WSI_RF model was 0.8088, while the VGE_RF model had a Kappa coefficient of 0.8755, indicating a strong agreement between the model's classification results and the actual data. In the validation of the bare land class, 173 sample points were correctly classified, resulting in an OA of 96.11%. These results further demonstrate the rationality of the newly constructed indices and thresholds.

Table 4. Model performance: OA and Kappa.

Metadata	OA	Kappa
WSI_RF	85.50%	0.8088
VGE_RF	93.77%	0.8755
Bare	96.11%	-

The GF-2 sample points selected in this study were used to validate both our classification results and those of the ESA product. Table 5 presents the number of correctly classified samples for each land cover category. Our classification achieved an overall accuracy of 85.24%, compared to 72.85% for the ESA product. Notably, our approach demonstrated higher accuracy in the classification of vegetation and water bodies. The number of correctly classified samples for the ice/snow and bare land categories was comparable to that of the ESA product. These findings further confirm the reliability of our classification results.

Table 5. Validation of classification results based on GF-2 sample dataset: correctly classified samples per class.

LULC Type	Proposed Classification	ESA Product
Water	27	23
Snow/Ice	24	24
Forest	25	17
Grassland	28	22
Cropland	24	41
Orchard	25	
Bare	26	26
OA	85.24%	72.85%

3.3. LULC Mapping of Wensu County

Extracting impervious surfaces is challenging due to spectral variability, the complex nature of buildings and structures, and the fragmented distribution of road networks. To overcome these challenges, this study integrated impervious surface data from the ESA WorldCover 10m v200 and ESRI LULC 10m datasets. A combined dataset was used to delineate impervious surfaces within the study region (Figure 8).

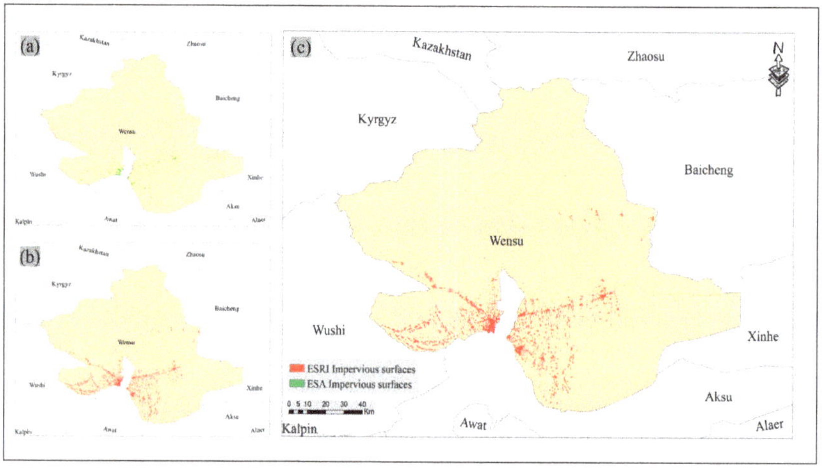

Figure 8. Impervious surface of ESRI and ESA datasets. (**a**) ESA2021. (**b**) ESRI2021. (**c**) ESRI merged with ESA.

An LULC classification map of Wensu County for the summer of 2021 was generated (Figure 9). Wensu County displays marked spatial variations in LULC between its northern and southern regions, reflecting distinct geographical and climatic conditions. The northern area is dominated by the Tianshan Mountains, characterized by complex topography and high elevations. This region is predominantly covered by extensive snow and ice, particularly at higher altitudes. The valleys in this area feature a mosaic of grasslands and forests. Grasslands are primarily found in valley bottoms and flatter terrains, supporting pastoral activities, while forests are concentrated on steeper slopes and rugged terrains. The water bodies in this region are largely sustained by glacial meltwater, with the Kumalak River flowing in from the east along the county's boundary and several minor streams winding through the central region.

In stark contrast, the southwestern and central–southern regions are dominated by oasis landscapes. These areas are predominantly covered by arable land, orchards, and impervious surfaces, reflecting intensive agricultural and urban activities. The topography here is relatively flat compared to the northern mountainous region, and water bodies are scattered within the oasis areas, primarily fed by irrigation systems rather than natural glacial sources.

The summer area statistics for each land cover type are presented in Table 6, under the "Sum_Area" column. Among these, the largest area is bare land, covering 11,750.89 km^2, followed by snow/ice at 3131.83 km^2. For vegetation types, grassland and cropland occupy the largest areas, at 1416.52 km^2 and 1373.01 km^2, respectively. The orchard area is 860.51 km^2, the forest area is 219.32 km^2, the water body area is 90.31 km^2, and the area of impervious surfaces is 302.94 km^2.

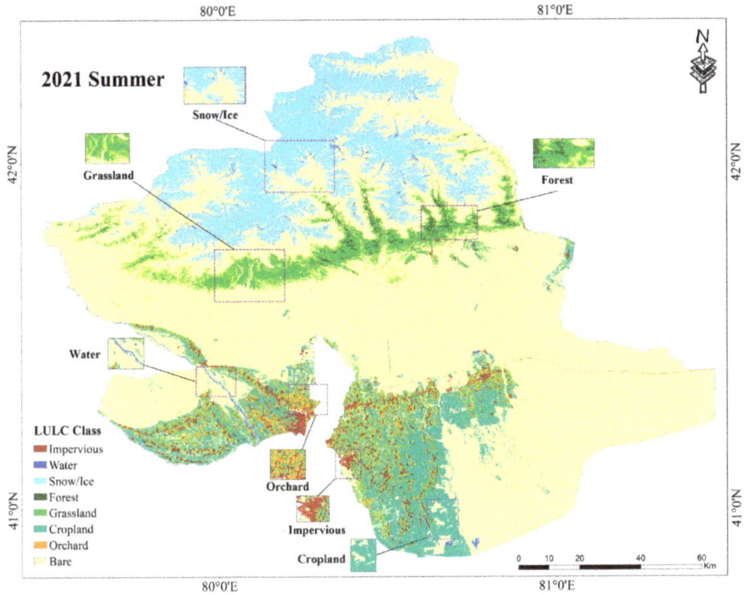

Figure 9. An LULC thematic map for the Wensu region in the summer of 2021.

Table 6. Seasonal LULC change matrix (Km^2).

	Winter								
Summer	**Water**	**Forest**	**Cropland**	**Grassland**	**Orchard**	**Snow/Ice**	**Bare**	**Impervious**	**Sum_Area**
Water	28.19	0	0	0	0	58.84	3.28	0	90.31
Forest	1.30	48.44	0	0	0	169.58	0	0	219.32
Cropland	63.09	0	1103.51	0	0	206.41	0	0	1373.01
Grassland	10.65	0	0	491.47	0	914.40	0	0	1416.52
Orchard	6.87	0	0	0	844.23	9.41	0	0	860.51
Snow/Ice	0.55	0	0	14.39	0	3111.84	5.05	0	3131.83
Bare	220.65	0	0	0	0	5040.53	6489.71	0	11,750.89
Impervious	0	0	0	0	0	0	0	302.94	302.94
Win_Area	331.3	48.44	1103.51	505.86	844.23	9511.01	6498.04	302.94	-
Changed Area	240.99	−170.88	−269.5	−910.66	−16.28	6379.18	−5252.85	0	-

3.4. Seasonal Variation Analysis

The seasonal variations in snow/ice and water bodies within the study area, based on the Otsu thresholding method and the WS_RF model, are shown in Figure 10. The spatial distribution of these features is influenced by meteorological factors such as temperature and precipitation, resulting in similar patterns between spring and autumn. In autumn, river water bodies expand compared to those in spring, while snow and ice in the northern regions are more extensive and denser in spring. Summer is characterized by the largest coverage of water bodies and minimal snow and ice. Snow and ice coverage decreases significantly from north to south, with some snow persisting in the southern oasis regions.

To further quantify the seasonal LULC changes between winter and summer, an LULC change matrix (Table 6) was constructed, illustrating the transitions in land cover types from summer to winter. The results reveal a significant increase in both water bodies and snow/ice coverage, while vegetated areas and bare land experienced relatively minor changes. Specifically, the area of water bodies expanded by 240.99 km^2. Snow/ice coverage, on the other hand, increased by 6379.18 km^2, with 5252.85 km^2 of bare land and 910.66 km^2 of grassland being covered by snow/ice. Furthermore, small portions of other vegetated areas were also partially covered by snow. This pattern indicates a notable shift in land cover, with snow and ice encroaching upon previously non-snow-covered areas.

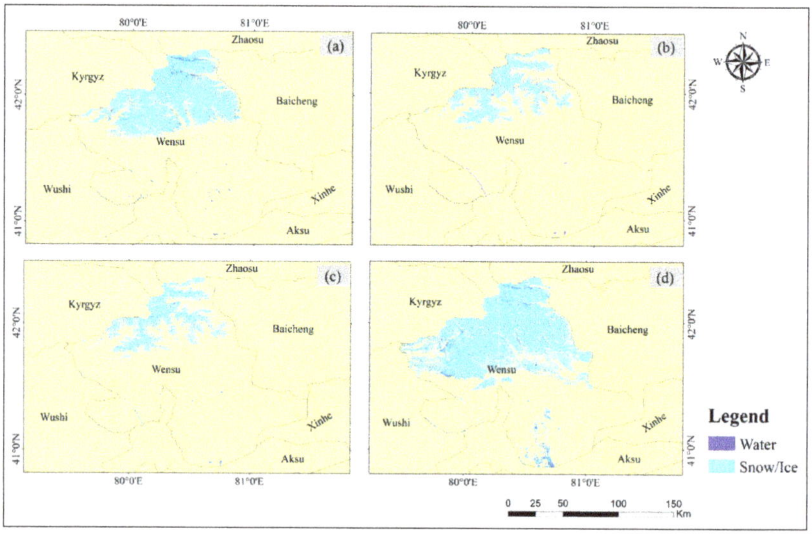

Figure 10. Water and snow/ice seasonal changes. (**a**) Spring. (**b**) Summer. (**c**) Autumn. (**d**) Winter.

Seasonal land use/land cover (LULC) changes were monitored using classification maps from winter and summer, with the spatial variations in land cover presented in Figure 11. The LULC changes in the study area are primarily driven by the coverage of snow during the winter season. In the southern region, where snow cover is limited, snow does not significantly impact other land cover types, resulting in minimal changes in LULC types. In contrast, in the northern mountainous areas, snow cover increases significantly, extensively covering forest and grassland areas. This snow cover advances southward, impacting parts of the oasis region and the bare land in the northern forest and grassland areas. These changes lead to a clear north–south gradient in the spatial distribution of LULC types.

Based on the land cover change data obtained, we spatially overlaid meteorological data to generate Figure 12. This figure illustrates the climatic characteristics of Wensu County from December 2021 to February 2022. Figure 12a shows the winter temperature variations in the study area, where the lowest temperature reached −31.4 °C and the highest temperature was −1.5 °C, resulting in a temperature range of 29.9 °C. The low-temperature zones are primarily concentrated in the high-altitude mountainous areas in the north. As the altitude decreases, temperatures in the valleys rise. Overall, the temperature shows an increasing trend from north to south, with the highest temperatures observed in the southernmost parts of the study area. This temperature distribution directly impacts human activity patterns, with southern areas, due to their higher temperatures and favorable environmental conditions, becoming the main regions for human activities and showcasing a wider variety of land cover types.

Figure 12c presents the winter precipitation levels, with the lowest value recorded at 0.6 mm and the highest at 4.3 mm. Higher-precipitation areas are mainly located in the northern regions, with a clear decreasing trend from north to south. Regions with higher temperatures and precipitation experience minimal land cover changes, where snow and glaciers effectively cover the area. However, in the transition zones between the northern and southern regions, the winter snow-covered area significantly expands, covering forests, grasslands, and bare lands. Figure 12b displays changes in snow cover, primarily concentrated in the low-temperature valleys and bare lands, extending southward. This phenomenon further reflects the impact of climatic factors on land cover types. In Figure 12d, a significant positive correlation is shown between precipitation and changes in snow cover. Specifically, areas with higher precipitation often have larger snow-covered areas, indicating that precipitation plays a crucial role in the formation and variation in snow cover. Additionally, the main rivers in the southern regions experience freezing due to temperatures falling below zero degrees, providing important information for understanding regional hydrological processes.

To quantitatively characterize these changes, we calculated the number of pixels representing snow and ice changes at different temperature and precipitation levels, as shown in Figure 13. By deeply analyzing pixel change patterns, we revealed the distribution characteristics of changed and unchanged pixels under various temperature and precipitation segments. This analysis helped identify the significance of changes under specific climate conditions and provided crucial insights into the impact of climatic factors on land cover changes.

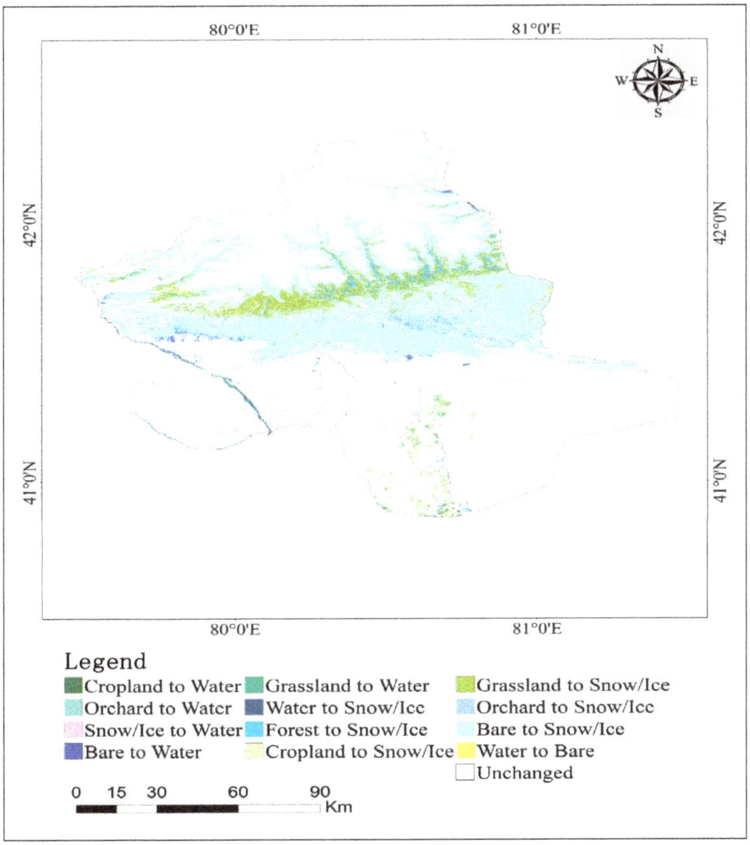

Figure 11. Spatial distribution of seasonal LULC changes.

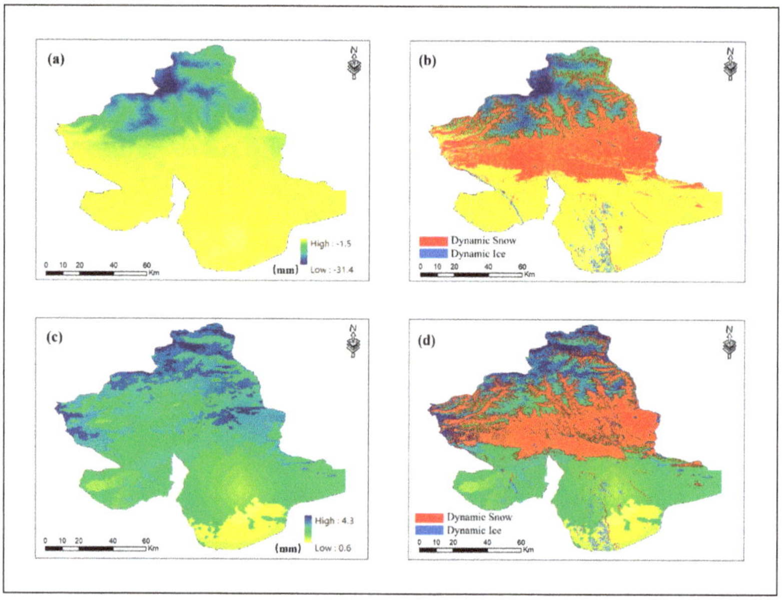

Figure 12. Spatial overlay of changing snow and ice with temperature and precipitation patterns. (**a**) Winter temperature, (**b**) spatial distribution of snow and ice with temperature, (**c**) winter precipitation, and (**d**) spatial distribution of snow and ice with precipitation.

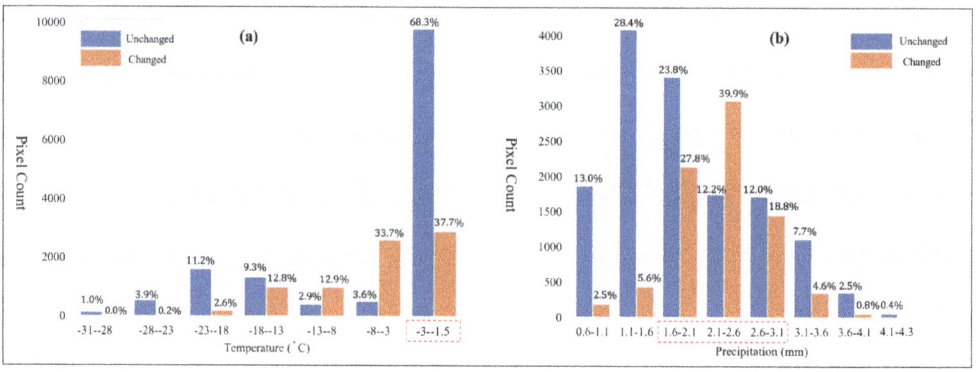

Figure 13. Statistical analysis of changed and unchanged pixels across temperature and precipitation segments. (**a**) Temperature. (**b**) Precipitation.

In Figure 13a, we observe that the changed pixels are mainly concentrated in the highest temperature range of −3 °C to −1.5 °C. A relatively high number of changed pixels is also found in the temperature range of −8 °C to −1.5 °C. Notably, there are considerable changes in the temperature ranges of −18 °C to −13 °C and −13 °C to −8 °C, indicating that land cover changes are minimal in areas with the lowest temperatures. In the temperature range of −13 °C to −8 °C, 39.9% of the pixels are changed, while in the range of −18 °C to −13 °C, 23.8% of the pixels are changed. Hence, temperature significantly influences land cover change patterns, especially within specific temperature ranges.

Figure 13b shows the distribution characteristics of changed pixels concerning precipitation levels. Changed pixels are heavily concentrated in the precipitation range of

1.61–2.6 mm, which falls in the middle of the overall precipitation levels. Additionally, changed pixels are primarily found in areas with precipitation below 1.6 mm. As precipitation increases, the number of changed land cover types gradually rises, peaking in the 2.1–2.6 mm range. In the precipitation range of 1.6–2.1 mm, 27.9% of the pixels are changed. As precipitation increases to 2.1–2.6 mm, the proportion of changed pixels decreases by 4.6%. However, the number of changed pixels decreases as precipitation continues to increase beyond this range. No changed pixels are observed in the highest precipitation range, highlighting the crucial role of precipitation in promoting land cover changes.

4. Discussion

The results of this study provide significant scientific evidence and practical guidance for land cover classification, particularly in regions with complex terrain and pronounced seasonal variations. By constructing the VGE_RF and WSI_RF classification models, classification accuracies of 93.77% and 85.50% were achieved, respectively, which robustly demonstrated the effectiveness of integrating Otsu's thresholding method with the Random Forest algorithm based on multi-dimensional feature data for land cover classification. This method significantly enhanced the differentiation between various land cover types, especially in the classification of complex surface types, offering a new technological approach for land cover classification research. The integration of Otsu's thresholding method with Random Forest further emphasized the importance of both spectral and terrain data in achieving high accuracy, particularly when applied to environments that exhibit significant heterogeneity in land cover types.

The relatively low user accuracy for orchards (77.16%) can be attributed to several factors. First, the spatial similarity between orchards and croplands: In the study area, the spatial distribution of orchards and croplands overlaps significantly, particularly in areas with concentrated farmland or scattered orchards, where the boundaries between the two are blurred. This resulted in minimal spectral differences, making them easily confused. Therefore, relying solely on spectral information for classification could be affected by spatial distribution similarities, leading to suboptimal classification results. Second, the diversity of tree species and complexity of tree ages: Orchards in the study area contain a wide variety of tree species and ages, resulting in a highly complex vegetation structure. This complexity caused significant spectral variation in the reflectance characteristics of orchards across different species and growth stages, increasing the difficulty of classification. Even with the inclusion of texture features, these differences remained difficult to fully eliminate, especially when spectral overlap occurred, making it challenging for the classification model to distinguish between categories. To address this issue, future studies may need to update sampling methods and incorporate more complex spatial information to improve classification accuracy.

One of the key contributions of this study is the systematic feature selection process. Using the RFE algorithm, the VGE_RF model selected 23 key features from the initial 48 image features, including nine spectral band features, 12 index features, and two terrain features; the WSI_RF model selected 20 key features, which included seven spectral band features, five index features, four texture features, and four terrain features. This feature selection process not only optimized the model's computational efficiency but also significantly improved classification accuracy, providing a reliable basis for the precise identification of land cover types. The results indicated that the spectral features (e.g., red-edge 1, shortwave infrared 1, and green bands) and index features (e.g., MNDWI, NDBI, MSI, and ARVI) played a crucial role in distinguishing vegetation types and water/ice cover, while terrain features (e.g., elevation and elevation standard deviation) further enhanced the model's adaptability to complex terrain regions. This feature selection process highlighted the importance of balancing spectral and spatial information to ensure comprehensive land cover characterization. It also emphasized the necessity of integrating multiple data types to fully capture the complex variability inherent in the study area.

Moreover, the inclusion of texture features in the WSI_RF model demonstrated the value of capturing fine-scale spatial variability, particularly in distinguishing water bodies from snow or ice. The texture variables, such as the correlation of the first principal component (pc1_corr) and the angular second moment (asm), underscored the utility of spatial information in enhancing the discriminative power of the model. The absence of significant texture variables in the VGE_RF model suggested that vegetation types, unlike water and snow/ice, exhibited relatively uniform spatial characteristics when viewed through optical remote sensing imagery. This distinction between the two models' feature sets reflected the differing physical properties and spatial arrangements of the land cover types, further illustrating the complex interaction between spectral and spatial data in remote sensing applications.

The seasonal land cover change detection results provided important insights into the dynamic variations in water and snow coverage. The study found that land cover in the study area exhibited significant seasonal fluctuations. By analyzing pixel change patterns under varying temperature and precipitation conditions, the non-linear impact mechanism of climatic factors on land cover change was revealed. Specifically, the change pixels were primarily concentrated within temperature ranges of −3 °C to −1.5 °C and precipitation ranges of 1.61–2.6 mm, indicating that land cover change responded to climatic factors with a distinct threshold effect. This finding not only deepens the understanding of the climate-land cover interaction mechanisms but also offers valuable references for the prediction and simulation of land cover changes at the regional scale. The results highlight that temperature and precipitation thresholds may serve as useful indicators for forecasting land cover dynamics, particularly in regions with marked seasonal variations. Understanding these climatic thresholds can inform resource management strategies, especially in areas that rely on water resources and agricultural productivity.

Further analysis of the relationship between temperature and land cover change revealed a complex, non-linear pattern, with temperature playing a significant role in regulating land cover shifts within specific ranges. For instance, the concentration of changing pixels within the temperature interval of −3 °C to −1.5 °C suggested that moderate temperature changes may have triggered transitions between land cover types. In contrast, extreme low temperatures, such as those in the −18 °C to −13 °C and −13 °C to −8 °C ranges, resulted in minimal land cover changes, emphasizing the relative stability of land cover types under harsher climatic conditions. This insight contributes to a more nuanced understanding of how land cover responds to extreme climatic conditions, offering practical implications for predicting land cover stability in regions prone to cold weather.

Precipitation, on the other hand, was found to have a threshold effect on land cover change. The concentration of changing pixels in the 1.61–2.6 mm precipitation range highlighted the sensitivity of land cover to moderate levels of precipitation, which could trigger vegetation growth and influence the dynamics of water bodies and snow cover. As precipitation levels increased beyond this range, the number of changing pixels began to decrease, suggesting that excessive precipitation may have had a dampening effect on land cover change, possibly due to saturation or other environmental factors. This finding provides a clearer understanding of how precipitation patterns affect land cover changes, particularly in areas with high seasonal variability in rainfall.

5. Conclusions

This study introduces an enhanced classification strategy that significantly boosts LULC classification accuracy, particularly in regions with distinct land cover boundaries but complex internal distributions. The proposed approach demonstrates notable effectiveness in classifying orchards and arable land, providing more accurate data to support regional agricultural development and resource management. Additionally, this method offers valuable insights for similar applications in comparable regions.

However, challenges remain in adapting to highly heterogeneous land cover types and performing fine-grained classification. Future research should explore integrating deep learning with traditional threshold-based methods to enhance classification adaptability. Deep learning models, such as Long Short-Term Memory (LSTM) networks and Convolutional Neural Networks (CNN), can automatically extract spatiotemporal features from time series remote sensing data, improving the detection of land cover changes. The fusion of multi-source remote sensing data, particularly the combination of high-resolution and multispectral imagery, is another promising direction. Leveraging multimodal learning techniques in deep learning allows for the effective fusion of data features from diverse remote sensing platforms, enhancing the model's adaptability to various land cover types. In addition, future research should focus on the dynamic monitoring of long-term time series to more comprehensively reveal the spatiotemporal evolution patterns of land cover changes. Analyzing long-term seasonal variations can help identify trends and anomalous fluctuations in land cover, thereby deepening our understanding of environmental changes and providing scientific support for regional ecological management and sustainable development.

In conclusion, this study validates the effectiveness of combining threshold methods with the Random Forest algorithm for land cover classification in Wensu County. The proposed approach provides a solid foundation for future improvements through more flexible threshold selection, deep learning integration, and multi-source data fusion. As climate change and ecological conservation continue to shape land use management, further advancements in classification techniques will play a crucial role in supporting sustainable resource management and environmental monitoring.

Author Contributions: Conceptualization, X.S.; Methodology, X.S.; Software, X.S., X.L. and L.W.; Validation, B.T. and J.G.; Formal analysis, S.X. All authors have read and agreed to the published version of the manuscript.

Funding: This research was supported by the Key R&D Task Special Project of the Xinjiang Uygur Autonomous Region for the establishment and demonstration of the application of the meteorological disaster monitoring and early warning service platform for forestry and fruit (Grant No. 2023B02004-3) and the Research Fund of the Chinese Academy of Forestry for the study on the remote sensing response mechanism of carbon sources and sinks in the forest management cycle (Grant No. CAFYBB2023MA013).

Data Availability Statement: The original contributions presented in the study are included in the article, further inquiries can be directed to the corresponding author.

Conflicts of Interest: The authors declare no conflicts of interest.

References

1. Abbass, K.; Qasim, M.Z.; Song, H.; Murshed, M.; Mahmood, H.; Younis, I. A review of the global climate change impacts, adaptation, and sustainable mitigation measures. *Environ. Sci. Pollut. Res.* **2022**, *29*, 42539–42559. [CrossRef]
2. IPCC. *Climate Change: The Physical Science Basis, Contribution of Working Group I to the Sixth Assessment Report of the Intergovernmental Panel on Climate Change*; Cambridge University Press: Cambridge, UK, 2021.

3. Buchhorn, M.; Lesiv, M.; Tsendbazar, N.-E.; Herold, M.; Bertels, L.; Smets, B. Copernicus global land cover layers—Collection 2. *Remote Sens.* **2020**, *12*, 1044. [CrossRef]
4. Aljenaid, S.S.; Kadhem, G.R.; AlKhuzaei, M.F.; Alam, J.B. Detecting and assessing the spatio-temporal land use land cover changes of Bahrain Island during 1986–2020 using remote sensing and GIS. *Earth Syst. Environ.* **2022**, *6*, 787–802. [CrossRef]
5. Salehi, H.; Shamsoddini, A.; Mirlatifi, S.M.; Mirgol, B.; Nazari, M. Spatial and temporal resolution improvement of actual evapotranspiration maps using Landsat and MODIS data fusion. *Front. Environ. Sci.* **2021**, *9*, 795287. [CrossRef]
6. Wiatkowska, B.; Słodczyk, J.; Stokowska, A. Spatial-temporal land use and land cover changes in urban areas using remote sensing images and GIS analysis: The case study of Opole, Poland. *Geosciences* **2021**, *11*, 312. [CrossRef]
7. Bhagat, S.; Prasad, P.R.C. Assessing the impact of spatio-temporal land use and LULC changes on land surface temperature, with a major emphasis on mining activities in the state of Chhattisgarh, India. *Spat. Inf. Res.* **2024**, *32*, 339–355. [CrossRef]
8. Umwali, E.D.; Kurban, A.; Isabwe, A.; Mind'je, R.; Azadi, H.; Guo, Z.; Udahogora, M.; Nyirarwasa, A.; Umuhoza, J.; Nzabarinda, V.; et al. Spatio-seasonal variation of water quality influenced by land use and land cover in Lake Muhazi. *Sci. Rep.* **2021**, *11*, 17376. [CrossRef] [PubMed]
9. Wang, H.; Wigneron, J.-P.; Ciais, P.; Yao, Y.; Fan, L.; Liu, X.; Li, X.; Green, J.K.; Tian, F.; Tao, S.; et al. Seasonal variations in vegetation water content retrieved from microwave remote sensing over Amazon intact forests. *Remote Sens. Environ.* **2022**, *285*, 113409. [CrossRef]
10. Mustapha, M.; Zineddine, M. Assessing the Impact of climate change on seasonal variation in agricultural land use using Sentinel-2 and machine learning. *Environ. Sci. Proc.* **2023**, *29*, 51. [CrossRef]
11. Sinha, R.K.; Eldho, T.I.; Subimal, G. Assessing the impacts of land use/land cover and climate change on surface runoff of a humid tropical river basin in Western Ghats, India. *Int. J. River Basin Manag.* **2020**, *21*, 141–152. [CrossRef]
12. Chen, D.; Zhou, Z.; Liu, Y.; Li, T.; Liu, J. Simulating and mapping the spatial and seasonal effects of future climate and land-use changes on ecosystem services in the Yanhe watershed, China. *Environ. Sci. Pollut. Res.* **2017**, *25*, 1115–1131. [CrossRef] [PubMed]
13. Zhao, W.; Du, S.; Emery, W.J. Object-based convolutional neural network for high-resolution imagery classification. *IEEE J. Sel. Top. Appl. Earth Obs. Remote Sens.* **2017**, *10*, 3386–3396. [CrossRef]
14. Wu, J.; He, Y.; Xu, C.; Jia, X.; Huang, Y.; Chen, Q.; Huang, C.; Eslamlou, A.D.; Huang, S. Interpretability Analysis of Convolutional Neural Networks for Crack Detection. *Buildings* **2023**, *13*, 3095. [CrossRef]
15. He, Z.; Zhang, C.; Wang, S.; Huang, J.; Zheng, X.; Jiang, W.; Bo, J.; Yang, Y. A Grad-CAM and capsule network hybrid method for remote sensing image scene classification. *Front. Earth Sci.* **2024**, *18*, 538–553. [CrossRef]
16. Saarela, M.; Jauhiainen, S. Comparison of feature importance measures as explanations for classification models. *SN Appl. Sci.* **2021**, *3*, 272. [CrossRef]
17. Yuan, X.; Liu, S.; Feng, W.; Dauphin, G. Feature Importance Ranking of Random Forest-Based End-to-End Learning Algorithm. *Remote Sens.* **2023**, *15*, 5203. [CrossRef]
18. Belgiu, M.; Drăguţ, L. Random forest in remote sensing: A review of applications and future directions. *ISPRS J. Photogramm. Remote Sens.* **2016**, *114*, 24–31. [CrossRef]
19. Nie, Y.; Huang, H.; Long, C.; Zhang, Q.; Maji, P.; Cai, H. Interleaving One-Class and Weakly-Supervised Models with Adaptive Thresholding for Unsupervised Video Anomaly Detection. *arXiv* **2024**, arXiv:2401.13551.
20. Otsu, N. A threshold selection method from gray-level histograms. *IEEE Trans. Syst. Man Cybern.* **1979**, *9*, 62–66. [CrossRef]
21. Xu, X.; Xu, S.; Jin, L.; Song, E. Characteristic analysis of Otsu threshold and its applications. *Pattern Recognit. Lett.* **2011**, *32*, 956–961. [CrossRef]
22. Yu, Y.; Bao, Y.; Wang, J.; Chu, H.; Zhao, N.; He, Y.; Liu, Y. Crop row segmentation and detection in paddy fields based on treble-classification otsu and double-dimensional clustering method. *Remote Sens.* **2021**, *13*, 901. [CrossRef]
23. Tang, W.; Zhao, C.; Lin, J.; Jiao, C.; Zheng, G.; Zhu, J.; Pan, X.; Han, X. Improved spectral water index combined with Otsu algorithm to extract muddy coastline data. *Water* **2022**, *14*, 855. [CrossRef]
24. Ren, J.; Wang, J.; Chen, R.; Li, H.; Xu, D.; Yan, L.; Song, J. Remote sensing identification of shallow landslide based on improved otsu algorithm and multi feature threshold. *Front. Earth Sci.* **2024**, *12*, 1473904. [CrossRef]
25. Huang, Y.; Zhao, C.; Yang, H.; Song, X.; Chen, J.; Li, Z. Feature selection solution with high dimensionality and low-sample size for land cover classification in object-based image analysis. *Remote Sens.* **2017**, *9*, 939. [CrossRef]
26. Xie, Z.; Chen, Y.; Lu, D.; Li, G.; Chen, E. Classification of land cover, forest, and tree species classes with ZiYuan-3 multispectral and stereo data. *Remote Sens.* **2019**, *11*, 164. [CrossRef]
27. Dian, Y.; Li, Z.; Pang, Y. Spectral and texture features combined for forest tree species classification with airborne hyperspectral imagery. *J. Indian Soc. Remote Sens.* **2014**, *43*, 101–107. [CrossRef]
28. Zeng, Y.; Hao, D.; Huete, A.; Dechant, B.; Berry, J.; Chen, J.M.; Joiner, J.; Frankenberg, C.; Bond-Lamberty, B.; Ryu, Y.; et al. Optical vegetation indices for monitoring terrestrial ecosystems globally. *Nat. Rev. Earth Environ.* **2022**, *3*, 477–493. [CrossRef]
29. Phiri, D.; Simwanda, M.; Salekin, S.; Nyirenda, V.R.; Murayama, Y.; Ranagalage, M. Sentinel-2 data for land cover/use mapping: A review. *Remote Sens.* **2020**, *12*, 2291. [CrossRef]

30. Abdi, A.M. Land cover and land use classification performance of machine learning algorithms in a boreal landscape using Sentinel-2 data. *GIScience Remote Sens.* **2019**, *57*, 1650447. [CrossRef]
31. Huang, S.; Tang, L.; Hupy, J.P.; Wang, Y.; Shao, G.F. A commentary review on the use of normalized difference vegetation index (NDVI) in the era of popular remote sensing. *J. For. Res.* **2021**, *32*, 1–6. [CrossRef]
32. Panaskar, S.; Nagarajan, K.; Narwade, R. Analysis of Changes in LULC of Western Ghat by Comparing NDVI and NDWI. *GIScience Remote Sens.* **2019**, *8*, 2.
33. Mishra, V.N.; Prasad, R.; Rai, P.K.; Vishwakarma, A.K.; Arora, A. Performance evaluation of textural features in improving land use/land cover classification accuracy of heterogeneous landscape using multi-sensor remote sensing data. *Earth Sci. Inform.* **2018**, *12*, 71–86. [CrossRef]
34. Tassi, A.; Vizzari, M. Object-oriented lulc classification in google earth engine combining snic, glcm, and machine learning algorithms. *Remote Sens.* **2020**, *12*, 3776. [CrossRef]
35. Hale, S.R.; Rock, B.N. Impact of topographic normalization on land-cover classification accuracy. *Photogramm. Eng. Remote Sens.* **2003**, *69*, 785–791. [CrossRef]
36. Friedl, M.A.; Sulla-Menashe, D.; Tan, B.; Schneider, A.; Ramankutty, N.; Sibley, A.; Huang, X. MODIS Collection 5 global LULC: Algorithm refinements and characterization of new datasets. *Remote Sens. Environ.* **2010**, *114*, 168–182. [CrossRef]
37. Homer, C.; Dewitz, J.; Yang, L.; Jin, S.; Danielson, P.; Xian, G.; Megown, K. Completion of the 2011 National LULC Database for the conterminous United States–representing a decade of LULC change information. *Photogramm. Eng. Remote Sens.* **2015**, *81*, 345–354.
38. Verbesselt, J.; Hyndman, R.; Zeileis, A.; Culvenor, D. Phenological change detection while accounting for abrupt and gradual trends in satellite image time series. *Remote Sens. Environ.* **2010**, *114*, 2970–2980. [CrossRef]
39. Hansen, M.C.; Potapov, P.V.; Moore, R.; Hancher, M.; Turubanova, S.A.; Tyukavina, A.; Thau, D.; Stehman, S.V.; Goetz, S.J.; Loveland, T.R.; et al. High-resolution global maps of 21st-century forest cover change. *Science* **2013**, *342*, 850–853. [CrossRef] [PubMed]
40. Drusch, M.; Del Bello, U.; Carlier, S.; Colin, O.; Fernandez, V.; Gascon, F.; Bargellini, P. Sentinel-2: ESA's optical high-resolution mission for GMES operational services. *Remote Sens. Environ.* **2012**, *120*, 25–36. [CrossRef]
41. Immitzer, M.; Vuolo, F.; Atzberger, C. First experience with Sentinel-2 data for crop and tree species classifications in central Europe. *Remote Sens.* **2016**, *8*, 166. [CrossRef]
42. Zhang, X.; Liu, L.; Chen, X.; Gao, Y.; Xie, S.; Mi, J. GLC_FCS30: Global land-cover product with fine classification system at 30 m using time-series Landsat imagery. *Earth Syst. Sci. Data* **2021**, *13*, 2753–2776. [CrossRef]
43. Zanaga, D.; Van De Kerchove, R.; Daems, D.; De Keersmaecker, W.; Brockmann, C.; Kirches, G.; Arino, O. ESA WorldCover 10 m 2021 v200. 2022. Available online: https://pure.iiasa.ac.at/id/eprint/18478/ (accessed on 12 August 2023).
44. Karra, K.; Kontgis, C.; Statman-Weil, Z.; Mazzariello, J.C.; Mathis, M.; Brumby, S.P. Global land use/LULC with Sentinel 2 and deep learning. In Proceedings of the 2021 IEEE International Geoscience and Remote Sensing Symposium (IGARSS), Brussels, Belgium, 11–16 July 2021; p. 4704.
45. Peng, S.; Ding, Y.; Wen, Z.; Chen, Y.; Cao, Y.; Ren, J. Spatiotemporal change and trend analysis of potential evapotranspiration over the Loess Plateau of China during 2011–2100. *Agric. For. Meteorol.* **2017**, *233*, 183–194. [CrossRef]
46. Ma, G.; Yue, X. An improved whale optimization algorithm based on multilevel threshold image segmentation using the Otsu method. *Eng. Appl. Artif. Intell.* **2022**, *113*, 104960. [CrossRef]
47. Loh, W.Y. Classification and regression trees. *Wiley Interdiscip. Rev. Data Min. Knowl. Discov.* **2011**, *1*, 14–23. [CrossRef]
48. Liaw, A.; Wiener, M. Classification and regression by random Forest. *R News* **2002**, *2*, 18–22.
49. Chen, Q.; Meng, Z.; Liu, X.; Jin, Q.; Su, R. Decision variants for the automatic determination of optimal feature subset in RF-RFE. *Genes* **2018**, *9*, 301. [CrossRef] [PubMed]
50. Wong, T.-T.; Yeh, P.-Y. Reliable accuracy estimates from *k*-fold cross validation. *IEEE Trans. Knowl. Data Eng.* **2019**, *32*, 1586–1594. [CrossRef]
51. Pohjankukka, J.; Pahikkala, T.; Nevalainen, P.; Heikkonen, J. Estimating the prediction performance of spatial models via spatial *k*-fold cross validation. *Int. J. Geogr. Inf. Sci.* **2017**, *31*, 2001–2019. [CrossRef]

remote sensing

MDPI

Article

A High-Performance Automated Large-Area Land Cover Mapping Framework

Jiarui Zhang [1,2], Zhiyi Fu [3], Yilin Zhu [1,2], Bin Wang [1,2], Keran Sun [1,2] and Feng Zhang [1,2,*]

1 School of Earth Sciences, Zhejiang University, 866 Yuhangtang Rd., Hangzhou 310058, China; jiaruihz@zju.edu.cn (J.Z.); yilinzhu@zju.edu.cn (Y.Z.); wbzju@zju.edu.cn (B.W.); krsun@zju.edu.cn (K.S.)
2 Zhejiang Provincial Key Laboratory of Geographic Information Science, 866 Yuhangtang Rd., Hangzhou 310058, China
3 Institute of Remote Sensing and Geographical Information Systems, School of Earth and Space Sciences, Peking University, Beijing 100871, China; zhiyif@pku.edu.cn
* Correspondence: zfcarnation@zju.edu.cn; Tel.: +86-571-8827-3287

Abstract: Land cover mapping plays a pivotal role in global resource monitoring, sustainable development research, and effective management. However, the complexity of the mapping process, coupled with significant computational and data storage requirements, often leads to delays between data processing and product publication, thereby bringing challenges to creating multi-timesteps large-area products for monitoring dynamic land cover. Therefore, improving the efficiency of each stage in land cover mapping and automating the mapping process is currently an urgent issue to be addressed. This study proposes a high-performance automated large-area land cover mapping framework (HALF). By leveraging Docker and workflow technologies, the HALF effectively tackles model heterogeneity in complex land cover mapping processes, thereby simplifying model deployment and achieving a high degree of decoupling between production models. It optimizes key processes by incorporating high-performance computing techniques. To validate these methods, this study utilized Landsat imagery data and extracted samples using GLC_FCS and FROM_GLC, all of which were acquired at a spatial resolution of 30 m. Several 10° × 10° regions were chosen globally to illustrate the viability of generating large-area land cover using the HALF. In the sample collection phase, the HALF introduced an automated method for generating samples, which overlayed multiple prior products to generate a substantial number of samples, thus saving valuable manpower resources. Additionally, the HALF utilized high-performance computing technology to enhance the efficiency of the sample–image matching phase, thereby achieving a speed that was ten times faster than traditional matching methods. In the mapping stage, the HALF employed adaptive classification models to train the data in each region separately. Moreover, to address the challenge of handling a large number of classification results in a large area, the HALF utilized a parallel mosaicking method for classification results based on the concept of grid division, and the average processing time for a single image was approximately 6.5 s.

Keywords: land cover; high-performance computing; remote sensing; workflow; automation

1. Introduction

Global land cover (GLC) plays a critical role in global resource monitoring and sustainable development research, as it provides vital insights into ecological diversity, carbon cycling, and human–environment relationships [1–3]. In recent years, with the rapid development of remote sensing technology, land cover classification based on remote sensing images has become the mainstream method for modern land cover mapping.

Unlike satellite images, which are usually obtained in near-real-time, GLC products are typically associated with substantial lag times between the processing of images and the release of data [4]. The shortest production cycle of a GLC product is usually one year, and seasonal or monthly datasets are relatively rare [5]. In addition, the higher the product resolution is, the longer the production cycle. Currently, there are many GLC products that are updated annually, such as the Moderate Resolution Imaging Spectroradiometer (MODIS) Land Cover Type product (MCD12Q1), each with a resolution of 500 m provided by National Aeronautics and Space Administration (NASA) from 2001 to present [6]; the Climate Change Initiative (CCI) program of the European Space Agency (ESA) provides a 300 m dataset (1992–2018), and the Copernicus Global Land Service (CGLS) [7] provides a 100 m dataset for land cover. Chen et al. [8] produced the GlobeLand30 with a resolution of 30 m from Landsat and China's HJ-1 satellite images for 2000 and 2010. The latest version was launched for updating in 2017 and was officially released in 2020. GLC with yearly or even higher intervals limits the monitoring capability of land cover dynamic changes. Until 2021, the ESA provided the WorldCover [9], a GLC product based on Sentinel data, with a resolution of 10m, with the potential to be produced within approximately 3 months of the acquisition of the last image. In 2022, Google released the Dynamic World [4], which is a near real-time LC product. It utilizes cloud computing platforms to acquire and predict the classification probabilities of Sentinel images in real-time. This marks a new era for the fast mapping and updating of GLC within the realm of big data, and an increasing number of researchers have started conducting related studies. Hence, improving the computational efficiency of various stages in land cover mapping using big data and automating the mapping process currently have become the key issues that need to be addressed. Larger area, higher resolution, and faster update frequency have exponentially increased the amount of data, thereby posing new challenges for storage and computing power. At the same time, the land cover mapping process is complex, and each stage depends on different models. The process of improving the efficiency of each link in land cover mapping, systematizing and standardizing the mapping process, and achieving the automated mapping of large-area land cover has gradually become one of the key issues.

As supervised classification algorithms are widely applied in remote sensing image classification, obtaining a large number of accurate samples is crucial for training classification models. Sample collection is considered the most time-consuming and labor-intensive process in producing GLC [10]. Existing methods for extracting classification samples still rely primarily on field investigations and visual interpretation of the imagery. This approach not only demands extensive manual labor, but also restricts the study area's scale significantly. To address this problem, some scholars have studied the automatic extraction of classification samples by combining prior products and auxiliary data [11–15]. This method uses classification rules extracted from prior products to quickly obtain a large number of spatially distributed and high-quality samples, thereby improving the efficiency of sample extraction and increasing the update frequency of land cover products. Zhang and Liu [16] used a total of 27,858,258 training samples in the production of a GLC product, which is several hundred times higher than the number of training samples traditionally selected manually, thereby occupying a large amount of storage and computing space. ESRI [17] even used a super-large dataset of Sentinel-2 with a total of more than five billion samples in the production of 2020 LandCover, which has a resolution of 10 m. In the process of matching samples and remote sensing images, the extraction of spectral features and indices of remote sensing images will consume a large amount of resources. In addition, auxiliary data such as DEM [18] and climate data [19] are vital for land cover classification. To incorporate these features, it is necessary to match the samples with these data, which further increases the computational burden on the matching step.

Remote sensing image mosaicking is the process of stitching multiple remote sensing images into a single geometric alignment composite scene for a wide field of view. Large-area image mosaicking is a computationally intensive process [20]. Creating a national or global level product by mosaicking tens of thousands of image scenes that cover tens of terabytes of data could take several days or even weeks to complete [21]. Using high-performance computing methods to process each task in parallel can improve the efficiency of image mosaicking. Researchers have studied parallel mosaicking methods.

Zhang et al. [22] combined aerial digital photogrammetry principles with parallel computing techniques to propose a parallel mosaicking method for massive aerial digital images. Chen et al. [23] improved an algorithm used for video image mosaicking to make it suitable for remote sensing images, thereby overcoming the problem of multi-images having to be divided into one-to-one mosaics. Ma et al. [24] improved parallel mosaicking efficiency by creating a dynamic task tree to divide the processing order of remote sensing images. Remote sensing data, intermediate data, and result data undergo frequent I/O operations during the computation process, which can lead to I/O bottlenecks. Jing et al. [25] customized Resilient Distributed Dataset (RDD) operators in Spark to accelerate the parallel mosaicking process through in-memory computation. Ma et al. [21] introduced a memory-based file management system called Alluxio, which was combined with computing-intensive remote sensing image mosaicking to effectively alleviate the I/O bottleneck in the mosaicking process. The process of mosaicking classification results is similar to that of remote sensing image mosaicking, but mapping after all data within the study area are collected may decrease the efficiency of dataset updating. Therefore, when mosaicking the classification results, it is necessary to consider that the acquisition time may vary between different regions.

The earliest remote sensing models were limited by the performance of desktop softwares running on a single machine. With the development of high-performance computing (HPC), some HPC-based methods have been considered as effective solutions for solving computational challenges, such as MPI/OpenMP [26], Hadoop [27], Spark [28], and CUDA programming supported by GPUs [29]. Many HPC frameworks targeting remote sensing and geographic information computing scenarios have also been proposed, such as SpatialHadoop [30], Hadoop GIS [31], GeoFlink [32], etc. Although using these frameworks and migrating computing to the cloud have alleviated the pressure of local computing, the sharing of data and models still presents difficulties due to different users' data storage and organizational formats [33]. The models for various stages of GLC mapping use different programming languages and run in different environments, which leads to researchers spending a lot of time and effort on setting up and maintaining the environment. Nüst et al. [34] encapsulated runtime environment dependencies and analysis components in containers, therby improving the reusability, accessibility, and transparency of programs. Wang et al. [35] implemented the automatic encapsulation of spatial information processing operators in different environments based on containers. Atomic-level geographic processing services have limited capacity and may not suffice to support large-area and complex processing tasks [36,37]. Thus, it is necessary to construct complex workflows for efficient resource allocation to facilitate the data and model orchestration of complex GLC products mapping [38].

To address the aforementioned challenges, this study designed a high-performance automated large-area land cover mapping framework(HALF) using distributed architecture. The HALF incorporates high-performance computing technology into different stages of the mapping process and presents three novel approaches: an automated method for generating samples, a parallel matching method for samples and images, and a parallel mosaicking method for classification results. These advancements significantly enhance the computational efficiency of the framework. Furthermore, the models in each stage of the mapping process have been encapsulated based on container technology and organized by Airflow, thereby resolving the heterogeneity of the operating environment and facilitating data reuse. The HALF was empirically tested in several $10^\circ \times 10^\circ$ regions worldwide, thus providing theoretical and technical support for automated GLC production.

2. Materials and Methods

2.1. Experimental Data

2.1.1. Landsat8

Landsat 8 is the eighth satellite in the United States Landsat program, and it carries the Operational Land Imager (OLI) and Thermal Infrared Sensor (TIRS). The remote sensing images obtained by OLI include nine bands, with spatial resolutions of 30 m for all bands except the panchromatic band, which has a spatial resolution of 15 m. The remote sensing images obtained by TIRS include only two separate thermal infrared bands, with a spatial resolution of 100 m. The data used in this article is the Level-2 Science Products of Landsat Collection 2, which are products obtained after radiometric correction and system-level geometric correction. The data for year 2015 was selected, and, for each scene, the image with the smallest cloud cover within a year was chosen. Figure 1 shows the number of images in each 10° × 10° grid, with darker colors indicating a greater abundance of remote sensing images in the corresponding area. The required remote sensing images were selected through conditional queries and downloaded in batches using a machine-to-machine (M2M) API. Landsat data were used in the study to match with the initial spatial points and generate samples with spectral features for training the classifier model. The final image classification results were then used for mosaic to create the land cover products.

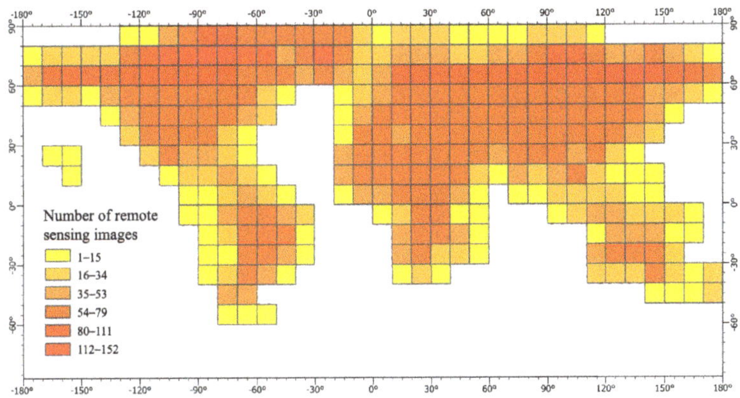

Figure 1. The quantity and distribution of of Landsat8 images of 2015 used in the study.

2.1.2. GDEM

The DEM data are the third version of The Advanced Spaceborne Thermal Emission and Reflection Radiometer (ASTER) Global Digital Elevation Model (GDEM) program [39], with a resolution of 30 m. It was jointly developed by the Ministry of Economy, Trade, and Industry (METI) of Japan and NASA of the United States, and it is available for free download and use. Three versions of data were released in 2009, 2011, and 2019, and the third version was chosen for the study. A DEM was used to match the classification samples and provide elevation information as a feature for the samples.

2.1.3. GLC Products

FROM_GLC [40] is a global 30 m land cover product produced by Gong et al. using Landsat Thematic Mapper (TM) and Enhanced Thematic Mapper Plus (ETM+) data. It classified over 6600 Landsat TM data scenes after 2006 and over 2300 Landsat TM and ETM+ data scenes before 2006, which were all selected from the green season. GLC_FCS [16] was produced by Zhang and Liu et al. It provides a refined classification system, including 16 global land cover classification system types and 14 regional land cover types, with high classification accuracy at 30 m resolution. The accuracy of FROM_GLC in 2015 ranges from 57.71% to 80.36%, while the accuracy of GLC_FCS in 2015 ranges from 65.59% to 84.33%. GLC products were used to obtain prior classification information, which was then used to automatically generate classification samples.

2.2. Framework Design

The HALF provides solutions for each stage of land cover mapping, as shown in Figure 2. It takes remote sensing image data, auxiliary data, and prior GLC products as input. Initially, a large number of initial samples are automatically generated. Then, training features are obtained by overlaying the samples with remote sensing and auxiliary data. Then, the generated samples are used to adaptively train the classifier. Finally, the final land cover product is generated by mosaicking the classification results of remote sensing images. The HALF accelerates the execution speed of the model in the mapping process through Spark, which encapsulates the models for each stage using Docker, which organizes the process using the Airflow workflow engine to read and process remote sensing image data using GDAL, which stores metadata information and performs spatial operator operations using Postgres SQL. The HALF's mapping process mainly includes the following key stages shown in the figure below and described in the following subsections.

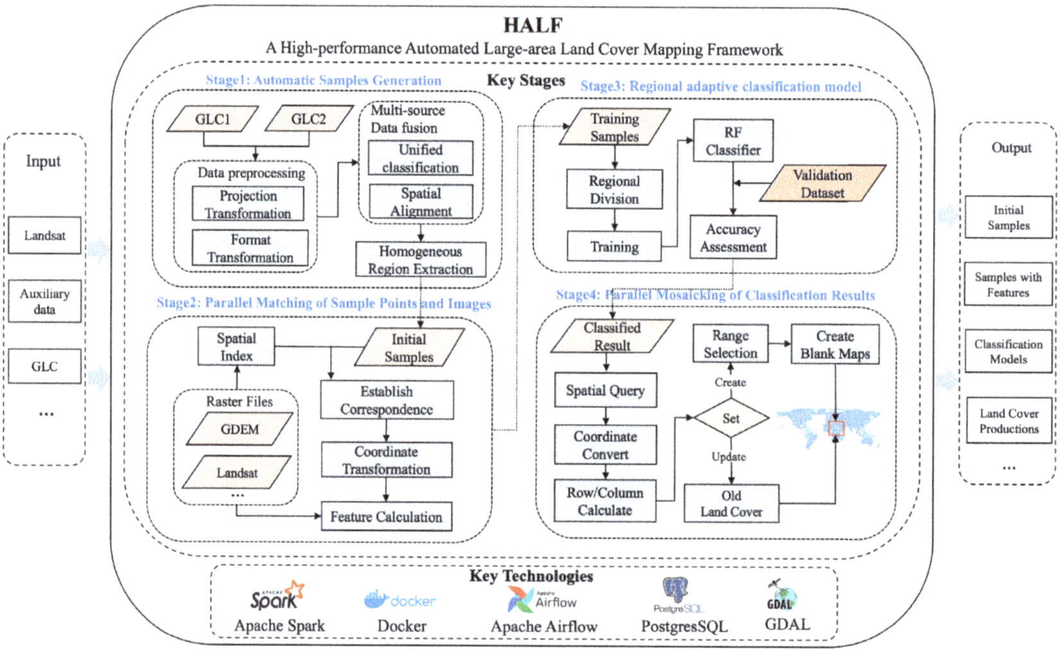

Figure 2. Framework of HALF.

2.2.1. Stage 1: Automatic Sample Generation

Collecting samples is one of the most important factors in land cover classification problems, and it can have a significant impact on classification results, which is sometimes even greater than the impact of the classifier [41]. The collection of sample data can be categorized into manual interpretation and derivation of training samples from existing land cover products. Many researchers have proven the benefits of exploring existing classification knowledge from available products [42–45]. The method is capable of producing a substantial amount of spatial samples that are evenly distributed without any manual intervention.

Several land cover products have been created based on remote sensing imagery using different classification systems. However, there is no universally accepted system among researchers and the international community. Table 1 lists the existing GLC classification systems. Although the classification systems constructed by different products and organizations are not exactly the same, there are certain similarities in category delineation.

Therefore, it is feasible to synthesize and generate new samples from existing products. The GLC_FCS classifies all land cover data into nine primary classes, including cropland, forest, grassland, shrubland, wetlands, water, impervious surfaces, bare areas, and ice and snow. FROM_GLC classifies all land cover into 11 primary classes, which includes the nine primary classes mentioned before (cropland, forest, grassland, shrubland, wetlands, water, impervious surfaces, bare areas, and ice and snow) and the additional classes of cloud and tundra.

Table 1. Comparison of existing GLC classification systems.

USGS [46] 1972	FAO [47] 1996	FROM_GLC 2013	GlobalLand30 2014	GLC_FCS 2021
Forest	Cultivated and Managed Terrestrial Areas	Forest	Forest	Forest
Agricultural	Natural and Semi-Natural Terrestrial Vegetation	Crop	Cultivated land	Cropland
	Cultivated Aquatic or Regularly Flooded Areas	Shrub	Shrubland	Shrubland
Range	Natural and Semi-Natural Aquatic or Regularly Flooded Vegetation	Grass	Grassland	Grassland
Wetlands		Wetland	Wetland	Wetlands
Urban or Built-Up	Artificial Surfaces and Associated Areas	Impervious	Artificial Surfaces	Impervious Surfaces
Barren	Bare Areas	Bareland	Bareland and Tundra	Bare Areas
Water	Artificial Waterbodies, Snow and Ice	Water	Water Bodies	Water Body
Perennial Snow and Ice	Natural Waterbodies, Snow and Ice	Snow/Ice	Permanent Snow/Ice	Permanent Ice and Snow
Tundra		Tundra Cloud		

The essence of sample selection is to obtain the spatial position of a certain pixel and its corresponding land cover type, which can be directly derived from existing products. Selecting high-precision training samples from land cover products is the key and prerequisite to obtain high-precision classification results. In order to ensure the reliability of samples, this section proposes a method for extracting homogenous areas from multi-source product data. The idea of homogenous area extraction is shown in Figure 3. First, select two land cover products with identical spatial range, resolution, and data acquisition time. Next, search for a (7 × 7) pixel range with an identical classification label in the same spatial position within a single product and between the two products. Finally, consider the middle (3 × 3) area as a homogenous region and quantify the number and distribution of these areas. The central pixel is selected as a high-confidence sample for extraction, thereby eliminating spatial deviations between the two land cover products and improving the reliability of samples.

2.2.2. Stage 2: Parallel Matching of Samples and Images

Initial samples from existing land cover products only include longitude, latitude, and land cover labels, and they lack the features required for classifier input. The initial samples need to be matched with remote sensing images and auxiliary data in order to generate the final samples. There is a significant computational burden when matching samples and remote sensing images. The purpose of the process is to correspond the spatial positions of samples with the pixel positions of remote sensing images in order to read band information and perform calculations. The process of image matching includes the following: obtaining the image corresponding to the point through a spatial index; converting the longitude and latitude of the sample point to the plane coordinate system of the image; transforming it into the pixel row and column position through affine six-parameter transformation;

and obtaining the average value in the window as the desired value. In the production process of GLC products, the number of remote sensing images reaches more than tens of thousands, and the number of spatial points in the training set can reach billions. The data volume of remote sensing images and auxiliary data can reach tens of terabytes. Traditional desktop software and unoptimized matching methods are unable to complete this process, so a more efficient matching method is needed to handle it. The HALF utilizes a parallel matching method for samples and images, as illustrated in Figure 4. The key to improving efficiency is to first establish an efficient correspondence relationship between samples and images through a spatial index, which is followed by distributing the reading tasks to each node. For each image, all matching points are read at once, thus avoiding the significant disk I/O pressure caused by frequent image reading. The method is illustrated as follows:

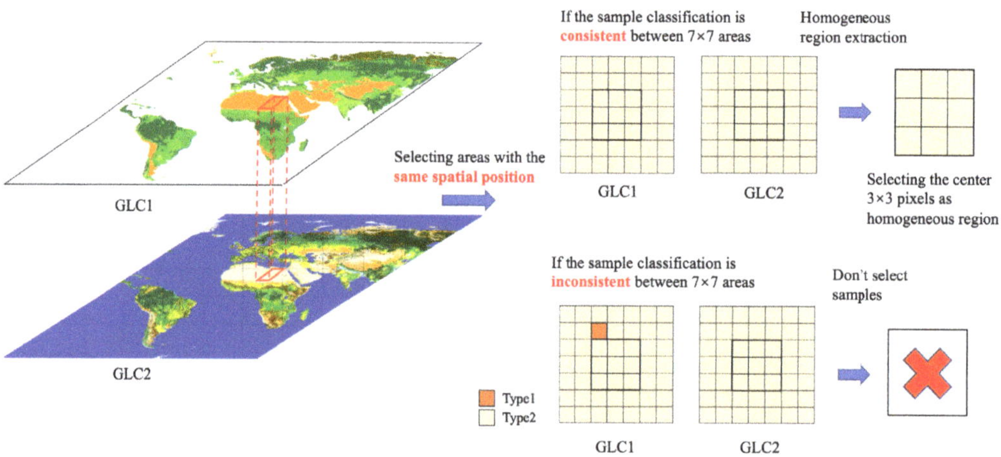

Figure 3. Schematic diagram of homogeneous region extraction.

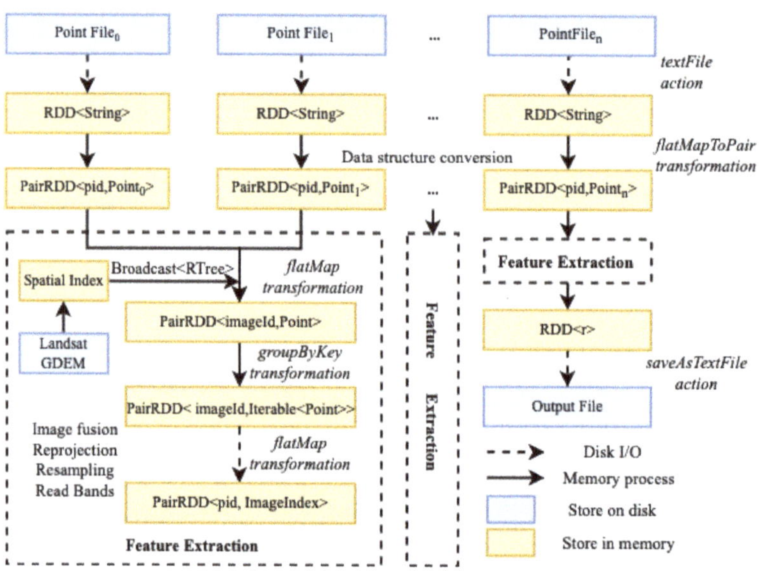

Figure 4. Parallel matching method for samples and remote sensing images.

- Read all the center point data to be matched into memory; use the longitude and latitude of the point to form a unique identifier pid; read its spatial and attribute information to form a spatial point object; and compose a tuple <pid, Point>;
- Read the Landsat image metadata information; use the boundary coordinates to construct a spatial index STRtree; and use the broadcast variable in Spark to transmit it to each node. The broadcast variable will save a copy on each node, thus saving the network transmission cost for multiple calls;
- Distribute the computing tasks. To address the issue of data skew resulting from unequal allocation of images to different computing nodes during task distribution, a custom partitioning strategy was developed, and the groupByKey operator was used to allocate data evenly across all computing nodes as much as possible;
- For each image, parallel operations are performed using GDAL. Then, the image is subjected to image fusion, reprojection, and resampling, and the row and column numbers of the sample point's longitude and latitude on the remote sensing image are calculated. If a sample is covered by clouds for most of the year or its quality is poor, it is discarded. The feature indicators are calculated for samples that meet the quality criteria, and both the data, along with their corresponding match results, are outputted.

2.2.3. Stage 3: Regional Adaptive Classification Model

The regional adaptive classification model was used to train the data in each region separately. Typically, the entire research area is divided into small areas, and the corresponding samples are used to train the model in each area adaptively. The classification results from all areas are merged to generate the final experimental outcomes. Although this approach may not be as straightforward in concept and implementation as global modeling, it overcomes the issue of overfitting due to small areas or underfitting due to large areas by sampling and training based on the unique characteristics of each region. Hankui [11] quantitatively compared the effectiveness of global modeling and adaptive modeling in the application of land cover in the United States, and their findings demonstrated that local modeling achieved greater classification accuracy than global modeling. In this study, the research area was segmented into a $10° \times 10°$ grid.

The classifier provided by the HALF is a random forest model. Gómez et al. [48] compared various classification models, such as artificial neural networks, decision trees, support vector machines, random forests, and clustering in the mapping of land cover, and they found that the random forest algorithm has the advantages of insensitivity to noise, good anti-overfitting performance, and good robustness to data reduction. In addition, it is relatively simple to set the parameters, as only the number of trees and the number of features for decision tree branching need to be set. It is suitable for training data with different distribution features in multiple different partitions. Traditional accuracy metrics are derived from a confusion matrix such as overall accuracy (OA), producer accuracy (PA), user accuracy (UA), and kappa coefficient [49].

2.2.4. Stage 4: Parallel Mosaicking of Large-Area Classification Results

The size of a Landsat satellite image is 185 km $\times$ 185 km. Therefore, when producing large-area land cover products, multiple scenes need to be stitched together. The main purpose of image mosaicking is to convert the trained remote sensing images to the map and to output the final classification results. In this process, there were a huge number of images. If the images are stitched in serial order, the computational pressure will be enormous. Thus, a certain strategy needs to be adopted to solve the problem of low efficiency in the image merging process.

This process involves coordinate system transformation, image registration, and overlap area processing. Traditional image mosaicking methods first specify the coordinate system of one image as the reference and then register other images with it to determine the attribute values of corresponding coordinate points. If images are processed sequentially in this process, the computational efficiency is slow. The HALF utilized an image mosaicking method, and Figure 5 shows its schematic diagram. Traditional image processing operates on a per-image basis, and parallelism is not sufficient. In this study, the study area was divided into a grid of 10° × 10° cells. The spatial index was used to query the remote sensing images corresponding to each cell. Each cell was then subdivided into several smaller sub-images, with a specified resolution and edge length. After obtaining the spatial information of the sub-images, a blank image slice file corresponding to each sub-image was created through GDAL. The spatial relationship between each sub-image and the remote sensing image was established, and the row and column numbers on the remote sensing image were transformed to the corresponding row and column numbers on the sub-image through affine coordinate transformation. The specific values were written to the image slice file, and finally, all sub-images were merged into a complete classification result through the gdalbuildvrt and gdal_translate tools provided by GDAL. The spatial resolution of the customized product can also be used to complete the resampling work during the mosaicking process.

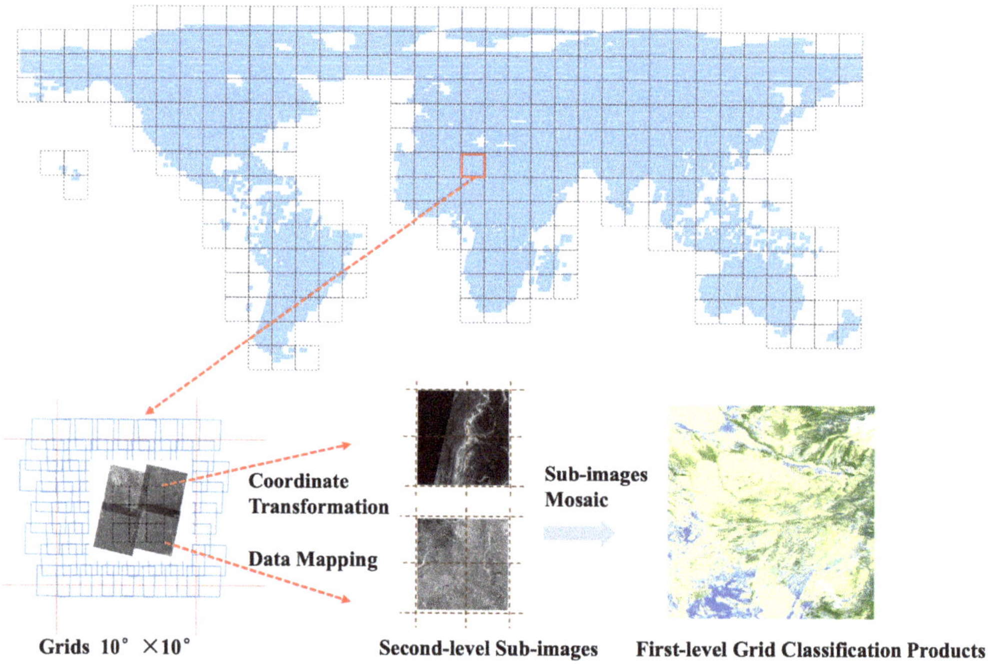

Figure 5. Schematic diagram of the mosaicking method for classification results. The world is partitioned into a grid system with each grid spanning 10° × 10°, and each grid is further divided into second-level sub-images. The blue border indicates the extent of the Landsat image, and when there are classification results within the grid, the pixels of the second-level sub-images within the spatial extent of the classification result are updated, and the final mosaic product is generated.

The parallel mosaicking method is demonstrated in Figure 6. The method first establishes the necessary parameters and determines the relationship between the sub-images and the remote sensing images. It then distributes tasks by using the sub-images as the parallel unit. When a sub-image corresponds to multiple remote sensing images, it first determines whether it is a no-data value. Otherwise, there may be a scenario where the invalid value will overwrite the valid value. Finally, it determines whether to output the latest results or all results based on the established rules. If the program is set to output multiple results, a blank view frame for multiple times is created to fill the corresponding sub-image with the corresponding results.

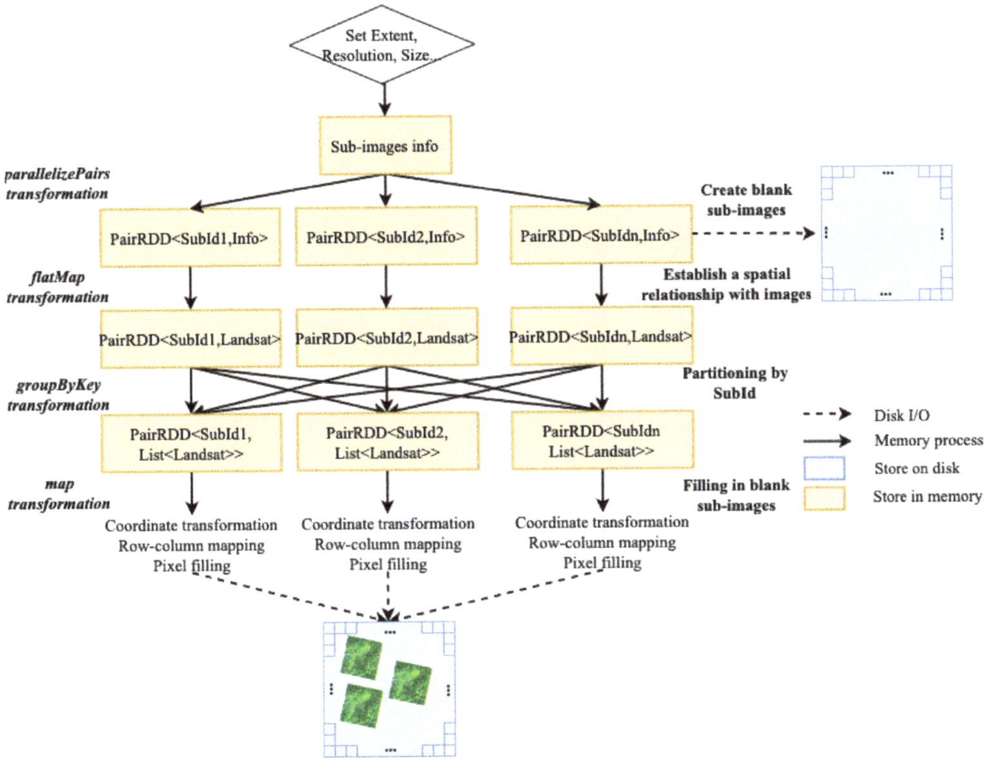

Figure 6. The parallel mosaicking method for classification results.

- Set the spatial extent, resolution, number of bands, and other parameters for different application scenarios. Then, create blank map sheets. For example, a latitude and longitude range of 10° × 10° corresponds to a grid of geographic extent of approximately 1100 km in length and width. If the sub-image resolution is set to 100 m, and each sub-image size is 256 × 256 pixels, then there are approximately 44 × 44 sub-images in this spatial range. The metadata information of the sub-images is stored in binary format, including spatial location and description, and files are created based on whether they already exist;
- Read the image information, and use the flatMap operator to match the spatial position of the sub-image with the metadata of the remote sensing image to establish a spatial correspondence. In this step, the key-value pairs of SubId and remote sensing images are obtained;
- Use the unique identification code pertaining to the SubId of the sub-image as the key to call the groupByKey operator to obtain the complete set of remote sensing images corresponding to each sub-image; the task is then distributed. Each task unit performs the sub-image filling task;
- In each task computing unit, perform coordinate transformation and row/column calculation for the image classification result, and fill in the corresponding pixel points for the sub-image. If no additional settings are made, the result from the latest remote sensing image corresponding to the sub-image is selected for writing.

2.3. HALF Workflow

The HALF provides workflow modules that encapsulate the key stages above through containers and organize each link using the workflow engine. The workflow model involves organizing remote sensing data, sample data, and processing functions into a structured and configurable set of steps. This approach enables practical problems to be resolved through automation or semi-automation. To encapsulate the model, the required basic environment image is searched and downloaded from the Docker Hub repository. Then, the necessary runtime environment is added, and the Docker commit command is used to create a new container image for storage. Relevant parameters of each model are recorded in XML description documents, while the basic model information is described in metadata files during registration. When adding a model, key details such as the model name, description, type, input parameters, output parameters, executable file path, dependent image, add time, user name, and test case are set. Metadata information of the operator is added to the database to complete the final model information storage and registration.

When organizing the workflow, use the Docker execution command as the Airflow workflow execution statement; access the database configured by Airflow; and monitor the running status of workflow nodes and instances. Airflow [50] is a lightweight workflow manager developed by AirBnB, Inc. Airflow provides a broad range of tools for managing the execution of workflows, such as pausing and resuming workflow execution, stopping and restarting individual workflow steps, and restarting workflows from a certain step. Airflow treats each workflow as a directed acyclic graph (DAG) of tasks with directional, non-cyclic dependencies. The DAG describes the relationships between tasks and defines their execution order. The Apache Foundation lists over 300 existing workflow languages and systems [51], and commonly used workflow systems in different fields have defined various organizational standards [52]. The international academic community has proposed the FAIR principles (Findable, Accessible, Interoperable, Reusable) for open and shared scientific data [53]. To this end, researchers have proposed the Common Workflow Language (CWL) [54], which entails a specification and standard for describing the creation of workflows that provides a universal standard for addressing the main issues in sharing workflows. The CWL describes the input and output of workflows in JSON or YAML format and has been applied in fields such as bioinformatics, hydrology, geospatial analysis, and physics. CWL-Airflow is one of the first workflow managers to support the CWL standard. It creates CWL-DAG instances based on workflow structure and executes them, followed by saving the results to an output folder.

The complete mapping process is presented in Figure 7. The research process consists of several stages encompassing sample generation, data matching, model training, image mosaicking, and data updating. The model description file specifies the parameters for each workflow node, and the input and output nodes for each workflow are designed accordingly.

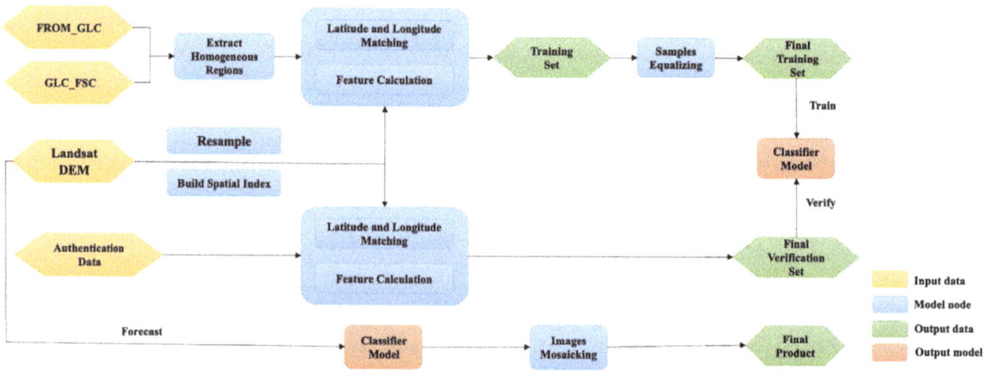

Figure 7. Conceptual design of the workflow. Describes the cartographic phase of HALF, with yellow nodes representing external data, green nodes representing computation results, blue nodes representing models, and red nodes representing classification models.

In the HALF, users can establish models by describing the metadata information of workflow nodes in the description file and setting the input and output parameters of the nodes according to different application requirements. The HALF quickly builds workflows according to different application requirements to achieve customized management and monitoring. For the combined workflow model, the system provides a save function that records node parameter information and spatial location on the interface. After entering the workflow name and comments, the workflow model can be saved to the server for future use. It also allows users to monitor the running status of each step with multiple states such as success, running, waiting, and failure. Set the input and output of each node; the input can select the output of the previous node, and the input correction is carried out in this link to ensure that the output format of the previous node is filled in correctly.

2.4. Experimental Environment

The experimental environment consisted of a distributed cluster composed of three Dell Power Edge R720 servers, each with an Intel Xeon E5-2630 v2 2.60 GHz CPU and 30GB of RAM. The three machines shared a NAS storage of 20T. The experimental software and operating environment were Hadoop 2.7.6, Spark 2.3.4, Scala 2.11.8, JDK 1.8, Python 3.8.3 and GDAL 3.0.3(released by the Open Source Geospatial Foundation, Chicago, US).

The research creates container images for the models. Table 2 shows some of the images used for the runtime environment. The initial images were obtained from Docker Hub and further encapsulated to ensure that each model was portable and reusable.

Table 2. Images environment configuration.

Images	Description	Base Images	Models
landuse-py	Used to run models written in Python	python:3.8.3	Sample Generation Classifier Training Image Classification Prediction
gdal-spark	Used to run models written in Java Spark	osgeo/gdal:ubuntu-full-3.6.3	Sample–Image Matching Image Mosaicking
pg12-citus-postgis	Used to storage metadata and spatial queries	postgres:12.14-alpine3.17	Building Spatial Index Spatial Querying
end-point	Used to launch the back-end project and provide network services	java:openjdk-8	Back-end Project

3. Results

3.1. Model Performance Testing

We compared the deployment workload and execution time of the operators' running tasks in containers and hosts. Deploying conventional applications on a physical machine may depend on multiple development and runtime environments, and the complexity of configuration depends on environment dependencies. For example, sample generation only requires a Python basic environment and GDAL to read images; the sample–image matching model is written in Java and also depends on Spark and GDAL; image training requires GDAL in the Python environment to fuse and resample images. The deployment of GDAL in the Python environment operating on the physical machine can be installed through conda, but the GDAL tool in the Java environment involves gcc compilation, which is extremely cumbersome. By pulling images provided on Docker Hub, the existing environment can be directly pulled and run, and the deployment workload is only one. Therefore, compared with the physical machine environment, the deployment efficiency can be effectively improved in the Docker container environment. Figure 8 shows the total time spent in production. The reason why the running time was longer in the Docker environment than in the physical machine environment is that the startup and destruction of Docker containers took some time, but the impact was very small in seconds. In addition, Docker containers themselves have the characteristic of light weight and will not occupy additional memory resources. In summary, deploying the model through Docker did not bring any losses in execution performance, but, compared to deploying the environment on a physical machine, it could reduce the configuration workload, thus making it convenient to migrate and reuse programs when deploying the production environment on new machines.

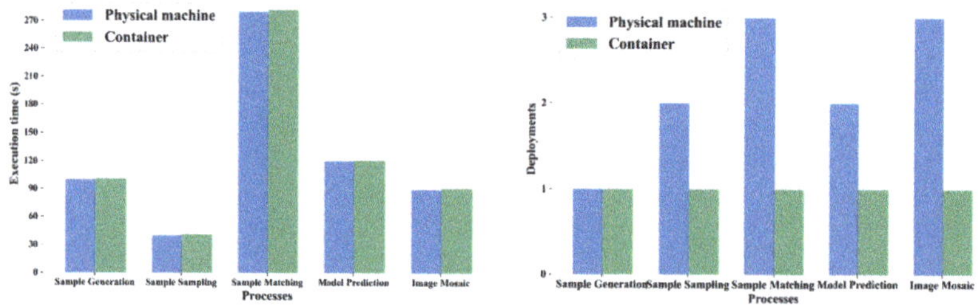

(**a**) Comparison of time consumption (**b**) Comparison of deployment

Figure 8. Performance comparison between physical machine and container. (**a**) shows the comparison of time consumption, with various models on the horizontal axis and their corresponding running time on the vertical axis. (**b**) shows the comparison of deployment amount, with various models on the horizontal axis and an abstract estimate of deployment workload on the vertical axis.

3.2. Automatically Generated Samples

Generating training samples based on existing land cover products makes it possible to obtain uniformly distributed samples at a large scale. The generated samples are visualized for several regions globally in Figure 9. The method guarantees complete automation of the entire classification process without any human intervention and ensures a large sample size and relatively uniform distribution in the sample space. Land cover classes such as water, bare area, or snow/ice tend to have higher discriminability during classification, thus resulting in denser sample points. On the other hand, classes such as cropland or grassland, which may have less distinct boundaries in different products, may exhibit more areas with "no samples" during classification. The classification system used for samples was inherited from previous products, thus eliminating the need for expert knowledge.

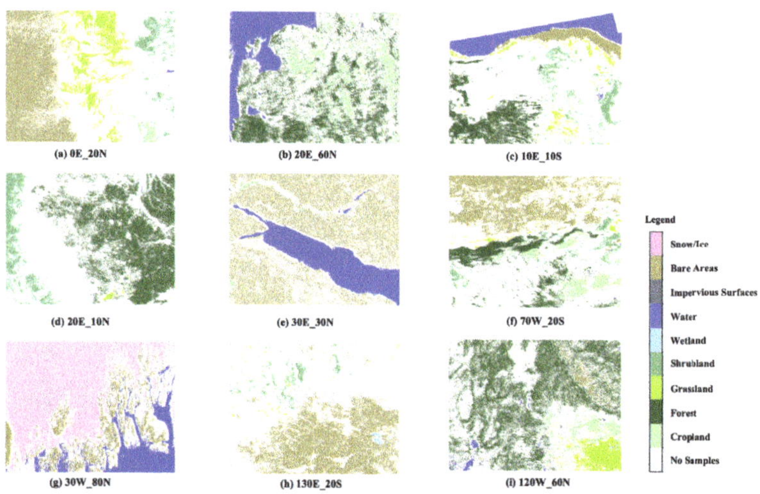

Figure 9. Automatically generated samples. The title of each subfigure indicates the latitude and longitude of the lower-left corner of the region. The legend is shown on the right side of each figure, and the white color indicates that there are no samples in this area.

After matching, the generated sample set exhibited the characteristics illustrated in Table 3, which included seven spectral features (bands), four remote sensing index features—Normalized Difference Vegetation Index (NDVI), Normalized Difference Water Index (NDWI), Enhanced Vegetation Index (EVI), Normalized Burn Ratio (NBR)—the latitude and longitude of the center point, the acquisition time of the image, and auxiliary indicators such as DEM, slope, and aspect.

Table 3. Descriptions of samples.

Feature	Data Source	Characteristic
Spatial Characteristics	Landsat	Longitude and Latitude
Temporal Characteristics		Image Acquisition Time
Spectral Characteristics		Band1, Band2...Band7
RS Index		NDVI, NDWI, EVI, NBR
Topographic Feature	DEM	DEM, Slope, Aspect
Land Cover Type	GLC	First-Level Type

3.3. Performance of Sample and Image Matching

In this section, we validated the method and conducted performance tests on different numbers of samples and remote sensing images.The experimental regions are shown in Figure 10, which include parts of Australia, Africa, and America. The longitude and latitude ranges for the Australian region were (110°E 40°S, 160°E 10°S), for the African region were (20°E 30°S, 50°E 10°N), and for the American region were (110°W 10°N, 50°W 40°N) and (80°W 50°S, 60°W 0°N). The method was able to complete the matching task within a short period of time when dealing with millions of samples and remote sensing images to be matched within a region. The execution time increased with the increase in samples and remote sensing images to be matched within a region.

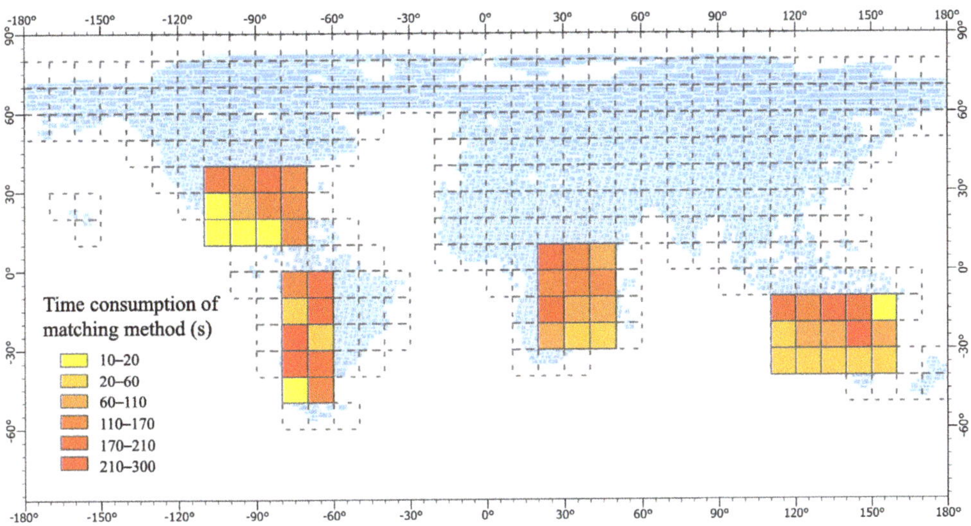

Figure 10. Time consumption statistics of the matching method.

To further validate the performance of the proposed method, the HALF matching method was set to Spark's local computing mode and compared with the method based on GDAL under the same computing resources. The experimental task was to perform sample point–image matching on randomly selected 10,000 to 500,000 points from 131 images and to calculate the corresponding total time. Figure 11 shows that the proposed method not only had a significant advantage in running speed, but also was not affected by the sharp increase in point data volume in terms of running time stability. This is because the spatial relationship between points and images was first established using spatial indexing; as the number of points increased, the I/O overhead of the remote sensing images was relatively fixed.

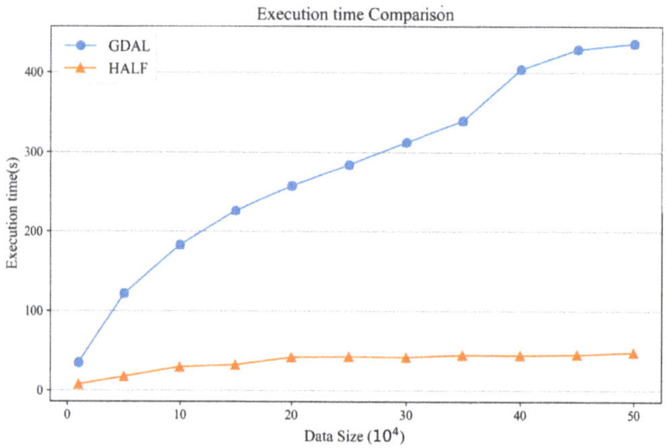

Figure 11. Performance comparison of matching method. The orange line indicates the time consumption of the matching method of HALF, and the blue line indicates the time consumption of the matching method of GDAL.

In order to further investigate the performance of each step of the method, Table 4 provides a detailed breakdown of the time taken for each stage of the two methods at sample sizes of 100,000. The results show that the HALF matching method took significantly less time than the GDAL-based method in the stages of reading points and establishing spatial relationships, as well as reading image values and performing calculations, thereby demonstrating superior computational performance. At the same time, for different sample sizes, the time taken for the Spark-based matching method to distribute tasks to computing nodes increased, while the time taken to read image values and perform calculations remained relatively stable. This indicates that the custom partitioning strategy developed by the method effectively addressed the problem of uneven distribution of node images during task distribution, thus ensuring that the task completion time of each node was as close as possible, which maximized the use of computing resources and reduced computational time.

Table 4. Time consumption analysis for each step of the matching method when data size is 100,000.

Stages	Time Cost(s)	GDAL	HALF
1	Build spatial index	0.7	0.7
2	Read data and create spatial relationships	4.2	2.6
3	Distribute tasks	/	4.9
4	Compute feature values	174.1	24.6
5	Total	179.2	32.8

3.4. *Performance of Large-Area Classification Result Mosaicking*

Exploration of the performance of the proposed large-area image parallel mosaic method was conducted by selecting some 10° × 10° grids globally for experiments. The experimental regions are shown in Figure 12, which are the same as Figure 10. Furthermore, a thematic map was constructed based on the matching time of each grid area, where darker colors indicate longer algorithmic processing time.

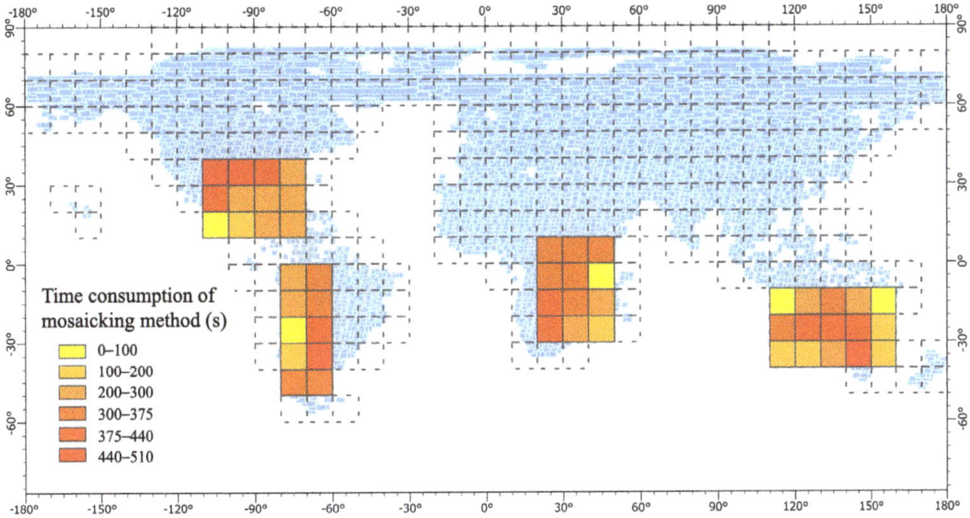

Figure 12. Time consumption of the mosaicking method.

A detailed analysis of the method performance in terms of processing time for the mosaic results of selected regions has been presented in Table 5. It was calculated that the average processing time for one image was around 6.5 s. For regions with around 60 images, such as regions 4 and 5, the calculation could be completed in around 400 s, while, for regions with around 30 images, such as region 11, it took over 200 s. As shown in the experimental results, with the increase in the number of images, the processing time of the mosaic method also gradually increased. Region 7 required only a little over 2 s to process one image on average, because only some parts of the region were land, and the sub-images not covered by any input images were not involved in the calculation. Among the selected regions, region 2 had the longest processing time, which was around 17% longer than that of region 4 and 5 with the same number of images. This is because Landsat images have deformation problems in different latitude regions, with larger image sizes in higher latitude regions, leading to longer processing time in high latitude regions. Further experiments were conducted to investigate the performance of the method under different regions and data volumes.

The study selected five regions with different latitudes that were all located on land. Figure 13 shows the changes in method runtime as the number of remote sensing images increased. To facilitate comparison, all these regions were located entirely on land. It was found that, regardless of the region, the method's processing time showed a linear trend as the number of images increased. This is because, during the calculation, a correspondence was established between the sub-images and the remote sensing images, and the calculation tasks were distributed based on the sub-images. For each calculation task, the remote sensing images were traversed and calculated. Therefore, as the number of images increased, the method's processing time increased almost linearly. In addition, the higher the latitude of the region, the more time it took to process the same number of images, because Landsat images at higher latitudes undergo distortion due to different projections, thereby resulting in larger image sizes and longer read times.

Table 5. Data volume and time consumption for partial regions.

Region	Spatial Extent	Number of Images	Number of Samples	Mosaic Time (s)	Match Time (s)
Region 1	(50°S–40°S, 70°W–60°W)	43	9,276,097	361	166.12
Region 2	(40°S–30°S, 70°W–60°W)	64	7,557,768	481	204.07
Region 3	(40°S–30°S, 140°E–150°E)	59	6,770,182	468	31.20
Region 4	(30°S–20°S, 20°E–30°E)	64	3,663,073	413	99.72
Region 5	(30°S–20°S, 140°E–150°E)	64	5,716,990	408	233.32
Region 6	(30°S–20°S, 130°E–140°E)	63	6,657,017	417	109.19
Region 7	(20°S–10°S, 110°E–120°E)	7	6,698,112	16	175.16
Region 8	(20°S–10°S, 20°E–30°E)	63	1,675,984	401	182.95
Region 9	(20°S–10°S, 30°E–40°E)	55	4,159,308	324	80.81
Region 10	(−10°S–0°N, 40°E–50°E)	13	11,522,471	72	169.84
Region 11	(30°N–40°N, 80°W–70°W)	31	16,008,303	222	143.02
Region 12	(0°N–10°N, 30°E–40°E)	62	1,022,583	359	119.86
Region 13	(10°N–20°N, 80°W–70°W)	35	9,698,301	250	134.51
Region 14	(20°N–30°N, 90°W–80°W)	39	12,549,213	253	146.13
Region 15	(20°N–30°N, 80°W–70°W)	34	11,688,087	224	153.62
Region 16	(30°N–40°N, 90°W–80°W)	63	12,793,479	439	259.34

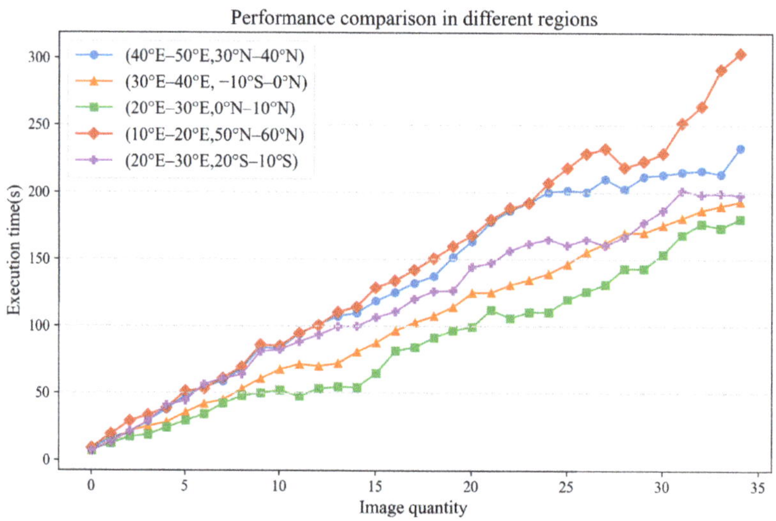

Figure 13. Performance analysis of mosaicking methods.

3.5. Result of Mapping

Figure 14 shows some regions around the world to present the classification results of the images. The results of the high-performance mosaicking method proposed were demonstrated by stitching together the classification results from regions such as Australia, Africa, and the Americas. The approach proposed in this study was effective for land mapping and could provide technical support to automate global land mapping.

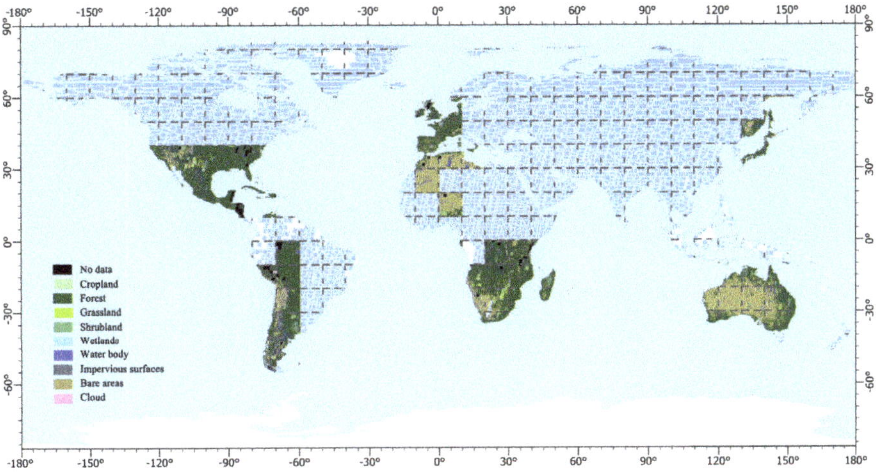

Figure 14. Large-area mosaic classification results.

Updating the classification results at the regional level could be performed on new incoming remote sensing images based on the results of the large-area mosaic method. As shown in Figure 15, there were missing data in some regions in the initial mapping result for the Korean Peninsula region. By querying the corresponding image row and column numbers based on spatial location information, new images could be inputted to fill in the missing areas. Figure 16 shows the process of updating existing products in the Japan region. In some areas, the product was affected by clouds, but higher-quality images could be used as input to update the obscured areas.

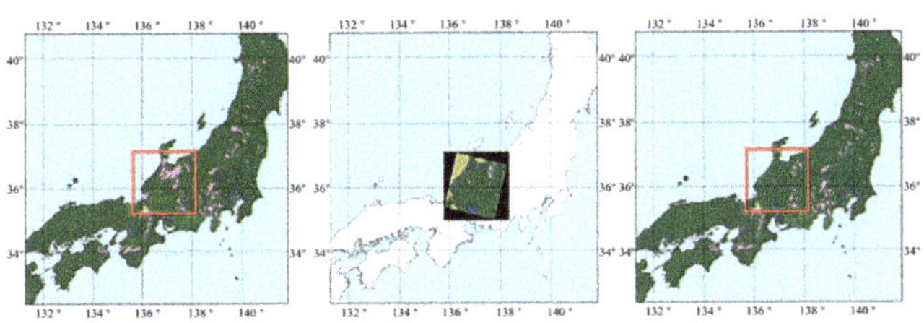

Figure 15. Update of regional classification results of the Korean Peninsula.

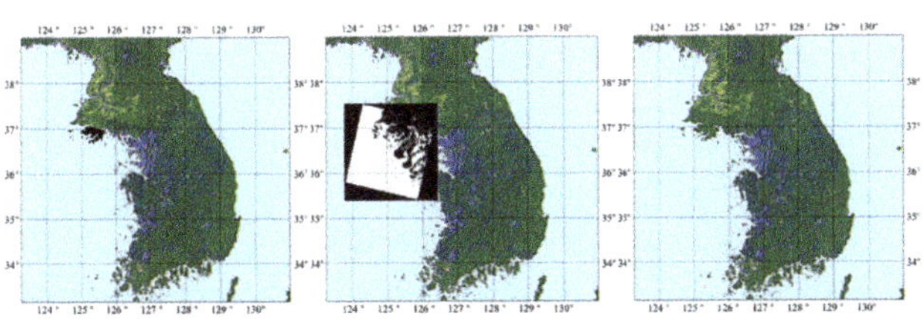

Figure 16. Update of regional classification results of Japan.

4. Discussion

The HALF framework presented in this study offers an automated and high-performance solution for large-area land cover mapping. By encapsulating each process's models using container technology, the HALF addresses the issue of model heterogeneity between different processes, thus resulting in a significant reduction of deployment workload without sacrificing operational performance. The HALF integrated various stages in land cover mapping by using the CWL-Airflow workflow to organize and arrange the models, thus increasing the automation and flexibility. While the proposed method was experimentally verified for large-area and efficient mapping, there was a need for service publishing and sharing functions for workflows. Collaborative efforts between experts from various fields and countries are crucial for effective GLC mapping. The real-time sharing of models, processes, and data can have significant implications for land cover mapping.

The HALF provides a method of extracting a large number of samples from multiple sources of prior products. This method facilitates the generation of a large number of spatially distributed and evenly balanced samples, thereby significantly reducing the human and material resources that are utilized in the sample selection process. However, the accuracy of the samples generated by this method is affected by the prior products. In this study, many samples were generated by overlaying classification products automatically, and the first-level classes in the FROM_GLC and GLC_FCS products were extracted. The classification system chosen was relatively coarse, and the samples produced by this method were constrained by the accuracy of the prior products, thereby affecting the training of the classification model. In future research, the quality of the produced samples can be more rigorously controlled by adopting a more refined classification system. In addition, techniques such as change detection methods and transfer learning can be utilized to create a substantial number of uniformly distributed and high-precision samples.

The HALF optimized and accelerated sample feature matching and classification result mosaicking using high-performance computing technology. The specific effects of the method and its performance under different data volumes were discussed and explained in Sections 3.4 and 3.5. The experimental results show that the matching method proposed in this study could quickly establish spatial relationships between points and images, and it could read image attributed to data in parallel, thereby effectively improving the computational efficiency of sample–image matching and feature extraction. Its efficiency was more than 10 times faster than conventional matching methods, and it performed well and stably under different data volume scenarios, thus making it applicable to other remote sensing applications that require large-area vector and raster data matching. The parallel mosaicking method for classification results used image slices as basic units, thereby achieving real-time mapping of large-area remote sensing image classification results to produce products and improving mapping efficiency. The total time required for resampling and mosaic of a single Landsat image in a 10° × 10° grid was about 6.5 s on average, thereby solving the problem of low computational efficiency when synthesizing multiple images. However, there may be practical difficulties in expanding the study area to the entire world, and the computational burden will increase exponentially with increasing research scope and resolution. Therefore, data fusion with other spatio-temporal resolution remote sensing data can be utilized to generate more real-time products if a higher update frequency is desired.

5. Conclusions

The process for producing large-area land cover products is complex, and as the demand for higher-resolution satellite remote sensing data and more frequent product updates increases, there is a growing need for automated tools and high-performance computing methods. The HALF aims to improve the operational efficiency of each link in the traditional process of land cover mapping in big data by utilizing high-performance computing technology. The HALF presents a method for large-area land cover mapping, which can serve as a valuable inspiration for researchers working on the development of remote sensing data processing or mapping tools. This is particularly beneficial for those who have limited access to paid computing power or require customization beyond existing platforms. To address the heterogeneity of operating systems, running environments, and programming languages between different links, container technology is used to encapsulate models such as sample generation, sample–image matching, model training and prediction, and classification result mosaicking, thereby supporting model reuse and data sharing, which greatly reduces the workload of deployment. Additionally, a general workflow language was introduced to model the land cover mapping process, organize data and models for each link, and decouple the production model from the overall process, thereby enhancing the automation and flexibility of the mapping process. In the future, we will further explore how to obtain higher-quality samples from multiple land cover products, encapsulate them into online data services with network technology, and integrate them into an integrated production portal to better support researchers in the field. By continuously improving the HALF, we can enhance the efficiency of large-area land cover mapping and better support global resource monitoring and sustainable development efforts.

Author Contributions: Conceptualization, J.Z. and Y.Z.; methodology, J.Z., Z.F. and Y.Z.; software, J.Z.; validation, Z.F.; formal analysis, Y.Z.; resources, Y.Z.; data curation, K.S.; writing—original draft preparation, J.Z.; writing—review and editing, Z.F., B.W. and F.Z.; visualization, J.Z. and B.W.; supervision, F.Z.; project administration, J.Z.; funding acquisition, F.Z. All authors have read and agreed to the published version of the manuscript.

Funding: This research was funded by National Natural Science Foundation of China grant number 42050105.

Data Availability Statement: The Level-2 Science Products of Landsat Collection 2 are sourced from the USGS website https://www.usgs.gov/core-science-systems/nli/landsat/landsat-data-access?qt-science_support_page_related_con=0#qt-science_support_page_related_con (accessed on 8 June 2023). The Global Digital Elevation Model (GDEM) is available from the NASA website https://terra.nasa.gov/news/aster-digital-elevation-model-version-3-released (accessed on 8 June 2023). GLC_FCS can be accessed at https://zenodo.org/record/3986872 (accessed on 8 June 2023) and FROM_GLC can be accessed at at http://data.starcloud.pcl.ac.cn/zh (accessed on 8 June 2023).

Conflicts of Interest: The authors declare no conflict of interest.

References

1. Sterling, S.M.; Ducharne, A.; Polcher, J. The impact of global land-cover change on the terrestrial water cycle. *Nat. Clim. Chang.* **2013**, *3*, 385–390. [CrossRef]
2. Feddema, J.J.; Oleson, K.W.; Bonan, G.B.; Mearns, L.O.; Buja, L.E.; Meehl, G.A.; Washington, W.M. The Importance of Land-Cover Change in Simulating Future Climates. *Science* **2005**, *310*, 1674–1678. [CrossRef]
3. Ban, Y.; Gong, P.; Giri, C. Global land cover mapping using Earth observation satellite data: Recent progresses and challenges. *ISPRS J. Photogramm. Remote Sens.* **2015**, *103*, 1–6. [CrossRef]
4. Brown, C.F.; Brumby, S.P.; Guzder-Williams, B.; Birch, T.; Hyde, S.B.; Mazzariello, J.; Czerwinski, W.; Pasquarella, V.J.; Haertel, R.; Ilyushchenko, S.; et al. Dynamic World, Near real-time global 10 m land use land cover mapping. *Sci. Data* **2022**, *9*, 251. [CrossRef]
5. Yu, L.; Du, Z.; Dong, R.; Zheng, J.; Tu, Y.; Chen, X.; Hao, P.; Zhong, B.; Peng, D.; Zhao, J.; et al. FROM-GLC Plus: Toward near real-time and multi-resolution land cover mapping. *Giscience Remote Sens.* **2022**, *59*, 1026–1047. [CrossRef]
6. Sulla-Menashe, D.; Gray, J.M.; Abercrombie, S.P.; Friedl, M.A. Hierarchical mapping of annual global land cover 2001 to present: The MODIS Collection 6 Land Cover product. *Remote Sens. Environ.* **2019**, *222*, 183–194. [CrossRef]
7. Buchhorn, M.; Lesiv, M.; Tsendbazar, N.E.; Herold, M.; Bertels, L.; Smets, B. Copernicus Global Land Cover Layers—Collection 2. *Remote Sens.* **2020**, *12*, 1044. [CrossRef]
8. Chen, J.; Chen, J.; Liao, A.; Cao, X.; Chen, L.; Chen, X.; He, C.; Han, G.; Peng, S.; Lu, M.; et al. Global land cover mapping at 30m resolution: A POK-based operational approach. *ISPRS J. Photogramm. Remote Sens.* **2015**, *103*, 7–27. [CrossRef]
9. ESA WorldCover 10 m 2020 v100. Available online: https://zenodo.org/record/7254221 (accessed on 7 June 2023).
10. Li, X.; Gong, P. An "exclusion-inclusion" framework for extracting human settlements in rapidly developing regions of China from Landsat images. *Remote Sens. Environ.* **2016**, *186*, 286–296. [CrossRef]

11. Zhang, H.K.; Roy, D.P. Using the 500m MODIS land cover product to derive a consistent continental scale 30m Landsat land cover classification. *Remote Sens. Environ.* **2017**, *197*, 15–34. [CrossRef]
12. Radoux, J.; Lamarche, C.; Bogaert, E.V.; Bontemps, S.; Brockmann, C.; Defourny, P. Automated Training Sample Extraction for Global Land Cover Mapping. *Remote Sens.* **2014**, *6*, 3965–3987. [CrossRef]
13. Yu, L.; Wang, J.; Li, X.; Li, C.; Zhao, Y.; Gong, P. A multi-resolution global land cover dataset through multisource data aggregation. *Sci. China Earth Sci.* **2014**, *57*, 2317–2329. [CrossRef]
14. Wessels, K.J.; Van den Bergh, F.; Roy, D.P.; Salmon, B.P.; Steenkamp, K.C.; MacAlister, B.; Swanepoel, D.; Jewitt, D. Rapid Land Cover Map Updates Using Change Detection and Robust Random Forest Classifiers. *Remote Sens.* **2016**, *8*, 888. [CrossRef]
15. Zhang, X.; Liu, L.; Chen, X.; Xie, S.; Gao, Y. Fine Land-Cover Mapping in China Using Landsat Datacube and an Operational SPECLib-Based Approach. *Remote Sens.* **2019**, *11*, 1056. [CrossRef]
16. Zhang, X.; Liu, L.; Chen, X.; Gao, Y.; Xie, S.; Mi, J. GLC_FCS30: Global land-cover product with fine classification system at 30m using time-series Landsat imagery. *Earth Syst. Sci. Data* **2021**, *13*, 2753–2776. [CrossRef]
17. Venter, Z.S.; Barton, D.N.; Chakraborty, T.; Simensen, T.; Singh, G. Global 10 m Land Use Land Cover Datasets: A Comparison of Dynamic World, World Cover and Esri Land Cover. *Remote Sens.* **2022**, *14*, 4101. [CrossRef]
18. Shirani, K.; Solhi, S.; Pasandi, M. Automatic Landform Recognition, Extraction, and Classification using Kernel Pattern Modeling. *J. Geovis. Spat. Anal.* **2023**, *7*, 2. [CrossRef]
19. Gong, P.; Yu, L.; Li, C.; Wang, J.; Liang, L.; Li, X.; Ji, L.; Bai, Y.; Cheng, Y.; Zhu, Z. A new research paradigm for global land cover mapping. *Ann. GIS* **2016**, *22*, 87–102. [CrossRef]
20. Camargo, A.; Schultz, R.R.; Wang, Y.; Fevig, R.A.; He, Q. GPU-CPU implementation for super-resolution mosaicking of Unmanned Aircraft System (UAS) surveillance video. In Proceedings of the 2010 IEEE Southwest Symposium on Image Analysis & Interpretation (SSIAI), Chicago, IL, USA, 23–25 May 2010; pp. 25–28. [CrossRef]
21. Ma, Y.; Song, J.; Zhang, Z. In-Memory Distributed Mosaicking for Large-Scale Remote Sensing Applications with Geo-Gridded Data Staging on Alluxio. *Remote Sens.* **2022**, *14*, 5987. [CrossRef]
22. Zhang, J.; Ke, T.; Sun, M. Parallel processing of mass aerial digital images base on cluster computer—The application of parallel computing in aerial digital photogrammetry. *Comput. Eng. Appl.* **2008**, *44*, 12–15. [CrossRef]
23. Chen, C.; Tan, Y.; Li, H.; Gu, H. A Fast and Automatic Parallel Algorithm of Remote Sensing Image Mosaic. *Microelectron. Comput.* **2011**, *28*, 59–62.
24. Ma, Y.; Wang, L.; Zomaya, A.Y.; Chen, D.; Ranjan, R. Task-Tree Based Large-Scale Mosaicking for Massive Remote Sensed Imageries with Dynamic DAG Scheduling. *IEEE Trans. Parallel Distrib. Syst.* **2014**, *25*, 2126–2137. [CrossRef]
25. Jing, W.; Huo, S.; Miao, Q.; Chen, X. A Model of Parallel Mosaicking for Massive Remote Sensing Images Based on Spark. *IEEE Access* **2017**, *5*, 18229–18237. [CrossRef]
26. Rabenseifner, R.; Hager, G.; Jost, G. Hybrid MPI/OpenMP Parallel Programming on Clusters of Multi-Core SMP Nodes. In Proceedings of the 2009 17th Euromicro International Conference on Parallel, Distributed and Network-based Processing, Washington, DC, USA, 18–20 February 2009; pp. 427–436. [CrossRef]
27. Apache Hadoop. Available online: https://hadoop.apache.org (accessed on 20 April 2023).
28. Zaharia, M.; Chowdhury, M.; Franklin, M.J.; Shenker, S.; Stoica, I. Spark: Cluster computing with working sets. *HotCloud* **2010**, *10*, 95. [CrossRef]
29. Garland, M.; Le Grand, S.; Nickolls, J.; Anderson, J.; Hardwick, J.; Morton, S.; Phillips, E.; Zhang, Y.; Volkov, V. Parallel Computing Experiences with CUDA. *IEEE Micro* **2008**, *28*, 13–27. [CrossRef]
30. Eldawy, A.; Mokbel, M. SpatialHadoop: A MapReduce framework for spatial data. In Proceedings of the 2015 IEEE 31st International Conference on Data Engineering, ICDE 2015. IEEE Computer Society, Proceedings—International Conference on Data Engineering, Seoul, Korea, 13–16 April 2015; pp. 1352–1363. [CrossRef]
31. Aji, A.; Wang, F.; Vo, H.; Lee, R.; Liu, Q.; Zhang, X.; Saltz, J. Hadoop GIS: A High Performance Spatial Data Warehousing System over Mapreduce. *Proc. VLDB Endow.* **2013**, *6*, 1009–1020. [CrossRef]
32. Shaikh, S.A.; Mariam, K.; Kitagawa, H.; Kim, K. GeoFlink: A Framework for the Real-time Processing of Spatial Streams. *arXiv* **2004**, arXiv.2004.03352. [CrossRef]
33. Kopp, S.; Becker, P.; Doshi, A.; Wright, D.J.; Zhang, K.; Xu, H. Achieving the Full Vision of Earth Observation Data Cubes. *Data* **2019**, *4*, 94. [CrossRef]
34. Nüst, D.; Konkol, M.; Pebesma, E.; Kray, C.; Schutzeichel, M.; Przibytzin, H.; Lorenz, J. Opening the publication process with executable research compendia. *D-Lib Mag.* **2017**, *23*, 451. [CrossRef]
35. Wang, B.; Zhang, M.; Huang, Q.; Yue, P. A Container-Based Service Publishing Method for Heterogeneous Geo-processing Operators. *J. Geomat.* **2021**, *46*, 174–177. [CrossRef]
36. Huffman, J.; Forsberg, A.; Loomis, A.; Head, J.; Dickson, J.; Fassett, C. Integrating advanced visualization technology into the planetary Geoscience workflow. *Planet. Space Sci.* **2011**, *59*, 1273–1279. [CrossRef]
37. Yue, P.; Zhang, M.; Tan, Z. A geoprocessing workflow system for environmental monitoring and integrated modelling. *Environ. Model. Softw.* **2015**, *69*, 128–140. [CrossRef]
38. Chen, Y.; Lin, H.; Xiao, L.; Jing, Q.; You, L.; Ding, Y.; Hu, M.; Devlin, A.T. Versioned geoscientific workflow for the collaborative geo-simulation of human-nature interactions—A case study of global change and human activities. *Int. J. Digit. Earth* **2021**, *14*, 510–539. [CrossRef]

39. Gesch, D.; Oimoen, M.; Danielson, J.; Meyer, D. Validation of the aster global digital elevation model version 3 over the conterminous united states. *Int. Arch. Photogramm. Remote Sens. Spat. Inf. Sci.* **2016**, *XLI-B4*, 143–148. [CrossRef]
40. Gong, P.; Wang, J.; Yu, L.; Zhao, Y.; Zhao, Y.; Liang, L.; Niu, Z.; Huang, X.; Fu, H.; Liu, S.; et al. Finer resolution observation and monitoring of global land cover: First mapping results with Landsat TM and ETM+ data. *Int. J. Remote Sens.* **2013**, *34*, 2607–2654. [CrossRef]
41. Foody, G.M.; Arora, M.K. An evaluation of some factors affecting the accuracy of classification by an artificial neural network. *Int. J. Remote Sens.* **1997**, *18*, 799–810. [CrossRef]
42. Du, P.; Lin, C.; Chen, Y.; Wang, X.; Zhang, W.; Guo, S. Training Sample Transfer Learning from Multi-temporal Remote Sensing Images for Dynamic and Intelligent Land Cover Classification. *J. Tongji Univ. (Nat. Sci. Ed.)* **2022**, *50*, 955–966. [CrossRef]
43. Huang, Y.; Liao, S. Automatic collection for land cover classification based on multisource datasets. *J. Remote Sens.* **2017**, *21*, 757–766. [CrossRef]
44. Liu, K.; Yang, X.; Zhang, T. Automatic Selection of Clasified Samples with the Help of Previous Land Cover Data. *J. -Geo-Inf. Sci.* **2012**, *14*, 507–513. [CrossRef]
45. Tianjun, W.; Jiancheng, L.; Liegang, X.; Haiping, Y.; Zhanfeng, S.; Xiaodong, H. An Automatic Sample Collection Method for Object-oriented Classification of Remotely Sensed Imageries Based on Transfer Learning. *Acta Geod. Cartogr. Sin.* **2014**, *43*, 908. [CrossRef]
46. Anderson, J.R. *A Land Use and Land Cover Classification System for Use with Remote Sensor Data*; US Government Printing Office: Washington, DC, USA, 1976; Volume 964.
47. Gregorio, A.D.; Jansen, L.J.M. *Land Cover Classification System (LCCS): Classification Concepts and User Manual*; FAO: Geneva, Switzerland, 2000; ISBN 92-5-104216-0.
48. Gómez, C.; White, J.C.; Wulder, M.A. Optical remotely sensed time series data for land cover classification: A review. *ISPRS J. Photogramm. Remote. Sens.* **2016**, *116*, 55–72. [CrossRef]
49. Chaaban, F.; Khattabi, J.E.; Darwishe, H. Accuracy Assessment of ESA WorldCover 2020 and ESRI 2020 Land Cover Maps for a Region in Syria. *J. Geovisualization Spat. Anal.* **2022**, *6*, 31. [CrossRef]
50. Apache Airflow. Available online: https://airflow.apache.org (accessed on 20 April 2023).
51. Amstutz, P.; Mikheev, M.; Crusoe, M.R.; Tijanić, N.; Lampa, S. Existing Workflow Systems. Available online: https://s.apache.org/existing-workflow-systems. (accessed on 18 April 2023).
52. Leipzig, J. A review of bioinformatic pipeline frameworks. *Briefings Bioinform.* **2017**, *18*, 530–536. . w020. [CrossRef] [PubMed]
53. Schultes, E.; Wittenburg, P. *FAIR Principles and Digital Objects: Accelerating Convergence on a Data Infrastructure*; Springer International Publishing: Berlin/Heidelberg, Germany, 2019; pp. 3–16.
54. Common Workflow Language. Available online: http://www.commonwl.org (accessed on 20 April 2023).

MDPI AG
Grosspeteranlage 5
4052 Basel
Switzerland
Tel.: +41 61 683 77 34

Remote Sensing Editorial Office
E-mail: remotesensing@mdpi.com
www.mdpi.com/journal/remotesensing

www.ingramcontent.com/pod-product-compliance
Lightning Source LLC
Chambersburg PA
CBHW061616220926
51850CB00005B/520
* 9 7 8 3 7 2 5 8 5 9 1 9 1 *